Postharvest Nanotechnology for Fresh Horticultural Produce

Food scarcity and insecurity is an alarming issue throughout the world. Postharvest loss due to both mechanical damage and microbial spoilage significantly influences the shelf life and hence the availability of agricultural produce. Once initiated, microbial spoilage can make a bulk quantity of a given agricultural product unacceptable for human consumption, and several methods have already been used to try to manage this. Considering the limited success of the available methods, there is an increasing interest in exploring nanotechnological methods. These methods are being considered for both developing various platforms for antimicrobial/ barrier packaging applications that minimize the contact of agricultural produce with the external environment and designing sensors to ensure food safety and quality.

The impact of various nanosystems developed through material engineering on the shelf-life enhancement and storage of fresh horticultural produce will have revolutionary effects on postharvest management in the coming years. Hence, ***Postharvest Nanotechnology for Fresh Horticultural Produce*** has been edited to advance understanding of material development, intelligent selection of nanomaterials to ensure the nontoxic nature, and future perspectives of nanotechnology on postharvest produce. This includes various types of nanoparticles exploited for the postharvest management; their mechanism of action; various applications and material engineering, along with natural products including essential oils and plant bioactives; modeling of various tailor-made materials to meet the required properties of the packed food; advancements in the nanotechnological applications for the minimally processed food; and toxicity concerns.

Key Features:

- Describes advances in nanotechnology for postharvest management
- Includes extensive details on the applications of material engineering for postharvest applications using nanotechnology and future aspects
- Provides extensive data on the types of nanomaterials used and the fabrication methods employed for the design of tailor-made products for the postharvest management

This book reviews the current scientific advancements and future prospects of nanotechnological interventions in meeting the quality and quantity standards of the horticultural produce and minimally processed food and will be a valuable reference for beginners, researchers, subject experts, and industrialists.

Innovations in Postharvest Technology Series

Series Editor Sunil Pareek

Department of Agriculture and Environmental Sciences National Institute of Food Technology Entrepreneurship and Management Kundli, Sonepat, Haryana, India

Postharvest Ripening Physiology of Crops (2016)
Edited by Sunil Pareek

Fresh-Cut Fruits and Vegetables: Technology, Physiology, and Safety (2016)
Edited by Sunil Pareek

Novel Postharvest Treatments of Fresh Produce (2017)
Edited by Sunil Pareek

Postharvest Physiological Disorders in Fruits and Vegetables (2019)
Edited by Sergio Tonetto de Freitas, and Sunil Pareek

Postharvest Nanotechnology for Fresh Horticultural Produce: Innovations and Applications (2024)
Edited by Radhakrishnan E.K., Ashitha Jose, and Sunil Pareek

Postharvest Nanotechnology for Fresh Horticultural Produce

Innovations and Applications

Edited by Radhakrishnan E.K., Ashitha Jose, and Sunil Pareek

CRC Press is an imprint of the
Taylor & Francis Group, an **informa** business

First edition published 2024
by CRC Press
2385 NW Executive Center Drive, Suite 320, Boca Raton FL 33431

and by CRC Press
4 Park Square, Milton Park, Abingdon, Oxon, OX14 4RN

CRC Press is an imprint of Taylor & Francis Group, LLC

ISBN: 978-0-367-67501-1 (hbk)
ISBN: 978-0-367-69558-3 (pbk)
ISBN: 978-1-003-14228-7 (ebk)

DOI: 10.1201/9781003142287

Typeset in Century Old Style Std
by Apex CoVantage, LLC

Contents

Preface

The current economy and population growth has accelerated the need for eliminating food loss through the shelf-life enhancement and protection from microbial agents during the various stages of growth, harvest, transportation, and storage. Nanotechnology has an emerging role in maintaining the postharvest quality of both fresh produce and processed food alike.

The origin, toxicity, and level of accumulation of nanoparticles in living forms determine their suitability for use in the food sector; hence, proper screening is mandatory. Owing to the minimal toxicity and better performance, various organic nanoparticles, including the cellulose nanoparticles and chitosan nanoparticles, are widely used for the postharvest storage. However, the enhanced antimicrobial property of the inorganic/metallic nanoparticles, such as silver nanoparticles, cannot be neglected. Here comes the importance of toxicity screening, accumulation, and degradation studies to ensure the safe use of nanomaterials.

The multimodular antimicrobial approaches of the engineered nanoparticles provide an added advantage in combating microorganisms. This is exploited in the development of the various nano-based packaging materials. The multitude of packaging materials developed using both nanoparticles and natural compounds such as essential oils are being exploited recently toward the shelf-life enhancement of both unprocessed and minimally processed horticultural produces.

Among the metallic nanoparticles, zinc-oxide nanoparticles have been exploited greatly in the food industry owing to it GRAS (generally recognized as safe) status. This, in association with various phytoformulations, can significantly ensure delayed oxidative damage and enhanced antimicrobial potency, which are the hallmarks of an ideal packaging material.

A major achievement made so far in the packaging industry is the development of materials with gas barrier property which can maintain the atmosphere and thereby prevent the gaseous exchange and thus prolong the shelf life. Nanosensors also play a crucial role in postharvest management. Research is being carried out to identify ethylene gas production/accumulation and detect harmful pathogens on the food surfaces. All these advancements in the postharvest sensor can ensure the quality of produce.

Even when the nanoparticles possess a greater advantage in the food industry, their toxicity concerns and life-cycle assessments play a critical role in ensuring its safety. However, the future scope of the nano-based materials in the postharvest technologies and shelf-life enhancement of the minimally processed food is unquestionable.

This book, ***Postharvest Nanotechnology for Fresh Horticultural Produce: Innovations and Applications***, reviews all aspects related to nanotechnological intervention on postharvest technology. The chapters are contributions from worldwide experts in the respective subject areas. Together, these make the book worth appreciation; a detailed index is also provided for reference.

About the Editors

Dr. Radhakrishnan E.K., a prominent academician and scientist in the biological sciences, started his career with a PhD in biotechnology from Rajiv Gandhi Centre for Biotechnology, Kerala, India, followed by the postdoctoral studies at the University of Tokyo, Japan. He joined the School of Biosciences, Mahatma Gandhi University, Kottayam, India, as an assistant professor in 2010. Currently, he continues as an associate professor. In addition to this, he currently holds other positions, including director of the Business Innovation and Incubation Centre and joint director of the Inter-University Centre for Organic Farming and Sustainable Agriculture, member of the member board of directors of Mahatma Gandhi University Innovation Foundation, nodal officer of the Innovation and Entrepreneurship Development Centre of Kerala Start Up Mission (KSUM), institutional facilitator of the Kerala Development and Innovation Strategic Council, and president of the Institution Innovation Council of Ministry of Education, Govt. of India. Dr. Radhakrishnan E.K. has advised 15 PhDs, 12 M.Phil and more than 100 other trainings. He has also completed 14 projects funded by various national and state government agencies as the principal investigator and 5 projects as co-investigator. His research areas include the functional biology of plant microbiomes, antimicrobial resistance evolution, and applications of bionanocomposites. He has 150 publications in various journals, with an *h*-index of 35; 50 book chapters; and edited 10 books. He has also been listed in the top 2% scientists as per the list prepared by Elsevier and Standford University in 2022. He has also been awarded with the Kairali Gaveshana puraskaram of Govt of Kerala in 2022. He has already filed five patents and technology transfer has been made with industry and his team developed the product nanopower for agricultural applications.

ABOUT THE EDITORS

Ms. Ashitha Jose is currently pursuing her PhD in biosciences (biochemistry) at the School of Biosciences, Mahatma Gandhi University Kottayam, Kerala, India. She has authored several book chapters and research publications with scrivener and springer. She is currently editing seven books (with Elsevier, CRC Press–Taylor and Francis, Scrivener publishing partnering with Wiley and Sons, USA, and Cambridge Scholars Publishing). Her area of research is the development and characterization of novel active bionanocomposites for various applications emphasizing the packaging applications of the developed materials for postharvest protection and shelf-life enhancement.

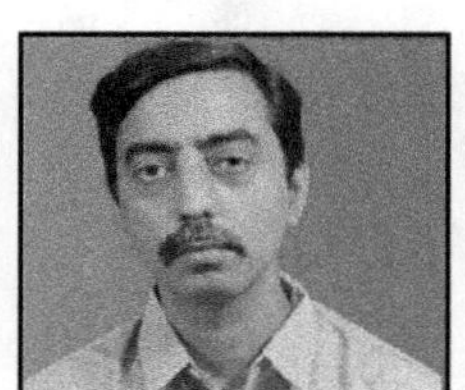

Dr. Sunil Pareek is serving as Professor (Horticulture PHT); Head, Department of Agriculture and Environmental Sciences, and Director (IQAC) at National Institute of Food Technology Entrepreneurship & Management, Kundli, India. He is also an Adjunct Faculty at Maharana Pratap Horticulture University, Karnal, India. He has specialization in postharvest management of fruits and vegetables. He has completed several research projects supported by ICAR, World Bank, DST, DBT, USAID etc. He generated technologies for extension of shelf life of fruits & vegetables, fresh-cuts and their processed products. Dr. Pareek has published more than 100 research papers, 10 books, 3 manuals, 6 technical bulletins, 50 popular articles, and 52 book chapters to his credit. Dr. Pareek is a series editor of 'Innovations in Postharvest Technology' from CRC Press, Taylor & Francis, USA. He is Editor of five SCI journals, namely, Scientific Reports, Journal of Food Quality, Journal of Food Biochemistry, Journal of Food Processing and Preservation, and Folia Horticulturae. Recently he has founded new journal "Food Safety and Health" as Associate Editor, and "Future Postharvest and Food" as Co-Editor-in-Chief from Wiley. He is the founder life member of Indian Society of Arid Horticulture and life member of several scientific societies. Dr. Sunil Pareek contributed immensely in agriculture and allied sector policies of Government of India. He is the member of various statutory bodies, notable few are Director on Board of Directors, National Horticulture Board, GoI; Board of Advisors, Institute of Technical and Professional Studies, Chattisgarh; Mandate review committee of Life Sciences Research Board of DRDO; Food and Agriculture Divisional Council, Bureau of Indian Standards; Research Advisory Committee of ICAR-National Research Centre on Banana, ICAR-National Research Centre on Pomegranate, and ICAR-Indian Institute of Spices Research; ICAR University Ranking Committee; Steering Committee of Good Agricultural Practices, Quality Council of India; Member, Working Group of NITI Ayog; Member, Agriculture and Natural Resources Commission, IOED, etc. He is an expert member of many scientific evaluation committees. He is involved in the formulation of various standards in Agri-Food commodities as Member of Scientific Panel on Fruits and Vegetables of Food Safety and Standards Authority of India; BIS FAD 9, FAD 10 and FAD 13; and Quality Council of India. He is also the member of Board of Governors, and Senate of NIFTEM. He is the recipient of Fellow of the Society for Horticultural Research & Development, India; Amity Academic Excellence Award 2020; Academic Excellence Award of Mobilization Society; IOED Award 2019 and 2021; Dr. RS Paroda Award 2019; University Best Team Research Award 2014; University Outstanding Services Award 2014; Young Scientist Award 2012; HS Mehta Young Scientist Award 2013; Fellow Award of Confederation of Horticultural Associations of India 2012, and JISL Fellowship 2014 and 2018.

Contributors

Lizet Aguirre-Güitrón
Universidad Politecnica del Estado de Nayarit
Nayarit, Mexico

Aiswarya P.
Cochin University of Science and Technology
Kochi, Kerala, India

María R. Ansorena
National University of Mar del Plata
Mar del Plata, Argentine

Fazilah Ariffin
Universiti Sains Malaysia
Penang, Malaysia

Manuel Toledano Ayala
Facultad de Ingeniería
Centro Universitario
Universidad Autónoma de Querétaro
Queretaro, Mexico

Jithasha Balan
Sree Sankara College
Kalady; Kerala, India

Kalyan Barman
Banaras Hindu University
Varanasi, Uttar Pradesh, India

Sarita G. Bhat
Cochin University of Science and Technology
Kochi, Kerala, India

Bharat Bhushan
ICAR-Indian Institute of Maize Research
Ludhiana, Punjab, India

Francisco J. Blancas-Benitez
Tecnológico Nacional de México/I. T. Tepic
Nayarit, Mexico

Alma Karen Burgos Araiza
Universidad Autónoma de Querétaro
Santiago de Querétaro, Mexico

Betsie Martínez Cano
Universidad Autónoma de Querétaro
Santiago de Querétaro, Mexico

M. Chandrasekar
Hindustan Institute of Technology & Science
Kelambakkam, Tamilnadu, India

Héctor J. Cortés-Rivera
Tecnológico Nacional de México/I. T. Tepic
Nayarit, Mexico

CONTRIBUTORS

Nathana L. Cristofoli
University of Algarve
Faro, Portugal

Ana Angélica Feregrino Pérez
Universidad Autónoma de Querétaro
Santiago de Querétaro, Mexico

Yolanda González García
Universidad Autónoma Agraria Antonio Narro
Coahuila, México

Sumi Mary George
Sree Sankara College
Kalady; Kerala, India

Ramsés R. González-Estrada
Tecnológico Nacional de México/I. T. Tepic
Nayarit, Mexico

González-Reza R.M.
National Autonomous University of Mexico
Cuautitlan Izcalli, Estado de Mexico, Mexico

Gutiérrez-Martínez P.
Tecnológico Nacional de México/I. T. Tepic
Nayarit, Mexico

Luis Guillermo Hernández-Montiel
Centro de Investigaciones Biológicas del Noroeste
Baja California Sur, Mexico

Dona V. Idicula
Sree Sankara College
Kalady; Kerala, India

Christopher Igwe Idumah
Nnamdi Azikiwe University
Awka, Nigeria

Kirti Jalgaonkar
ICAR-Central Institute for Research on Cotton Technology
Mumbai, Maharashtra, India

Ashitha Jose
Mahatma Gandhi University
Kottaym, Kerala, India

Minu Joys
Assistant Professor,
Department of Chemistry,
Union Christian College,
Aluva

Karim A.A.
Universiti Sains Malaysia
Penang, Malaysia

Fabián Pérez Labrada
Universidad Estatal de Sonora
Sonora, México

Alexandre R. Lima
University of Algarve
Faro, Portugal

Mari Carmen López Pérez
Universidad Estatal de Sonora
Sonora, México

Archana Mahapatra
ICAR-Central Institute for Research on Cotton Technology
Mumbai, Maharashtra, India

Manoj Kumar Mahawar
ICAR-Central Institute for Research on Cotton Technology
Mumbai, Maharashtra, India

Antonio Juárez Maldonado
Universidad Autónoma Agraria Antonio Narro
Coahuila, México

Girish N. Mathad
National Pingtung University of Science and Technology
Pingtung, Taiwan

Maya Mathew
Mahatma Gandhi University
Kottaym, Kerala, India

Cristian Josué Mendoza Meneses
Universidad Autónoma de Querétaro
Santiago de Querétaro, Mexico

Midhun Dominic C.D.
Sacred Heart College
Kochi, Kerala, India

Shatakashi Mishra
Banaras Hindu University
Varanasi, Uttar Pradesh, India

Cristina Moreno-Hernández
Tecnológico Nacional de México/I. T. Tepic
Nayarit, Mexico

Abdorreza Mohammadi Nafchi
Universiti Sains Malaysia
Penang, Malaysia

Ajalesh B. Nair
Union Christian College
Aluva, Kerala, India

Sreeja Narayanan
Cochin University of Science and Technology
Kochi, Kerala, India

Neenu K.V.
Cochin University of Science and Technology
Kerala, India

Shrikrishna Nishani
Karunya Institute of Technology and Sciences
Coimbatore, Tamil Nadu, India

Nazila Oladzadabbasabadi
Universiti Sains Malaysia
Penang, Malaysia

Jyotishkumar Parameswaranpillai
Mar Athanasios College for Advanced Studies Tiruvalla
Pathanamthitta, Kerala, India

Sherin Joy Parappilly
Sree Sankara College
Kalady, Kerala, India

Sharmila Patil
ICAR-Central Institute for Research on Cotton Technology
Mumbai, Maharashtra, India

Gregorio Cadenas Pliego
Centro de Investigación en Química Aplicada
Coahuila, México

G.L. Praveen
Wimpey Laboratories, Ras Al Khor, Industrial Area
Dubai—United Arab Emirates

Radhakrishnan E.K.
Mahatma Gandhi University
Kottaym, Kerala, India

Sabarish Radoor
King Mongkut's University of Technology North Bangkok
Bangkok, Thailand

Rajesh R.
St Albert's College (Autonomous)
Kochi, Kerala, India

Surelys Ramos-Bell
Tecnológico Nacional de México/I. T. Tepic
Nayarit, Mexico

P.M. Sabura Begum
Cochin University of Science and Technology
Kochi, Kerala, India

Mohammad Reza Saeb
Gdańsk University of Technology
Gdańsk, Poland

Jonathan Michel Sanchez-Silva
Universidad Autónoma de San Luis Potosí
San Luis Potosí, Mexico

CONTRIBUTORS

Sandhya C.
Assistant Professor
Department of Biochemistry
Kuriakose Elias College Mannanam

Satish Kumar
ICAR- National Institute of Abiotic Stress Management
Pune, Maharashtra, India

K. Senthilkumar
King Mongkut's University of Technology North Bangkok
Bangkok, Thailand

T. Senthil Muthu Kumar
Kalasalingam Academy of Research and Education
Krishnankoil, Tamil Nadu, India

Simple Kumar
AmityUniversity
Noida, Uttar Pradesh, India

Swati Sharma
ICAR-Indian Institute of Vegetable Research
Varanasi, Uttar Pradesh, India

Shasiya P.S.
Union Christian College
Aluva, Kerala, India

Suchart Siengchin
King Mongkut's University of Technology North Bangkok
Bangkok, Thailand

Jagdish Singh
ICAR-Indian Institute of Vegetable Research
Varanasi, Uttar Pradesh, India

Gobika Thiripuranathar
Institute of Chemistry Ceylon
Rajagiriya, Sri Lanka

Neethumol Varghese
Union Christian College
Aluva, Kerala, India

Gabriela E. Viacava
National University of Mar del Plata
Mar del Plata, Argentine

C. Vibha
Mar Athanasios College for Advanced Studies Tiruvalla (MACFAST)
Pathanamthitta, Kerala, India

Margarida C. Vieira
University of Algarve
Faro, Portugal

Raji Vijayamma
Mahatma Gandhi University
Kottayam, Kerala, India

Udari Wijesinghe
Institute of Chemistry Ceylon
Rajagiriya, Sri Lanka

Upekshya Welikala
Institute of Chemistry Ceylon
Rajagiriya, Sri Lanka

Zambrano-Zaragoza M.L.
National Autonomous University of Mexico
Cuautitlan Izcalli, Estado de Mexico, Mexico

Chapter 1

Introduction to Nanotechnological Applications in the Postharvest Processing of Fresh Horticultural Produce

González-Reza R.M., Zambrano-Zaragoza M.L., and Gutiérrez-Martínez P.

1.1 Introduction

Horticultural produce is of great importance worldwide and contributes to the maintenance of human health; it provides most of the energy necessary to develop specific functions in the human organism and is attractive due to its color, flavor, and juiciness. However, horticultural produce continues the metabolic processes after it is harvested, so treatments are a must to preserve and help transform it into ready-to-eat products. Therefore, all operations involving its handling will result in security and safety risks that can be prevented, and treatments are needed to avoid foodborne disease consumption (Mahajan et al., 2017). Some of the changes that should be considered to improve horticultural processing strategies are changes in pH, sugar content, volatile compounds, vitamins, susceptibility to endogenous and exogenous ethylene, and external O_2 that influence respiratory and metabolic activity. In addition, ethylene can influence whether the product is climacteric or non-climacteric, the type of organ, the state of maturity, and development conditions. The alternatives being looked for

DOI: 10.1201/9781003142287-1

are those that, pre- and postharvest, promote increased shelf life (Aggarwal et al., 2018). Numerous strategies have been tested postharvest to keep products fresh as long as possible; this includes active nano-packaging with a photocatalytic effect capable of absorbing O_2 and ethylene, inhibiting or controlling microbial growth, and diminishing metabolic activity (Liu et al., 2020). However, minimally processed horticultural produce had a decreased shelf life due to cutting, which released cytoplasm content and produced changes in sensory, nutritional, and physicochemical properties. In addition, microbial development depends on the plant part because the green parts like the leaves and stems are more susceptible to bacterial growth, and in the fruits, fungi growth, without forgetting that the products stored in refrigeration conditions, has a susceptibility to cold damage and being handling in cold stores (Vivek et al., 2019).

Nanotechnology is an emerging technology that focuses on designing, characterizing, modeling, and evaluating systems with dimensions less than 1000 nm, preferably between 100–500 nm (Zambrano-Zaragoza et al., 2018). In the last two decades, new and novelty developments allow incorporating natural active components as additives, and this contributes to improving the processing, handling, storage, and distribution of horticultural produce, including fresh fruits and minimally processed storage on refrigeration conditions. The synthesis of nanostructures has shown a possibility that marks a breaking point between the different technologies previously used. Nanosystems are prepared in two ways: (1) top down, which implies the use of high energy, homogenization pressure, ultrasound, and others, to achieve the reduction of the size of components, and (2) bottom up, in which nanostructures are constructed from simple features to accomplish polymerization reactions to add to ingredients that will integrate these. A great advancement in developing antioxidant nanosystems to avoid the formation of reactive oxygen species in the cells of horticultural produce and limit the effects at the membrane level has been developed. The active components in the nanosystems interact with the cell membrane and restrict the diffusion of gases and lipoperoxidation reactions, thus contributing to conservation and increasing shelf life (Yadollahi et al., 2010).

Nanosystems also have excellent possibilities in the control of postharvest diseases, and these could encapsulate natural and synthetic antifungal compounds (essential oils, extracts obtained from plants and their by-products, and other secondary metabolites) to reduce postharvest rot and losses. Moreover, these can be loaded with antibacterial actives and contribute to ensuring the safety of horticultural products. The principal nanostructures are polymeric nanoparticles (nanospheres and nanocapsules), solid lipid nanoparticles, lipid nanocarriers, metallic nanoparticles, nanoemulsions, Pickering nanoemulsions, nanofibers, and nanotubes. The principal active components with antifungal and antibacterial power come from essential oils and natural extracts obtained from algae, tree bark, and by-products of industrial processes. Essential oils have low water solubility, are volatile, and are sensitive to ambient conditions such as light, temperature, and O_2. Thus, the nanostructure represents a form for increasing the compatibility between this and the tissue of fruits and vegetables with high water content and an approach to the composition and characteristics of the biopolymers used to form the nanostructure that protects and regulates the release of active components (Lazar et al., 2010).

Moreover, nanosystems can be harnessed to maximize the shelf life of horticultural produce by using scavenging, absorbing, and oxide substances that decrease the ethylene and O_2 absorption. For example, the TiO_2 nanoparticles and cellulose nanocrystals help scavenge O_2 across active packaging to increase shelf life (Gaikwad et al., 2020; Hussain et al., 2011). Another application of nanotechnology is to guarantee the safety and security of horticultural produce, reducing microorganisms and viruses present due to the risk of contamination from product handling, transport, and packaging; these aspects are addressed in later sections of this chapter.

1.2 Postharvest Losses Global Overview

World production of tropical and subtropical fruits and vegetables in 2015 was more than 200 million tons (FAO, 2016). According to Food and Agriculture Organization of the United Nations (FAO), 90 % of tropical and subtropical fruits were produced in developing countries, and more than 50 % were consumed as fresh fruits in the European Union countries. In addition, fruit production and export are vital to several countries' economies and work generation (Altendorf, 2017).

Horticultural products are essential for a balanced and nutritious diet, which results in the preservation and maintenance of health. Fruits and vegetables are the main sources of vitamins A and C and are excellent sources of calcium, iron, and other micronutrients. For example, bananas, roots, and tubers are important sources of calories and some protein. All the previously mentioned horticultural crops play an important role in developing countries. These provide nutrients and phytochemicals, which justifies that it is necessary to do what is required to reduce the high losses in these fruits (Kader and Yahia, 2011).

According to FAO reports, about one-third of the food destined to feed the world's population is lost, accounting for more than 1.3 billion tons per year. A significant part of the losses of these products occurs in the postharvest stage, originating from harvesting, packaging, transportation, storage, and marketing centers. The causes of the postharvest losses are due to various circumstances, inadequate infrastructure, a lack of knowledge in the management and technology of postharvest on the conservation of fresh products, and attacks by pathogenic fungi (Kitinoja and Kader, 2015; FAO, 2016; Fabi and English, 2019).

During their growth and development in the garden, horticultural products are susceptible to being infected by pathogenic fungi; in some cases, fungi use natural wounds or openings, stomata, or lenticels or are directed by enzymes produced by the pathogen, being established below the cuticle in a latent form. When the ripening of horticultural products begins, the latent fungi are activated and attack the fruit, causing deterioration and a loss of external and internal quality (Alkan et al., 2015). Hence, to control these pathogenic fungi, synthetic fungicides have traditionally been applied pre-harvest and postharvest; the pathogen is controlled, but the environment is contaminated and damaged, and there is a latent danger to human health (Barkai-Golan, 2001; Adaskaveg and Förster, 2009). Some of the substances commonly applied are, for example, imazalil, thiabendazole, and pyrimethanil. Although currently it has been shown that these fungi develop resistance to these fungicides (Sánchez-Torres and Tuset, 2011), nanotechnology used in conjunction with natural substances can be a great alternative to this problem, as well as represent a potential risk to living beings, the environment, and human health (Romanazzi et al., 2014; Nicolopoulou-Stamati et al., 2016).

According to the evidence obtained, synthetic fungicides are polluting and dangerous. In some parts of the world, such as the European Union, there is already the disposal of zero residues in horticultural products (Horská et al., 2020). Alternative technologies and systems for fungicides control the postharvest deterioration of fruit and vegetable products caused by pathogenic fungi. Losses caused by applying products of natural origin, such as generally-recognized-as-safe (GRAS) compounds and plant extracts, are excellent alternatives to treatments with substances synthetics (Wisniewski et al., 2016; Gutiérrez-Martínez et al., 2018; Chávez-Magdaleno et al., 2018; Palou, 2018; Cortes-Rivera et al., 2019). Currently, the research in postharvest pathology focuses on improving the host fruit and vegetable product's individual potential to respond to the attack of pathogens through resistance inducers (Romanazzi et al., 2016; Gutiérrez-Martínez et al., 2018; Ramos-Guerrero et al., 2018). Nanotechnology is an alternative for controlling pathogenic fungi in postharvest fruits and vegetables. In combination with other materials and fungicides, nanomaterials represent a viable alternative to decrease postharvest loss.

1.3 Postharvest Treatment in Fruit and Vegetable Preservation

Numerous technologies are used to reduce the postharvest losses of horticultural produce. For example, drying, osmotic dehydration, pickling, and crystallization are some conservation methods for decreasing water activity and moisture content to delay metabolic reactions and microbial growth, and canning is a way to transform horticultural produce into a product with a considerably increased of shelf life. However, these produce irreversible changes in the sensory quality and losses the characteristics of fresh produce. Therefore, the world trend toward fresh fruit and vegetables is the best approach for consuming phytonutrients, antioxidants, and functional properties. In this sense, the phytonutrients present in horticultural products contribute to controlling chronic degenerative diseases (hypertension, diabetes, atherosclerosis); moreover, these have anti-inflammatory, anticancer, and immunologic properties (Olaimat et al., 2020; Yadollahi et al., 2010), so it is necessary to look to nanotechnology as a viable alternative for maintaining these properties and increase the time to market.

Different chemical, physical, and biological methods have been studied and applied to preserve fresh horticultural produce as long as possible. Chemical processes include using different fungicides, antibacterial treatments, and antioxidants, such as ascorbic acid, citric acid, peroxide, and calcium chloride, to mention a few. However, eliminating chemical fungicides and chemical additives when preserving horticultural produce is necessary due to their high toxicity indexes, adverse effects on human health, and water contamination. Moreover, microorganisms have developed resistance to these fungicides, lowering pathogen control. Physical methods include heat treatment and emerging technologies like edible coatings and irradiation (UV-C, modified atmospheres, ozone; Helmy et al., 2020; Mahajan et al., 2017).

The principal pathogens in fruits and vegetables are the *Penicillium digitatum*, *Botrytis cinerea*, *Colletrotichium gloeosporioides*, *Alternaria alternata*, *Fusarium spp*., and *Geotrichum candidum*, and among others. The development of fungi resistance to chemical treatments carried through to combining chemical treatments with physical methods, such as hydrothermal treatments, to modify the metabolism of horticultural products and diminish the use of chemical fungicides by changing the absorption and improving their effectiveness (Liu et al., 2018). Heat treatment at 40–60° C has been considered an alternative for disease and infestation control in the world, mainly for tropical and subtropical fruits; this treatment increases the resistance of these fruits to cold damage and modifies their metabolism by slowing their respiration rate due to the denaturation of enzymes. Another treatment widely used to eliminate sprouts and delay senescence is γ-irradiation in low doses up to 1 kG because doses between 1 at 10 kG reduce the microbial load, and high doses have not considered due to undesirable metabolic changes (Deng et al., 2020; Mahajan et al., 2014).

In the last decade, the increased use of emerging eco-friendly technologies has promoted the commercialization of healthy products, limiting the use of synthetic chemical substances and leading to new and innovative food protection systems. Nanosystems are used as vehicles to deliver compounds capable of maintaining horticultural products' quality characteristics and contribute to maintaining their safety and security. Perhaps currently, due to the SARS-CoV-2 pandemic situation, society has entered a new adaptation stage at which nanotechnology will have an excellent place regarding food safety.

Nanosystems provide ample possibilities for preserving horticultural produce since they show greater homogeneity and surface areas than larger systems, such as microparticles and emulsions. In addition, fresh and minimally processed products have had antimicrobial substances incorporated into them to maintain the quality of the products with a control system to inhibit or control the growth of bacteria and other pathogenic microorganisms that reduce shelf life (Ceylan et al., 2020; Liu et al., 2020).

Nanosystems used in the disinfection of horticultural products mainly include nanoemulsions and polymeric nanoparticles, while other nanostructures are under study. These can better the controlled release of active components, protect other processing steps, and increase the compatibility of active components with the cell membrane. The reduction in particle size is a critical point that diminishes the number of active components to achieve good effectiveness; for example, silver and copper are some of the most potent antimicrobials and have viricidal potential; however, they are highly toxic in concentration. In addition, their use for disinfection can generate high waste discharge into water. Therefore, the nanometric size, with a smaller quantity of active components, will require reducing the risk of contamination from effluents (Rajwade et al., 2020).

Essential oils are used because they own antimicrobial potential principally in disinfection by aspersion of horticultural produce. Their activity depends on their composition since the disinfecting effect is attributable to terpenes, phenols, aldehydes, and ketones (Deng et al., 2020; Gago et al., 2019). Nanosystems provide a membrane or protective wall that is formed by the polymers used; improve the solubility, stability, and functionality of essential oils; and, for their size, can be used at low concentrations. Furthermore, efforts should be made to establish instant release systems (burst release) for the best approach to disinfection properties. The nanoemulsions represent an excellent way to apply natural disinfectants with rapid-release profiles compared to other nanostructures. For example, a nanoemulsion with citral and linalool was used to inhibit *L. monocytogenes*, and it was shown that 5% of the citral was enough to decrease 83.52% of bacterial growth (Prakash and Vadivel, 2020). Another study with thyme essential oil in a nanoemulsion coating with chitosan and whey protein isolates was tested on a surface to inhibit *E. coli* and *S. aureus* and found that the essential oil was more effective against *E. coli* prolonged their effectivity during storage (Li et al., 2021). Regarding the effectiveness of disinfecting horticultural produce, considering the origin and optimal concentration to use, another study probed the effects of cinnamon, rosemary, and oregano essential oils in a nanoemulsion on decontaminating fresh celery and found a reduction of 5 log in *E. coli* and *L. monocytogenes* in less than 60 minutes (Dávila-Rodríguez et al., 2019). In addition, an ultrasound process and thyme essential oil nanoemulsion application effectively inhibit the growth of *E. coli* O157:H7, reducing bacteria in the wastewater (He et al., 2021).

1.4 Nanotechnology Applications in Postharvest Horticultural Produce

Nanotechnology, an emerging technology, has impacted the conservation of horticultural produce. In addition, the SARS-CoV-2 pandemic has dramatically changed consumer needs, including a demand for fruit and vegetable products with beneficial effects and the highest nutritional and functional properties that act on the immune system. Figure 1.1 shows some nanotechnology advantages to help conserve postharvest fruit and vegetable produce.

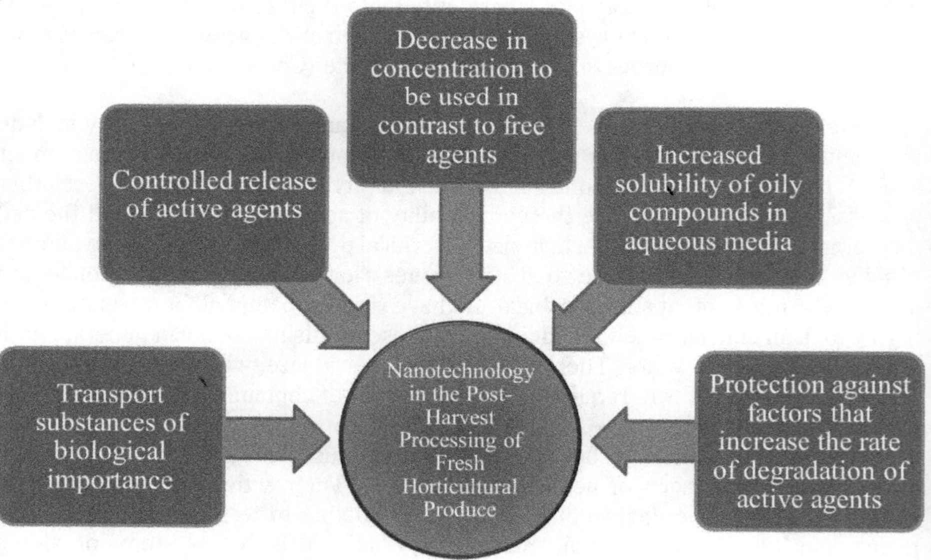

Figure 1.1 Beneficial effects of nanotechnology in postharvest fruits and vegetables.

1.4.1 The Structural Component of Nanosystems

Natural and synthetic polymers are used in nanosystem preparation; the selection depends on preference and the active components' function, antioxidant capacity, antimicrobial capacity, and controlled release characteristics. According to need, compatibility, and functionality, natural biopolymers obtained from by-products, wastes, and other sources from animals and vegetables are preferred. The principal biopolymers from these sources are proteins and polysaccharides (Sivakanthan et al., 2020). Examples of the polysaccharides used as the matrix in nanostructures are chitosan, gellan gum, cellulose, arabic gum, agave fructans, and pectin. Among the proteins commonly used in the formation of nanostructured systems are zein, gelatin, bovine serum albumin, and proteins obtained from amaranth and soy, among others. An example of the formation of nanostructures with a mixture of polysaccharides and a protein was developed by Liang et al. (2017), who formed zein-coated chitosan nanoparticles as a targeted delivery system for epigallocatechin gallate. One of the most relevant aspects of this nanostructured system was the controlled release obtained. The release profiles showed an initial explosion effect, followed by a slow release of the active component. A release mechanism is an approach for increasing the shelf life of horticultural produce. An example of an approach using biopolymers for controlled release was the development of an edible coating with ZnO nanoparticles (10 a 40 nm) prepared by the hydrothermal method using chitosan and arabic gum as wall biopolymers showed that the best antimicrobial effect was obtained with 0.3% of ZnO on avocado at ambient conditions, indicating that the use of polysaccharides allows control of the release of ZnO (Le et al., 2021).

The encapsulation in a biopolymeric matrix of active components considers the compatibility between components, stability, and desired functionality, based on the specific action site in the horticultural produce. Therefore, just a few biodegradable polymers are considered. An example of using these polymers with potential application in fruit and vegetable was developed by González-Reza et al. (2020), who encapsulated thyme essential oil in the polymers poly-ε-caprolactone and ethylcellulose using pluronic® F-127 and polyvinyl alcohol as stabilizing polymers. The main findings that may significantly impact applications to fruit and vegetable products are that the nanosystems present controlled release, describing it as Fickian. In addition, these showed excellent physical stability and high antioxidant capacity.

Lipids are other biological substances used in addition to polysaccharides and proteins. These are excellent carriers of active components and an excellent alternative for preserving horticultural produce, with significant advantages in the release of active components and cell membrane compatibility. Solid lipid nanoparticles, nanoliposomes, and lipid nanocarriers are examples of nanostructures with lipids. The lipid most used for nanosystem formation are waxes, phospholipids, butter, oils, and glycerides (Zambrano-Zaragoza et al., 2018). Table 1.1 provides some novel studies on different nanostructured systems with potential applications in horticultural products, highlighting antimicrobial and antioxidant activities. Several examples were also reported using different matrices (polysaccharides, proteins, lipids) to deliver important active components.

Figure 1.2 showed the principal nanostructures used to preserve postharvest fruit and vegetable products. These types of nanostructures aim to improve the functional properties of support materials and enhance the different activities of the active components.

Table 1.1 Nanostructured Matrices and Potential Application in Horticultural Produce

Active compound	*Matrix*	*Findings*	*Authors*
Curcumin	Zein/carboxymethyl dextrin	The photothermal stability and antioxidant activity of curcumin were significantly improved after loading onto nanoparticles (212 nm), indicating that these nanoparticles can be excellent vehicles for bioactive food applications.	Meng et al. (2021)
Ursolic acid	Gelatin/odium dodecyl sulfate/ ammonium oxalate	Nanoparticles with a mean diameter of approximately 702 nm and encapsulation efficiency of 87% were obtained by spray-drying. In addition, the release profiles revealed that the nanostructuring of ursolic acid improved bioaccessibility by 94.58%.	Karimi et al. (2020)
Gallic acid	Chitosan/zinc oxide	The nanostructured systems' functionalization with gallic acid in films improved the mechanical properties, the permeability to water vapor and oxygen, and the solubility in water. Therefore, the authors suggest that the films obtained may have potential applications in food packaging.	Yadav et al. (2021)
–	Sulfur nanoparticles/ alginate	The films showed bacteriostatic activity against *E. coli* and a bactericidal effect against *L. monocytogenes*. Its potential application in frozen food packaging with high moisture content was reported.	Priyadarshi et al. (2021)
β-carotene	Poly-ε-caprolactone/ pluronic F-127	Nanocapsules' physical stability with a size smaller than 300 nm was established using a biodegradable synthetic encapsulating polymer, finding good physical stability during storage for 28 days. In addition, the controlled release of the active ingredient attributed to color changes was suggested, making it attractive for food preservation.	González-Reza et al. (2018)

Active compound	*Matrix*	*Findings*	*Authors*
β-lactoglobulin/ ferulic acid/ tocopherol	Glyceryl tristearate	The adsorption of lactoglobulin increased with adding tocopherol in the nanostructured system, in contrast to the decrease in ferulic acid. Average particle sizes less than 300 nm were reported in all cases. In this article, the concept of "protein corona" is adapted from pharmaceutical technology to a food engineering claim	Oehlke et al. (2019)

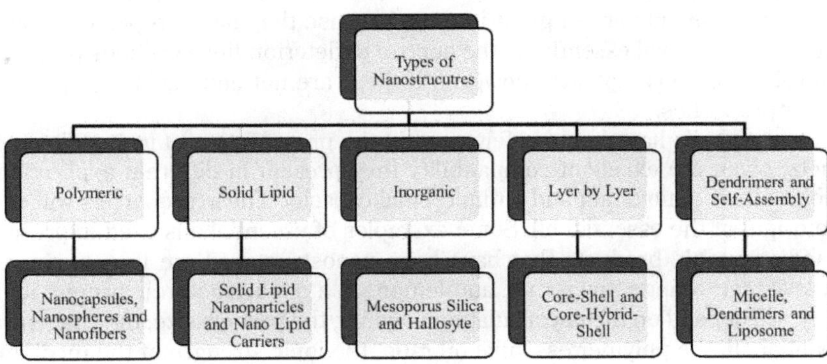

Figure 1.2 Nanostructures types for preservation of posfarvest horticultural produce.

1.4.2 Inorganic Materials in Nanostructures Preparation

The conservation of horticultural produce across the worldwide require different strategies and diverse materials to prepare nanosystems; these include mainly proteins, lipids, and polysaccharides, with specific functions. However, there is a tendency to incorporate inorganic materials into these matrices to reinforce and better the properties that the polymers and lipids do not possess; their use as an additive in the formulation of edible films and films modify the transport properties (e.g., water vapor, CO_2, and O_2). The incorporation of inorganic materials considers only GRAS materials. Some examples of these materials reported in nanosystems preparation are montmorillonite, nano-SiOx, nano-TiO_2, and nano-ZnO, as well as silver (Ag) nanoparticles. Its most significant impact on the conservation of fruit and vegetable products is the potentiation of antimicrobial capacity (Ajmal et al., 2019; Zhu et al., 2019). Notably, Ag and ZnO nanoparticles are effective antibacterial agents widely used in food packaging materials (Wang et al., 2021). In addition, TiO_2 and SiO_2 nanoparticles have a photocatalytic and reduce the ethylene presence. Therefore, TiO_2 and SiO_2 mesoporous mixture were tested, and the best combination of proportions was TiO_2 80%:SiO_2 20%, producing the degradation and total elimination of ethylene; the application in packaging of green tomatoes resulted in the increase of shelf life (de Chiara et al., 2015). ZnO nanoparticles are important due to their effective antifungal that helps reduce the contamination of fresh fruit produce. In a study, the efficacy of ZnO nanoparticles was evaluated on *Penicillium expansum*, *Alternaria alternata*, *Botrytis cinerea*, and *Rhizopus stolonifer*. Inhibition was measured by the growth in the diameter of mycelium at concentrations between 0 mmol to 15 mM of ZnO nanoparticles, and all fungi tested were eliminated with a concentration >6 mmol; then the effectivity depends on inorganic material concentration (Sardella et al., 2017).

1.4.3 Active Components in Nanosystems for Horticultural Produce

The advantage of nanosystems to preserve horticultural produce is that the nanosystems acts as a vector for active component delivery. These allow for active compound releases in a sustained, instant, or retarded manner during the packaging, storage, or commercialization according to the specific needs of horticultural produce. Generally, the active components are of great interest because they have properties of antioxidant and antimicrobial essential to the control of deterioration reactions or inhibition or control of microbial growth; however, if these are not encapsulated, they quickly lose their properties.

Essential oils have been widely used in the preservation of fruit and vegetable products; given the excellent compatibility they present in different applications, in addition to their antioxidant and antimicrobial capacity. These properties will depend on the origin of the essential oil. Some examples of essential oils used in preserving fruit and vegetable products that have been nanostructured are thyme, cinnamon, clove, lavender, orange, tangerine, and lemon. In a nanostructured system, its functionality will depend on the encapsulating polymer, the particle size, the concentration of essential oil, the polymer:essential oil ratio, the ionic strength of the medium, the presence of soluble solids, stabilizing agents, and others.

Phenolic compounds natively found in fruit and vegetables have been nanostructured using natural, synthetic polymers and lipids. The nanostructured properties of curcumin, catechin, epicatechin, quercetin, naringin, gallic acid, and ferulic acid were studied to functionalize antimicrobial and antioxidant characteristics (Matrose et al., 2020). Some examples of natural extracts with potential applications in horticultural produce preservation and nanostructure type are shown in Table 1.2.

Table 1.2 Nanoparticles Functionalized with Natural Extracts on the Preservation of Horticultural Produce

Active compound	*Matrix/ nanostructure*	*Findings*	*Authors*
Extracts from blackberries (*Rubus spp.*)	Nanoparticles based on pectin and lysozyme self-assembling	90% of the crude extract obtained was constituted by cyanidin-3-O-glucoside. Under the optimal conditions obtained by the response surface methodology, nanostructures with a size of 198.5 nm and a polydispersity index of 0.18 were obtained. The added pectin was on the surface of the nanostructures. The encapsulation efficiency was 78%.	Rosales et al. (2021)

Active compound	*Matrix/ nanostructure*	*Findings*	*Authors*
Opuntia humifusa fruit extract	ZnO nanoparticles	Functionalized nanoparticles with an average diameter of 25 nm with a hexagonal wurtzite structure were obtained by green synthesis. The nanostructures functionalized showed significant antimicrobial activity (zone of inhibition) against gram-positive bacteria (*Staphylococcus aureus* [17 mm], *Bacillus cereus* [18mm]) and gram-negative bacteria (*Shigella* [12 mm], *Escherichia coli* [13 mm], and *Pseudomonas* [12 mm]). The analysis of the phytochemicals present in the extract denotes the presence of saponins, phenols, flavonoids, tannins, terpenoids, and steroids, which are directly linked to antimicrobial activities.	Chennimalai et al. (2021)
Lemon balm extract (*Melissa officinalis* L.)	Azelaic acid-chitosan nanoparticles	Spherical nanoparticles with an average size of 814 nm with monomodal distribution were obtained. Inhibitory capacity was evaluated in 9 different species of gram-positive/gram-negative bacteria and 1 species of fungus. The nanostructures functionalized with the lemon balm extract presented an antibacterial effect against *B. cereus*, *S. aureus*, and *S. epidermidis*, highlighting a more significant inhibition effect against *E. faecalis* in contrast to gentamicin (positive control).	Tarassoli et al. (2021)
Extract of rosemary (*Rosmarinus officinalis* L.)	Chitosan nanoparticles	Chitosan nanoparticles were cross-linked with γ-polyglutamic acid (γ-PGA) to enhance the antimicrobial activity of rosemary extract. Functionalized nanoparticles with an average diameter between 212.2 and 605 nm were obtained depending on the concentration of chitosan used (0.1–1.8 mg/mL). The nanosystems obtained presented antimicrobial activity significantly higher than when the extract was used individually, denoted in reducing the viable population of *Bacillus subtilis* (0.5 to 3.6 CFU/mL).	Lee et al. (2019)

(Continued)

Table 1.2 (Continued)

Active compound	*Matrix/ nanostructure*	*Findings*	*Authors*
Safflower (*Carthamus tinctorius* L.) waste extract	Silver nanoparticles	Functionalized silver nanoparticles were obtained by green synthesis, with an average particle size of 8.67 nm. The nanoparticles inhibited the growth of both types of bacteria from the lowest concentration evaluated (0.9 µg/mL). The inhibition tests denoted that the minimum inhibitory concentration (MIC) and minimum lethal concentration (MLC) of silver nanoparticles were (a) *Staphylococcus aureus* (1.9 and 3.9 µg/mL) and (b) *Pseudomonas fluorescens* (7.8 and 15.6 µg/mL), respectively.	Rodríguez-Félix et al. (2021)
Pomegranate (*Punica granatum* L.) peel extract	Chitosan nanoparticles	The nanoparticles obtained (208.2 nm) at a 2:1 ratio (chitosan, peel extract) presented good physical stability (zeta potential = 26.5 mV) with an encapsulation efficiency of 50.5%. In addition, the inhibition test for *S. aureus* was successful for the extract individually and nanostructure. Fourier Transform Infrared (FTIR) and gas chromatography-mass spectroscopy analysis attribute these properties to phenolic compounds, such as tannic acid, ellagic acid, and gallic acid, which are present in the extract.	Soltanzadeh et al. (2021)

1.5 Nanotechnology Treatments in Postharvest Horticultural Produce

Recently, innovative treatment that includes emerging technologies to increase the shelf life of perishable products has been tested and applied. Nanotechnology is one of these; an example of this is the treatment by electric pulses combined with nisin-pectin nanostructures used to inactivate microorganisms in different drinks and fruit juices that showed effects against *S. typhimurium* and *L. innocua* about growth inhibition, revealing that the combination of these treatments improved the sensitivity of gram-negative bacteria to nisin (Novickij et al., 2020).

Among the potential applications of nanotechnology in the conservation of fresh postharvest horticultural products are the development of nanobiosensors to detect pesticides, allergens, and pathogenic microorganisms in the products. These are placed in different polymeric materials and can potentially apply in nano-packaging, which also has a high potential for applications in this type of product. Table 1.3 shows

Table 1.3 Applications of Nanotechnology in the Conservation of Horticultural Produce

Nanostructure	*Compound*	*Matrix*	*Findings*	*Reference*
Nanocrystals	—	Cellulose/chitosan/ polyvinyl alcohol (PVA)	The antifungal activity against postharvest pathogens, *Colletotrichum gloeosporioides* and *Lasiodiplodia theobromae*, was established, reporting that films with chitosan nanoparticles presented 70% inhibition against *C. gloeosporioides* and films containing PVA presented 51% inhibition against *L. theobromae*. The analysis in mangoes packed with films with chitosan nanoparticles did not present postharvest decomposition for 20 days at 26°C.	Dey et al., 2021
Nanotubes Nanocomposites	Horseradish peroxidases	Multiwalled carbon/ sodium alginate/ carboxymethyl chitosan films	A novel technique for the mass production of disposable immunosensors for rapid detection of *Escherichia coli* O157:H7 and *Enterobacter sakazakii* inexpensively is reported.	Dou et al., 2013
Nanoparticles	—	Nanocellulose and activated carbon	A novel film was developed for smart food packaging. Films with nanocellulose and activated carbon at 50% and, 30% have good electrical properties for their function as a biosensor.	Sobhan et al., 2019
Inclusion complex	Cinnamon essential oil	β-cyclodextrin inclusion complex/electrospun polylactic acid nanofilm	The inclusion complexes to the polylactic acid nanofibers significantly improved thermal stability and antimicrobial activity against *Escherichia coli* and *Staphylococcus aureus*.	Wen et al., 2016
Inclusion complex	Quercetin	β-cyclodextrin inclusion complex/electrospun zein nanofilm	Quercetin contained in the inclusion complexes incorporated into the zein nanofibers was shown to have a good inhibitory effect on *E. coli* and *S. aureus*, in addition to the slow release of the compound, ideal for its potential application in food packaging.	Wang et al., 2021
Nanoclusters	—	Gold nanoclusters/ chitosan induced by ultraviolet irradiation	The developed colorimetric method effectively detected *S. aureus*, *E. coli*, and *Bacillus subtilis*. Revealing an excellent alternative for the rapid in situ detection of pathogenic bacteria and detecting enterotoxin in food.	Xie et al., 2019
Nanoparticles	Polietilenimina	Magnetic quantum dot nanoparticles	The sensibility on quantification and simultaneous detection of two toxins (botulinum neurotoxin type A and staphylococcal enterotoxin B) was performed with a wide dynamic range. The biosensor is a potential test tool for the sensitive detection of trace toxins.	Wang et al., 2019

some examples of said novel treatments that have currently been developed to preserve food and have potential application in fruit and vegetable products, given their improved properties such as antimicrobial and antioxidant.

Nanotechnology has implications in the control of diseases of horticultural produce as well as safety and security; in the following two sections, some applications are described, and the potential uses of these nanosystems are highlighted.

1.5.1 Nanosystems in Disease Control of Horticultural Produce

Horticultural produce is susceptible to fungi development, and these cause about 70–80% of postharvest losses. The infection is produced from pre-harvest, and when the biotic and abiotic conditions are present in postharvest, then the fungi growth causes damage and losses throughout the pericarp. The losses in developing countries from this cause are approximately 25% and up to 50 % in undeveloped countries (Zhang et al., 2020). The result of decay occurs in different stages of the commercialization of fruits and vegetables, dependent on maturity and growth factors. Besides, the decrease in quality due to the development of pathogenic fungi induces the production of secondary metabolites from the horticultural produce as a defense mechanism, and some of these can potentially be toxic for consumption and therefore cause diseases (Atiq et al., 2020). In this sense, and to reduce the effects of contamination by fungi and other pathogens, nanotechnology is now being considered a helpful emerging technology in controlling fungi that cause diseases, deterioration, and rotting in fruits and vegetables. However, to increase the effectiveness of active components in nanosystems, sometimes it is necessary more than one component. In such a way, essential oils, plant extracts, or their pure components are used as organic active components and SiO_2, CuO, ZnO, and CaO as active inorganic components (Rajwade et al., 2020). The use of nanosystems charged with natural fungicides represents a beneficial form to reduce the chemical fungicides harness. Moreover, the small size contributes to reducing the quantity of functional ingredients needed to obtain an effective antifungal in horticultural produce (Atiq et al., 2020).

Studies have been carried out combining substances with different sites of action on fungi cells; thus, Cu nanoparticles combined with commercial fungicides $Cu(OH)_2$, fluazinam, mancozeb, thiophanate methyl, carbendazim, fenhexamid, zoxamide, fludioxonil, and difenoconazole were tested on the inhibition of isolates of *Botrytis cinerea*, highlighting thiophanare methyl and benizmadoles showed synergistic effect when these were studied *in vitro* and on the surfaces of apples, establishing that the use of Cu nanoparticles can contribute effectively to the control of the pathogen and can also be considered a sustainable and environmentally safe method of fungi control (Malandrakis et al., 2020).

The antifungal activity of zein nanoparticles functionalized with natamycin-D prepared by the phase separation method and with a particle size of 223 nm, an efficiency encapsulation of 68.3%, a zeta potential of –50.7 mV, and a coating with carboxymethyl chitosan was tested on the growth of *Botrytis cinerea*, achieving a 64.4% of micellar growth and inhibiting the spore germination when the concentration of natamycin-D was of 10 mg/L in zein nanoparticles (Lin et al., 2020). Different types of nanosystems have been used to control fungi growth, including inorganic nanoparticles such as nano-zinc, nano-silver, nano-magnesium, nano-calcium, and other compounds generally in their oxide form.

Nano-zeolites and nanoclays have been used as an absorbent of the active component to potentiate the antifungal effect since these components increase the absorption and retention of active components and modify the release of essential oils and extracts with the advantage of reducing water in the process. This modification confers biopolymer molecular flexibility that changed the mechanical and thermal resistance and their functionality during the application in fresh produce and the chain of distribution. The packaging material, edible coatings, is an excellent form for applying nanosystems on the internal surface for the active component delivery during storage and distribution. Another form is infiltrating the fruit by ultrasound, packaging, or well sprinkled on the surface to control the development of fungi.

1.5.2 Human Pathogen Control to Security and Safety Products

Due to the SARS-CoV-2 pandemic and the necessity to consume healthy foods with beneficial effects on the population, horticultural produce represents the best alternative to obtain nutrients and functional food containing active components. It is also important to guarantee the security and safety of these products, including those minimally processed fruits and vegetables (Anelich et al., 2020). Furthermore, the lack of sanitary control gives rise to foodborne diseases that represent a high cost of public health, considering that this contamination will depend mainly on processing (Mahajan et al., 2017). Nanosystems are viable alternatives that have been considered and studied in the last two decades to encapsulate different active components like phenols, flavonols, terpenoids, sesquiterpenes, and other components of essential oils and plant extracts. The beneficial effects obtained are an increase in solubility, biocompatibility, antioxidant capacity, and antimicrobial effect.

Moreover, incorporating material that is nanometric in size will allow the proportion of active components to be lowered with the same beneficial effect with respect to those larger sizes; nanosystems have greater surface areas with higher interactions between the nanosystem and fresh produce, increasing antioxidant and antimicrobial effectiveness (Ijaz et al., 2020). Natural additives (e.g., essential oils, plant extracts, and biotechnological products) effectively reduce and control microbial growth during the storage, commercialization, and consumption of fresh and minimally processed fruits and vegetables. The antibacterial activity of essential oils is principally due to their volatile presence; then, the nanoencapsulation exerts a protective effect on the volatile presence, allowing a controlled release. A good encapsulation enables the release of volatile and other active components in the functioning of biotic and abiotic conditions.

The choice between different nanosystems and their composition depends on the application, structure, and tissue composition. The specific function of the products as antimicrobials, antioxidants, and metabolic modifiers is considered regarding the active component characteristics needed for encapsulation and controlled-release conditions. The principal nanosystems used in horticultural produce are nanospheres, nanocapsules, nanofibers, and nanostructured lipid carriers; the considerations to take into account are (a) the source of the active component with antimicrobial activity; (b) the minimum concentration necessary of the active component to inhibit or retard the microbial activity; (c) the compatibility of biopolymer, lipid, or oil used as encapsulating with the horticultural produce; (d) the characteristics of release as a function of the action site; and (e) the interaction of components with the cellular membrane. Encapsulation with natural biopolymers has the advantage of not being toxic. For example, chitosan from marine sources has been used to prepare biopolymeric nanoparticles to entrap an active component that has antimicrobial potential itself, increasing the antimicrobial effect of the extract, essential oils, or other natural antimicrobial used (Zheng and Zhu, 2003). Some essential oils encapsulated in chitosan with antibacterial effectivity have been studied; examples are cloves, cinnamon, rosemary, thyme, and lime lemon, which have shown effectiveness against *Escherichia coli*, *Listeria monocytogenes*, *Staphylococcus aureus*, *Erwinia carotovora*, and *Pseudomonas fluorescens* (Hadidi et al., 2020; Mohammadi et al., 2020; Sotelo-Boyás et al., 2017). The effect of chitosan, allyl isothiocyanate, and nisin was evaluated on the development of *Salmonella spp*. in fresh melon, concluding their use as disinfecting in subsequent applications was effective to decrease the recontamination of product due to the handling or product during the process (Chen et al., 2012). Nanoparticles from polylactic acid, positively charged by modification with polyethyleneimine, were studied for encapsulating carvacrol and tested to control the growth of *E. coli*, *L. monocytogenes*, *Salmonella enterica*, and *Staphylococcus aureus*, establishing that the surface charge allowed the best add-in the cell and that the composition effectively inhibited the growth of these microorganisms (Niza et al., 2020).

Nanosystems with mixtures of materials, organic and inorganic, have been studied. Silver nanoparticles have been the type of nanoparticles most studied to control microbial growth. The silver nanoparticles had a higher effect on gram-negative bacteria than gram-positive when tested on *E. coli*, *Salmonella typhimurium*, *Klebsiella pneumoniae*, and *Staphylococcus aureus* (Hungund et al., 2015). Thus, the nanosystem can potentially be used in the pathogen control of gram-negative microorganisms during the handling of fresh fruits and vegetables. According to a review by Keat et al. 2015, silver nanoparticles inhibit the growth of bacteria due to the small size of nanoparticles, which adhere to the cell wall of the microorganisms; moreover, the adhesion of silver nanoparticles causes damage to DNA and the proteins that contain sulfur, causing an outbreak of cell wall.

However, Ag is a toxic component, and its use is limited to packaging materials in contact with foods even though it has been highly debated; thus, ZnO nanoparticles have been tested for fungi and bacterial control. Studies carried out combined chitosan, and ZnO nanoparticles for gram-positive and gram-negative bacterial control were most effective against *S. aureus* than against *E. coli* (Yusof et al., 2019). In other studies with TiO_2 nanoparticles (12.5 at 100 µg/mL) combined with a by-product from kiwi, peach, or plum, the results showed that organic–inorganic components reduced the growth effectively gram-negative bacteria, such as *E. coli* (Ajmal et al., 2019). The action mechanism of these oxides is associated with the generation of reactive oxygen species that interact with the cell membrane producing the alteration of metabolism and causing the loss of their integrity, thereby having an antibacterial effect (Kumar et al., 2020).

1.6 Horticultural Produce Nano-Packaging

Continuously packaging systems have an improvement process to include biodegradable and eco-friendly materials to reduce effluents and waste; significant efforts to develop packaging minimize petroleum-derived use. Although the new materials are prepared with by-products and the nanotechnology represents an alternative to the development of new and innovative packaging prepared with natural-origin biopolymers with high mechanical resistance, they are able to control the release of a natural active component, offering a gas and water vapor barrier as well as an excellent physiological and microbial control (Han et al., 2018; Singh and Kalia, 2019). Moreover, the packaging of fresh and minimally processed horticultural products is the final stage before refrigerated storage and represents a way of communicating the characteristics and presentation of products to the consumer. At the beginning of this century, packaging functionality improved thermal resistance, mechanical properties, and barrier properties and reduced migration. However, the active packaging was developed with the incorporation of active components such as essential oils (clove, rosemary, cinnamon, lemon, among others) and plant extracts (e.g., phenols, anthocyanins, and flavonols), generally encapsulated in natural biopolymers, and could be reinforced with nanoclays and inorganic nanoparticles (e.g., TiO_2, Ag, CuO, ZnO, SiO_2), with antimicrobial, antioxidant effect, respiration, and ethylene control and controlled release of the active compound during the storage time. Including nanoclays in packaging improves the resistance and interaction with the cell structures and allows a sustained release that remains during the storage of horticultural produce, contributing to maintaining freshness for a longer time (Ebrahimi et al., 2021).

The interactions between the nano-packaging environment and horticultural product take advantage of the modified properties and metabolism of horticultural produce, which achieves ethylene control since the packaging has the capacity to absorb and scavenge oxide or block the receptors, thus achieving an extended shelf life (Sahraei Khosh Gardesh et al., 2016; Shakeri et al., 2019). The nanosystems most used for ethylene control are absorber and oxidation agents; the most important photocatalytic compounds are based on TiO_2 nanoparticles. These nanoparticles act in the presence of water and light, producing the photodegradation of ethylene and a reduction in metabolic activity (Hussain et al., 2011). Another scavenger widely used is $KMnO_4$, when these is absorbed into nano-silica and nanoclays to form a new film based on a polyolefin elastomer, finding that the higher concentration of $KMnO_4$ had a more significant ethylene entrapment, which managed to increase the shelf life of bananas up to 15 days (Ebrahimi et al., 2021).

1.7 Nanosystems as an Edible Coating for Fresh Produce

Edible coatings are an emerging technology that helps preserve the sensory, nutritional, physicochemical, and microbiological characteristics of fresh and minimally processed horticultural produce. Edible coatings containing nanostructures are called "nanocoatings"; their unique condition is that almost one component preserves the nanometric size. These nanocoatings can entrap active and functional substances that fulfill different functions that regulate metabolic activity and therefore contribute to controlling oxidation reactions in membrane cells and increasing shelf life (Flores-López et al., 2016; Zambrano-Zaragoza et al., 2018). In addition, the characteristics of the nanocoating layer on fruits or vegetables generate a modified atmosphere that limits gas exchange and loss of weight. These depend on the composition and nanostructure as nanoemulsions, polymeric nanoparticles, solid lipid nanoparticles, nanostructured lipid carriers, metallic nanoparticles, nanofibers, nanoclays, and nano-zeolites, among others, that contribute to different nanocoating functions, such as modifying the thermal properties, reducing the cold damage, and protecting them from environmental conditions. Nanocomposites developed as coatings have infinite possibilities, as well as all the active components necessary to be incorporated.

Moreover, green methods for producing edible coatings represent an alternative sustainable to protect horticultural produce, resulting in an eco-friendly technology (Cerrillo et al., 2017; Flores-López et al., 2016; Gago et al., 2019). In the study on controlling citrus green mold (*Penicillium digitatum*), Ag-zeolite nanoparticles, considering two zeolite pore sizes and silver concentration, were studied; these nanoparticles were prepared by ion exchange. The Ag nanoparticles show that a high concentration of Ag inhibits the growth of fungi. However, the Ag concentration causes physiological damage in orange; then, the zeolite:Ag ratio contributes to a controlled release on the surface of citrus (Cerrillo et al., 2017). Chitosan nanoparticles prepared by ionic gelation showed an effect on *Aspergillus niger*, *Botrytis cinerea*, and *Rhizopus stolonifera* when an edible coating was formed on strawberries previously inoculated with these microorganisms, with the conclusion that chitosan nanocomposites were able to reduce the severity of the infection caused by the fungus studied (Melo et al., 2020).

1.8 Future Trends

With better knowledge of the interactions between cell structures, biocompatibility, and bioaccessibility, the modification and best integration of active components into a nanosystem will be possibly achieved, for example, a controlled release with the specific site of action. Furthermore, studies related to the cellular response to the stimuli of active components will allow nanosystems that modified the signalization into the cell.

The study of microorganisms' cellular response to nanosystems and their active components enables the integration of substances that promote the modification of the defense metabolism of fresh fruits and vegetables. It will also be possible to obtain intelligent materials capable of detecting and inhibiting the growth of pathogen microorganisms that decrease food security and safety.

Another important parameter will be the control release mechanisms and the vectorization of specific cell parts. It will also be essential to study the effect of signaling between the cellular components attributed to nanostructures' interaction from the cell wall and membrane to the interior.

Nanobiosensors, used in conjunction with controlled-release nanosystems (independently or integrated), will be an excellent tool for controlling different pathogens, as well as preserving the components of biological importance in the horticultural produce (maximizing properties such as antioxidants), while monitoring the quality changes. Furthermore, these systems can be vectorized as a nano–edible coating or for active nano-packaging.

1.9 Conclusion

The nanosystems developed with potential applications in the conservation of horticultural produce have represented a viable and safe alternative that leads to the exploration of industrial applications. A reduction in postharvest microbial load and an increase in antioxidant capacity with a long shelf life are attributable to the incorporation of bioactive substances in combination with different polymers (polysaccharides and proteins of natural origin), lipid matrices, and inorganic compounds (with

GRAS components) in nanosystems produced by various methods; this is due to the controlled release of active components. Furthermore, the targeted delivery of essential oils and natural extracts (as active components of interest) makes the integration of nanostructured systems a viable and reliable alternative, in contrast to other substances with undesired health effects (e.g., pesticides or synthetic origin). The applications will depend on the requirements of each product and the compatibility of the materials that make up the nanostructured system with the selected fruit and vegetable product.

References

Adaskaveg, J.E., and Förster, H. 2009. New developments in postharvest fungicide registrations for edible horticultural crops and use strategies in the United States. In *Postharvest pathology*. Springer. 107–117.

Aggarwal, S., Mohite, A.M., and Sharma, N. 2018. The maturity and ripeness phenomenon with regard to the physiology of fruits and vegetables: A review. *Bulletin of the Transylvania University of Brasov, Series II: Forestry, Wood Industry, Agricultural Food Engineering*, 11(60), 77–88.

Ajmal, N., Saraswat, K., Bakht, M.A., Riadi, Y., Ahsan, M.J., and Noushad, M. 2019. Cost-effective and eco-friendly synthesis of titanium dioxide (TiO_2) nanoparticles using fruit's peel agro-waste extracts: Characterization, in vitro antibacterial, antioxidant activities. *Green Chemistry Letters and Reviews*, 12(3), 244–254.

Alkan, N., Friedlander, G., Ment, D., Prusky, D., and Fluhr, R. 2015. Simultaneous transcriptome analysis of *Colletotrichum gloeosporioides* and tomato fruit pathosystem reveals novel fungal pathogenicity and fruit defense strategies. *New Phytologist*, 205(2), 801–815.

Altendorf, S. 2017. Global prospects for major tropical fruits. The food outlook: Biannual report on global food markets. The Food and Agriculture Organization of the United Nations (FAO), Rome, 68–81.

Anelich, L.E.C.M., Lues, R., Farber, J.M., and Parreira, V.R. 2020. SARS-CoV-2 and risk to food safety. *Frontiers in Nutrition*, 7, 1–8.

Atiq, M., Naeem, I., Sahi, S.T., Rajput, N.A., Haider, E., Usman, M., Shahbaz, H., Fatima, K., Arif, E., and Qayyum, A. 2020. Nanoparticles: A safe way towards fungal diseases. *Archives of Phytopathology and Plant Protection*, 53(17–18), 1–12.

Barkai-Golan, R. 2001. *Postharvest diseases of fruits and vegetables*. Development and Control, Elsevier Science.

Cerrillo, J.L., Palomares, A.E., Rey, F., Valencia, S., Palou, L., and Pérez-Gago, M.B. 2017. Ag-zeolites as fungicidal material: Control of citrus green mold caused by *Penicillium digitatum*. *Microporous and Mesoporous Materials*, 254, 69–76.

Ceylan, Z., Meral, R., and Cetinkaya, T. 2020. Relevance of SARS-CoV-2 in food safety and food hygiene: Potential preventive measures, suggestions and nanotechnological approaches. *Virus Disease*, 31(2), 154–160.

Chávez-Magdaleno, M.E., González-Estrada, R.R., Ramos-Guerrero, A., Plascencia-Jatomea, M., and Gutiérrez-Martínez, P. 2018. Effect of pepper tree (*Schinus molle*) essential oil-loaded chitosan bio-nanocomposites on postharvest control of *Colletotrichum gloeosporioides* and quality evaluations in avocado (*Persea americana*) cv. Hass. *Food Science and Biotechnology*, 27(6), 1871–1875.

Chen, W., Jin, T.Z., Gurtler, J.B., Geveke, D.J., and Fan, X. 2012. Inactivation of Salmonella on whole cantaloupe by application of an antimicrobial coating containing chitosan and allyl isothiocyanate. *International Journal of Food Microbiology*, 155(3), 165–170.

Chennimalai, M., Vijayalakshmi, V., Senthil, T.S., and Sivakumar, N. 2021. One-step green synthesis of ZnO nanoparticles using *Opuntia humifusa* fruit extract and their antibacterial activities. *Materials Today: Proceedings*, 47, 1842–1846.

Cortes-Rivera, H. J., Blancas-Benítez, F. J., Romero-Islas, L., Gutiérrez-Martínez, P., and González-Estrada, R. 2019. In vitro evaluation of residues of coconut (*Cocos nucifera* L.) aqueous extracts, against the fungus *Penicillium italicum*. *Emirates Journal of Food and Agriculture*, 31(8), 613–617.

Dávila-Rodríguez, M., López-Malo, A., Palou, E., Ramírez-Corona, N., and Jiménez-Munguía, M.T. 2019. Antimicrobial activity of nanoemulsions of cinnamon, rosemary, and oregano essential oils on fresh celery. *LWT-Food Science and Technology*, 112, 108247.

de Chiara, M.L.V., Pal, S., Licciulli, A., Amodio, M.L., and Colelli, G. 2015. Photocatalytic degradation of ethylene on mesoporous TiO_2/SiO_2 nanocomposites: Effects on the ripening of mature green tomatoes. *Biosystems Engineering*, 132, 61–70.

Deng, L.Z., Mujumdar, A.S., Pan, Z., Vidyarthi, S.K., Xu, J., Zielinska, M., and Xiao, H.W. 2020. Emerging chemical and physical disinfection technologies of fruits and vegetables: A comprehensive review. *Critical Reviews in Food Science and Nutrition*, 60(15), 2481–2508.

Dey, D., Dharini, V., Periyar Selvam, S., Rotimi Sadiku, E., Mahesh Kumar, M., Jayaramudu, J., and Nath Gupta, U. 2021. Physical, antifungal, and biodegradable properties of cellulose nanocrystals and chitosan nanoparticles for food packaging application. *Materials Today: Proceedings*, 38, 860–869.

Dou, W., Tang, W., and Zhao, G. 2013. A disposable electrochemical immunosensor arrays using 4-channel screen-printed carbon electrode for simultaneous detection of *Escherichia coli* O157: H7 and *Enterobacter sakazakii*. *Electrochimica Acta*, 97, 79–85.

Ebrahimi, A., Zabihzadeh Khajavi, M., Mortazavian, A.M., Asilian-Mahabadi, H., Rafiee, S., Farhoodi, M., and Ahmadi, S. 2021. Preparation of novel nano—based films impregnated by potassium permanganate as ethylene scavengers: An optimization study. *Polymer Testing*, 93, 106934.

Fabi, C., and English, A. 2019. Methodological proposal for monitoring SDG target 12.3. Sub-indicator 12.3. 1. The food loss index design, data collection methods and challenges. In FAO Statistics Working Paper Series. FAO publisher. 1–59.

FAO. 2016. Pérdidas y desperdicios de alimentos en América Latina y el Caribe. www.fao.org/publications.

Flores-López, M.L., Cerqueira, M.A., de Rodríguez, D.J., and Vicente, A.A. 2016. Perspectives on utilization of edible coatings and nano-laminate coatings for extension of postharvest storage of fruits and vegetables. *Food Engineering Reviews*, 8(3), 292–305.

Gago, C.M.L., Artiga-Artigas, M., Antunes, M.D.C., Faleiro, M.L., Miguel, M.G., and Martín-Belloso, O. 2019. Effectiveness of nanoemulsions of clove and lemongrass essential oils and their major components against *Escherichia coli* and *Botrytis cinerea*. *Journal of Food Science and Technology*, 56(5), 2721–2736.

Gaikwad, K.K., Singh, S., and Negi, Y.S. 2020. Ethylene scavengers for active packaging of fresh food produce. *Environmental Chemistry Letters*, 18(2), 269–284.

González-Reza, R., García-Betanzos, C., Sánchez-Valdes, L., Quintanar-Guerrero, D., Cornejo-Villegas, M., and Zambrano-Zaragoza, M. 2018. The functionalization of nanostructures and their potential applications in edible coatings. *Coatings*, 8(5), 160.

González-Reza, R.M., Hernandez-Sanchez, H., Zambrano-Zaragoza, M.L., Gutierrez-Lopez, G.F., Del Real, A., Quintanar-Guerrero, D., and Velasco-Bejarano, B. 2020. Influence of stabilizing and encapsulating polymers on antioxidant capacity, stability, and kinetic release of thyme essential oil nanocapsules. *Foods*, 9(12), 1884.

González-Reza, R.M., Quintanar-Guerrero, D., Del Real-López, A., Piñon-Segundo, E., and Zambrano-Zaragoza, M.L. 2018. Effect of sucrose concentration and pH onto the physical stability of β-carotene nanocapsules. *LWT–Food Science and Technology*, 90, 354–361.

Gutiérrez-Martínez, P., Ramos-Guerrero, A., Rodríguez-Pereida, C., Coronado-Partida, L., Angulo-Parra, L., and González-Estrada, R. 2018. Chitosan for postharvest disinfection of fruits and vegetables. In: *Postharvest disinfection of fruits and vegetables*. Academic Press-Elsevier. 231–241.

Hadidi, M., Pouramin, S., Adinepour, F., Haghani, S., and Jafari, S.M. 2020. Chitosan nanoparticles loaded with clove essential oil: Characterization, antioxidant and antibacterial activities. *Carbohydrate Polymers*, 236, 116075.

Han, J.W., Ruiz-Garcia, L., Qian, J.P., and Yang, X.T. 2018. Food packaging: A comprehensive review and future trends. *Comprehensive Reviews in Food Science and Food Safety*, 17(4), 860–877.

He, Q., Guo, M., Jin, T.Z., Arabi, S.A., and Liu, D. 2021. Ultrasound improves the decontamination effect of thyme essential oil nanoemulsions against *Escherichia coli* O157: H7 on cherry tomatoes. *International Journal of Food Microbiology*, 337, 108936.

Helmy, K.G., Partila, A.M., and Salah, M. 2020. Gamma radiation and polyvinyl pyrrolidone mediated synthesis of zinc oxide/zinc sulfide nanoparticles and evaluation of their antifungal effect on pre and post harvested orange and pomegranate fruits. *Biocatalysis and Agricultural Biotechnology*, 29, 101728.

Horská, T., Kocourek, F., Stará, J., Holý, K., Mráz, P., Krátký, F., Kocourek, V., and Hajšlová, J. 2020. Evaluation of pesticide residue dynamics in lettuce, onion, leek, carrot and parsley. *Foods*, 9, 680.

Hungund, B.S., Dhulappanavar, G.R., and Ayachit, N.H. 2015. Comparative evaluation of antibacterial activity of silver nanoparticles biosynthesized using fruit juices. *Journal of Nanomedicine & Nanotechnology*, 6(2), 1.

Hussain, M., Bensaid, S., Geobaldo, F., Saracco, G., and Russo, N. 2011. Photocatalytic degradation of ethylene emitted by fruits with TiO_2 nanoparticles. *Industrial and Engineering Chemistry Research*, 50(5), 2536–2543.

Ijaz, M., Zafar, M., Afsheen, S., and Iqbal, T. 2020. A review on Ag-nanostructures for enhancement in shelf time of fruits. *Journal of Inorganic and Organometallic Polymers and Materials*, 30(5), 1475–1482.

Kader, A.A., and Yahia, E.M. 2011. Postharvest biology of tropical and subtropical fruits. In *Postharvest biology and technology of tropical and subtropical fruits*. Woodhead Publishing. 79–111.

Karimi, A., Askari, G., Yarmand, M.S., Salami, M., and EmamDjomeh, Z. 2020. Development, modification and characterization of ursolic acid-loaded gelatin nanoparticles through electrospraying technique. *Food and Bioproducts Processing*, 124, 329–341.

Keat, C.L., Aziz, A., Eid, A.M., and Elmarzugi, N.A. 2015. Biosynthesis of nanoparticles and silver nanoparticles. *Bioresources and Bioprocessing*, 2(1).

Kitinoja, L., and Kader, A.A. 2015. Measuring postharvest losses of fresh fruits and vegetables in developing countries. *PEF White Paper*, 15, 26.

Kumar, S., Mukherjee, A., and Dutta, J. 2020. Chitosan based nanocomposite films and coatings: Emerging antimicrobial food packaging alternatives. *Trends in Food Science and Technology*, 97, 196–209.

Lazar, E.E., Jobling, J.J., and Benkeblia, N. 2010. Postharvest disease management of horticultural produce using essential oils: Today's prospects. *Stewart Postharvest Review*, 6(3), 1–9.

Le, K.H., Nguyen, M.D.B., Dai Tran, L., Thi, H.P.N., Van Tran, C., Van Tran, K., . . . La, D.D. 2021. A novel antimicrobial ZnO nanoparticles-added polysaccharide edible coating for the preservation of postharvest avocado under ambient conditions. *Progress in Organic Coatings*, 158, 106339.

Lee, K.H., Lee, J.S., Kim, E.S., and Lee, H.G. 2019. Preparation, characterization, and food application of rosemary extract-loaded antimicrobial nanoparticle dispersions. *LWT*, 101, 138–144.

Li, S., Sun, J., Yan, J., Zhang, S., Shi, C., McClements, D.J., Liu, X., and Liu, F. 2021. Development of antibacterial nanoemulsions incorporating thyme oil: Layer-by-layer self-assembly of whey protein isolate and chitosan hydrochloride. *Food Chemistry*, 339, 128016.

Liang, J., Yan, H., Wang, X., Zhou, Y., Gao, X., Puligundla, P., and Wan, X. 2017. Encapsulation of epigallocatechin gallate in zein/chitosan nanoparticles for controlled applications in food systems. *Food Chemistry*, 231, 19–24.

Lin, M., Fang, S., Zhao, X., Liang, X., and Wu, D. 2020. Natamycin-loaded zein nanoparticles stabilized by carboxymethyl chitosan: Evaluation of colloidal/chemical performance and application in postharvest treatments. *Food Hydrocolloids*, 106, 105871.

Liu, J., Sui, Y., Wisniewski, M., Xie, Z., Liu, Y., You, Y., Zhang, X., Sun, Z., Li, W., Li, Y., and Wang, Q. 2018. The impact of the postharvest environment on the viability and virulence of decay fungi. *Critical Reviews in Food Science and Nutrition*, 58(10), 1681–1687.

Liu, W., Zhang, M., and Bhandari, B. 2020. Nanotechnology—A shelf life extension strategy for fruits and vegetables. *Critical Reviews in Food Science and Nutrition*, 60(10), 1706–1721.

Mahajan, P.V., Caleb, O.J., Gil, M.I., Izumi, H., and Colelli, G. 2017. Quality and safety of fresh horticultural commodities: Recent advances and future perspectives. *Food Packaging and Shelf Life*, 14, 2–11.

Mahajan, P.V., Caleb, O.J., Singh, Z., Watkins, C.B., and Geyer, M. 2014. Postharvest treatments of fresh produce. *Philosophical Transactions of the Royal Society A: Mathematical, Physical and Engineering Sciences*, 372(2017).

Malandrakis, A.A., Kavroulakis, N., and Chrysikopoulos, C.V. 2020. Synergy between Cu-NPs and fungicides against Botrytis cinerea. *Science of the Total Environment*, 703, 135557.

Matrose, N.A., Obikeze, K., Belay, Z.A., and Caleb, O.J. 2020. Plant extracts and other natural compounds as alternatives for postharvest management of fruit fungal pathogens: A review. *Food Bioscience*, 100840.

Melo, N.F.C.B., de Lima, M.A.B., Stamford, T.L.M., Galembeck, A., Flores, M.A.P., de Campos Takaki, G.M., da Costa Medeiros, J.A., Stamford-Arnaud, T.M., and Montenegro Stamford, T.C. 2020. In vivo and in vitro antifungal effect of fungal chitosan nanocomposite edible coating against strawberry phytopathogenic fungi. *International Journal of Food Science and Technology*, 55(11), 3381–3391.

Meng, R., Wu, Z., Xie, Q.T., Cheng, J.S., and Zhang, B. 2021. Preparation and characterization of zein/carboxymethyl dextrin nanoparticles to encapsulate curcumin: Physicochemical stability, antioxidant activity and controlled release properties. *Food Chemistry*, 340, 127893.

Mohammadi, A., Hosseini, S.M., and Hashemi, M. 2020. Emerging chitosan nanoparticles loading-system boosted the antibacterial activity of *Cinnamomum zeylanicum* essential oil. *Industrial Crops and Products*, 155, 112824.

Nicolopoulou-Stamati, P., Maipas, S., Kotampasi, C., Stamatis, P., and Hens, L. 2016. Review chemical pesticides and human health: The urgent need for a new concept in agriculture. *Front Public Health*, 4, 148.

Niza, E., Božik, M., Bravo, I., Clemente-Casares, P., Lara-Sanchez, A., Juan, A., Klouček, P., and Alonso-Moreno, C. 2020. PEI-coated PLA nanoparticles to enhance the antimicrobial activity of carvacrol. *Food Chemistry*, 328.

Novickij, V., Stanevičienė, R., Staigvila, G., Gruškienė, R., Sereikaitė, J., Girkontaitė, I., Novickij, J., and Servienė, E. 2020. Effects of pulsed electric fields and mild thermal treatment on antimicrobial efficacy of nisin-loaded pectin nanoparticles for food preservation. *LWT-Food Science and Technology*, 120.

Oehlke, K., Keppler, J.K., Milsmann, J., Mayer-Miebach, E., Greiner, R., and Steffen-Heins, A. 2019. Adsorption of β-lactoglobulin to solid lipid nanoparticles (SLN) depends on encapsulated compounds. *Journal of Food Engineering*, 247, 144–151.

Olaimat, A.N., Shahbaz, H.M., Fatima, N., Munir, S., and Holley, R.A. 2020. Food safety during and after the era of COVID-19 pandemic. *Frontiers in Microbiology*, 11.

Palou, L. 2018. Postharvest treatments with GRAS salts to control fresh fruit decay. *Horticulturae*, 4, 46.

Prakash, A., and Vadivel, V. 2020. Citral and linalool nanoemulsions: Impact of synergism and ripening inhibitors on the stability and antibacterial activity against *Listeria monocytogenes*. *Journal of Food Science and Technology*, 57(4), 1495–1504.

Priyadarshi, R., Kim, H.J., and Rhim, J.W. 2021. Effect of sulfur nanoparticles on properties of alginate-based films for active food packaging applications. *Food Hydrocolloids*, 110, 106155.

Rajwade, J.M., Chikte, R.G., and Paknikar, K.M. 2020. Nanomaterials: New weapons in a crusade against phytopathogens. *Applied Microbiology and Biotechnology*, 104(4), 1437–1461.

Ramos-Guerrero, A., Gonzalez-Estrada, R., Hanako-Rosas, G., Bautista-Baños, S., Acevedo-Hernandez, G., Tiznado, M., and Gutiérrez-Martinez, P. 2018. Use of inductors in the control of *Colletotrichum gloeosporioides* and *Rhizopus stolonifer* isolated from soursop fruits: In vitro tests. *Food Science and Biotechnology*, 27, 755–763.

Rodríguez-Félix, F., López-Cota, A.G., Moreno-Vásquez, M.J., Graciano-Verdugo, A.Z., Quintero-Reyes, I.E., Del-Toro-Sánchez, C.L., and Tapia-Hernández, J.A. 2021. Sustainable-green synthesis of silver nanoparticles using safflower (*Carthamus tinctorius* L.) waste extract and its antibacterial activity. *Heliyon*, 7(4), e06923.

Romanazzi, G., and Feliziani, E. 2014. *Post-harvest decay* (Botrytis cinerea (Gray Mold)). Academic Press. 131–146.

Romanazzi, R., Sanzani, M., Bi, Y., Tian, S., Gutiérrez-Martínez, P., and Alkan, N. 2016. Induced resistance to control postharvest decay of fruit and vegetables. *Postharvest Biology and Technology*, 122, 82–94.

Rosales, T.K.O., da Silva, M.P., Lourenco, F.R., Hassimotto, N.M.A., and Fabi, J.P. 2021. Nanoencapsulation of anthocyanins from blackberry (*Rubus* spp.) through pectin and lysozyme self-assembling. *Food Hydrocolloids*, 114, 106563.

Sahraei Khosh Gardesh, A., Badii, F., Hashemi, M., Ardakani, A.Y., Maftoonazad, N., and Gorji, A.M. 2016. Effect of nanochitosan based coating on climacteric behavior and postharvest shelf-life extension of apple cv. Golab Kohanz. *LWT–Food Science and Technology*, 70, 33–40.

Sánchez-Torres, P., and Tuset, J.J. 2011. Molecular insights into fungicide resistance in sensitive and resistant *Penicillium digitatum* strains infecting citrus. *Postharvest Biology and Technology*, 59, 159–165.

Sardella, D., Gatt, R., and Valdramidis, V.P. 2017. Physiological effects and mode of action of ZnO nanoparticles against postharvest fungal contaminants. *Food Research International*, 101, 274–279.

Shakeri, M., Razavi, S., and Shakeri, S. 2019. Carvacrol and astaxanthin co-entrapment in beeswax solid lipid nanoparticles as an efficient nanosystem with dual antioxidant and anti-biofilm activities. *LWT–Food Science and Technology*, 107, 280–290.

Singh, G., and Kalia, A. 2019. Nano-enabled technological interventions for sustainable production, protection, and storage of fruit crops. *Nanoscience for Sustainable Agriculture*, 299–322.

Sivakanthan, S., Rajendran, S., Gamage, A., Madhujith, T., and Mani, S. 2020. Antioxidant and antimicrobial applications of biopolymers: A review. *Food Research International*, 136, 109327.

Sobhan, A., Muthukumarappan, K., Cen, Z., and Wei, L. 2019. Characterization of nanocellulose and activated carbon nanocomposite films' biosensing properties for smart packaging. *Carbohydrate Polymers*, 225, 115189.

Soltanzadeh, M., Peighambardoust, S.H., Ghanbarzadeh, B., Mohammadi, M., and Lorenzo, J.M. 2021. Chitosan nanoparticles as a promising nanomaterial for encapsulation of pomegranate (*Punica granatum* L.) peel extract as a natural source of antioxidants. *Nanomaterials*, 11(6), 1439.

Sotelo-Boyás, M.E., Correa-Pacheco, Z.N., Bautista-Baños, S., and Corona-Rangel, M.L. 2017. Physicochemical characterization of chitosan nanoparticles and nanocapsules incorporated with lime essential oil and their antibacterial activity against food-borne pathogens. *LWT–Food Science and Technology*, 77, 15–20.

Tarassoli, Z., Najjar, R., and Amani, A. 2021. Formulation and optimization of lemon balm extract loaded azelaic acid-chitosan nanoparticles for antibacterial applications. *Journal of Drug Delivery Science and Technology*, 102687.

Vivek, K., Suranjoy Singh, S., Ritesh, W., Soberly, M., Baby, Z., Baite, H., Mishra, S., and Pradhan, R.C. 2019. A review on postharvest management and advances in the minimal processing of fresh-cut fruits and vegetables. *Journal of Microbiology, Biotechnology and Food Sciences*, 8(5), 1178–1187.

Wang, C., Xiao, R., Wang, S., Yang, X., Bai, Z., Li, X., Rong, Z., Shen, B., and Wang, S. 2019. Magnetic quantum dot based lateral flow assay biosensor for multiplex and sensitive detection of protein toxins in food samples. *Biosensors and Bioelectronics*, 146, 111754.

Wang, M., Li, Y., Yang, J., Shi, R., Xiong, L., and Sun, Q. 2021. Effects of food-grade inorganic nanoparticles on the probiotic properties of *Lactobacillus plantarum* and *Lactobacillus fermentum*. *LWT*, 139, 110540.

Wang, Z., Zou, W., Liu, L., Wang, M., Li, F., and Shen, W. 2021. Characterization and bacteriostatic effects of β-cyclodextrin/quercetin inclusion compound nanofilms prepared by electrospinning. *Food Chemistry*, 338, 127980.

Wen, P., Zhu, D.H., Feng, K., Liu, F.J., Lou, W.Y., Li, N., Zong, M.H., and Wu, H. 2016. Fabrication of electrospun polylactic acid nanofilm incorporating cinnamon essential oil/β-cyclodextrin inclusion complex for antimicrobial packaging. *Food Chemistry*, 196, 996–1004.

Wisniewski, M., Droby, S., Norelli, J., Liu, J., and Schena, L. 2016. Alternative management technologies for postharvest disease control: The journey from simplicity to complexity. *Postharvest Biology and Technology*, 122, 3–10.

Xie, X., Tan, F., Xu, A., Deng, K., Zeng, Y., and Huang, H. 2019. UV-induced peroxidase-like activity of gold nanoclusters for differentiating pathogenic bacteria and detection of enterotoxin with colorimetric readout. *Sensors and Actuators, B: Chemical*, 279, 289–297.

Yadav, S., Mehrotra, G.K., and Dutta, P.K. 2021. Chitosan based ZnO nanoparticles loaded gallic-acid films for active food packaging. *Food Chemistry*, 334, 127605.
Yadollahi, A., Arzani, K., and Khoshghalb, H. 2010. The role of nanotechnology in horticultural crops postharvest management. *Acta Horticulturae*, 875, 49–56.
Yusof, N.A.A., Zain, N.M., and Pauzi, N. 2019. Synthesis of ZnO nanoparticles with chitosan as stabilizing agent and their antibacterial properties against Gram-positive and Gram-negative bacteria. *International Journal of Biological Macromolecules*, 124, 1132–1136.
Zambrano-Zaragoza, M.L., González-Reza, R., Mendoza-Muñoz, N., Miranda-Linares, V., Bernal-Couoh, T.F., Mendoza-Elvira, S., and Quintanar-Guerrero, D. 2018. Nanosystems in edible coatings: A novel strategy for food preservation. *International Journal of Molecular Sciences*, 19(3).
Zhang, X., Li, B., Zhang, Z., Chen, Y., and Tian, S. 2020. Antagonistic yeasts: A promising alternative to chemical fungicides for controlling postharvest decay of fruit. *Journal of Fungi*, 6(3), 1–15.
Zheng, L.Y., and Zhu, J.F. 2003. Study on antimicrobial activity of chitosan with different molecular weights. *Carbohydrate Polymers*, 54(4), 527–530.
Zhu, Z., Zhang, Y., Shang, Y., and Wen, Y. 2019. Electrospun nanofibers containing TiO_2 for the photocatalytic degradation of ethylene and delaying postharvest ripening of bananas. *Food and Bioprocess Technology*, 12, 281–287.

Chapter 2

Synthesis and Characterization Methods of Nano-Based Materials for the Postharvest Storage of Horticultural Produce

Sherin Joy Parappilly, Raji Vijayamma, Dona V. Idicula, Jithasha Balan, and Sumi Mary George

2.1 Introduction

Postharvested fruits and vegetables harbor a large number of living organisms. During storage and transportation, the properties of agricultural products, such as flavor, nutritional content, and their appearance, deteriorate or get damaged, owing to the loss of moisture content, decay and browning process, microbial activity, and so on (Lin & Zhao, 2007). It will lead to the wastage of agricultural products and a reduction in their commercial value with a serious impact on producers. To reduce the farmer's crisis and extend the shelf life of the postharvested produce, various methods have been used like irradiation, coating, modified atmosphere packaging, and low-temperature storage (Jianglian & Shaoying, 2013). Among the different measures, one of the promising procedures is edible coatings using nanoparticles, which will protect from aroma and water loss and restrict microbial growth by inhibiting oxygen penetration into plant tissues (Mantilla et al., 2013). Accordingly, the application of nanotechnology has a significant role in reducing the postharvest wastage of food items (Singh, 2020).

DOI: 10.1201/9781003142287-2

Nanotechnology is a modern, inventive, and versatile scientific attitude with superior effectiveness when compared with previously existing techniques. The fundamental components of nanotechnology are nanoparticles and nanostructured materials with unique physicochemical properties, used for the designing, development, and application of devices and materials. This emerging scientific area will not leave any field untouched with its fascinating scientific approach, and the agriculture sector is no exception (Ali et al., 2014).

Recently, nanotechnology has become the most popular hot spot in the scientific world, due to its beneficial impact on humankind, especially in areas like biomedical, biotechnology, and food technology (Khan et al., 2019a). The widespread application of nanoparticles in food technology includes the nanoencapsulation of food nutrients (Coles & Frewer, 2013; Eleftheriadou et al., 2017) and the development of food packaging materials, food additives, and food nanosensors (Kumar et al., 2020). Nowadays, the use of nanoparticles has greatly increased in the food packaging industry, where they are used mainly as antimicrobial agents against foodborne pathogens (Eleftheriadou et al., 2017; Kumar & Gautam, 2019; Kumar et al., 2020). Nanotechnology has been used widely in the field of horticulture for boosting plant growth and development and improving the productivity and health of plant species (Pramanik et al., 2020; Egirani et al., 2020) in the form of nanoformulated agrochemicals (Nair & Kumar, 2013; Pérez-de-Luque & Hermosín, 2013; Gopal et al., 2011; Zhao et al., 2012), nanosensors for monitoring the agrochemical residues (Omanović-Mikličanina & Maksimović, 2016; Kaushal & Wani, 2017; Yılmaz et al., 2017; Srivastava et al., 2018), catalyzing agents to diagnose the plant diseases (Manchester & Singh, 2006; Elbeshehy et al., 2015; Khan et al., 2019b), nanodevices for the genetically engineered plants (Ghidan & Al Antary, 2019), bioremediation of contaminated soil (Song et al., 2019; Vázquez-Núñez et al., 2020; Neethu et al., 2021), and as materials to enhance plant defense mechanisms against the oxidative stress (Prasad et al., 2017; Hussain et al., 2019; Khanna et al., 2015).

The potential beneficial effects of nanomaterials in agriculture and plant science have not been fully explored. Due to the increased demand of consumers for good-quality and highly nutritious fresh fruits and vegetables, the food industry has been motivated to develop better methods for maintaining the quality and shelf life of food (Rico et al., 2007). The potential application of nanotechnology and nanoparticles were illustrated in Figures 2.1 and 2.2. Among the different postharvest approaches, the use of palatable coatings and nano-overlay coatings (Flores-López et al., 2016) have received considerable attraction due to their substantial protection to agricultural products (Arnon et al., 2015; Silva-Weiss et al., 2013; Galus & Kadzińska, 2015; Xing et al., 2020). One of the prime advantages of the coatings is that they can be gobbled with the wrapped products, hence every constituent used in such nanoformulations should be distinguished as generally recognized as safe (GRAS). The barrier and mechanical characteristics of the edible coatings are strongly dependent on the physiochemical properties of the polysaccharides, lipids, and proteins used in it (Embuscado & Huber, 2009; Flores-López et al., 2016). The natural polysaccharide chitosan, with its antimicrobial activity, mechanical strength, and film-forming properties, has important applications as a coating material. In 2016, Kaya et al. (2016) have already disclosed chitosan as an effective food coating to extend the shelf life of the kiwifruit. Cellulose nanocrystals can also pop up the storage capacity of chitosan coverings employed for postharvest application on pears at different storage conditions (Deng et al., 2017). However, chitosan is less effective in managing moisture and preserving the fundamental arrangement (Xing et al., 2020). Other nanomaterials that have been

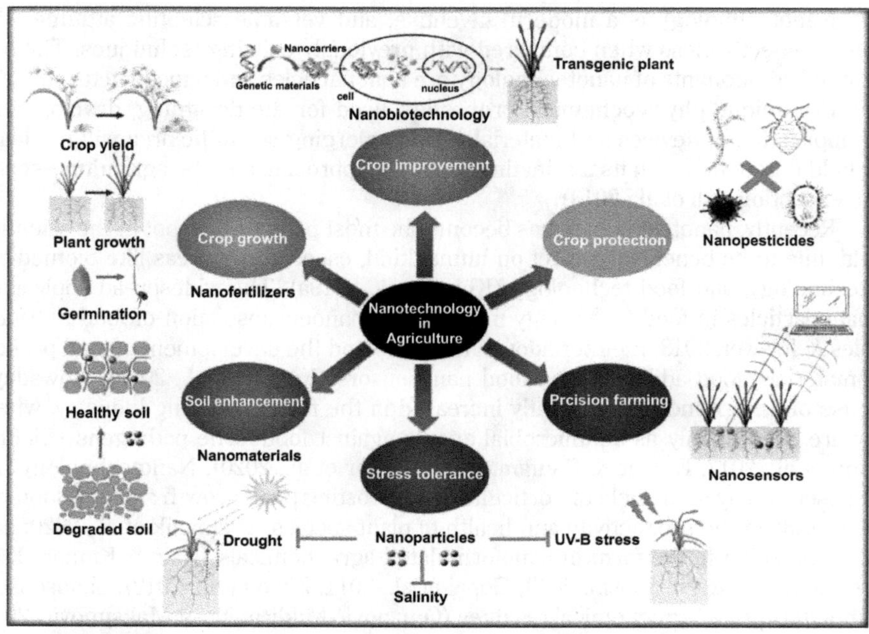

Figure 2.1 Applications of nanotechnology in agriculture (Shang et al., 2019).

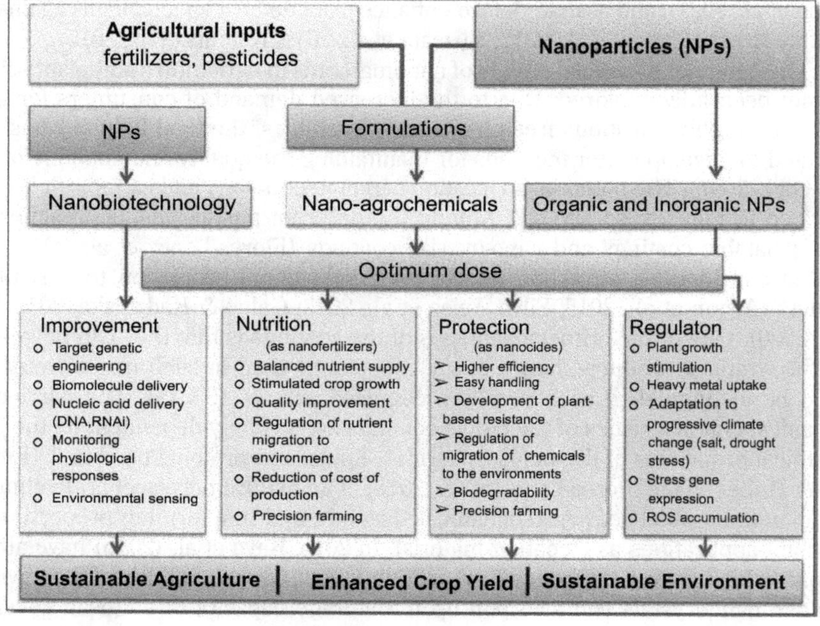

Figure 2.2 Potential applications of nanomaterials in sustainable agriculture production (Shang et al., 2019).

used as nanofillers in polymer coatings for packaging applications are nanoparticles of silver, TiO_2, silica, and montmorillonite. This chapter mainly discusses the synthesis and characterization methods used for the development of nano-based materials for the postharvest storage of horticultural produce.

2.2 Classification of Nanoparticles

Based on the chemical composition, morphology, dimensionality, size, and state (Gleiter, 2000), nanoparticles are classified into various groups. Based on dimensionality and shape, nanoparticles can be of (1) zero-dimensional nanomaterials (0D; i.e., sized smaller than100nm), which comprises spherical nanomaterials, nanorods, core-shell, cubes, hollow spheres, polygons, metal nanomaterials, and quantum dots (QDs); (2) one-dimensional nanomaterials (1D), which encompasses nanotubes, metallic and ceramic nanofibers, polymeric nanowires, nanorod fibers, or filaments; (3) two-dimensional nanomaterial (2D), which includes thin films, single-layered and multilayered nanoplates and nanocoating, crystalline or amorphous; and (4) three-dimensional nanomaterials (3D) includes carbon nanobuds, pillars, nanotubes, foams, fullerenes, fibers, polycrystals, layer skeletons, and honeycombs (Pokropivny & Skorokhod, 2007; Asghari et al., 2016; Aversa et al., 2018; Saleh, 2020). Dimension based classification of nanoparticles was illustrated in Figure 2.3. Morphologically different nanomaterials include nanorods, nano zigzags, nano hooks, nanostars, nanocubes, nano helices, and nanoplates.

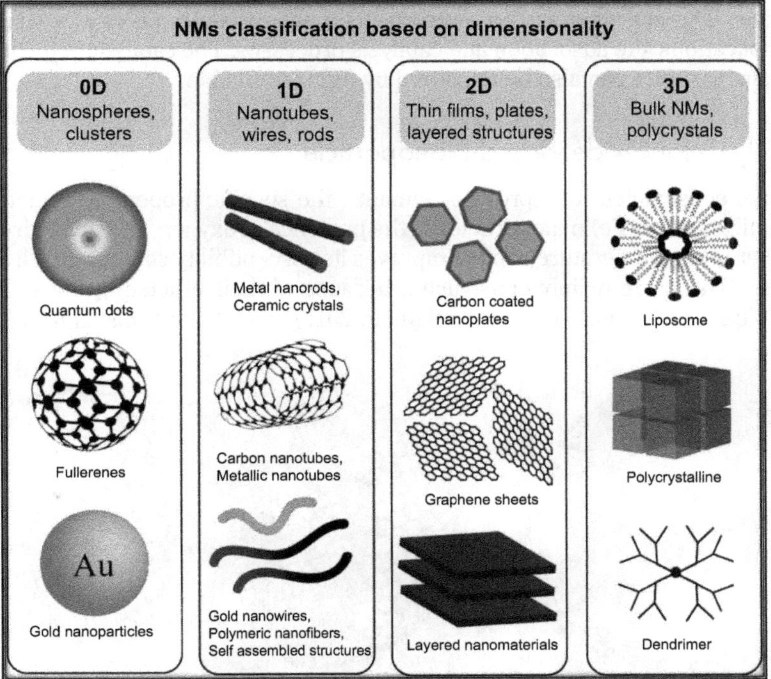

Figure 2.3 Diagrammatic illustration of dimension-based classification of nanoparticles (Poh et al., 2018).

2.2.1 Organic Nanoparticles

Organic nanoparticles are nontoxic and biodegradable. The commonly known organic particles include ferritin, dendrimers, liposomes, and micelles (hollow core; Figure 2.4; Tiwari et al., 2008). Organic nanoparticles have wide applications as drug delivery systems (in the biomedical area), because of their high efficiency, and likewise are used for site-specific targeted drug delivery.

2.2.2 Inorganic Nanoparticles

Inorganic nanoparticles include metal- and metal oxide–based particles.

2.2.2.1 Metal-Based Nanoparticles

These are prepared out of metals to nanometric size besides constructive or destructive methods (Salavati-Niasari et al., 2008). Metallic nanoparticles are assembled from metals like silver (Ag), copper (Cu), iron (Fe), aluminum (Al), zinc (Zn), titania (Ti), and silica (Saleh, 2020). The characteristics features of these metal-based nanoparticles include, surface charge, high surface area-to-volume ratio, amorphous structures, surface charge density, pore size, crystalline, etc. The shape of nanoparticles varies from cylindrical to spherical, and they show their susceptibility to climatic elements (moisture, heat, and sunlight, etc.) and color reactivity, which change from particle to particle. The activated surface area of metal nanoparticles plays a crucial role in their applications like adsorption and catalysis processes. The catalytic activities of the metal nanoparticles can also be improved by light irradiation.

2.2.2.2 Metal Oxide–Based Nanoparticle

These are constructed to improve or enhance the specific properties of metal-based nanoparticles. Iron (Fe) nanoparticles in the presence of oxygen oxidized to iron oxide (Fe_2O_3) at room temperature, which improves its susceptibility compared to iron nanomaterials. These are mainly constructed because of their efficiency and susceptibility (Tai et al., 2007; Ealias & Saravanakumar, 2017). Metal oxide–based nanoparticles

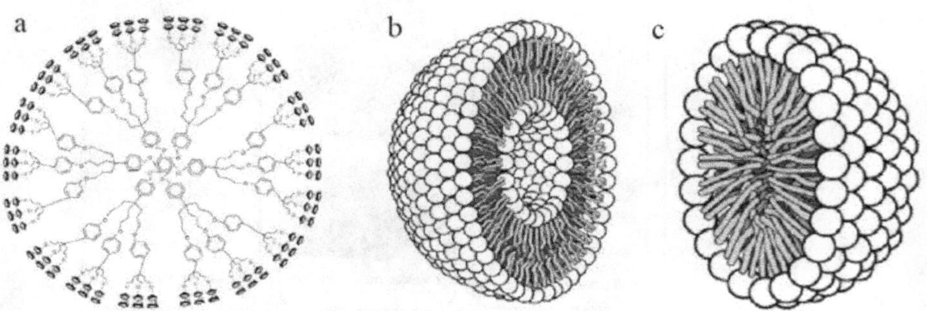

Figure 2.4 Organic nanoparticles: (a) dendrimers, (b) liposomes, and (c) micelles (Ealias & Saravanakumar, 2017).

like TiO_2, $SiO_2Fe_2O_3$, Al_2O_3, and ZnO are constructed by hydrothermal and sol-gel responses. The change in surface properties of the metal oxide–based nanoparticles affects the band gap energy of the particles and makes it have advantages in chemical sensors, catalysts, and semiconductor applications (Saleh and Fadillah, 2019; Saleh, 2020). The major characteristics of such materials are their biodegradability in supplying an extremely vital surface area. By adding doping metal ions, polymer chains, coupling agents (Qi et al., 2019; Das et al., 2020), and several organic compounds like thiols (Zeng et al., 2019), amines (Gaur & Banerjee, 2019), epoxies (Chu et al., 2017), anionic compounds (Hur et al., 2019), and others, the surface of nanoparticles can be altered.

2.2.3 Carbon-Based Nanoparticles

These are completely made of carbon (Bhaviripudi et al., 2007) and can be divided into carbon nanotubes (CNTs), grapheme, fullerenes, carbon black, nanofibers, and nano-sized activated carbon (Ealias & Saravanakumar, 2017; Figure 2.5). They play

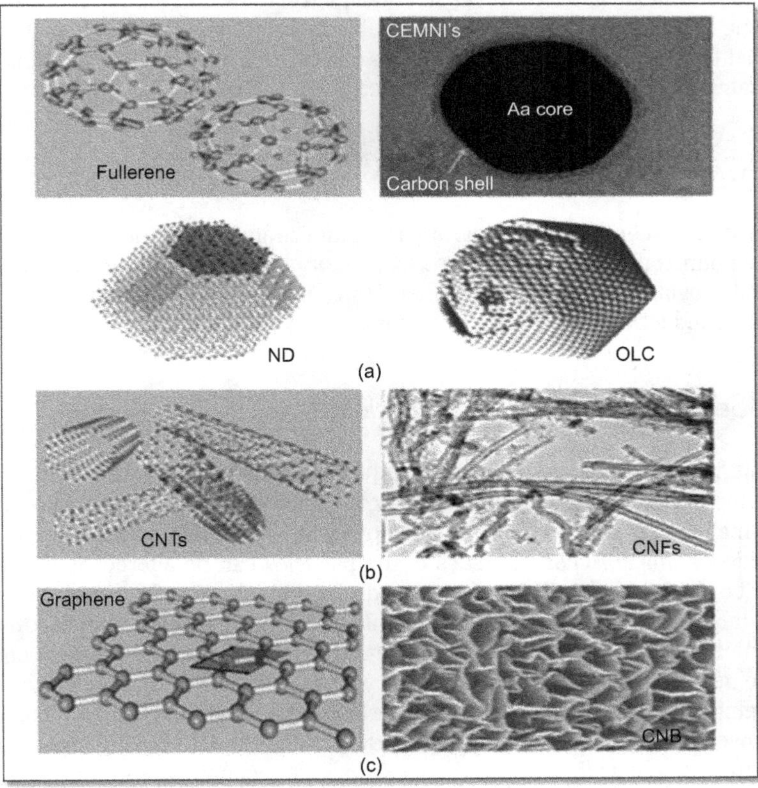

Figure 2.5 Schematic representation of crystal structures of the different low-dimensional carbon nanoparticles. From left to right: (a) fullerene, carbon-encapsulated metal nanoparticles, nanodiamond, and onion-like carbon, (0D); (b) carbon nanotube and carbon nanofibers (1D); and (c) graphene and carbon nanowalls (2D; Mostofizadeh et al., 2011).

a crucial role in different interdisciplinary areas because of their unique properties. Carbon-based nanoparticles have hybridized sp2 carbon atoms that have been prepared in different dimensions (Li et al., 2019; Nehra et al., 2019; Xie et al., 2019). They have received a great deal of attention for their various applications due to their physiochemical characteristics like their thermal and mechanical properties, conductivity, and chemical stability, among others.

2.2.4 Composite Nanoparticles

The multiphase solid materials have different phases, in which one of the phases having dimensions smaller than 100nm or nanoscale structures joining the various phases that fabricate the particles are called the nanocomposites (Lozhkomoev et al., 2019; Oh et al., 2018). The unprecedented flexibility and the dimensions in the nanometer size are explored in the synthesis of constructing the composite shapes. Other properties of nanocomposites include water absorption, wear and gloss retention, optical properties, and flexural strength (Zhang et al., 2019; Biswas et al., 2012; Abozaid et al., 2019). The production of nanocomposites by the combination of the materials discloses that the surface area of the materials can be improved and that such nanocomposite materials improve the adsorption capacity of the materials.

2.2.5 Ceramic Nanoparticles

The nanoparticles containing ceramic or nanoceramics are divided into heat-resistant inorganic and nonmetallic solids and heat-resistant inorganic, along with metal and nonmetal compounds having less than 100nm dimensions. Such materials exhibit improved structural, superconductive, ferroelectric or ferromagnetic, and electro-optical properties.

2.2.6 Semiconductor Nanoparticles

Germanium, silicon, gallium arsenide, and the compounds close to the "metalloid staircase" on the periodic tables exist as well-known semiconductors. It has a low band-gap energy smaller than 4eV and is composed of various elements from different classes. The physicochemical Acharacteristics of the particles can be altered by the quantum size effect or by improving the surface area during modification. There are two types of semiconductor nanoparticles: The first are intrinsic semiconductors, which contain only solid elements or compounds without doping. The major essence of this semiconductor is to have negative temperature coefficients of resistance, which means that by improving the temperature, the resistivity of the particles will be reduced and the conductivity will improve. The second are pure compounds or elements with doping in their shape to improve their conductivity to make them extrinsic semiconductors (Yang et al., 2020).

2.2.7 Polymeric Nanoparticles

The nano-sized solid particles containing synthetic or natural polymers are known as polymeric nanoparticles. They have a widespread application in the medical and pharmaceutical fields as drug-release controllers. They are broadly classified into two groups: First are polymeric micelles with the self-assembly of amphiphilic block copolymers in the specified solvents. Due to their unique nature, polymeric micelles of chitosan are appropriate for drug delivery, and their nanoform has less toxicity, stability, micellar association, and biocompatibility. Second are polymeric nanoparticles that contain biocompatible and biodegradable polymers with a size range from 10–100nm and are used to deliver drugs in the targeted area (Sur et al., 2019).

2.2.8 Lipid-Based Nanoparticle

These are appropriate for drug delivery and contain solid lipid nanoparticles (SLNs), liposomes, and nanostructured lipid carriers (NLCs). These are utilized to transport hydrophilic and hydrophobic particles; they have less toxicity and are used for controlled drug release at the target sites within humans (Zhong & Zhang, 2019). Physical and chemical stability, site-specific targeting, low cost, a lack of toxicity, and the possible control of both hydrophilic and hydrophobic molecules are the beneficial characteristics of the SLNs. Makalle (2012) has, however, reported that SLN materials also have negative properties, like restricted volume for drug loading and drug expulsion as the particles can crystallize throughout the storage procedure. Due to the increased stability of NLCs compared to the SLNs, they are utilized as drug-release regulating particles. Liposomes (50–100nm in size) are used for the delivery of cytotoxic drugs, because of their decreased toxicity and improved bioavailability.

2.2.9 Metal-Organic Frameworks

Metal-organic framework (MOF)–hybrid materials contain organic and inorganic metal ions or ligands. A well-organized structure, improved surface area, larger porosity, configuration, and easily modified surfaces are the major properties of MOFs. Due to the high reactivity provided by the nanoparticles distributed on the surfaces, MOFs are widely used as a support material. Iron-based metal-organic framework [*Fe*(*BTC*) (BTC: 1,3,5-benzenetricarboxylate)] MOFs materials are reported to have various reactivity and superior enzyme immobilization by undemanding interactions between the enzyme and the surface of the particles when used as a support for enzyme immobilization (Gascón et al., 2018).

2.2.10 Core–Shell Nanoparticles

The different layers of the nanoparticles here are (1) the core of the particle (the central region of nanoparticles); (2) the surface layer (branched with various metal ions, surfactants, polymeric branches, and small moieties); and (3) the shell layer (different from the particle of the core). Different types of core–shell nanoshapes, such as (a) a metal core along with various metals, which contain continuous layers of tiny spheres; (b) a metal core along with a nonmetal shell; (c) a metal core containing a polymer shell; (d) a nonmetal core containing a nonmetal shell; € nonmetal shell containing a polymer core; and (f) a polymer core with a shell designed with various polymers or in addition with multiple shells, have already been studied (Su et al., 2019).

2.3 Synthesis and General Characterization Methods

There are different techniques to prepare nanoparticles like (1) bottom-up and (2) top-down methods, and they are simply represented in Figure 2.6.

Bottom-up method: This method is also called a constructive method. Here, nanoparticles are first acquired from atoms in clusters and then integrated into the desired nanoparticles. The commonly used bottom-up method for producing nanoparticles is the formation of powders from sol-gel, chemical vapor deposition (CVD), colloidal dispersion, atomic layer deposition, sedimentation, biochemical synthesis, deoxyribonucleic acid scaffolding of nanoelectronics, reduction, molecular self-assembly, spinning, pyrolysis, and biosynthesis (Biswas et al., 2012; Ealias & Saravanakumar, 2017).

Top-down method: Top-down method is also called a destructive method in which bulk material (macroscopic level) is trimmed down to nanoparticles in various ways. The most widely used methods include laser ablation, soft lithography mechanical milling, thermal decomposition, ball milling, sputtering, grinding/milling, etching, CVD, optical lithography, block copolymer lithography, e-beam lithography, nanoimprint lithography, physical vapor deposition (PVD), and scanning probe lithography.

Due to their unique characteristics, nanoparticles have widespread applications. It is very crucial to distinguish the physiochemical and other properties of the nanoparticles, according to the type of applications. The characterization of nanoparticles can be done by using various tools as listed in Table 2.1.

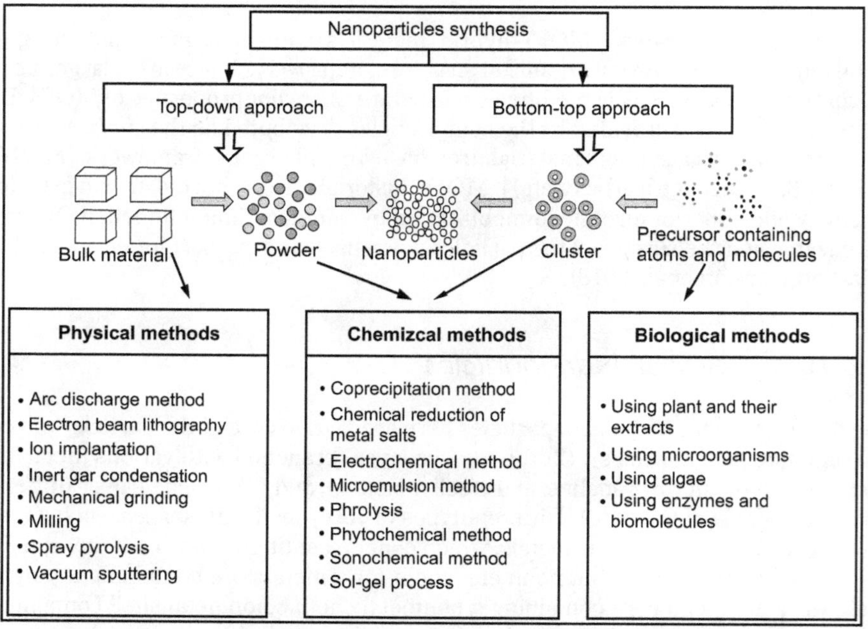

Figure 2.6 Process of nanoparticle synthesis (Patra & Baek, 2015).

Table 2.1 Techniques Used for the Characterization of Nanoparticles

Techniques for analysis	*Characterization of nanoparticles*
Scanning tunneling microscopy (STM), Scanning electron microscopy (SEM), Field-emission SEM (FESEM)	Size and shape, size distribution, aggregation
Atomic force microscopy (AFM), Dynamic light scattering electron microscopy	Surface properties, size and shape, structure
Differential scanning colorimetry (DSC)	Physiochemical state and possible interactions of the polymer and the drug
High-performance liquid chromatography (HPLC)	Contents of a nanoparticles
Mass spectrometry (MS)	Structure, composition, molecular weight, and surface properties
In vivo and in vitro cell viability and molecular colony viability	Biological properties of the particles
Dynamic light scanning (DLS)	Hydrodynamic size distribution pattern
Raman and infrared spectroscopy (IR), Surface-enhanced Raman spectroscopy (SERS)	Structure and function, analysis of functional groups
X-ray photoelectron spectroscopy (XPS)	Elemental and surface composition of the surface of the particles
Nuclear magnetic resonance (NMR)	Composition and structure and purity of the particles
Ultraviolet (UV)-Visible spectroscopy Electron dispersive X-ray spectroscopy	Chemical properties

2.4 Applications of Nanoparticles

The application of nanoparticles spread in different areas of our life such as electronics, cosmetics, catalytic activity, medicinal, construction, food and environmental remediation, and renewable energy. Nanoparticles (cadmium sulfide, nanocrystalline lead telluride, sulfide, zinc selenide) are used inside the computer monitor and the television displays as light-emitting diodes (LEDs; Teng et al., 2008), metal hybrids and nanocrystalline nickel batteries in computers and mobile phones (Lu et al., 2010). For the detection of gases like NO_2 and NH_3, nanoparticles are used (Liu et al., 2011). Inorganic nanoparticles (zinc oxide and titanium dioxide) and organic nanoparticles (tris-biphenyl triazine [nano] and methylene bis-benzotriazolyl tetramethylbutylphenol [MBBT; nano]) are unique choices of components for the sunscreen formulation (Couteau et al., 2015; Doak, 2016); nanoparticles of iron oxide are used in lipsticks (Wiechers & Musee, 2010); nanosilica is used in cosmetics (Fytianos et al., 2020); carbon Black, CI 77266, is used in eye decorative cosmetic products (Sahu et al., 2014); nano-hydroxyapatite particles are used in dentifrices and mouthwashes (Coelho et al., 2019; Bernauer, 2018; Ramis et al., 2018), and gold and silver nanoparticles are used in deodorants and antiaging creams (Fytianos et al., 2020). Biodegradable and biocompatible nanoliposomes are used in antiperspirants, body-spray deodorants, and lipsticks (Rigano & Lionetti, 2016; Santos et al., 2019). Platinum nanoparticles in automotive catalytic converters act as a systematic catalyst in chemical production, because of their very large surface-to-volume ratio (Crooks et al., 2001). The use of polymeric nanoparticles in drug delivery to specific cells or sites offers many superiorities over other systems (Rahman et al., 2012; Ganesh & Archana, 2013; Ealias & Saravanakumar, 2017). Polymeric nanoparticles (polylactic acid, chitosan, poly(lactic-co-glycolic acid) [PLGA]), sodium alginate, employed for drug delivery (Krishnamurthy et al., 2015), PLGA nanoparticles loaded with doxycycline hydrochloride for eliminating lymphatic filarial parasites (Singh et al., 2016), and carboplatin containing CNTs used for inhibition of bladder cancer (Hampel et al., 2008), among others, are some of the applications of nanoparticles in the medicine that have been reported. Food packaging with nanocomposite coating with antimicrobial substances (Laad & Jatti, 2018) and nanodrops, to fetch minerals and vitamins in the food, (Ealias & Saravanakumar, 2017), have also been described to be more effective in the food industry. Nano-silica (SiO_2) (Nazari & Riahi, 2011), hematite (Fe_2O_3; Laanaiya & Zaoui, 2020) in the concrete, and nano-sized steel cables are used in bridge construction (Akhnoukh, 2020); titanium dioxide (TiO_2) nanoparticles in glass (Xu et al., 2004) provide an effective application of nanotechnology in constructions (Machado et al., 2015). Soil contamination caused by heavy metals, toxic industrial waste, and the like could also be cleared by nanotechnological applications (Vázquez-Núñez et al., 2020; Mallikarjunaiah et al., 2020; Okoh et al., 2020).

2.5 General Methods for the Postharvest Storage of Grains

Traditional warehouse operations use big or small storage houses, outdoor or indoor, individual/community, temporary or lifelong storage designs, having open, semi-open, or closed systems (Gwinner et al., 1996). Such practices have been used for a long time with or without modification. The choice of such storage systems is related to the climate and is influenced by the customs and type of regional natural resources. Having unique storage practices inside a society that possibly differed from other localities and villages and among countries, traditional knowledge of storage progressed into the community and was passed on from one age group to another (Natarajan & Govind, 2006). The storage practices arose also from the connections with specific environmental conditions. Food storage in developing countries is carried out in traditional structures in unthreshed or threshed forms at home. Farmers take measures to decrease food loss after harvesting by drying crops properly before storage and using moisture-proof and adequately aired storage structures (Nduku et al., 2013). Other storage practices include the use of earthen pots or baskets, gourds, metal container platforms, plastic containers, and cribs; hessian, polyethylene, or plant fibers and jute bags; and herbs like Mexican marigold and hot pepper. Some commonly used postharvest storage practices include the following:

1. **Solarization:** One of the age-old storage practices by the farmers before storing the grains or pulses by heating grains in the sun to inhibit insects is known as solarization, which may reduce the seed viability (Chua & Chou, 2003).

2. **Open fireplace:** In rural areas, farmers store the grains near the kitchen, where heat and smoke from the flaming wood perforate the food. This helps keep it free from pests or insects (Sarangi et al., 2009), using specially raised barns for large quantities of food or above the farm hut kitchen fire or open into direct solar radiation.
3. **Open-air/aerial storage:** In the open-air or aerial storage process, the shelled maize cobs and cereals are air-dried using sunlight (Ofor, 2011) until it is needed by the farmers for marketing or consumption. The major disadvantage of this technique is that the grains are always exhibited to the pests and environment.
4. **Storage with diatomized earth:** Storage with diatomized earth is a natural method used for the preservation of agricultural food products. Diatomaceous particles adhere to the insects and attack the cuticles of the insects by absorption and abrasion of hydrocarbon, which results in the death of the insects because of dehydration (Korunić, 2013).
5. **Gourds:** They are prepared from the dried hard outer layers of fruits and are used to preserve small amounts of food grains by depositing them on a platform to protect them from moisture absorption. It is economical and preserves grains a lot better than unlined pits (Makalle, 2012).
6. **Metal or plastic drums:** First, the grains are sun-dried to minimize the moisture content, and the large containers are filled with the food grains by sealing them with the grease screw cap and are placed on a pallet under shade. The large containers must remain sealed; otherwise, the flies are susceptible to resuming the physiological activity at the tiny holes of oxygen when opened carelessly is a disadvantage of using large containers as storage systems (Makalle, 2012).
7. **Storage bags:** Storage bags (prepared by sisal, cotton, woven jute, cotton) are used for the small-time preservation of food grains in sacks and are used in villages, commercial storage areas, and farms, depending on the materials available in the area. This is used to guarantee moisture resistance and entirely sealed storage conditions (Mutungi et al., 2015).
8. **Storage with table salt:** It is utilized for short-term storage. Due to the grinding effect of the salt, it protects from insect motion inside the package container; population buildup and insect infestation are also defeated (Jeeva et al., 2006).
9. **Camphor:** It is employed for the small, short-time preservation of food grains for the coming season of planting. Sun-dried grains or paddies are preserved in bags or pots by placing camphor inside the storage containers or bags. Camphor may act either as an antifeedant, a repellent, or a fumigant, ascribed to the strong odor emerging from it (Karthikeyan et al., 2009).

2.5.1 Role of Nanomaterials in Food Storage

Almost 40% of vegetables and fruits are wasted yearly because of inappropriate handling, preservation, packing, or shipment. Reducing postharvest mislaying is very crucial; to ensure enough food, both in quality and in quantity at a convenient level. The biopreservation of food is a great challenge in the food industry as tons of food is wasted every year due to damage by microorganisms and reduced shelf life. Innovative ideas in the area of development of decay-preventing techniques or shelf-life-increasing nanocomposites have recently got much attention. A food coating, cover, or protective film is an immediate layer of material applied directly to the food product. It typically needs to be edible in order to avoid removal before food consumption. Multifunctional nanoformulations can act as edible coatings for vegetables and fruits that extend the life span of food and function as efficient carriers for pharmaceuticals and phytochemicals. A polymer coating also reduces the damage to food by interacting with microorganisms that affect quality and leads to economic losses during postharvest storage. Nanotechnology emerges as a potentially innovative and different process to solve problems with regular packaging, also used in edible coatings. The mixture of lipids and polymers along with edible coatings improves the storage of fruits and vegetables because they work as semipermeable obstacles to water and gas. The nanoformulation of an edible coating was found to be able to protect fresh vegetables and fruits for more than 15 days.

Nanotechnology has played a vital role in the food industry, making certain moderations of color, flavor, and nutritional content; improving the life span of food products; and watching the integrity of food by barcodes (Aigbogun et al., 2017). Some applications of nanotechnology in food packaging are listed in Figure 2.7.

Microbial contamination is the major problem faced by the food industry throughout the production, processing, transport, and preservation of food. The use of nanoparticles with antimicrobial activity has a crucial role in protecting the decaying of food products, and as a result, the life span of the food products could also be

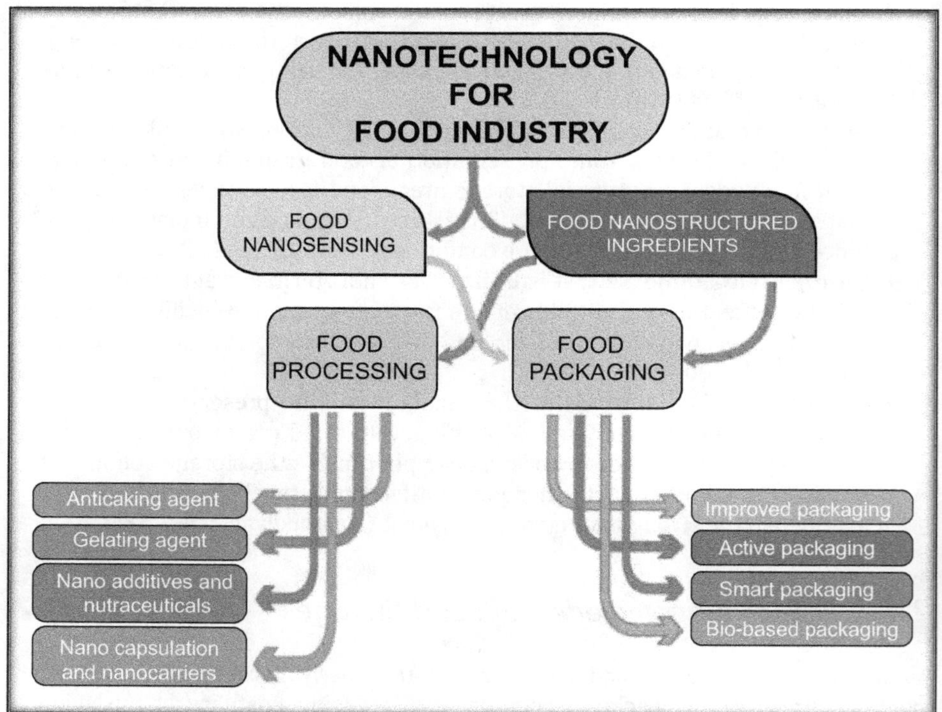

Figure 2.7 Application of nanotechnology in the food industry (Primožič et al., 2021).

expanded (Mitura & Zarzycki, 2018). Nanoparticles made of metals and metal oxides were observed to be good as antimicrobials, responsible for the uncontrolled production of reactive oxygen species (ROS), leading to oxidative stress and cell damage of microbial pathogens (Wu et al., 2014; Fu et al., 2014). Nanocomposites with metal or metal oxide are utilized for food coating, packaging, or as an ingredient in food products. Among them, silver nanocomposites and their nanoparticles were widely used as food antimicrobials in the food industry (He & Hwang, 2016), silver-containing zeolites are also accepted by the United States Food and Drug Administration (USFDA) for decontamination (Duncan, 2011).

Many researchers have reported the silver nanocomposites to be safe for packaging, due to no detectable or insignificant level of silver nanoparticles released (Addo Ntim et al., 2015) and it also enhances stability. The widely used polymers in nanocomposites include polylactic acid, gelatin, low-density polyethylene (LDPE), and isotactic polypropylene. The antimicrobial property of nanocomposites depends on the organoclay types, microbial species, and polymer matrix utilized for nanocomposite formulation (Hong & Rhim, 2008). Various biopolymers are incorporated with cellulose nanoparticles (CNPs) to prepare efficient bio-based nanocomposites, with improved mechanical and physical properties (De Paula et al., 2011). Other biopolymers used to prepare the bio-based nanocomposites are all-cellulose nanocomposite (ACNC) films from sugarcane bagasse (Ghaderi et al., 2014), regenerated cellulose/montmorillonite (RC/MMT) nanocomposite films (Mahmoudian et al., 2012), regenerated cellulose/sepiolite bio-nanocomposite films (Soheilmoghaddam et al., 2014), chitosan-based biodegradable films (Khan et al., 2012), and bacterial cellulose nanocrystals (BCNCs; George, 2012), among others.

Nanotechnology-enabled sensors (Dobrucka, 2020) are also applied as labels or coatings, for adding an intelligent function in food packaging, for detecting the leak signal of time-temperature changes, or for microbial food safety (the deterioration of foodstuffs) in the form of nanosensors (Mahalik & Nambiar, 2010; World Health Organization, 2010; Watson et al., 2011). They are also helpful in detecting pathogens, chemical compounds, toxins in food (like aflatoxins; Myndrul et al., 2020) and providing the real-time status of food freshness, eliminating the need for inaccurate expiration dates.

2.5.2 Nanocomposites for Fruits and Vegetable Coatings

A wide range of nanomaterials has been used for food packaging like silver nanoparticles (AgNPs), nano clay, nano-zinc oxide (nano-ZnO; Hirvikorpi et al., 2010), nano-titanium dioxide (nano-TiO_2), titanium nitride nanoparticles (nano-TiN), and more (Chaudhary et al., 2020). Nanomaterials like titanium dioxide and zinc oxide are utilized as photocatalyst agents to degrade organic molecules and inhibit microbes by contributing to ROS formation. This results in cytoplasm oxidation of microbial cells and thereby the cell death (Bodaghi et al., 2013). However, AgNPs, layered silicates, and nanoclays are utilized as antimicrobial materials (Majeed et al., 2013). ZnO is more attractive and efficient than AgNPs, because of its lower toxicity and cost-effectiveness (Silvestre et al., 2011). They have a wide application in medical devices, atmospheric processed (MAP) packaging, medication delivery, and cosmetics. When exposed to UV light, ZnO can produce a large quantity of hydrogen peroxide, which leads to oxidative stress in microbial cells. Hence, nanotechnology will have a revolutionary impact on postharvest applications in the coming years.

2.6 Summary and Conclusion

Nanotechnology and nanoparticles have a major implementation in postharvest management, because of their excellent nano-based properties. It can also have a vital role in increasing the shelf life of highly perishable fruits and vegetables. Several investigational reports support nanocomposite use to have quite beneficial results on the physiological and physical and chemical properties compared with normal packaging. The application of nanoparticles in packaging may provide a promising replacement to improve the storability and preservation qualities of vegetables and fruits. However, further research is needed to clarify the exact nano-packing mechanisms for stored food products and the road to successful nanoparticles for postharvest storage of food is still long.

Acknowledgments

The authors acknowledged DIST-FIST, KSCSTE—SARD, for the financial assistance in the form of a project grant. Management—Sree Sankara College, Kalady for providing research facilities in the department. Mrs. Sherin Joy Parappilly is grateful to University Grants Commission (UGC)—Maulana Azad National Fellowship for minority students, for providing fellowship for her research work.

References

Abozaid, R. M., Lazarević, Z. Ž., Radović, I., Gilić, M., Šević, D., Rabasović, M. S., & Radojević, V. 2019. Optical properties and fluorescence of quantum dots CdSe/ZnS-PMMA composite films with interface modifications. *Optical Materials, 92*, 405–410.

Addo Ntim, S., Thomas, T. A., Begley, T. H., & Noonan, G. O. 2015. Characterisation and potential migration of silver nanoparticles from commercially available polymeric food contact materials. *Food Additives & Contaminants: Part A, 32*(6), 1003–1011.

Aigbogun, I. E., Mohammed, S. S. D., Orukotan, A. A., & Tanko, J. D. 2017. The role of nanotechnology in food industries: A review. *Journal of Advances in Microbiology*, 1–9.

Akhnoukh, A. K. 2020. The use of micro and nano-sized particles in increasing concrete durability. *Particulate Science and Technology, 38*(5), 529–534.

Ali, M. A., Rehman, I., Iqbal, A., Din, S., Rao, A. Q., Latif, A., Samiullah, T. R., Azam, S., & Husnain, T. 2014. Nanotechnology, a new frontier in agriculture. *Advancements in Life Sciences, 1*(3), 129–138.

Arnon, H., Granit, R., Porat, R., & Poverenov, E. 2015. Development of polysaccharides-based edible coatings for citrus fruits: A layer-by-layer approach. *Food Chemistry, 166*, 465–472.

Asghari, F., Jahanshiri, Z., Imani, M., Shams-Ghahfarokhi, M., & Razzaghi-Abyaneh, M. 2016. Antifungal nanomaterials: Synthesis, properties, and applications. In *Nanobiomaterials in antimicrobial therapy* (pp. 343–383). William Andrew Publishing.

Aversa, R., Modarres, M. H., Cozzini, S., Ciancio, R., & Chiusole, A. 2018. The first annotated set of scanning electron microscopy images for nanoscience. *Scientific Data*, 5, 180172.

Bernauer, U. 2018. Opinion of the Scientific Committee on Consumer Safety (SCCS)—revision of the opinion on hydroxyapatite (nano) in cosmetic products. *Regulatory Toxicology and Pharmacology, 98*, 274–275.

Bhaviripudi, S., Mile, E., Steiner, S. A., Zare, A. T., Dresselhaus, M. S., Belcher, A. M., & Kong, J. 2007. CVD synthesis of single-walled carbon nanotubes from gold nanoparticle catalysts. *Journal of the American Chemical Society, 129*(6), 1516–1517.

Biswas, A., Bayer, I. S., Biris, A. S., Wang, T., Dervishi, E., & Faupel, F. 2012. Advances in top—down and bottom—up surface nanofabrication: Techniques, applications & future prospects. *Advances in Colloid and Interface Science, 170*(1–2), 2–27.

Bodaghi, H., Mostofi, Y., Oromiehie, A., Zamani, Z., Ghanbarzadeh, B., Costa, C., . . . & Del Nobile, M. A. 2013. Evaluation of the photocatalytic antimicrobial effects of a TiO_2 nanocomposite food packaging film by in vitro and in vivo tests. *LWT–Food Science and Technology, 50*(2), 702–706.

Chaudhary, P., Fatima, F., & Kumar, A. 2020. Relevance of nanomaterials in food packaging and its advanced future prospects. *Journal of Inorganic and Organometallic Polymers and Materials, 30*(12), 5180–5192.

Chu, P., Zhang, H., Zhao, J., Gao, F., Guo, Y., Dang, B., & Zhang, Z. 2017. On the volume resistivity of silica nanoparticle filled epoxy with different surface modifications. *Composites Part A: Applied Science and Manufacturing, 99*, 139–148.

Chua, K. J., & Chou, S. K. 2003. Low-cost drying methods for developing countries. *Trends in Food Science & Technology, 14*(12), 519–528.

Coelho, C. C., Grenho, L., Gomes, P. S., Quadros, P. A., & Fernandes, M. H. 2019. Nano-hydroxyapatite in oral care cosmetics: Characterization and cytotoxicity assessment. *Scientific Reports, 9*(1), 1–10.

Coles, D., & Frewer, L. J. 2013. Nanotechnology applied to European food production—A review of ethical and regulatory issues. *Trends in Food Science & Technology, 34*(1), 32–43.

Couteau, C., Paparis, E., Chauvet, C., & Coiffard, L. 2015. Tris-biphenyl triazine, a new ultraviolet filter studied in terms of photoprotective efficacy. *International Journal of Pharmaceutics, 487*(1–2), 120–123.

Crooks, R. M., Zhao, M., Sun, L., Chechik, V., & Yeung, L. K. 2001. Dendrimer-encapsulated metal nanoparticles: Synthesis, characterization, and applications to catalysis. *Accounts of Chemical Research, 34*(3), 181–190.

Das, L., Das, P., Bhowal, A., & Bhattachariee, C. 2020. Synthesis of hybrid hydrogel nano-polymer composite using graphene oxide, chitosan and PVA and its application in waste water treatment. *Environmental Technology & Innovation, 18*, 100664.

Deng, Z., Jung, J., Simonsen, J., Wang, Y., & Zhao, Y. 2017. Cellulose nanocrystal reinforced chitosan coatings for improving the storability of postharvest pears under both ambient and cold storages. *Journal of Food Science, 82*(2), 453–462.

De Paula, E. L., Mano, V., & Pereira, F. V. 2011. Influence of cellulose nanowhiskers on the hydrolytic degradation behavior of poly (d, l-lactide). *Polymer Degradation and Stability, 96*(9), 1631–1638.

Doak, S. 2016. Opinion of the Scientific Committee on consumer safety (SCCS)-Opinion on the use of 2, 2′-methylene-bis-(6-(2H-benzotriazol-2-yl)-4-(1, 1, 3, 3-tetramethylbutyl) phenol) (nano)—S79—In cosmetic products. *Regulatory Toxicology and Pharmacology, 76*, 215–216.

Dobrucka, R. 2020. Application of nanotechnology in food packaging. *Journal of Microbiology, Biotechnology and Food Sciences*, 353–359.

Duncan, T. V. 2011. Applications of nanotechnology in food packaging and food safety: Barrier materials, antimicrobials and sensors. *Journal of Colloid and Interface Science*, *363*(1), 1–24.

Ealias, A. M., & Saravanakumar, M. P. 2017. A review on the classification, characterisation, synthesis of nanoparticles and their application. *IOP Conference Series: Materials Science and Engineering*, *263*, 032019.

Egirani, D. E., Shehata, N., & Khedr, M. H. 2020. A review of nano materials in agriculture and allied sectors: Preparation, characterization, applications, opportunities, and challenges. *Materials International*, *2*(3), 421–432.

Elbeshehy, E. K., Elazzazy, A. M., & Aggelis, G. 2015. Silver nanoparticles synthesis mediated by new isolates of *Bacillus* spp., nanoparticle characterization and their activity against Bean Yellow Mosaic Virus and human pathogens. *Frontiers in Microbiology*, *6*, 453.

Eleftheriadou, M., Pyrgiotakis, G., & Demokritou, P. 2017. Nanotechnology to the rescue: Using nano-enabled approaches in microbiological food safety and quality. *Current Opinion in Biotechnology*, *44*, 87–93.

Embuscado, M. E., & Huber, K. C. 2009. *Edible films and coatings for food applications* (Vol. 222). Springer.

Flores-López, M. L., Cerqueira, M. A., de Rodríguez, D. J., & Vicente, A. A. 2016. Perspectives on utilization of edible coatings and nano-laminate coatings for extension of postharvest storage of fruits and vegetables. *Food Engineering Reviews*, *8*(3), 292–305.

Fu, P. P., Xia, Q., Hwang, H. M., Ray, P. C., & Yu, H. 2014. Mechanisms of nanotoxicity: Generation of reactive oxygen species. *Journal of Food and Drug Analysis*, *22*(1), 64–75.

Fytianos, G., Rahdar, A., & Kyzas, G. Z. 2020. Nanomaterials in cosmetics: Recent updates. *Nanomaterials*, *10*(5), 979.

Galus, S., & Kadzińska, J. 2015. Food applications of emulsion-based edible films and coatings. *Trends in Food Science & Technology*, *45*(2), 273–283.

Ganesh, K., & Archana, D. 2013. Review article on targeted polymeric nanoparticles: An overview. *American Journal of Advanced Drug Delivery*, *3*(3), 196–215.

Gascón, V., Jiménez, M. B., Blanco, R. M., & Sanchez-Sanchez, M. 2018. Semi-crystalline Fe-BTC MOF material as an efficient support for enzyme immobilization. *Catalysis Today*, *304*, 119–126.

Gaur, P., & Banerjee, S. 2019. CN cross coupling: Novel approach towards effective aryl secondary amines modification on nanodiamond surface. *Diamond and Related Materials*, *98*, 107468.

George, J. 2012. High performance edible nanocomposite films containing bacterial cellulose nanocrystals. *Carbohydrate Polymers*, *87*(3), 2031–2037.

Ghaderi, M., Mousavi, M., Yousefi, H., & Labbafi, M. 2014. All-cellulose nanocomposite film made from bagasse cellulose nanofibers for food packaging application. *Carbohydrate Polymers*, *104*, 59–65.

Ghidan, A. Y., & Al Antary, T. M. 2019. Applications of nanotechnology in agriculture. In *Applications of nanobiotechnology*. IntechOpen.

Gleiter, H. 2000. Nanostructured materials: Basic concepts and microstructure. *Acta Materialia*, *48*(1), 1–29.

Gopal, M., Gogoi, R., Srivastava, C., Kumar, R., Singh, P. K., Nair, K. K., Yadav, S., and Goswami, A. 2011. Nanotechnology and its application in plant protection. *Plant Pathology in India: Vision*, *2030*, 224–232.

Gwinner, J., Harnisch, R., & Mück, O. 1996. Manual on the prevention of post-harvest grain losses. Manual on the prevention of post-harvest grain losses.346 pp.

Hampel, S., Kunze, D., Haase, D., Krämer, K., Rauschenbach, M., Ritschel, M., Leonhardt, A., et al. 2008. Carbon nanotubes filled with a chemotherapeutic agent: A nanocarrier mediates inhibition of tumor cell growth. *Nanomedicine, 3*(2), 175–182. https://doi.org/10.2217/17435889.3.2.175

He, X., & Hwang, H. M. 2016. Nanotechnology in food science: Functionality, applicability, and safety assessment. *Journal of Food and Drug Analysis, 24*(4), 671–681.

Hirvikorpi, T., Vähä-Nissi, M., Mustonen, T., Iiskola, E., & Karppinen, M. 2010. Atomic layer deposited aluminum oxide barrier coatings for packaging materials. *Thin Solid Films, 518*(10), 2654–2658.

Hong, S. I., & Rhim, J. W. 2008. Antimicrobial activity of organically modified nanoclays. *Journal of Nanoscience and Nanotechnology, 8*(11), 5818–5824.

Hur, J. U., Han, J. S., Shin, J. R., Park, H. Y., Choi, S. C., Jung, Y. G., & An, G. S. 2019. Fabrication of SnO_2-decorated Fe_3O_4 nanoparticles with anionic surface modification. *Ceramics International, 45*(17), 21395–21400.

Hussain, A., Rizwan, M., Ali, Q., & Ali, S. 2019. Seed priming with silicon nanoparticles improved the biomass and yield while reduced the oxidative stress and cadmium concentration in wheat grains. *Environmental Science and Pollution Research, 26*(8), 7579–7588.

Jeeva, S. R. D. N., Laloo, R. C., & Mishra, B. P. 2006. Traditional agricultural practices in Meghalaya, North East India. *Indian Journal of Traditional Knowledge, 5*(1), 7–18.

Jianglian, D., & Shaoying, Z. 2013. Application of chitosan based coating in fruit and vegetable preservation: A review. *Journal of Food Processing & Technology, 4*(5), 227.

Karthikeyan, C., Veeraragavathatham, D., Karpagam, D., & Firdouse, S. A. 2009. Traditional storage practices. *Indian Journal of Traditional Knowledge, 8*(4), 564–568.

Kaushal, M., & Wani, S. P. 2017. Nanosensors: Frontiers in precision agriculture. In *Nanotechnology* (pp. 279–291). Springer.

Kaya, M., Česonienė, L., Daubaras, R., Leskauskaitė, D., & Zabulionė, D. 2016. Chitosan coating of red kiwifruit (*Actinidia melanandra*) for extending of the shelf life. *International Journal of Biological Macromolecules, 85*, 355–360.

Khan, A., Khan, R. A., Salmieri, S., Tien, C. L., Riedl, B., Bouchard, J., Chauve, G., Tan, V., Kamal, M. R., & Lacroix, M. 2012. Mechanical and barrier properties of nanocrystalline cellulose reinforced chitosan based nanocomposite films. *Carbohydrate Polymers, 90*(4), 1601–1608.

Khan, I., Saeed, K., & Khan, I. 2019a. Nanoparticles: Properties, applications and toxicities. *Arabian Journal of Chemistry, 12*(7), 908–931. https://doi.org/10.1016/j.arabjc.2017.05.011

Khan, M. R., Ahamad, F., & Rizvi, T. F. 2019b. Effect of nanoparticles on plant pathogens. In *Advances in phytonanotechnology* (pp. 215–240). Academic Press.

Khanna, P., Ong, C., Bay, B. H., & Baeg, G. H. 2015. Nanotoxicity: An interplay of oxidative stress, inflammation and cell death. *Nanomaterials, 5*(3), 1163–1180.

Korunić, Z. 2013. Diatomaceous earths: Natural insecticides. Pesticides. *Phytomedicine (Belgrade), 28*(2), 77–95.

Krishnamurthy, S., Vaiyapuri, R., Zhang, L., & Chan, J. M. 2015. Lipid-coated polymeric nanoparticles for cancer drug delivery. *Biomaterials Science, 3*(7), 923–936. https://doi.org/10.1039/C4BM00427B

Kumar, P., & Gautam, S. 2019. Developing ZnO nanoparticle embedded antimicrobial starch biofilm for food packaging. arXiv:1909.05083.

Kumar, P., Mahajan, P., Kaur, R., & Gautam, S. 2020. Nanotechnology and its challenges in the food sector: A review. *Materials Today Chemistry, 17*, 100332.

Laad, M., & Jatti, V. K. S. 2018. Titanium oxide nanoparticles as additives in engine oil. *Journal of King Saud University-Engineering Sciences*, *30*(2), 116–122.

Laanaiya, M., & Zaoui, A. 2020. Preventing cement-based materials failure by embedding Fe_2O_3 nanoparticles. *Construction and Building Materials*, *260*, 120466.

Li, Z., Wang, L., Li, Y., Feng, Y., & Feng, W. 2019. Carbon-based functional nanomaterials: Preparation, properties and applications. *Composites Science and Technology*, *179*, 10–40.

Lin, D., & Zhao, Y. 2007. Innovations in the development and application of edible coatings for fresh and minimally processed fruits and vegetables. *Comprehensive Reviews in Food Science and Food Safety*, *6*(3), 60–75.

Liu, X., Zhang, J., Wang, L., Yang, T., Guo, X., Wu, S., & Wang, S. 2011. 3D hierarchically porous ZnO structures and their functionalization by Au nanoparticles for gas sensors. *Journal of Materials Chemistry*, *21*(2), 349–356.

Lozhkomoev, A. S., Pervikov, A. V., Chumaevsky, A. V., Dvilis, E. S., Paygin, V. D., Khasanov, O. L., & Lerner, M. I. 2019. Fabrication of Fe-Cu composites from electroexplosive bimetallic nanoparticles by spark plasma sintering. *Vacuum*, *170*, 108980.

Lu, Y. C., Xu, Z., Gasteiger, H. A., Chen, S., Hamad-Schifferli, K., & Shao-Horn, Y. 2010. Platinum-gold nanoparticles: A highly active bifunctional electrocatalyst for rechargeable lithium-air batteries. *Journal of the American Chemical Society*, *132*(35), 12170–12171.

Machado, S., Pacheco, J. G., Nouws, H. P. A., Albergaria, J. T., & Delerue-Matos, C. 2015. Characterization of green zero-valent iron nanoparticles produced with tree leaf extracts. *Science of the Total Environment*, *533*, 76–81.

Mahalik, N. P., & Nambiar, A. N. 2010. Trends in food packaging and manufacturing systems and technology. *Trends in Food Science & Technology*, *21*(3), 117–128.

Mahmoudian, S., Wahit, M. U., Ismail, A. F., & Yussuf, A. A. 2012. Preparation of regenerated cellulose/montmorillonite nanocomposite films via ionic liquids. *Carbohydrate Polymers*, *88*(4), 1251–1257.

Majeed, K., Jawaid, M., Hassan, A., Bakar, A. A., Khalil, H. A., Salema, A. A., & Inuwa, I. 2013. Potential materials for food packaging from nanoclay/natural fibres filled hybrid composites. *Materials & Design*, *46*, 391–410.

Makalle, A. M. 2012. Post harvest storage as a rural household food security strategy in Tanzania. *ARPN Journal of Science and Technology*, *12*(9), 814–821.

Mallikarjunaiah, S., Pattabhiramaiah, M., & Metikurki, B. 2020. Application of nanotechnology in the bioremediation of heavy metals and wastewater management. In *Nanotechnology for food, agriculture, and environment* (pp. 297–321). Springer.

Manchester, M., & Singh, P. 2006. Virus-based nanoparticles (VNPs): Platform technologies for diagnostic imaging. *Advanced Drug Delivery Reviews*, *58*(14), 1505–1522.

Mantilla, N., Castell-Perez, M. E., Gomes, C., & Moreira, R. G. 2013. Multilayered antimicrobial edible coating and its effect on quality and shelf-life of fresh-cut pineapple (Ananas comosus). *LWT–Food Science and Technology*, *51*(1), 37–43.

Mitura, K. A., & Zarzycki, P. K. 2018. Biocompatibility and toxicity of allotropic forms of carbon in food packaging. In *Role of materials science in food bioengineering* (pp. 73–107). Academic Press.

Mostofizadeh, A., Li, Y., Song, B., & Huang, Y. 2011. Synthesis, properties, and applications of low-dimensional carbon-related nanomaterials. *Journal of Nanomaterials*, *2011*, Article ID 685081-1–685081-21.

Mutungi, C., Affognon, H. D., Njoroge, A. W., Manono, J., Baributsa, D., & Murdock, L. L. 2015. Triple-layer plastic bags protect dry common beans (Phaseolus vulgaris) against damage by Acanthoscelides obtectus (Coleoptera: Chrysomelidae) during storage. *Journal of Economic Entomology, 108*(5), 2479–2488.

Myndrul, V., Coy, E., Bechelany, M., & Iatsunskyi, I. 2020. Photoluminescence label-free immunosensor for the detection of Aflatoxin B1 using polyacrylonitrile/zinc oxide nanofibers. *Materials Science and Engineering: C, 118*, 111401.

Nair, R., & Kumar, D. S. 2013. Plant diseases—control and remedy through nanotechnology. In *Crop improvement under adverse conditions* (pp. 231–243). Springer.

Natarajan, M., & Govind, S. 2006. Indigenous agricultural practices among tribal women. *Indian Journal of Traditional Knowledge*, *5*(1), 118–121.

Nazari, A., & Riahi, S. 2011. RETRACTED: The effects of SiO_2 nanoparticles on physical and mechanical properties of high strength compacting concrete. *Composites Part B: Engineering, 42*(3), 570–578. https://doi.org/10.1016/j.compositesb.2010.09.025

Nduku, T. M., De Groote, H., & Nzuma, J. M. 2013. Comparative analysis of maize storage structures in Kenya (No. 309-2016-5259).

Neethu, S., Midhun, S. J., Radhakrishnan, E. K., & Jyothis, M. 2021. Microbially synthesized nanomaterials for remediation of contaminated soil and water environment. In *Microbe mediated remediation of environmental contaminants* (pp. 157–176). Woodhead Publishing.

Nehra, M., Dilbaghi, N., Hassan, A. A., & Kumar, S. 2019. Carbon-based nanomaterials for the development of sensitive nanosensor platforms. In *Advances in nanosensors for biological and environmental analysis* (pp. 1–25). Elsevier.

Ofor, M. O. 2011. Traditional methods of preservation and storage of farm produce in Africa. *New York Science Journal*, *4*(3), 58–62.

Oh, Y., Lee, J., & Lee, M. 2018. Fabrication of Ag-Au bimetallic nanoparticles by laser-induced dewetting of bilayer films. *Applied Surface Science, 434*, 1293–1299.

Okoh, E., Yelebe, Z. R., Oruabena, B., Nelson, E. S., & Indiamaowei, O. P. 2020. Clean-up of crude oil-contaminated soils: Bioremediation option. *International Journal of Environmental Science and Technology, 17*(2), 1185–1198.

Omanović-Mikličanina, E., & Maksimović, M. 2016. Nanosensors applications in agriculture and food industry. *Bulletin of the Chemists and Technologists of Bosnia and Herzegovina*, *47*, 59–70.

Patra, J. K., & Baek, K. H. 2015. Green nanobiotechnology: Factors affecting synthesis and characterization techniques. *Journal of Nanomaterials, 2014*, 219–219.

Pérez-de-Luque, A., & Hermosín, M. C. 2013. Nanotechnology and its use in agriculture. In *Bio-nanotechnology: A revolution in food, biomedical and health sciences* (pp. 383–398). Wiley.

Poh, T. Y., Ali, N. A. T. B. M., Mac Aogáin, M., Kathawala, M. H., Setyawati, M. I., Ng, K. W., & Chotirmall, S. H. 2018. Inhaled nanomaterials and the respiratory microbiome: Clinical, immunological and toxicological perspectives. *Particle and Fibre Toxicology, 15*(1), 1–16.

Pokropivny, V. V., & Skorokhod, V. V. 2007. Classification of nanostructures by dimensionality and concept of surface forms engineering in nanomaterial science. *Materials Science and Engineering: C, 27*(5–8), 990–993.

Pramanik, P., Krishnan, P., Maity, A., Mridha, N., Mukherjee, A., & Rai, V. 2020. Application of nanotechnology in agriculture. In *Environmental nanotechnology* (Vol. 4, pp. 317–348). Springer.

Prasad, R., Gupta, N., Kumar, M., Kumar, V., Wang, S., & Abd-Elsalam, K. A. 2017. Nanomaterials act as plant defense mechanism. In *Nanotechnology* (pp. 253–269). Springer.

Primožič, M., Knez, Ž., & Leitgeb, M. 2021. (Bio) nanotechnology in food dcience—food packaging. *Nanomaterials, 11*(2), 292.

Qi, Y., Ye, J., Zhang, S., Tian, Q., Xu, N., Tian, P., & Ning, G. 2019. Controllable synthesis of transition metal ion-doped CeO_2 micro/nanostructures for improving photocatalytic performance. *Journal of Alloys and Compounds, 782*, 780–788.

Rahman, M., Ahmad, M. Z., Kazmi, I., Akhter, S., Afzal, M., Gupta, G., Ahmed, F. J., and Anwar, F. 2012. Advancement in multifunctional nanoparticles for the effective treatment of cancer. *Expert Opinion on Drug Delivery, 9*(4), 367–381.

Ramis, J. M., Coelho, C. C., Córdoba, A., Quadros, P. A., & Monjo, M. 2018. Safety assessment of nano-hydroxyapatite as an oral care ingredient according to the EU cosmetics regulation. *Cosmetics, 5*(3), 53.

Rico, D., Martin-Diana, A. B., Barat, J. M., & Barry-Ryan, C. 2007. Extending and measuring the quality of fresh-cut fruit and vegetables: A review. *Trends in Food Science & Technology, 18*(7), 373–386.

Rigano, L., & Lionetti, N. 2016. Nanobiomaterials in galenic formulations and cosmetics. In *Nanobiomaterials in galenic formulations and cosmetics* (pp. 121–148). William Andrew Publishing.

Sahu, D., Kannan, G. M., & Vijayaraghavan, R. 2014. Carbon black particle exhibits size dependent toxicity in human monocytes. *International Journal of Inflammation, 2014*, Article ID 827019, 10 pages. https://doi.org/10.1155/2014/827019

Salavati-Niasari, M., Davar, F., & Mir, N. 2008. Synthesis and characterization of metallic copper nanoparticles via thermal decomposition. *Polyhedron, 27*(17), 3514–3518.

Saleh, T. A. 2020. Nanomaterials: Classification, properties, and environmental toxicities. *Environmental Technology & Innovation*, 101067.

Saleh, T. A., & Fadillah, G. 2019. Recent trends in the design of chemical sensors based on graphene—metal oxide nanocomposites for the analysis of toxic species and biomolecules. *TrAC Trends in Analytical Chemistry, 120*, 115660.

Santos, A. C., Morais, F., Simões, A., Pereira, I., Sequeira, J. A. D., Pereira-Silva, M., Veiga, F., and Ribeiro, A. 2019. Nanotechnology for the development of new cosmetic formulations. *Expert Opinion on Drug Delivery, 16*(4), 313–330.

Sarangi, S. K., Singh, R., & Singh, K. A. 2009. Indigenous method of rat proof grain storage by Adi tribes of Arunachal Pradesh. *Indian Journal of Traditional Knowledge*, *8*(2), 230–233.

Shang, Y., Hasan, M., Ahammed, G. J., Li, M., Yin, H., & Zhou, J. 2019. Applications of nanotechnology in plant growth and crop protection: A review. *Molecules, 24*(14), 2558.

Silva-Weiss, A., Ihl, M., Sobral, P. D. A., Gómez-Guillén, M. C., & Bifani, V. 2013. Natural additives in bioactive edible films and coatings: Functionality and applications in foods. *Food Engineering Reviews, 5*(4), 200–216.

Silvestre, C., Duraccio, D., & Cimmino, S. 2011. Food packaging based on polymer nanomaterials. *Progress in Polymer Science, 36*(12), 1766–1782.

Singh, K. 2020. Nanotechnology: Advanced technique in postharvest management of horticultural crops. *Biotica Research Today, 2*(6), 456–458.

Singh, Y., Srinivas, A., Gangwar, M., Meher, J. G., Misra-Bhattacharya, S., & Chourasia, M. K. 2016. Subcutaneously administered ultrafine PLGA nanoparticles containing doxycycline hydrochloride target lymphatic filarial parasites. *Molecular Pharmaceutics, 13*(6), 2084–2094.

Soheilmoghaddam, M., Wahit, M. U., Yussuf, A. A., Al-Saleh, M. A., & Whye, W. T. 2014. Characterization of bio regenerated cellulose/sepiolite nanocomposite films prepared via ionic liquid. *Polymer Testing, 33*, 121–130.

Song, B., Xu, P., Chen, M., Tang, W., Zeng, G., Gong, J., Zhang, P., and Ye, S. 2019. Using nanomaterials to facilitate the phytoremediation of contaminated soil. *Critical Reviews in Environmental Science and Technology, 49*(9), 791–824.

Srivastava, A. K., Dev, A., & Karmakar, S. 2018. Nanosensors and nanobiosensors in food and agriculture. *Environmental Chemistry Letters, 16*(1), 161–182.

Su, H., Sun, H., Zhang, Y., Yang, Y., Shi, X., & Guo, Z. 2019. Luminescence of perovskite light-emitting diodes with quasi-core/shell structure enhanced by Al-TiO_2-Ag bimetallic nanoparticle. *Superlattices and Microstructures, 136*, 106323.

Sur, S., Rathore, A., Dave, V., Reddy, K. R., Chouhan, R. S., & Sadhu, V. 2019. Recent developments in functionalized polymer nanoparticles for efficient drug delivery system. *Nano-Structures & Nano-Objects, 20*, 100397.

Tai, C. Y., Tai, C. T., Chang, M. H., & Liu, H. S. 2007. Synthesis of magnesium hydroxide and oxide nanoparticles using a spinning disk reactor. *Industrial & Engineering Chemistry Research, 46*(17), 5536–5541.

Teng, W. Y., Jeng, S. C., Kuo, C. W., Lin, Y. R., Liao, C. C., & Chin, W. K. 2008. Nanoparticles-doped guest-host liquid crystal displays. *Optics Letters, 33*(15), 1663–1665.

Tiwari, D. K., Behari, J., & Sen, P. 2008. Application of nanoparticles in waste water treatment. *World Applied Sciences Journal, 3*(3), 417–433.

Vázquez-Núñez, E., Molina-Guerrero, C. E., Peña-Castro, J. M., Fernández-Luqueño, F., & de la Rosa-Álvarez, M. 2020. Use of nanotechnology for the bioremediation of contaminants: A review. *Processes, 8*(7), 826.

Watson, S. B., Gergely, A., & Janus, E. R. 2011. Where is agronanotechnolgoy heading in the United States and European Union. *Natural Resources & Environment, 26*, 8.

Wiechers, J. W., & Musee, N. 2010. Engineered inorganic nanoparticles and cosmetics: Facts, issues, knowledge gaps and challenges. *Journal of Biomedical Nanotechnology, 6*(5), 408–431.

World Health Organization. 2010. *FAO/WHO expert meeting on the application of nanotechnologies in the food and agriculture sectors: Potential food safety implications: Meeting report*. World Health Organization.

Wu, H., Yin, J. J., Wamer, W. G., Zeng, M., & Lo, Y. M. 2014. Reactive oxygen species-related activities of nano-iron metal and nano-iron oxides. *Journal of Food and Drug Analysis, 22*(1), 86–94.

Xie, F., Yang, M., Jiang, M., Huang, X. J., Liu, W. Q., & Xie, P. H. 2019. Carbon-based nanomaterials—A promising electrochemical sensor toward persistent toxic substance. *TrAC Trends in Analytical Chemistry, 119*, 115624.

Xing, Y., Yang, H., Guo, X., Bi, X., Liu, X., Xu, Q., Wang, Q., et al. 2020. Effect of chitosan/Nano-TiO_2 composite coatings on the postharvest quality and physicochemical characteristics of mango fruits. *Scientia Horticulturae, 263*, 109135.

Xu, X., Stevens, M., & Cortie, M. B. 2004. In situ precipitation of gold nanoparticles onto glass for potential architectural applications. *Chemistry of Materials, 16*(11), 2259–2266.

Yang, X., Lian, K., Tan, Y., Zhu, Y., Liu, X., Zeng, Y., Yu, T., Meng, T., Yuan, H., & Hu, F. 2020. Selective uptake of chitosan polymeric micelles by circulating monocytes for enhanced tumor targeting. *Carbohydrate Polymers, 229*, 115435.

Yılmaz, F., Saylan, Y., Akgönüllü, S., Çimen, D., Derazshamshir, A., Bereli, N., & Denizli, A. 2017. Surface plasmon resonance based nanosensors for detection of triazinic pesticides in agricultural foods. In *New pesticides and soil sensors* (pp. 679–718). Academic Press.

Zeng, T., Zhang, P., Li, X., Yin, Y., Chen, K., & Wang, C. 2019. Facile fabrication of durable superhydrophobic and oleophobic surface on cellulose substrate via thiol-ene click modification. *Applied Surface Science, 493*, 1004–1012.

Zhang, Q., Gao, Q., Qian, W., Zhang, H., Tian, W., & Li, Z. 2019. A C-coated and Sb-doped SnO_2 nanocompsite with high surface area and low charge transfer resistance as ultrahigh capacity lithium ion battery anode. *Materials Today Energy*, *13*, 93–99.

Zhao, M., Liu, L., Her, R., Kalantar, T., Schmidt, D., Mathieson, T., . . . & Zettler, M. 2012. Nano-sized delivery for agricultural chemicals. In *Nano formulation* (pp. 256–265). Royal Society of Chemistry.

Zhong, Q., & Zhang, L. 2019. Nanoparticles fabricated from bulk solid lipids: Preparation, properties, and potential food applications. *Advances in Colloid and Interface Science*, *273*, 102033.

Chapter 3

Organic Nanoparticles in Postharvest Management and the Mechanisms Involved

Aiswarya P., Sarita G. Bhat, and Sreeja Narayanan

List of Abbreviations

AAO	*Artemisia annua* oil
AFM	Atomic force microscopy
BCN	Bacterial cellulose nanofibrils
BSN	Banana starch nanoparticles
CHNP-ECEO	Nanochitosan encapsulated with *Eryngium campestre L.* essential oil
CMC	Carboxymethyl cellulose
CNC	Cellulose nanocrystal
CNF	Cellulose nanofibers
CNP	Cellulose nanoparticle
CNW	Cellulose nanowhiskers
CS	Chitosan
CSN	Crystalline silk nanodisc
CSNP	Chitosan nanoparticle
EC	Edible coating
EO	Essential oil
GRAS	Generally recognized as safe
HDM pectin	High-density methyl pectin
LDM pectin	Low-density methyl pectin
LLD	Liquid–liquid dispersion

DOI: 10.1201/9781003142287-3

LRNC	Lignin-rich nanocellulose
MFC	Microfibrillated cellulose
NCC	Nanocrystalline cellulose
NCS	Nanochitosan
NFC	Nanofibrillated cellulose
P	Pectin
P:NCS	Pectin: nanochitosan
PDI	Polydispersity index
PE	Polyethylene
PET	Polyethylene terephthalate
PLA	Polylactic acid
PP	Polypropylene
SEM	Scanning electron microscopy
SLN	Solid lipid nanoparticle
SNC	Silk nanocrystal
SNP	Starch nanoparticle
Starch-BTCAD-NHS–SBN	Starch-butanetetracarboxylic acid dianhydride-N-hydroxysuccinimide)
TE	*Thymus vulgaris L.* extract
WPI	Whey protein isolate
XA	Xylonic acid
ZN	Zein nanoparticle

3.1 Introduction

Deterioration of fruits and vegetables occurs at every point, from pre-harvest and harvest to postharvest, transportation, and consumption. The decline is attributed to damage caused by environmental factors or human handlers alone or in combination. Once harvested, fresh produce starts to lose water and shrinks, causing weight loss at varying degrees, which directly depends on the surface area-to-volume ratio of the product. In order to extend shelf life, the rate of water loss should be minimal. The vapor pressure of water inside the product and the atmosphere surrounding it are important factors governing the rate of water loss. Careless handling can cause skin breaks and internal bruising in fruits and vegetables, further increasing the rate of moisture loss, which can also cause postharvest decay, mainly attributed to oxidative reactions and microbial deterioration (Madani et al., 2019).

Most infections are due to fungi, followed by bacteria; microscopic spores distributed in the air, water, and soil; and decayed plant material that spreads the disease, leading to considerable quality loss. Enzymes for various biological activities are present inherently in fruits and vegetables. Tissue dissolution, product softening, and excessive maturation are all integral parts of ripening. Some of the primary biochemical changes are oxidative damage of important pigments like chlorophyll, anthocyanin, oxidation of vitamins like ascorbic acid, and starch-to-sugar conversion. Fruit ripening accelerates the loss of firmness, weight loss, flavor, poor appearance, color, nutrition, and texture of fresh produce (Basumatary et al., 2020). Increased demand for a year-round supply of fresh food produces, and the globalization of fruit and vegetable marketing necessitates the requirement of storage and transport for a long period. Specifically, for postharvest fresh food produce, packaging becomes essential. Materials that are widely used for packing are plastics (42%), paper/cardboard (31%), metals (15%), ceramics/glass (7%), and other materials (5%; Jeevahan & Chandrasekaran, 2019). The materials used for packaging have to fulfill stringent criteria and are also dependent on the type of food production and the storage conditions. When food consumption is delayed or when food is transported over long distances, a demand for better packaging material arises since there are hardly any pure materials that equally exhibit all the desired thermal, physical, and biological properties relevant to the conceivable food packaging, polymer blends, composites, and complex multilayer films that are often used (Duncan, 2011).

Among the packaging materials, plastic has become the most dominant in use today. After use, the majority of the plastic packs are discarded, with only a limited quantity recycled. These nonrecycled plastics are either dumped in landfills, creating microplastics, or incinerated, which emits harmful gases into the atmosphere and causes pollution and harm to the environment. Biodegradable plastics are an attractive alternative to synthetic, non-decomposable plastics since they decompose and turn into carbon dioxide, methane, and water. However, for this process to occur, it requires aerated conditions supportive of active decomposition and significant land areas for landfills. Packaging materials developed from polysaccharides, proteins, and lipids utilize renewable organic sources, and they are quickly replacing nonrenewable fossil sources such as petroleum-based, non-degradable, and degradable polymers. Various organic nanoparticles applied in packaging mainly are of three different categories: polysaccharide-based, protein-based, and lipid-based (Figure 3.1). Film production uses nanoparticles alone or in combination with polymer matrices to form nanocomposite materials, together with other bioactive compounds, inorganic nanoparticles, and other organic nanomaterials, rendering better postharvest preservation of food produce (Huang et al., 2018). Organic nanoparticle-based packaging is safe to use under every intended condition. It is inert and can withstand extreme processing conditions, in addition to having properties such

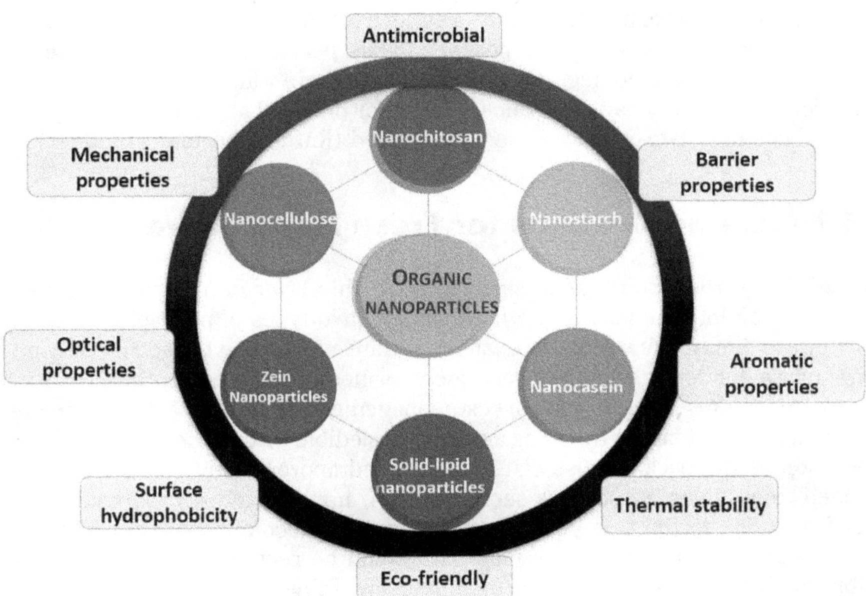

Figure 3.1 Types of organic nanoparticles applied in postharvest management and their significant properties.

as lightness in weight, producibility, cost-effectiveness, easy disposability or reusability, and resistance to external forces (Ameta et al., 2020).

3.2 Properties and Mechanisms of Organic Nanoparticles and Their Composites

Food packaging ensures food preservation from all types of physical, chemical, and biological contamination encountered during processing, harvest, storage, and consumption (Bajpai et al., 2018). It also secures food quality by maintaining a good atmosphere inside the coating, imparting mechanical strength, avoiding microbial interventions, separating the products from chemical contaminants (viz. oxygen, moisture, light), and providing stability under different temperatures and moisture conditions, thereby extending a product's shelf life. To improve properties such as biodegradability, mechanical, and antimicrobial attributes, nanoparticles are incorporated into the packaging materials without changing their density and transparency (Echegoyen, 2015; Mohamad Fauzi & Abu Hassan, 2017; Wesley et al., 2014).

A significant concern with packing material is its permeability to water vapor, atmospheric gases, and inherent substances from fruits and vegetables. In order to address such challenges, nanoparticles made of polysaccharides and protein package inclusions were built. The fundamental difference between polysaccharides and protein nanoparticles lies in their barrier properties. Proteins show better mechanical strength while polysaccharide provides a better gas barrier. However, both can form self-supporting films, unlike lipids, which are suitable water vapor barriers but cannot form self-supporting films. Hence, lipids are applied directly over the surface of fruits and vegetables as coatings. Lipids act as additives in films made of a polysaccharide or protein self-supporting emulsion (Jeevahan & Chandrasekaran, 2019). Nowadays, as an alternative to control postharvest diseases of fruit and vegetables, such as strawberries, grapes, apples, pears, mangos, peaches, citrus, peaches, guava, bananas, apricots, avocados, papayas, and pomegranates, nanomaterials are widely used. Single nanoparticles alone cannot fulfill all properties. Hence, for practical applications, composite in films and coatings, two or more materials are used (Ruffo Roberto et al., 2019).

3.3 Films and Coatings for Fresh Horticulture Produce

Packaging that rests on the food surface with a thin layer of material, if edible, can be eaten as an integral part of the product. The two types of packaging in the food industry are films and coatings. In coatings, a thin layer will be formed directly on the food surface, while the films are separately formed and placed on the food surface at later stages. Research in postharvest management based on packaging materials mainly focuses on making the films and coatings edible so that there will be no waste generation as the package also can be consumed appropriately. Films and coatings obtained from biological sources such as plants, animals, and microorganisms have gained attention in the past years. It will be a great accomplishment if all the agricultural waste, such as cellulosic materials, could be recycled as nanomaterials for various applications focusing on the food industry. Table 3.1 lists the organic and biologically relevant materials, such as polysaccharides, proteins, and lipids, used in postharvest packaging, owing to their various properties. Several studies support the use

Table 3.1 List of Various Organic Nanoparticles, Their Application in Postharvest Management and the Possible Mechanisms

Nanoparticles, polymer matrix and bioactive compounds	*Application in postharvest management*	*Functionality/Mechanism*	*References*
Chitosan			
1. Chitosan nanoparticles (CSNPs) and propolis (P)	Used for extending the shelf life of vegetables and fruits	Quality and antioxidant capacity of strawberries	Martínez-González et al., 2020
2. Active cellulose-based papers coated with chitosan–Ag/tio2 nanocomposite	Coating of walnut kernels	Improved barrier properties and antimicrobial properties increasing the shelf-life	Mihaly Cozmuta et al., 2018
3. Chitosan nanoparticles and cinnamon essential oil composite with zein films	Food packaging applications	Improved tensile strength and decreased the elongation of zein film composite, antimicrobial properties	Vahedikia et al., 2019
4. Chitosan nanoparticles with high- and low-methyl pectin films	Biodegradable biopolymeric films for packaging	Improved mechanical, thermal, and barrier properties	Lorevice et al., 2016
5. Chitosan nanoparticles with incorporated resveratrol	Improve the properties of polyethylene (PE) and polypropylene (PP) polymer foils suitable for applications in food packaging	Improved antioxidant and antimicrobial properties, reduced oxygen permeability	Glaser et al., 2019
Cellulose			
1. Nano filler incorporated (hydroxypropyl) methyl cellulose (HPMC)	Coating of nanoparticles incorporated HPMC polymer Gel to unripened mangoes	Delayed ripening and antimicrobial process of Mangifera indica L.	Kumar et al., 2018
2. Chitosan/cellulose nanofibril (CNF) nanocomposite	Coating of Strawberries with Bionanocomposites	Reduced 'Camarosa' strawberry mass loss and softening, mitigates the changes of color, soluble solids, fungal contamination, vitamin C, total phenolics, anthocyanins, PAL activity, and antioxidant capacity	Resende et al., 2018

(Continued)

Table 3.1 (Continued)

Nanoparticles, polymer matrix and bioactive compounds	*Application in postharvest management*	*Functionality/Mechanism*	*References*
Starch			
1. Biodegradable starch/clay nanocomposite films	Food packaging applications	Improved mechanical properties	Avella et al., 2005
2. Clay, silver, and starch nanostructured films	Food packaging applications	Enhanced mechanical, barrier, and antimicrobial properties	Abreu et al., 2015
3. Ag-starch nanocomposite	Food packaging applications	Increase their mechanical properties and physical resistance	Mohseni et al., 2020
4. Edible corn starch nanocomposite films	Biodegradable packaging films	Decreased water vapor permeability, tensile strength significantly improved	Liu et al., 2016
5. Biodegradable chitosan–rice-starch nanocomposite films	Nano-composite films for peach fruit	Antimicrobial activity and enhanced shelf life improved postharvest quality of stored peach fruit	Kaur et al., 2017
6. Banana starch nanoparticles and montmorillonite incorporated banana flour film	Nanoparticle reinforced biopolymer films for packaging	Improved mechanical and water barrier properties	Orsuwan & Sothornvit, 2017
Gums			
1. Xanthan gum alone and in different concentration with guar gum	Edible coatings on the cassava root	Affected ethylene production rate, respiration, and dry matter content of cassava with extended postharvest shelf life	Atieno et al., 2019
2. Starch-cashew tree gum nanocomposite films (montmorillonite (MMT) as nanofillers)	Cashew nuts coating	Increased tensile strength and elastic modulus and barrier properties and decreased oxidation rates	Pinto et al., 2015

Other common polysaccharides			
Pullulan			
1. Nanocomposite films composed of pullulan (PL) and lysozyme nanofibers (LNFs)	Eco-friendly edible films for active food packaging	Good mechanical properties, antibacterial and antioxidant activities	Silva et al., 2018
Carrageenan			
1. Carrageenan-zein and Mica bionanocomposite	Food packaging and coating applications for shelf-life extension of food products	Reduced water permeability and uptake, e ability to block the UV-vis radiation (absorbing ability of fillers)	Sanchez-Garcia et al., 2010
Zein			
1. Zein-sodium caseinate (NaCas) curcumin-loaded nanoparticles	Long term storage of food	Antioxidant film, significantly improve the tensile strength, decreased water vapor permeability	Wang et al., 2019
2. Natamycin-loaded zein nanoparticles stabilized by carboxymethyl chitosan	Encapsulate hydrophobic food preservatives for fruit preservation, application in postharvest treatments	Increased light stability of natamycin reduce the occurrence of rot and mildew in strawberries by inhibit the spore germination and mycelial growth of *B. cinerea*	Lin et al., 2020
Casein			
1. Natamycin encapsulated in zein/casein	Nanocomposite films	Antifungal activities against three spoilage molds	Mo et al., 2021
2. Eugenol-entrapped casein nanoparticles	Economical and simple-manufactured preservative for postharvest fruits against microbial spoilage	Enhanced antifungal activities against Anthracnose in postharvest fruits (Pear)	Xue et al., 2019

(Continued)

Table 3.1 (Continued)

Nanoparticles, polymer matrix and bioactive compounds	*Application in postharvest management*	*Functionality/Mechanism*	*References*
Solid Lipid Nanoparticle			
1. Beeswax solid lipid nanoparticles (BSLN) along with xanthan gum (XG) and propylene glycol	Conservation and increase of shelf life of strawberry stored in refrigeration	Antimicrobial activity, limited oxygen diffusion and water loss, limited respiration	Zambrano-Zaragoza et al., 2020
2. Solid lipid nanoparticles (SLN)	To control and reduce postharvest disease in small fruits during shipping and storage	Good mechanical, optical, and antifungal properties	McDaniel et al., 2019
3. Solid Lipid Nanoparticles (SLN) containing 1% w/w cinnamaldehyde and pullulan coating	Antimicrobial packaging films on strawberries	Maintains quality and reduces fungal decay during refrigerated storage	Trinetta et al., 2020
4. Solid lipid nanoparticles (SLNs) and xanthan gum	Coatings for the preservation of guava	Improved shelf life of guava	Zambrano-Zaragoza et al., 2013 Gad & Zagzog, 2017
5. Candeuba wax solid lipid nanoparticles, xanthan gum (XG) as coatings on guava	Postharvest life preservation	Delayed biochemical and physiological changes	García-Betanzos et al., 2017
Lignin			
Lignin nanoparticles in polyvinyl alcohol/chitosan films	Active packaging of food items	Antioxidant and antibacterial properties	Yang et al., 2016

Table 3.2 Methods of Nanoparticle Synthesis and Packaging Techniques Employed

Nanoparticle	*Method of preparation*	*Packaging method*	*References*
1. Solid lipid nanoparticles (SLNs)	Hot lipid dispersion Hot high-shear stirring method	Dip coating Dip coating	Zambrano-Zaragoza et al., 2013 García-Betanzos et al., 2017
2. Sodium caseinate (NaCas)-zein composite nanoparticles	Self-assembly method	Solution casting method	Wang et al., 2019
3. Natamycin was encapsulated in zein nanoparticles	Anti-solvent method	Spraying method	Lin et al., 2020
4. Chitosan nanoparticles	Ionotropic gelation	Solution casting method	Lorevice et al., 2016
5. Silk nanoparticles	Standard degumming process	Dip coating	Brenckle et al., 2016
6. Starch nanoparticles	Water-in-oil mini emulsion cross-linking method	Solution casting method	Orsuwan & Sothornvit, 2017
7. Chitosan nanoparticles	Ionic gelation technique	Roll printing	Glaser et al., 2019
8. Liposomes nanoparticles	Self-assembly	Dip coating	Makwana et al., 2014

of polysaccharides, mostly homopolymers such as starch and cellulose, while some advocate the use of proteins, which are heteropolymers consisting of different types of amino acids. Various packaging methods employed for improving the shelf life of the harvested products are briefed in Table 3.2. Additionally, to improve the packaging efficiency, anti-browning agents, nutraceuticals, antimicrobials, and other functional components can also be added (Saberi & Golding, 2018). The following sections of the chapter discuss, in length, the different organic materials that are nanofabricated and used for postharvest management of fruits and vegetables.

3.4 Polysaccharide-Based Organic Nanoparticles for Packaging

3.4.1 NCS

CS is a soluble form of chitin obtained after deacetylation (Kumar et al., 2020). The molecular weight and degree of deacetylation of these heterogeneous copolymers mainly influence the physicochemical properties such as viscosity, solubility, and biological properties. The wide acceptability of CS nanoparticle–incorporated films for coating fruit and vegetables is due to their biodegradability, selective permeability to gases (such as CO_2, O_2, and ethylene), biocompatibility, and antimicrobial activity. During the postharvest period, it can also act as a resistance elicitor and induces defense mechanisms in the produce, both locally and systemically (Martínez et al., 2018; Orzali et al., 2017; Sivakumar et al., 2016).

The CS nanoparticles are reported to show two general mechanisms of action, which account for their filmogenic properties as an EC material. First, CS nanoparticle–based coatings can slow down the respiration rates of fruits and act as a physical barrier against infectious pathogens (Mohammadi et al., 2015). The shelf life of cucumber was extended and stored for 21 days using *Zataria multiflora* EO–encapsulated CS nanoparticles. The nanoparticles effectively enhanced the antioxidant activity during food processing and storage by protecting the bioactive compounds from unfavorable environmental conditions (Mohammadi et al., 2016). Additionally, the antifungal activity of CS nanoparticles, owing to the induction of enzyme chitinases and its interference with fungal growth, is an attribute that renders CS nanoparticles good packaging material (Correa-Pacheco et al., 2017; Martínez et al., 2018; Melo, Lima, et al., 2020; Romanazzi & Feliziani, 2016; Sahraei Khosh Gardesh et al., 2016; Sivakumar et al., 2016; Zahid et al., 2013).

Concerning the preparation of CS nanoparticles, ionic gelation is the most common protocol for obtaining CS nanoparticles that are later embedded into biocompatible matrices to obtain films (Sharma et al., 2017; Lin et al., 2019). This matrix-reinforcing strategy has been proven to elicit efficient antimicrobial and barrier properties (Martínez et al., 2018). Several studies on fresh produce packaging using ECs based on NCS (Esyanti et al., 2019), gel enriched with edible NCS, CS nanocomposite (Melo, Lima, et al., 2020; Melo, Pintado, et al., 2020), NCS combined with EOs (Mohammadi et al., 2015) (Correa-Pacheco et al., 2017), NCS combined with plant extracts (González-Saucedo et al., 2019), chito-metallic nanoparticles (Eshghi et al., 2014; Saqib et al., 2020), CMC (Jannatyha et al., 2020), CMC and whey protein nanocomposites (Shojaei et al., 2019), and hydroxypropyl methylcellulose (Moura et al., 2008), with interesting conclusions about their effects on postharvest food preservation, have been reported.

An exciting study by Arabpoor et al. employed NCS CHNP-ECEO as a fruit coating material to prolong the fresh cherry fruits' shelf life. Cherries coated with a thin layer of CHNP-ECEO were stored for 21 days to check the microbial activity at a storage temperature of 4°C, and the coating demonstrated efficient microbial growth control. An assessment of different physicochemical characteristics showed a decrease in weight loss and titratable acidity while there was a moderate increase in the rate of respiration, pH, firmness, total phenolic contents, and antioxidant activity of the cherries. Additionally, the coating also did not alter the natural color of the fruits (Arabpoor et al., 2021). Similar studies conducted on fresh strawberries stored at 2°C for the same period showed decreased weight loss while firmness and titratable acidity remain conserved. The coating retarded the production of malondialdehyde, a marker of oxidative stress, and inhibited polyphenol oxidase activity significantly (Nguyen & Nguyen, 2020). In yet another study, different ratios of edible films based on P and NCS were characterized for their antimicrobial properties. The hydrophobic films exhibited efficient inhibition of fungal growth, including *Colletotrichum gloeosporioides*, *Saccharomyces cerevisiae*, and *Aspergillus niger*, and bacterial species *Escherichia coli*. A 50:50 proportion of P to NCS in films was found to be good for active packaging applications for extending shelf life (Ngo et al., 2020). Advanced packaging materials developed by adding CS nanoparticles to low- and high-methyl P biopolymer matrices to produce LDM/CSNP and HDM/CSNP nanocomposite films showed improved mechanical and barrier properties owing to the CS nanoparticles. Unlike HDM pectin/CSNP, LDM pectin/CSNP nanocomposite films also showed good thermal properties, preventing the warming up of the packed food produce. The study proved that biodegradable, eco-friendly, and renewable polysaccharides like P could be successfully reinforced with nanostructures, yielding effective alternatives for conventional packaging (Lorevice et al., 2016).

A combination of treatments, organic calcium foliar spraying, and NCS fruit coating during postharvest extended the storage life of fresh early swelling peach fruits. The combined treatment improved sensory properties, fruit weight, volume, firmness, total soluble solids, color attributes, and overall yield. CS is economical and cost-effective; its wide acceptability is possible for extended storage and shipping of peaches during off-seasons (Ali & Toliba, 2018). Dadvar et al. used an innovative and inexpensive technique in a case study on apples. They used three optimization methods based on two variables concerning all experimental replications and two other heuristics methods—the basis of the method is an algorithm known as the Lagrange interpolation optimization algorithm. The advantage of this new approach was that there was no additional material requirement; it was easy to perform and was healthful (Dadvar et al., 2021).

In another report, edible films were made using CS nanoparticles with tara gum. This approach improved the mechanical and physicochemical properties and antimicrobial activity of the film. The compact structure of the film after nanoparticle incorporation reduced the film's water solubility and hydrophilicity and reduced the available free volume of the polymer matrix, decreasing the film's moisture content (Antoniou et al., 2015). The storage life and quality of guava fruits (*Psidium guajava L.*) were investigated by making a film using xanthan gum and CS nanoparticle mix. Mixing 1% xanthan gum with 0.2% CS nanoparticles achieved the lowest decay percentage, maintained the level of vitamin C and fruit firmness, but reduced total sugar content compared to xanthan gum and control. However, an increased nanoparticle concentration of 0.4% increased the percentage of decay, firmness loss, and vitamin C (Gad & Zagzog, 2017)

A nanostructured EC based on EC-P-CSNPs with natural components, such as terpenes, was employed to preserve the postharvest quality of *Capsicum annuum L.* An EC incorporating α-pinene encapsulated CS nanoparticles inhibited the fungal pathogen *Alternaria alternata* and preserved the physicochemical quality of bell peppers (Hernández-López et al., 2020). CS nanoparticles have also been combined with starch in another study to prepare starch/CSNP films that were later evaluated for their in vitro and in vivo antimicrobial properties. Compared to the tested gram-negative bacteria, the gram-positive bacteria were inhibited more efficiently using the composite film. An inhibition zone of 15% and 20% was observed for the tested bacteria such as *Escherichia coli*, *Bacillus cereus*, *Staphylococcus aureus*, and *Salmonella typhimurium*. In vivo studies on microbial growth in cherry tomatoes were also conducted and supported the potential application of the starch/CSNP films as an antimicrobial food packaging (Shapi'i et al., 2020). CS nanoparticles have also been employed to improve the properties of PE and PP polymer foils. In this study, the CS-nanoparticle encapsulated resveratrol, an antioxidant polyphenol with multispectral biologically relevant activities. The proposed modification of the polymer foils with nanoparticles showed both an increase in antioxidant activity by a factor of 10 and antimicrobial properties against *Staphylococcus aureus* (90%) and *Escherichia coli* (77%) simultaneously, making the film suitable for packaging applications (Glaser et al., 2019).

3.4.2 Nanocellulose

Cellulose is the most abundant hydrophilic polymeric material obtained from natural sources such as cotton, jute, banana rachis, hemp, flax, oil palm, wheat, rice, wood, sugarcane bagasse, corncob, cassava bagasse, bamboo, sisal, and soy hulls (Silvério et al., 2013). Currently, cellulose in the form of nanoparticles is being considered a cost-effective alternative for synthetic materials in the food sector because of its properties such as biodegradability, edibility, transparency, flexibility, mechanical and barrier properties (oxygen, water vapor, oil, grease, liquid), antimicrobial attributes, high aspect ratio, biocompatibility, renewability, low density, low toxicity, low thermal expansion, and so on (Abdul Khalil et al., 2016). Hydrophilic groups in cellulose enable them to establish strong intermolecular hydrogen bonds, rendering the material mechanically strong and chemically stable (Kumar et al., 2014; de Amorim et al., 2020).

Incorporation of CNPs, including, CNFs (also known as NFC or MFC) and CNCs (also known as CNWs or NCCs) and bacterial nanocellulose into biopolymers matrices have shown improvements in the quality and shelf life of fruits and vegetables when the composite is used as packing material. The dense structures formed by polymer–CNC intercalation have sound reinforcement, significantly improving the mechanical and barrier properties of the film. Among the four crystalline polymorphs of cellulose (I, II, III, IV), cellulose II is of great interest for packaging applications due to its better functionalities that allow a better interaction with the surrounding matrix (Dhar et al., 2015). Nanocellulose involved in packaging is generally available as nanocomposite coatings and films, biodegradable and edible food packaging, hydrogels, blends, emulsion-based food packaging, and others (Mondal et al., 2019). Some active substances can further enhance packaging efficiency, for example, antioxidants, and antimicrobials (Manikkam, 2018). Additionally, a study demonstrated the use of nanocellulose and lemongrass EO to prolong the postharvest shelf life of blackberry fruits up to 6 days without any soluble solid changes (Silva et al., 2020).

Five percent of CNC (w/w)–reinforced CMC composite film for red chili packaging showed a 50.8% increase in tensile strength, improved UV barrier, a 53.4% decrease in water vapor permeability, and improved thermal stability and effectively reduced weight loss while maintaining the level of vitamin C in the chilies when compared with uncoated ones (H. Li et al., 2020). The combined effects of phosphorylated CNC-incorporated CS nanocomposite films were compared to neat CS and unmodified cellulose-filled CS. Phosphorylated CNC composites showed better mechanical, antibacterial, and thermal properties and interesting intercellular catalase activity comparatively. Moreover, cyclotriphosphazene-grafted cellulose nanocrystals as fillers showed better mechanical properties and biological properties than microcrystals. The film showed a significant increase in the adsorption of hemoglobin after 24 hours of incubation. Antibacterial activity was observed against *S. aureus* and *E. coli* (Blilid et al., 2020). These studies elucidated the role of phosphorylation in the rational designing of next-generation bioplastics based on nanocellulose. In another study, Luo et al., 2020 fabricated hybrid films combining LRNC from depectinized celery and P containing up to 20% XA. A pioneering study gave fundamental guidance about combining XA with LRNC to reconstruct hybrid films for food packaging. With the incorporation of higher concentrations of XA into the film, CNFs showed antimicrobial activity along with other properties, including transparency, flexibility, and hydrophilic properties (Luo et al., 2020). Reinforcing effects, such as UV blocking, antimicrobial, and antioxidant properties, of nanocellulose-based nanohybrids for prolonging food shelf life and recent advancements and synthesis methods have been reviewed in detail recently (Abdalkarim et al., 2021).

3.4.3 Nanostarch

Starch is second in abundance as an organic material produced biologically only after cellulose (Ashori, 2014). It is an attractive alternative to plastic-based packaging as it is edible and inexpensive, biodegradable, renewable, and retrievable from its raw nature for industrial and horticultural applications. The extraction of starch granules is chiefly from cereals and tubers as starch is mainly stored in seeds and roots as the chief energy reserve of the plant (Chivrac, 2009). In order to overcome the issues of compatibility during blending and lamination, different starch derivatives have been synthesized. These starch-based thermoplastic materials are biological and compostable, and therefore, their commercialized products are in high demand (Dhall, 2013).

Starch-based coatings are colorless, tasteless, and flavorless and are transparent or translucent. Along with these critical benefits, an add-on advantage of starch-based coating is their oil-free appearance and low permeability to oxygen due to the highly ordered H bonds in the starch configuration acting as a barrier for oxygen entry. This coating ability can significantly contribute to keeping the freshness of horticulture produce by decreasing the rate of respiration and senescence (Saberi & Golding, 2018). The newly developed pea starch-based nanocomposite films impregnated with potato SNPs of size 15–30 nm effectively showed an increase in the physicochemical properties of the starch film. About 6% of SNPs exhibited the highest tensile strength of 15MPa, decreased water vapor permeability and water solubility, and had a high melting temperature, indicating good thermal stability (Jiang et al., 2016). The incorporation of nanoparticles prepared from potato starch as a filler in composite films improved the film's properties, such as film thickness and biodegradability, and reduced the water vapor transmission rate (Gujral et al., 2021).

Type A crystals of starch with a crystallinity of 41.42% were prepared by citrate esterification and ultrasound treatment to test the influence of the same on the formation and properties of SNPs. SNPs with a higher crystallinity and efficiency indicate that the esterification followed by a 5-min ultrasonic treatment would yield particles with a mean size of 352.8 nm and a PDI of 0.292. The impact of the prepared SNPs as a filler on CS composite films was evaluated, which revealed a noteworthy enhancement in the film's mechanical properties, suggesting the application of SNPs as good reinforcement for nanocomposite packaging films (Yan et al., 2021). An ideal and edible choice to create films for packaging in the food industry was developed, where they used cornstarch-based nanoparticles and gelatin as potential sources. For this, a novel macromolecular crosslinker starch-BTCAD-NHS–SBN was reinforced to modify a gelatin film. The modified gelatin films were safe to use and exhibited improved properties such as more excellent opacity, poorer water vapor uptake, high tensile strength, low Young's modulus values, surface hydrophobicity, and better anti-degradation capacity. The SBN-gel crosslinked films showed a substantial reduction in corruption and could effectively extend the shelf life of coated peeled apples (Tao et al., 2018). In yet another study, the mechanical properties and water vapor permeability of a banana flour film improved when reinforced with a 1:1 ratio of BSNs and nanoclay at 5% and 10% (w/w), respectively. A 5% mixture of BSNs and nanoclay was sufficient as a good reinforcement agent (Orsuwan & Sothornvit, 2017)

3.5 Protein-Based Organic Nanoparticles for Packaging

3.5.1 ZNs

Corn zein, an essential protein in corn, is a prolamine with high proline amino acid and is water-insoluble. Zein obtained from maize has more than 50% residues of hydrophobic amino acid in them. Based on the difference in solubility and amino acid sequences, there are four different types of zein, namely, α, β, γ, and δ. with, α-zein being the most abundant among them. Zein films are prepared commonly by the LLD method, where an aqueous solution of zein and a lipophilic compound is dissolved in ethanol and then dispersed into the water to precipitate zein as nanoparticles; this enables the compound to be encapsulated into ZNs (Pan & Zhong, 2016). They have a suitable binding property, which contributes to their good coating- and film-producing abilities. They also act as good adhesive substances, and the film formation is believed to involve limited disulfide bonds and noncovalent interactions, such as hydrophobic and hydrogen bonds, between zein chains. The resulting matrix has weak mechanical properties, and therefore, plasticizers are added to make them more flexible. However, the use of crosslinking agents and plasticizers raises concerns about the edibility of such films (Raghav et al., 2016; Dhall, 2013).

In order to evade the issue of non-edibility rendered by the crosslinkers and for developing sustainable food packaging based on zein, a multifunctional nanocomposite was developed using ZNs incorporated in BCNs that are mechanically robust. Li et al. (2020) introduced a simple, fast, and environmentally friendly route for developing a nanocomposite where the BCN matrix and ZNs interacted strongly, causing interfacial adhesion and resulting in a nanocomposite with markedly improved thermal stability, tensile strength, and mechanical properties (Li et al., 2020). In another study, ZNs have been successfully reinforced to potato starch–olive oil edible films to study their physicochemical and mechanical properties. It was understood that by increasing the concentration of oil in the composite, the water vapor permeability and tensile strength were decreased, while an increase in the number of ZNs tended to increase the barrier properties and tensile strength. The color, contact angle, and UV transmittance increased with the addition of ZNs and olive oil (Farajpour et al., 2020).

Similar studies were conducted on nanocomposites made by spraying ZNs on starch films to study the sustained release of EOs for their application as ECs and biopreservatives in perishable food products. The predominant mechanism behind the release of clove EO from ZNs was based on Fickian diffusion, and the Peleg model described the sustained release behavior well. A 10% electro-sprayed ZN–clove EO film reduced elongation at break and improved the film's water vapor permeability, tensile strength, and other mechanical properties (Alinaqi et al., 2021). Solution casting of sodium caseinate–coated ZNs on WPI was done to obtain a nanocomposite film with uniform distribution of the nanoparticles. With the addition, the elongation of the composite films were not adversely affected. SEM and AFM images suggest that the hydrophilic and hydrophobic groups of sodium caseinate provide an efficient interface between the hydrophilic whey protein matrix and hydrophobic ZNs, respectively. This interface allows the homogeneous nanoparticle distribution on the composite films even at high concentrations. A decrease in the fraction of free volume and the hydrophobic nature of ZNs, when combined, contribute significantly to effective reinforcement and improved barrier properties (Oymaci & Altinkaya, 2016).

3.5.2 Casein Nanoparticles

Casein is the major protein component obtained from bovine milk. Interestingly, the native structure of the protein does not lose its essential integrity even at high temperatures (100°C) and pressure (100 MPa). Casein can form micellar structures of 10–400 nm in size as it has the ability to self-assemble due to the amphipathic nature that favors micelle formation. It forms micelles of alternating charge and hydrophobicity, and poor water-soluble components can be encapsulated easily into the hydrophobic core of casein micelle. Regarding its use as packaging material, bovine casein has been previously used to encapsulate eugenol, which proved to be an effective nanoformulation with antifungal activity against anthracnose disease. Nanoparticles of mean size 307.4 ± 2.5 nm were formed when the eugenol solution was dispersed into casein solution under stirring. A nanoparticle was developed with a 5:1 ratio of casein: eugenol showed the highest encapsulation efficiency. Casein is an eco-friendly and cost-effective food packaging alternative for vegetables and fruits (Xue et al., 2019). A new possibility of designing opaque coatings using casein nanoparticles in the field of packaging is mentioned by Zhang et al. (2016).

3.5.3 Nano Silk

Silk is a naturally obtained filamentous fibrous protein with a mechanical flexibility that unfolds in various applications, such as in tissue engineering, food packaging, and biomedical engineering. Silk fibroin is made up of repeating units of amino acid residues, such as glycine, alanine, and serine, linked by hydrogen bonding. Lateral H bonds formed between carboxyl groups and amino groups form a highly stable antiparallel β-pleated sheet (Marsh et al., 1955). The different forms of nano-silk fibroin are SNCs and CSNs. As a result of the acid hydrolysis of fibrous SNC to its powder form, the size is reduced, and the crystalline regions become identifiable (Mondal et al., 2019).

Crystalline silk nano-discs from Muga silk by acid hydrolysis yield particles of 50 nm diameter. Nano silk as a CSN is highly ordered with a thickness of 2–4 nm. CNCs are highly crystalline spherical crystals around 100 nm in size that are fabricated by acid hydrolysis. These have been used to modify synthetic polymers to suit packaging applications. In a study, a poly(lactic acid) (PLA)–based food packaging film, which is generally brittle, with higher gas permeability and slow crystallization, was modified with CSNs to obtain a bio-nanocomposite containing 1% w/w uniformly distributed CSNs. This composite showed an increase in toughness up to 65%, percentage elongation of about 40%, tensile strength of 10% and reduced water vapor permeability by 30%, and reduction in oxygen permeability by 70%. An overall improvement in mechanical, thermal, and structural properties along with a 12% increase in storage modulus was observed, which suggested that PLA/CSN-based nanocomposite is a promising candidate for food packaging applications (Patwa et al., 2018). The thermal stability, non-isothermal crystallization, and kinetic modeling of the nanocomposite were investigated, and a profound influence of the CSNs over film properties was observed (Patwa et al., 2019). Nanocomposite coating of silk nanodisc–based edible CS over banana fruits improved wettability, hydrophobicity, and thermal and mechanical stability, without any change in overall weight, firmness, opacity, and color, which makes them a suitable EC material on food products. The investigators identified that composite films based on CS and silk nanodiscs show antimicrobial properties against *E. coli* and *S. aureus*. However, silk nanofibers lack any antimicrobial properties but can act as carriers for antimicrobial compounds (Ghosh et al., 2021).

3.6 Lipid-Based Organic Nanoparticles for Packaging

3.6.1 SLNs

SLNs are a submicronic (range 50–1000 nm) colloidal system developed to encapsulate lipophilic compounds. They have a high potential application as packaging films, especially in postharvest management. SLNs are unique due to their peculiar surface chemistry, small particle size, solidity, and large specific surface area. They are usually spherical in shape and possess a water-insoluble core. Typically, SLN production is done through the hot homogenization process, where the product obtained is an oil–water nanoemulsion that is cooled to obtain the solid particles. The lipid phase is dispersed and homogenized into an aqueous phase solution at a temperature above the melting point of the lipid, along with suitable surfactants. These submicronic systems with several advantages have various uses and applications as they distribute homogeneously on the fruit skin. SLNs also allow the controlled release of bioactive compounds from the core (Zambrano-Zaragoza et al., 2018).

SLNs prepared with Candeuba® wax have been reported to effectively improve the shelf life of guava. The study used xanthan gum as the polysaccharide base to incorporate wax SLNs whose particle size was greater than 300 nm. The coating process employed in this study was dip coating, where the fruit was dipped in the coating material and refrigerated to obtain a continuous layer around it. The coating allowed adequate transpiration and metabolic activities, thereby extending the storage period and quality preservation. SLN-based ECs also can minimize senescence when applied to fruits and vegetables (Zambrano-Zaragoza et al., 2013). Nanocoatings of beeswax SLNs (10 g/L) with xanthan gum (4 g/L) and propylene glycol (5 g/L) were found to be the best concentrations for conserving fruit characteristics. The effect of the edible film coating on the O_2 consumption rate, juiciness, firmness, skin color changes, and other physical and chemical changes in strawberries (*Fragaria ananassa*) was evaluated. The coating decreased the decay rate, with a decay index of 31%, which reflected less weight loss, fungal growth, and physiological damage. However, a concentration of beeswax SLNs of 30 g/L caused physiological damage due to limited respiration, oxygen diffusion, and water loss through transpiration (Zambrano-Zaragoza et al., 2020). In another study, Candeuba wax SLN–xanthan gum films as a coating on guava fruit (*Psidium guajava L.*) were evaluated for their effect on physicochemical and nutritional parameters. It was understood that coatings with 65 g/L of SLNs maintained the texture and low P methylesterase and delayed biochemical and physiological changes.

However, a concentration of SLN greater than 65 g/L produced physiological damage in guava (García-Betanzos et al., 2017). In order to control postharvest fungal diseases during the shipping and storage of food produce, pullulan packaging films loaded with EO nanoemulsions were formulated. Along with good mechanical and physical properties, these films showed potent antifungal activity against *Rhizopus stolonifer*, *Alternaria spp.*, and *Aspergillus niger* (McDaniel et al., 2019).

3.6.2 Nanoliposomes

Nanoliposomes are nanoscale lipid vesicles with the phospholipid bilayer and the overall chemical composition of nanoliposomes is similar to their larger counterparts. The advantages of nanoliposomes are their biodegradability, small size, biocompatibility, prevention of unwanted interactions, and increased solubility. The hydrophilic and hydrophobic interaction between its component phospholipids and surrounding water molecules is the mechanism behind the formation of liposomes. Based on the method and source used to prepare the liposomes, their characteristics, such as size and number of bilayers, may vary. If the liposomes are unilamellar, the vesicles are spherical and have a single phospholipid bilayer, while in multilamellar liposomes, vesicles have more than one phospholipid layer with an onion structure. The fatty acid chains self-assemble so that the hydrophobic tail regions form the core while the hydrophilic head regions face the outside aqueous environment (Mozafari et al., 2008).

In general, four basic steps are involved in any method for liposome synthesis, namely, (1) drying of lipids obtained from organic solvent, (2) formation of particles by dispersing the lipid in aqueous media, (3) purification of the resultant product to concentrate liposome, and (4) analysis of the final product. Standard mechanical dispersion methods are sonication and extrusion, among others, while other methods are solvent dispersion using ethanol or ether and removing non-encapsulated materials through dialysis. Among these, sonication is the simplest method for making nanoliposomes (Nomani & Govinda, 2016). One of the major causes of the decay of fruits and vegetables is the mechanical injuries that may occur during the period from harvest to consumption. In order to detect the robustness of liposomal coatings for preventing decay due to mechanical injury, two separate formations were developed: (1) curcumin-loaded liposomes and (2) limonene-loaded liposomes (Dhital et al., 2017). One set of each type of liposomal coating was subjected to simulated in-transit vibration. The limonene formulation exhibited higher titratable acidity and total phenolic contents as well as allowed only resisted fungal growth on strawberries. Even though limonene is generally regarded as safe, its application as a coating material is scarcely reported due to its hydrophobicity and its property of getting degraded under oxidative conditions. As an extension of the previously mentioned work, a novel nanocoating treatment was developed to store fresh strawberries of the 'Chandler' variety at a storage temperature of 4°C. Lipid films of d-Limonene liposomes were prepared using the thin-film dehydration method. The novel coating applied on strawberries extended shelf life and maintained the quality parameters of the fruit suggesting its application in postharvest storage. A comparative study on the novel limonene liposomes and alginate coating with appropriate controls proved that limonene liposomes were most effective as a coating material because they have a low pH, a low respiration rate, and a higher anthocyanin content, which supported their usability (Dhital et al., 2018).

AAO, a volatile natural antibacterial agent against *Escherichia coli O157:H7* was encapsulated in nanoliposomes under optimal conditions and assessed for their size, zeta potential, and PDI. The AAO nanoliposome showed a zeta potential of greater than 30 mV, loading efficiency of 67.45%, and PDI of 0.463. The antibacterial activity of the AAO nanoliposomes incorporated in agar film with CS was evaluated against *E. coli O157:H7* on cherry tomatoes. The results indicate a bacteriostatic effect of the film that opens a broad prospect for the film as a potent candidate in the field of fresh produce preservation (Cui et al., 2017). WPI films incorporated with TE-loaded nanoliposomes were fabricated and subjected to study properties such as mechanical, antioxidant, antimicrobial, and release properties. The film containing TE-loaded nanoliposomes showed intense antimicrobial activity against *S. aureus* in comparison to *E. coli*. However, both antimicrobial and antioxidant activity of the films was diminished in nano-active films compared to the free TE–loaded samples, indicating a controlled-release profile possibly existing for the nano-TE when compared to the free TE (Aziz & Almasi, 2018). Nanoliposome-mediated encapsulation of EOs and phenolic compounds ameliorated the antimicrobial and antioxidant properties of the coatings, which, in turn, extended the shelf life of fruits, reduced the weight loss and deterioration due to the action of microorganisms (Lopez-Polo et al., 2021).

3.7 Conclusion

With their unique properties compared to their bulk counterparts, nanomaterials have revolutionized the agri-food industry lately, especially in improving the physical, mechanical, and biological properties of bulk packaging to assure food quality, improved nutrition, and extended shelf lives. Conventional food packaging is defective mainly due to its permeability, which could eventually compromise the quality of food. In order to resist water vapor and atmospheric gases, these general food packaging materials were reinforced with nanoparticles. Such composites were of improved functional characteristics, including improved barrier, mechanical, and thermal properties; this paved the way for more advances in packaging, such as intelligent and active packaging strategies. Later, biopolymer-based renewable or green packaging gained much attention, and edible packaging was promoted to generate zero waste. Organic nanoparticles are synthesized generally from renewable sources containing polysaccharides, proteins, lipids, and others. The widely used organic nanomaterials for food packaging applications included NCS, nanocellulose, nanostarches, ZNs, and SLNs. In addition, nanoliposomes, nano silk, nano casein, and nano gum are other emerging materials used in food packaging. Individually or in combination with other materials, they provide a platform for improved, smart, and active packaging applications in the era of green packaging. Other exciting strategies such as active agents encapsulated nanoparticle-based coatings have also advanced widely as active food packaging materials. From past studies, it can be understood that nanomaterials of biological origin are broadly nontoxic, biodegradable, biocompatible, and cost-effective. However, for successful commercialization, the safety of the reported nanoparticles should be assured before human consumption. Simultaneously, to avoid any adverse conditions, nanoparticle migration toward the food contact surface also must be investigated and understood.

References

Abdalkarim, S. Y. H., Chen, L.-M., Yu, H.-Y., Li, F., Chen, X., Zhou, Y., & Tam, K. C. (2021). Versatile nanocellulose-based nanohybrids: A promising-new class for active packaging applications. International Journal of Biological Macromolecules, 182, 1915–1930. https://doi.org/10.1016/j.ijbiomac.2021.05.169

Abdul Khalil, H. P. S., Davoudpour, Y., Saurabh, C. K., Hossain, M. S., Adnan, A. S., Dungani, R., Paridah, M. T., Islam Sarker, M. Z., Fazita, M. R. N., Syakir, M. I., & Haafiz, M. K. M. (2016). A review on nanocellulosic fibres as new material for sustainable packaging: Process and applications. Renewable and Sustainable Energy Reviews, 64, 823–836. https://doi.org/10.1016/j.rser.2016.06.072

Abreu, A. S., Oliveira, M., de Sá, A., Rodrigues, R. M., Cerqueira, M. A., Vicente, A. A., & Machado, A. V. (2015). Antimicrobial nanostructured starch based films for packaging. Carbohydrate Polymers, 129, 127–134. https://doi.org/10.1016/j.carbpol.2015.04.021

Ali, A. A., & Toliba, A. O. (2018). Effect of organic calcium spraying and nano Chitosan fruits coating on yield, fruit quality and storability of peach cv 'Early swelling'. Current Science International, 7, 737–749. https://doi.org/10.13140/RG.2.2.30115.60965

Alinaqi, Z., Khezri, A., & Rezaeinia, H. (2021). Sustained release modeling of clove essential oil from the structure of starch-based bio-nanocomposite film reinforced by electrosprayed zein nanoparticles. International Journal of Biological Macromolecules, 173, 193–202. https://doi.org/10.1016/j.ijbiomac.2021.01.118

Ameta, S. K., Rai, A. K., Hiran, D., Ameta, R., & Ameta, S. C. (2020). Use of nanomaterials in food science. In M. Ghorbanpour, P. Bhargava, A. Varma, & D. K. Choudhary (Eds.), Biogenic nano-particles and their use in agro-ecosystems (pp. 457–488). Springer. https://doi.org/10.1007/978-981-15-2985-6_24

Antoniou, J., Liu, F., Majeed, H., & Zhong, F. (2015). Characterization of tara gum edible films incorporated with bulk chitosan and chitosan nanoparticles: A comparative study. Food Hydrocolloids, 44, 309–319. https://doi.org/10.1016/j.foodhyd.2014.09.023

Arabpoor, B., Yousefi, S., Weisany, W., & Ghasemlou, M. (2021). Multifunctional coating composed of *Eryngium campestre L.* essential oil encapsulated in nano-chitosan to prolong the shelf-life of fresh cherry fruits. Food Hydrocolloids, 111, 106394. https://doi.org/10.1016/j.foodhyd.2020.106394

Ashori, A. (2014). Effects of graphene on the behavior of chitosan and starch nanocomposite films. Polymer Engineering & Science, 54(10), 2258–2263. https://doi.org/10.1002/pen.23774

Atieno, L., Owino, W., Ateka, E. M., & Ambuko, J. (2019). Influence of coating application methods on the postharvest quality of Cassava. International Journal of Food Science, 2019, 1–16. https://doi.org/10.1155/2019/2148914

Avella, M., De Vlieger, J. J., Errico, M. E., Fischer, S., Vacca, P., & Volpe, M. G. (2005). Biodegradable starch/clay nanocomposite films for food packaging applications. Food Chemistry, 93(3), 467–474. https://doi.org/10.1016/j.foodchem.2004.10.024

Aziz, S. G.-G., & Almasi, H. (2018). Physical characteristics, release properties, and antioxidant and antimicrobial activities of whey protein isolate films incorporated with thyme (*Thymus vulgaris L.*) extract-loaded nanoliposomes. Food and Bioprocess Technology, 11(8), 1552–1565. https://doi.org/10.1007/s11947-018-2121-6

Bajpai, V. K., Kamle, M., Shukla, S., Mahato, D. K., Chandra, P., Hwang, S. K., Kumar, P., Huh, Y. S., & Han, Y.-K. (2018). Prospects of using nanotechnology for food preservation, safety, and security. Journal of Food and Drug Analysis, 26(4), 1201–1214. https://doi.org/10.1016/j.jfda.2018.06.011

Basumatary, I. B., Mukherjee, A., Katiyar, V., & Kumar, S. (2020). Biopolymer-based nanocomposite films and coatings: Recent advances in shelf-life improvement of fruits and vegetables. Critical Reviews in Food Science and Nutrition, 62(7), 1912–1935. https://doi.org/10.1080/10408398.2020.1848789

Blilid, S., Kędzierska, M., Miłowska, K., Wrońska, N., El Achaby, M., Katir, N., Belamie, E., Alonso, B., Lisowska, K., Lahcini, M., Bryszewska, M., & El Kadib, A. (2020). Phosphorylated micro- and nanocellulose-filled chitosan nanocomposites as fully sustainable, biologically active bioplastics. ACS Sustainable Chemistry & Engineering, 8(50), 18354–18365. https://doi.org/10.1021/acssuschemeng.0c04426

Brenckle, M. A., Kaplan, D. L., Omenetto, F. G., & Marelli, B. (2016). Silk fibroin as edible coating for perishable food preservation. Scientific Reports. https://dspace.mit.edu/handle/1721.1/105177

Chivrac, F. (2009). Progress in nano-biocomposites based on polysaccharides and nanoclays. Semantic Scholar. www.semanticscholar.org/paper/Progress-in-nano-biocomposites-based-on-and-Chivrac-Pollet/30d9b5f81a09351d87c2ff78cdd9927fcb871fb1

Correa-Pacheco, Z. N., Bautista-Baños, S., Valle-Marquina, M. Á., & Hernández-López, M. (2017). The effect of nanostructured chitosan and chitosan-thyme essential oil coatings on colletotrichum gloeosporioides growth in vitro and on cv hass avocado and fruit quality. Journal of Phytopathology, 165(5), 297–305. https://doi.org/10.1111/jph.12562

Cui, H., Yuan, L., Li, W., & Lin, L. (2017). Edible film incorporated with chitosan and Artemisia annua oil nanoliposomes for inactivation of *Escherichia coli* O157:H7 on cherry tomato. International Journal of Food Science & Technology, 52(3), 687–698. https://doi.org/10.1111/ijfs.13322

Dadvar, A. A., Vahidi, J., Hajizadeh, Z., Maleki, A., & Reza Bayati, M. (2021). Experimental study on classical and metaheuristics algorithms for optimal nano-chitosan concentration selection in surface coating and food packaging. Food Chemistry, 335, 127681. https://doi.org/10.1016/j.foodchem.2020.127681

de Amorim, J. D. P., de Souza, K. C., Duarte, C. R., da Silva Duarte, I., de Assis Sales Ribeiro, F., Silva, G. S., de Farias, P. M. A., Stingl, A., Costa, A. F. S., Vinhas, G. M., & Sarubbo, L. A. (2020). Plant and bacterial nanocellulose: Production, properties and applications in medicine, food, cosmetics, electronics and engineering. A review. Environmental Chemistry Letters, 18(3), 851–869. https://doi.org/10.1007/s10311-020-00989-9

Dhall, R. K. (2013). Advances in edible coatings for fresh fruits and vegetables: A review. Critical Reviews in Food Science and Nutrition, 53(5), 435–450. https://doi.org/10.1080/10408398.2010.541568

Dhar, P., Tarafder, D., Kumar, A., & Katiyar, V. (2015). Effect of cellulose nanocrystal polymorphs on mechanical, barrier and thermal properties of poly(lactic acid) based bionanocomposites. RSC Advances, 5(74), 60426–60440. https://doi.org/10.1039/C5RA06840A

Dhital, R., Becerra Mora, N., Watson, D., Kohli, P., & Choudhary, R. (2018). Efficacy of limonene nano coatings on post-harvest shelf life of strawberries. LWT, 97. https://doi.org/10.1016/j.lwt.2018.06.038

Dhital, R., Joshi, P., Becerra-Mora, N., Umagiliyage, A., Chai, T., Kohli, P., & Choudhary, R. (2017). Integrity of edible nano-coatings and its effects on quality of strawberries subjected to simulated in-transit vibrations. LWT, 80, 257–264. https://doi.org/10.1016/J.LWT.2017.02.033

Duncan, T. (2011). Applications of nanotechnology in food packaging and food safety: Barrier materials, antimicrobials and sensors. Journal of Colloid and Interface Science, 363, 1–24. https://doi.org/10.1016/j.jcis.2011.07.017

Echegoyen, Y. (2015). Nano-developments for food packaging and labeling applications. In M. Rai, C. Ribeiro, L. Mattoso, & N. Duran (Eds.), Nanotechnologies in food and agriculture (pp. 141–166). Springer International Publishing. https://doi.org/10.1007/978-3-319-14024-7_7

Eshghi, S., Hashemi, M., Mohammadi, A., Badii, F., Mohammadhoseini, Z., & Ahmadi, K. (2014). Effect of nanochitosan-based coating with and without copper loaded on physicochemical and bioactive components of fresh strawberry fruit (*Fragaria x ananassa Duchesne*) during storage. Food and Bioprocess Technology, 7(8), 2397–2409. https://doi.org/10.1007/s11947-014-1281-2

Esyanti, R. R., Zaskia, H., Amalia, A., & Nugrahapraja, D. H. (2019). Chitosan nanoparticle-based coating as post-harvest technology in banana. Journal of Physics: Conference Series, 1204, 012109. https://doi.org/10.1088/1742-6596/1204/1/012109

Farajpour, R., Emam Djomeh, Z., Moeini, S., Tavakolipour, H., & Safayan, S. (2020). Structural and physico-mechanical properties of potato starch-olive oil edible films reinforced with zein nanoparticles. International Journal of Biological Macromolecules, 149, 941–950. https://doi.org/10.1016/j.ijbiomac.2020.01.175

Gad, M., & Zagzog, O. (2017). Mixing xanthan gum and chitosan nano particles to form new coating for maintaining storage life and quality of elmamoura guava fruits. International Journal of Current Microbiology and Applied Sciences, 6, 1582–1593. https://doi.org/10.20546/ijcmas.2017.611.190

García-Betanzos, C. I., Hernández-Sánchez, H., Bernal-Couoh, T. F., Quintanar-Guerrero, D., & Zambrano-Zaragoza, M. D. L. (2017). Physicochemical, total phenols and pectin methylesterase changes on quality maintenance on guava fruit (*Psidium guajava L.*) coated with candeuba wax solid lipid nanoparticles-xanthan gum. Food Research International, 101, 218–227. https://doi.org/10.1016/j.foodres.2017.08.065

García-Ramón, J. A., Carmona-García, R., Valera-Zaragoza, M., Aparicio-Saguilán, A., Bello-Pérez, L. A., Aguirre-Cruz, A., & Alvarez-Ramirez, J. (2021). Morphological, barrier, and mechanical properties of banana starch films reinforced with cellulose nanoparticles from plantain rachis. International Journal of Biological Macromolecules, 187, 35–42. https://doi.org/10.1016/j.ijbiomac.2021.07.112.

Gardesh, A. S. K., Badii, F., Hashemi, M., Ardakani, A. Y., Maftoonazad, N., & Gorji, A. M. (2016). Effect of nanochitosan based coating on climacteric behavior and postharvest shelf-life extension of apple cv. Golab Kohanz. LWT—Food Science and Technology, 70, 33–40.

Ghosh, T., Mondal, K., Giri, B. S., & Katiyar, V. (2021). Silk nanodisc based edible chitosan nanocomposite coating for fresh produces: A candidate with superior thermal, hydrophobic, optical, mechanical and food properties. Food Chemistry, 360, 130048. https://doi.org/10.1016/j.foodchem.2021.130048

Glaser, T. K., Plohl, O., Vesel, A., Ajdnik, U., Ulrih, N. P., Hrnčič, M. K., Bren, U., & Fras Zemljič, L. (2019). Functionalization of polyethylene (PE) and polypropylene (PP) material using chitosan nanoparticles with incorporated resveratrol as potential active packaging. Materials, 12(13), 2118. https://doi.org/10.3390/ma12132118

González-Saucedo, A., Barrera-Necha, L. L., Ventura-Aguilar, R. I., Correa-Pacheco, Z. N., Bautista-Baños, S., & Hernández-López, M. (2019). Extension of the postharvest quality of bell pepper by applying nanostructured coatings of chitosan with

Byrsonima crassifolia extract (L.) kunth. Postharvest Biology and Technology, 149, 74–82. https://doi.org/10.1016/j.postharvbio.2018.11.019

Gujral, H., Sinhmar, A., Nehra, M., Nain, V., Thory, R., Pathera, A. K., & Chavan, P. (2021). Synthesis, characterization, and utilization of potato starch nanoparticles as a filler in nanocomposite films. International Journal of Biological Macromolecules, 186, 155–162. https://doi.org/10.1016/j.ijbiomac.2021.07.005

Hassan, F. A. S., Ali, E. F., Mostafa, N. Y., & Mazrou, R. (2021). Shelf-life extension of sweet basil leaves by edible coating with thyme volatile oil encapsulated chitosan nanoparticles. International Journal of Biological Macromolecules, 177, 517–525. https://doi.org/10.1016/j.ijbiomac.2021.02.159

Hernández-López, G., Ventura-Aguilar, R. I., Correa-Pacheco, Z. N., Bautista-Baños, S., & Barrera-Necha, L. L. (2020). Nanostructured chitosan edible coating loaded with α-pinene for the preservation of the postharvest quality of *Capsicum annuum L.* and *Alternaria alternata* control. International Journal of Biological Macromolecules, 165, 1881–1888. https://doi.org/10.1016/j.ijbiomac.2020.10.094

Huang, Y., Mei, L., Chen, X., & Wang, Q. (2018). Recent developments in food packaging based on nanomaterials. Nanomaterials (Basel, Switzerland), 8(10), E830. https://doi.org/10.3390/nano8100830

Jannatyha, N., Shojaee-Aliabadi, S., Moslehishad, M., & Moradi, E. (2020). Comparing mechanical, barrier and antimicrobial properties of nanocellulose/CMC and nanochitosan/CMC composite films. International Journal of Biological Macromolecules, 164, 2323–2328. https://doi.org/10.1016/j.ijbiomac.2020.07.249

Jeevahan, J., & Chandrasekaran, M. (2019). Nanoedible films for food packaging: A review. Journal of Materials Science, 54(19), 12290–12318. https://doi.org/10.1007/s10853-019-03742-y

Jiang, S., Liu, C., Wang, X., Xiong, L., & Sun, Q. (2016). Physicochemical properties of starch nanocomposite films enhanced by self-assembled potato starch nanoparticles. LWT–Food Science and Technology, 69, 251–257. https://doi.org/10.1016/j.lwt.2016.01.053

Kaur, M., Kalia, A., & Thakur, A. (2017). Effect of biodegradable chitosan–rice-starch nanocomposite films on post-harvest quality of stored peach fruit. Starch-Stärke, 69(1–2). https://doi.org/10.1002/star.201600208

Kumar, A., Negi, Y. S., Choudhary, V., & Bhardwaj, N. (2014). Characterization of cellulose nanocrystals produced by acid-hydrolysis from sugarcane bagasse as agro-waste. Journal of Materials Physics and Chemistry, 2, 1–8. https://doi.org/10.1007/978-3-642-27758-0_1162-2

Kumar, M. C., Hajira Banu, S., Shamarao, N., & B V, L. (2018). Studies on effect of nano filler incorporated (hydroxypropyl) methyl cellulose (hpmc) polymer films on post-harvest mango (mangifera indica l.) Ripening. International journal of pharmacy and biological sciences, 8(4), 42–48. https://ijpbs.com/abstract.php?iid=1282#

Kumar, S., Mukherjee, A., & Dutta, J. (2020). Chitosan based nanocomposite films and coatings: Emerging antimicrobial food packaging alternatives. Trends in Food Science & Technology, 97, 196–209. https://doi.org/10.1016/j.tifs.2020.01.002

Li, H., Shi, H., He, Y., Fei, X., & Peng, L. (2020). Preparation and characterization of carboxymethyl cellulose-based composite films reinforced by cellulose nanocrystals derived from pea hull waste for food packaging applications. International Journal of Biological Macromolecules, 164, 4104–4112. https://doi.org/10.1016/j.ijbiomac.2020.09.010

Li, Q., Gao, R., Wang, L., Xu, M., Yuan, Y., Ma, L., Wan, Z., & Yang, X. (2020). Nanocomposites of bacterial cellulose nanofibrils and zein nanoparticles for food packaging. ACS Applied Nano Materials, 3(3), 2899–2910. https://doi.org/10.1021/acsanm.0c00159

Lin, L., Gu, Y., & Cui, H. (2019). Moringa oil/chitosan nanoparticles embedded gelatin nanofibers for food packaging against Listeria monocytogenes and Staphylococcus aureus on cheese. Food Packaging and Shelf Life, 19, 86–93. https://doi.org/10.1016/j.fpsl.2018.12.005

Lin, M., Fang, S., Zhao, X., Liang, X., & Wu, D. (2020). Natamycin-loaded zein nanoparticles stabilized by carboxymethyl chitosan: Evaluation of colloidal/chemical performance and application in postharvest treatments. Food Hydrocolloids, 106, 105871. https://doi.org/10.1016/j.foodhyd.2020.105871

Liu, C., Jiang, S., Zhang, S., Xi, T., Sun, Q., & Xiong, L. (2016). Characterization of edible corn starch nanocomposite films: The effect of self-assembled starch nanoparticles. Starch—Stärke, 68 (3–4), 239–248. https://doi.org/10.1002/star.201500252

Lopez-Polo, J., Monasterio, A., Cantero-López, P., & Osorio, F. A. (2021). Combining edible coatings technology and nanoencapsulation for food application: A brief review with an emphasis on nanoliposomes. Food Research International, 145, 110402. https://doi.org/10.1016/j.foodres.2021.110402

Lorevice, M. V., Otoni, C. G., Moura, M. R. D., & Mattoso, L. H. C. (2016). Chitosan nanoparticles on the improvement of thermal, barrier, and mechanical properties of high- and low-methyl pectin films. Food Hydrocolloids, 52, 732–740. https://doi.org/10.1016/j.foodhyd.2015.08.003

Luo, J., Huang, K., Zhou, X., & Xu, Y. (2020). Preparation of highly flexible and sustainable lignin-rich nanocellulose film containing xylonic acid (XA), and its application as an antibacterial agent. International Journal of Biological Macromolecules, 163, 1565–1571. https://doi.org/10.1016/j.ijbiomac.2020.07.281

Madani, B., Mirshekari, A., & Imahori, Y. (2019). Physiological responses to stress. In E. M. Yahia (Eds.), Postharvest physiology and biochemistry of fruits and vegetables (pp. 405–423). Woodhead Publishing. https://doi.org/10.1016/B978-0-12-813278-4.00020-8

Makwana, S., Choudhary, R., Dogra, N., Kohli, P., & Haddock, J. (2014). Nanoencapsulation and immobilization of cinnamaldehyde for developing antimicrobial food packaging material. LWT–Food Science and Technology, 57, 470–476. https://doi.org/10.1016/j.lwt.2014.01.043

Manikkam, V. (2018, February 6). What are the applications of nanocellulose in packaging?—PreScouter—custom intelligence from a global network of experts. PreScouter. www.prescouter.com/2018/02/nanocellulose-applications-packaging/

Marsh, R. E., Corey, R. B., & Pauling, L. (1955). An investigation of the structure of silk fibroin. Biochimica et Biophysica Acta, 16, 1–34. https://doi.org/10.1016/0006-3002(55)90178-5

Martínez, K., Ortiz, M., Albis, A., Gilma Gutiérrez Castañeda, C., Valencia, M. E., & Grande Tovar, C. D. (2018). The effect of edible chitosan coatings incorporated with thymus capitatus essential oil on the shelf-life of Strawberry (*Fragaria x ananassa*) during cold storage. Biomolecules, 8(4). https://doi.org/10.3390/biom8040155

Martínez-González, M. D. C., Bautista-Baños, S., Correa-Pacheco, Z. N., Corona-Rangel, M. L., Ventura-Aguilar, R., Río-García, J. C. D., & Ramos-García, M. D. L. (2020). Effect of nanostructured chitosan/propolis coatings on the quality and antioxidant capacity of strawberries during storage. Coatings, 10, 90. https://doi.org/10.3390/coatings10020090

McDaniel, A., Tonyali, B., Yucel, U., & Trinetta, V. (2019). Formulation and development of lipid nanoparticle antifungal packaging films to control postharvest disease. Journal of Agriculture and Food Research, 1, 100013. https://doi.org/10.1016/j.jafr.2019.100013

Melo, N. F. C. B., Lima, M. A. B. D., Stamford, T. L. M., Galembeck, A., Flores, M. A. P., Takaki, G. M. D. C., Medeiros, J. A. D. C., Stamford-Arnaud, T. M., & Stamford, T. C. M. (2020). In vivo and in vitro antifungal effect of fungal chitosan nanocomposite edible coating against strawberry phytopathogenic fungi. International Journal of Food Science & Technology, 55(11), 3381–3391. https://doi.org/10.1111/ijfs.14669

Melo, N. F. C. B., Pintado, M. M. E., Medeiros, J. A. D. C., Galembeck, A., Vasconcelos, M. A. D. S., Xavier, V. L., Lima, M. A. B. D., Stamford, T. L. M., Stamford-Arnaud, T. M., Flores, M. A. P., & Stamford, T. C. M. (2020). Quality of postharvest strawberries: Comparative effect of fungal chitosan gel, nanoparticles and gel enriched with edible nanoparticles coatings. International Journal of Food Studies, 9(2), Article 2. https://doi.org/10.7455/ijfs/9.2.2020.a9

Mo, X., Peng, X., Liang, X., Fang, S., Xie, H., Chen, J., & Meng, Y. (2021). Development of antifungal gelatin-based nanocomposite films functionalized with natamycin-loaded zein/casein nanoparticles. Food Hydrocolloids, 113, 106506. https://doi.org/10.1016/j.foodhyd.2020.106506

Mohamad Fauzi, S., & Abu Hassan, N. (2017). Nanomaterials—recent advancements in edible coating technology. Palm Oil Developments. http://palmoilis.mpob.gov.my/POD/index.php/2020/03/28/nanomaterials-recent-advancements-in-edible-coating-technology/

Mohammadi, A., Hashemi, M., & Hosseini, S. M. (2015). Chitosan nanoparticles loaded with *Cinnamomum zeylanicum* essential oil enhance the shelf life of cucumber during cold storage. Postharvest Biology and Technology, 110, 203–213. https://doi.org/10.1016/j.postharvbio.2015.08.019

Mohammadi, A., Hashemi, M., & Hosseini, S. M. (2016). Postharvest treatment of nanochitosan-based coating loaded with *Zataria multiflora* essential oil improves antioxidant activity and extends shelf-life of cucumber. Innovative Food Science & Emerging Technologies, 33, 580–588. https://doi.org/10.1016/j.ifset.2015.10.015

Mohseni, M. S., Khalilzadeh, M. A., Mohseni, M., Hargalani, F. Z., Getso, M. I., Raissi, V., & Raiesi, O. (2020). Green synthesis of Ag nanoparticles from pomegranate seeds extract and synthesis of Ag-Starch nanocomposite and characterization of mechanical properties of the films. Biocatalysis and Agricultural Biotechnology, 25, 101569. https://doi.org/10.1016/j.bcab.2020.101569

Mondal, K., Ghosh, T., Bhagabati, P., & Katiyar, V. (2019). Sustainable nanostructured materials in food packaging. In Dynamics of advanced sustainable nanomaterials and their related nanocomposites at the bio-nano interface (pp. 171–213). Elsevier. https://doi.org/10.1016/B978-0-12-819142-2.00008-2

Moura, M., Avena-Bustillos, R., Mchugh, T., Krochta, J. M., & Mattoso, L. H. C. (2008). Properties of novel hydroxypropyl methylcellulose films containing chitosan nanoparticles. Journal of Food Science, 73, N31–N37. https://doi.org/10.1111/j.1750-3841.2008.00872.x

Mozafari, M., Johnson, C., Hatziantoniou, S., & Demetzos, C. (2008). Nanoliposomes and their applications in food nanotechnology. Journal of Liposome Research, 18, 309–327. https://doi.org/10.1080/08982100802465941

Ngo, T. M. P., Nguyen, T. H., Dang, T. M. Q., Tran, T. X., & Rachtanapun, P. (2020). Characteristics and antimicrobial properties of active edible films based on pectin and nanochitosan. International Journal of Molecular Sciences, 21(6), 2224. https://doi.org/10.3390/ijms21062224

Nguyen, H. V. H., & Nguyen, D. H. H. (2020). Effects of nano-chitosan and chitosan coating on the postharvest quality, polyphenol oxidase activity and malondialdehyde content of strawberry (*Fragaria x ananassa* Duch.). Journal of Horticulture and Postharvest Research, 3(1), 11–24. https://doi.org/10.22077/jhpr.2019.2698.1082

Nomani, S., & Govinda, J. (2016). Nanoliposome: An alternative approach for drug delivery system. International Journal of Advances in Pharmacy Medicine and Bioallied Sciences, 2016, 1–10.

Orsuwan, A., & Sothornvit, R. (2017). Development and characterization of banana flour film incorporated with montmorillonite and banana starch nanoparticles. Carbohydrate Polymers, 174, 235–242. https://doi.org/10.1016/j.carbpol.2017.06.085

Orzali, L., Corsi, B., Forni, C., & Riccioni, L. (2017). Chitosan in agriculture: A new challenge for managing plant disease. Biological Activities and Application of Marine Polysaccharides, 17–36. https://doi.org/10.5772/66840

Oymaci, P., & Altinkaya, S. A. (2016). Improvement of barrier and mechanical properties of whey protein isolate based food packaging films by incorporation of zein nanoparticles as a novel bionanocomposite. Food Hydrocolloids, 54, 1–9. https://doi.org/10.1016/j.foodhyd.2015.08.030

Pan, K., & Zhong, Q. (2016). Low energy, organic solvent-free co-assembly of zein and caseinate to prepare stable dispersions. Food Hydrocolloids, 52, 600–606. https://doi.org/10.1016/j.foodhyd.2015.08.014

Patwa, R., Kumar, A., & Katiyar, V. (2018). Effect of silk nano-disc dispersion on mechanical, thermal, and barrier properties of poly(lactic acid) based bionanocomposites. Journal of Applied Polymer Science, 135(38), 46671. https://doi.org/10.1002/app.46671

Patwa, R., Singh, M., Kumar, A., & Katiyar, V. (2019). Kinetic modelling of thermal degradation and non-isothermal crystallization of silk nano-discs reinforced poly (lactic acid) bionanocomposites. Polymer Bulletin, 76. https://doi.org/10.1007/s00289-018-2434-7

Pinto, A. M. B., Santos, T. M., Caceres, C. A., Lima, J. R., Ito, E. N., & Azeredo, H. M. C. (2015). Starch-cashew tree gum nanocomposite films and their application for coating cashew nuts. LWT – Food Science and Technology, 62(1), 549–554. https://doi.org/10.1016/j.lwt.2014.07.028

Raghav, P. K., Agarwal, N., & Saini, M. (2016). Edible coating of fruits and vegetables: A review. Education, 1(2), 188–204.

Resende, N. S., Gonçalves, G. A. S., Reis, K. C., Tonoli, G. H. D., & Boas, E. V. B. V. (2018). Chitosan/cellulose nanofibril nanocomposite and its effect on quality of coated strawberries. Journal of Food Quality, 2018. www.hindawi.com/journals/jfq/2018/1727426/

Romanazzi, G., & Feliziani, E. (2016). Chapter 6—use of chitosan to control postharvest decay of temperate fruit: Effectiveness and mechanisms of action. In S. Bautista-Baños, G. Romanazzi, & A. Jiménez-Aparicio (Eds.), Chitosan in the preservation of agricultural commodities (pp. 155–177). Academic Press. https://doi.org/10.1016/B978-0-12-802735-6.00006-9

Ruffo Roberto, S., Youssef, K., Hashim, A. F., & Ippolito, A. (2019). Nanomaterials as alternative control means against postharvest diseases in fruit crops. Nanomaterials, 9(12), 1752. https://doi.org/10.3390/nano9121752

Saberi, B., & Golding, J. (2018). Postharvest application of biopolymer-based edible coatings to improve the quality of fresh horticultural produce. In Polymers for food applications (pp. 211–250). https://doi.org/10.1007/978-3-319-94625-2_9

Sahraei Khosh Gardesh, A., Badii, F., Hashemi, M., Ardakani, A. Y., Maftoonazad, N., & Gorji, A. M. (2016). Effect of nanochitosan based coating on climacteric behavior and postharvest shelf-life extension of apple cv. Golab Kohanz. LWT, 70, 33–40. https://doi.org/10.1016/j.lwt.2016.02.002

Sanchez-Garcia, M. D., Hilliou, L., & Lagaron, J. M. (2010). Nanobiocomposites of Carrageenan, Zein, and mica of interest in food packaging and coating applications. Journal of Agricultural and Food Chemistry, 58(11), 6884–6894. https://doi.org/10.1021/jf1007659

Saqib, S., Zaman, W., Ayaz, A., Habib, S., Bahadur, S., Hussain, S., Muhammad, S., & Ullah, F. (2020). Postharvest disease inhibition in fruit by synthesis and characterization of chitosan iron oxide nanoparticles. Biocatalysis and Agricultural Biotechnology, 28, 101729. https://doi.org/10.1016/j.bcab.2020.101729

Shapi'i, R. A., Othman, S. H., Nordin, N., Kadir Basha, R., & Nazli Naim, M. (2020). Antimicrobial properties of starch films incorporated with chitosan nanoparticles: In vitro and in vivo evaluation. Carbohydrate Polymers, 230, 115602. https://doi.org/10.1016/j.carbpol.2019.115602

Sharma, C., Dhiman, R., Rokana, N., & Panwar, H. (2017). Nanotechnology: An untapped resource for food packaging. Frontiers in Microbiology, 8. https://doi.org/10.3389/fmicb.2017.01735

Shojaei, M., Eshaghi, M., & Nateghi, L. (2019). Characterization of hydroxypropyl methyl cellulose—whey protein concentrate bionanocomposite films reinforced by chitosan nanoparticles. Journal of Food Processing and Preservation, 43(10). https://doi.org/10.1111/jfpp.14158

Silva, E., Almeida, M., & Ayub, R. (2020). Blackberry extend shelf life by nanocellulose and vegetable oil coating [Review of Blackberry extend shelf life by nanocellulose and vegetable oil coating, by T. Carvalho]. Horticulture International Journal, 4. https://doi.org/10.15406/hij.2020.04.00158

Silva, N. H. C. S., Vilela, C., Almeida, A., Marrucho, I. M., & Freire, C. S. R. (2018). Pullulan-based nanocomposite films for functional food packaging: Exploiting lysozyme nanofibers as antibacterial and antioxidant reinforcing additives. Food Hydrocolloids, 77, 921–930. https://doi.org/10.1016/j.foodhyd.2017.11.039

Silvério, H. A., Flauzino Neto, W. P., Dantas, N. O., & Pasquini, D. (2013). Extraction and characterization of cellulose nanocrystals from corncob for application as reinforcing agent in nanocomposites. Industrial Crops and Products, 44, 427–436. https://doi.org/10.1016/j.indcrop.2012.10.014

Sivakumar, D., Bill, M., Korsten, L., & Thompson, K. (2016). Chapter 5—integrated application of chitosan coating with different postharvest treatments in the control of postharvest decay and maintenance of overall fruit quality. In S. Bautista-Baños, G. Romanazzi, & A. Jiménez-Aparicio (Eds.), Chitosan in the preservation of agricultural commodities (pp. 127–153). Academic Press. https://doi.org/10.1016/B978-0-12-802735-6.00005-7

Tao, F., Shi, C., & Cui, Y. (2018). Preparation and physicochemistry properties of smart edible films based on gelatin-starch nanoparticles. Journal of the Science of Food and Agriculture, 98(14), 5470–5478. https://doi.org/10.1002/jsfa.9091

Trinetta, V., McDaniel, A., Batziakas, G. K., Yucel, U., Nwadike, L., & Pliakoni, E. (2020). Antifungal packaging film to maintain quality and control postharvest diseases in strawberries. Antibiotics, 9(9), 618. https://doi.org/10.3390/antibiotics9090618

Vahedikia, N., Garavand, F., Tajeddin, B., Cacciotti, I., Jafari, S. M., Omidi, T., & Zahedi, Z. (2019). Biodegradable zein film composites reinforced with chitosan nanoparticles and cinnamon essential oil: Physical, mechanical, structural and antimicrobial attributes. Colloids and Surfaces B: Biointerfaces, 177, 25–32. https://doi.org/10.1016/j.colsurfb.2019.01.045

Wang, L., Xue, J., & Zhang, Y. (2019). Preparation and characterization of curcumin loaded caseinate/zein nanocomposite film using pH-driven method. Industrial Crops and Products, 130, 71–80. https://doi.org/10.1016/j.indcrop.2018.12.072

Wesley, S., Raja, P., Raj, A., & Tiroutchelvamae, D. (2014). Review on-nanotechnology applications in food packaging and safety. International Journal of Engineering Research, 3, 645–651. https://doi.org/10.17950/ijer/v3s11/1105

Xue, Y., Zhou, S., Fan, C., Du, Q., & Jin, P. (2019). Enhanced antifungal activities of eugenol-entrapped casein nanoparticles against anthracnose in postharvest fruits. Nanomaterials (Basel, Switzerland), 9. https://doi.org/10.3390/nano9121777

Yan, X., Diao, M., Yu, Y., Gao, F., Wang, E., Wang, Z., & Zhang, T. (2021). Influence of esterification and ultrasound treatment on formation and properties of starch nanoparticles and their impact as a filler on chitosan based films characteristics. International Journal of Biological Macromolecules, 179, 154–160. https://doi.org/10.1016/j.ijbiomac.2021.03.004

Yang, W., Owczarek, J. S., Fortunati, E., Kozanecki, M., Mazzaglia, A., Balestra, G. M., Kenny, J. M., Torre, L., & Puglia, D. (2016). Antioxidant and antibacterial lignin nanoparticles in polyvinyl alcohol/chitosan films for active packaging. Industrial Crops and Products, 94, 800–811. https://doi.org/10.1016/j.indcrop.2016.09.061

Zahid, N., Alderson, P. G., Ali, A., Maqbool, M., & Manickam, S. (2013). In vitro control of *Colletotrichum gloeosporioides* by using chitosan loaded nanoemulsions. Acta Horticulturae, 1012, 769–774. https://doi.org/10.17660/ActaHortic.2013.1012.104

Zambrano-Zaragoza, M. L., González-Reza, R., Mendoza-Muñoz, N., Miranda-Linares, V., Bernal-Couoh, T. F., Mendoza-Elvira, S., & Quintanar-Guerrero, D. (2018). Nanosystems in edible coatings: A novel strategy for food preservation. International Journal of Molecular Sciences, 19(3). https://doi.org/10.3390/ijms19030705

Zambrano-Zaragoza, M. L., Mercado-Silva, E., Ramirez-Zamorano, P., Cornejo-Villegas, M. A., Gutiérrez-Cortez, E., & Quintanar-Guerrero, D. (2013). Use of solid lipid nanoparticles (SLNs) in edible coatings to increase guava (*Psidium guajava L.*) shelf-life. Food Research International, 51(2), 946–953. https://doi.org/10.1016/j.foodres.2013.02.012

Zambrano-Zaragoza, M. L., Quintanar-Guerrero, D., Del Real, A., González-Reza, R. M., Cornejo-Villegas, M. A., & Gutiérrez-Cortez, E. (2020). Effect of nano-edible coating based on beeswax solid lipid nanoparticles on strawberry's preservation. Coatings, 10(3), 253. https://doi.org/10.3390/coatings10030253

Zhang, F., Ma, J., Xu, Q., Zhou, J., Simion, D., Carmen, G., Wang, J., & Li, Y. (2016). Hollow casein-based polymeric nanospheres for opaque coatings. ACS Applied Materials & Interfaces, 8(18), 11739–11748. https://doi.org/10.1021/acsami.6b00611

Chapter 4

Cellulose Nanoparticles for Postharvest Management and the Mechanism Involved

Neenu K.V., P.M. Sabura Begum, Rajesh R., Jyotishkumar Parameswaranpillai, Mohammad Reza Saeb, Midhun Dominic C.D.

List of Abbreviations

BC	Bacterial cellulose
BNC	Bacterial nanocellulose
CHX	Chlorohexidine digluconate
CMC	Carboxymethyl cellulose
CNC	Cellulose nanocrystal
EO	Essential oil

4.1 Introduction

Horticulture is the study and practice of propagating, cultivating, processing, inspecting, storing, and distributing plants (Yadollahi et al., 2010). The benefits of horticulture include the development of new plant production technology, the introduction of new plant species, and high-nutrient crops. Regardless of the aforementioned positive aspects, the postharvest treatment of fruits, vegetables, and crops is a huge issue for the food industry. The food products will spoil unless they are properly treated.

DOI: 10.1201/9781003142287-4

Numerous changes that occur in fruits and vegetables after harvests, such as rots, molds, and so on, must be examined. As a result, adequate postharvest treatment is necessary to extend the shelf life of fruits, vegetables, and crops and reduce postharvest losses. Besides the intrinsic nutritional value of food, the quality of food after production is also important. Postharvest treatment is especially needed in the case of fruits and vegetables since they contain high moisture content and are easily attacked and destroyed by microorganisms and pathogens (James et al., 2017). The attack of pathogens/microorganisms results in postharvest diseases and a reduction in the quality of foods.

4.1.1 Postharvest Diseases

Because of the increasing demand for high-quality foods, advancements in postharvesting procedures are gaining attraction. Correspondingly, it became an issue to establish a cost-effective, simple, well-organized, and helpful strategy for postharvest conservation (Resende et al., 2018). The postharvest diseases in fruits and vegetables are mainly caused by (1) high-water content in fruits (Sanzani et al., 2009), (2) fungus and bacteria (Romanazzi et al., 2016), and (3) surface wounds caused by insects or mechanical forces (Lindy & Johnson, 1997). Different postharvest diseases, affected fruits or vegetables, and disease-causing pathogens are given in Table 4.1. From the table, different pathogens cause a variety of postharvest diseases in crops. Given the importance of this factor, one must maintain food safety by applying effective postharvest treatment.

4.1.2 Postharvest Techniques

Different postharvest treatments have been carried out using a variety of natural and scientific procedures for the preservation of fruits and vegetables (James et al., 2017). The widely accepted natural and scientific procedures are cleaning, controlled ripening, temperature control, coatings, and chemical treatments. Controlled ripening consists of controlling the levels of CO_2 and O_2 in the storage atmosphere thereby delaying the ripening process (Mahajan et al., 2014). Temperature control (Aghdam et al., 2013) is another option in which fruits, vegetables, and crops are exposed to varying degrees of heat, resulting in the elimination of microbes and control of decay. However, this technique has some drawbacks, such as increased labor and high energy costs. Coatings, also known as thin films (Silva et al., 2019a; Syafiq et al., 2020), are a scientific, environmentally friendly method that could be used to create a protective barrier for fruits and vegetables. Edible antimicrobial polymer films for food packaging are a good choice for improving the shelf life and maintaining the quality of the fruits and vegetables for a longer period, this can also be coated over the fruits and vegetables (Otoni et al., 2017; Suput et al., 2015). Chemical treatment is another option for postharvest treatment. Chemical treatments (Ramos et al., 2013; Sibomana et al., 2017) involve treating fruits and vegetables with chemicals such as chlorine, chlorine dioxide, sulfur dioxide, nitric oxide, ozone, and others to prevent postharvest decay, water loss, and respirational loss. However, there is a risk to health after the chemical treatments (Mahajan et al., 2014).

Table 4.1 Different Postharvest Diseases, Affected Plant Species, and Disease-Causing Pathogens

Postharvest disease	*Fruit or vegetable*	*Pathogen*	*References*
Bacterial soft rot	Tomato, spinach, cabbage, cauliflower, carrot, cucumber, Chinese cabbage, Chinese radish, bok choy, and asparagus	*Envini acarotovora*	Phokum et al., 2006
Bulb rot	Onion	*Pantoea agglomerans*	Vahling-Armstrong et al., 2016
Lenticel rot	Apple	*Gloeosporium album*	Edney et al., 1977
Black pit	Citrus fruits	*Pseudomonas syringae*	Abdellatif et al., 2017
Sour rot	Citrus fruits	*Geotrichum candidum*	Talibi et al., 2012; Thornton et al., 2010
Gray mold	Strawberry, grapes, and blackberry	*Botrytis cinerea*	Archbold et al., 1997
Dry rot	Potato	*Fusarium spp.*	Bojanowski et al., 2013, Khedher et al., 2021
Stem rot	Brassica crops	*Sclerotinia sclerotiorum*	Mei et al., 2012
Pink mold rot	Tomato	*Trichothecium roseum*	Han et al., 2012
Rot	Grapes	*Cladosporium cladosporioides and Cladosporium herbarum*	Latorre et al., 2011
Green mold	Citrus species	*Penicillium digitatum*	Papoutsis et al., 2019
Blue mold	Citrus species	*Penicillium italicum*	Papoutsis et al., 2019

4.1.3 Edible Films

Edible coatings are very thin films being applied to a variety of fruits and vegetables to protect them as part of active packing (Dhall, 2013). The use of biodegradable and edible films/coatings on fruits and vegetables is a useful way to prevent postharvest loss, pathogens, and spoilage, and keep them fresh, thus ensuring food safety (Bonilla

Figure 4.1 Grapes after 21 days of refrigerator storage without (1) and with (2) a cornstarch- gelatin coating (Fakhouri et al., 2015). (Reproduced with thanks from Elsevier, License Number: 5107611244908)

Lagos, 2013; Shahbazi, 2018). A fully biodegradable polymer matrix and a suitable filler are the two main components of an edible film. The addition of nanoscale fillers to films improves their characteristics. Various types of fruit- and vegetable-based films and coatings have been reported, such as corn starch, guava, orange, and beetroot (Basumatary et al., 2020; Fakhouri et al., 2015; Otoni et al., 2017). Figure 4.1 shows the photograph of edible corn starch-gelatin coating over grapes that were refrigerated for 21 days (Fakhouri et al., 2015). The advantages of edible coatings include the ability to reduce pollution caused by synthetic packaging, eventually maintain good barrier properties, reduce respiration and nutrient loss from crops, preservation of fruits and vegetables for a longer time (improved shelf life), maintain quality, protection from oxidation, antifungal properties, antibacterial properties, and novel packing and appearance (Basumatary et al., 2020; Dhall, 2013; Fakhouri et al., 2015; Ghosh et al., 2021; Tahir et al., 2019; Tavassoli-Kafrani et al., 2020).

Films and edible coatings are of three types: hydrocolloids, lipids, and composites (Dashipour et al., 2015). Hydrocolloid edible films are plant-based, include both polysaccharide- and protein-based, and have many hydroxyl groups (Raghav et al., 2016). Cellulose (Simsek et al., 2020), chitosan (Wang et al., 2011), starch (Bonilla Lagos, 2013), and pectin (Jantrawut et al., 2018) comprise the hydrocolloid types. Fatty acid–based (Sebti et al., 2002) and wax-based (Syahida et al., 2020) edible coatings are lipid types. This type of coating is hydrophobic, which ensures a glossy appearance and has good barrier properties (Raghav et al., 2016). The composite category includes a multipolymer matrix or a single polymer containing fillers to ensure better properties. Mixed composite films, films embedded with EO (Han et al., 2018; Noshirvani et al., 2017), and others are examples of composite films. Table 4.2 lists the different types of edible coatings/films that can be applied to fruits and vegetables to improve their shelf life. The advantages and disadvantages of the coatings are also listed.

Table 4.2 Edible Polymer Coatings and Their Advantages and Disadvantages

Type of polymer films for coating	*Advantages*	*Disadvantages*	*Reference*
Starch-gelatin polymer blend containing cinnamon, clove, and oregano EO	Enhanced transparency, barrier properties, and antifungal properties	—	Acosta et al., 2016
Fish gelatin films with palm wax	Improved tensile strength, elongation at break, ultraviolet (UV) barrier properties, and water resistance	Thicker and opaque films	Syahida et al., 2020
Chitosan/ cellulose nanofibril nano composite	Enhanced barrier and mechanical properties	High concentration of chitosan resulted in darker color.	Resende et al., 2018
CMC + EO	Improved tensile strength and antimicrobial activity, stronger	Water vapor permeability increased with the incorporation of *Santolina chamaecyparissus* and *Schinus molle* EO	Simsek et al., 2020
Chitosan + bergamot EO	Improved antibacterial and water vapor permeability resistance	Drop in tensile strength and less glossy	Sánchez-González et al., 2010
Hydroxypropyl methylcellulose + bergamot EO	Improved antimicrobial activity, inhibits respiration rates	—	Sánchez-González et al., 2011
cellulose acetate + EO	Improved elongation at break and water vapor permeability resistance	—	Bastos et al., 2016
CMC-chitosan + Cinnamon/ginger EO	Increased water contact angles, water vapor permeability resistance, antimicrobial properties, and elongation at break	Lower tensile strength and crystallinity index	Noshirvani et al., 2017
Nanofibrillated cellulose films	Reduced respiration rates and improved shelf life	—	Pacaphol et al., 2019

Type of polymer films for coating	*Advantages*	*Disadvantages*	*Reference*
Poly(sulfobetaine methacrylate)+ BNC	Homogeneous and transparent films, antimicrobial activity; thermally stable, good mechanical properties; exhibits proton conductivity; and enables UV protection	—	Vilela et al., 2019
Nanofibrillated cellulose + polyvinyl alcohol + polyacrylic acid	UV resistance, strong crosslinking, high oxygen barrier capacity, thermally stable, high tensile strength and increased shelf life	—	Poiini et al., 2020
Polyvinyl alcohol + CNC/chitosan nanoparticle	Improved tensile strength, antifungal activity and increased shelf life.	—	Dey et al., 2020
Sodium alginate-CMC + cinnamon EO	Improved antimicrobial activity and increased shelf life	Drop in tensile strength	Han et al., 2018
Cellulose nanofiber	Increased aroma and texture	—	Jung et al., 2018
Fatty acid–cellulose films	Improved moisture barrier properties	Lower mechanical resistance and antimicrobial activity	Sebti et al., 2002

Edible coatings are promising materials to store and protect a variety of fruits, vegetables, and meat. However, microbial attack is the most important problem one must address. The addition of antimicrobial agents to coatings could improve their antimicrobial activity and storage properties. Antimicrobial compounds such as nanoparticles, EO, and herbs, among others, can be safely applied to films to combat diseases (Kaewklin et al., 2018; Varghese et al., 2020; Zhang et al., 2018). One of the natural products with antimicrobial properties is EO. Several studies on the inclusion of EO- embedded polymer films have been reported. Table 4.3 represents the polymer films/coatings containing EO as an antimicrobial agent.

4.2 Why Cellulose?

Various biopolymers along with different active agents can be used for the preservation of postharvested fruits and vegetables. Among the various biopolymers, cellulose is widely preferred in the application of postharvest treatment because of the ability of cellulose and nanocellulose to form thin films or edible coatings. Also, cellulose is an environmentally friendly non-toxic packing material that ensures food safety. CMC

Table 4.3 Different Polymer Films/Coatings Containing EO as Antimicrobial Agent

Polymer coating/film	*Antimicrobial agent*	*References*
Plasticized banana flour	Garlic EO	Orsuwan & Sothornvit, 2018
Cornstarch	Orange EO	do Evangelho et al., 2019
Gelatin-chitosan	Clove EO Fennel EO Cypress EO Lavender EO Thyme EO Herb-of-the-cross EO Pine EO Rosemary EO	Gómez-Estaca et al., 2010
Chitosan/carboxy methyl cellulose (CMC)	Mentha spicata oil	Shahbazi, 2018
Starch/CMC	Rosemary EO	Mohsenabadi et al., 2018
Cellulose	Clove and oregano EO	Wieczyńska & Cavoski, 2018
Chitosan	Limonene Bergamot EO Cinnamon, clove bud, and star anise EO	Maleki et al., 2018 Sánchez-González et al., 2010 Wang et al., 2011
Nanocellulose	Lemongrass EO	Silva et al., 2019a
Cyclodextrin nanosponges	Coriander EO	Silva et al., 2019b
CMC-agar	Savory EO	Abdollahi et al., 2019
Hydroxypropyl methylcellulose	Bergamot EO	Sánchez-González et al., 2011
Pectin	Orange EO	Jantrawut et al., 2018, Chaiwarit et al., 2018

(Biswas et al., 2018), hydroxypropyl methylcellulose (Sánchez-González et al., 2011), cellulose esters (Bastos et al., 2016), cellulose nanofibers (Jung et al., 2018; Lee et al., 2009), CNCs (Criado et al., 2019), and BCs (Nguyen et al., 2008) are some of the widely used types of cellulose for food packaging.

4.2.1 Cellulose in Postharvest Management

Nanocellulose has been extensively researched for use in food packaging due to its properties such as lightweight, low cost, renewable, biodegradability, and transparency (Fotie et al., 2020). Different physical and chemical approaches can be used to isolate nanocellulose from native cellulose (Ali et al., 2020; Chen et al., 2016; Hu et al.,

2017; Johnson et al., 2009; Nie et al., 2018; Rajinipriya et al., 2018, Thomas et al., 2021a, 2021b). Nanocellulose is known for its remarkable strength and other physical properties, and it is a feasible alternative for replacing petroleum-based packaging materials (Hubbe et al., 2017). Cellulose is hydrophilic due to the presence of hydroxyl groups. Surface modification is an effective way to change its hydrophilic characteristics (Phanthong et al., 2018). Studies have shown that the nanocellulose based films/coatings can be utilized for postharvest treatments such as storability, disease control, oxygen and oil barrier properties, reduce respiration rate, and shelf-life extension (Fujisawa et al., 2011; Aulin et al., 2010; Pacaphol et al., 2019; Sánchez-González et al., 2011; Vilela et al., 2019). The incorporation of antimicrobial agents in cellulose films improved the antimicrobial activity, and studies suggest that these films are suitable for packaging applications that can protect food from pathogens (Han et al., 2018; Motelica et al., 2020; Simsek et al., 2020; Wieczyńska & Cavoski, 2018).

4.2.2 Cellulose Nanocomposites for Antimicrobial Applications

Antimicrobial packaging can control bacterial or fungal growth in fruits and vegetables in the food packaging industry. This packaging minimizes pathogen growth and improves the quality, safety, and shelf life of fruits and vegetables. Various strategies can be used to obtain antibacterial activity, such as encapsulating volatile antimicrobial compounds in nano/micro containers within the packaging, introducing active agents directly in biopolymer films or coatings, and/or employing antimicrobial papers or pads (Chiabrando et al., 2019; Varghese et al., 2020). Cellulose-based materials are typically used for reinforcing, but they can also be employed as antimicrobial films or coatings for food packaging (M. Hubbe et al., 2008). However, because cellulose lacks antibacterial activity, antimicrobial chemicals must be added to it. In addition, its low water resistance and mechanical properties limit its use in the packaging industry. Thus, the use of modified cellulose nanocomposites, on the other hand, can give good structural properties and on the other hand can provide good antibacterial properties, making them suitable for food packaging applications.

Yang et al. (2016) prepared polylactic acid films having CNCs and lignin. The composites showed good antibacterial activity against *Pseudomonas syringae* pv. *tomato* (Pst) due to the presence of CNCs and lignin. The study proposed the use of PLA-composite packaging films containing CNCs and LNP to extend the shelf life of tomatoes during the postharvest stage. In another study, Lu et al. (2018) reported that the coating of cellulose nanofibers loaded with nisin improves the properties of plasma-treated biaxially oriented polyproylene/low density polyethylene (BOPP/LDPE) film. The results showed that the nanofibrillated cellulose (NC)/nisin coating possessed good antimicrobial activity against the growth of *Listeria monocytogenes*, suggesting the potential use of composite for commercial food packaging applications. Similarly, a study conducted by (Salmieri et al., 2014a) used a film containing polylactic acid–reinforced CNC for slow and controlled release of nisin. The composite films showed a good antibacterial effect against *L. Monocytogenes*. Khan et al. (2016) conducted a study on antimicrobial efficiency of genipin-crosslinked chitosan/CNC films loaded with ethylenediaminetetraacetate (EDTA) and nisin. The composites showed improved antibacterial against *E. coli* and *L. monocytogenes*.

Introducing functional materials to cellulose enhances the antimicrobial properties of as-prepared bioactive films. A blend of alginate, CMC, carrageenan, and grapefruit seed extract was used to prepare an antibacterial coating on wrapping paper (Shankar & Rhim, 2018). biopolymer-coated paper exhibited strong antibacterial activity against bacteria *L. monocytogenes* and *E. coli* and increased the surface hydrophobicity and tensile properties. These multifunctional features can provide a solution for potential eco-friendly packaging materials. Lavoine et al. (2014) conducted a study on the efficacy of a microfibrillated cellulose/antibacterial agent CHX coating on paper. The release studies showed that CHX-impregnated paper reached a maximum release value after 2 h, whereas CHX-impregnated + MFC coating samples required 19 h for the maximum release of CHX. Thus, the coating of MFC exhibited a significant role in extending the antibacterial activity of composite coatings.

In a study conducted by Shahmohammadi et al. (2016), ZnO-BC nanocomposite films were prepared by incorporating 5 wt.% ZnO nanoparticles. The nanocomposite film showed antibacterial activity toward *E. coli* (gram-negative bacteria) and *S. Aureus* (gram-positive bacteria). An effective release of ZnO particles was attained within at least 96 h, suggesting ZnO-BC nanocomposite as a potential food packaging material with the regulated release of antimicrobial agents. Jipa et al. (2012) worked on innovative mono- and multilayer films based on BC, incorporating sorbic acid (SA) as an antimicrobial agent. The nanocomposite films showed antibacterial activity toward *E. coli*. Antimicrobial efficacy was mainly controlled by the SA release rate and by the water solubility of films. SA release rate studies showed that monolayer films exhibited a diffusion coefficient in the range of 10^{-12} m^2s^{-2} and multilayer films showed in the range of 10^{-13} m^2s^{-1}. The results revealed that these films can function as novel antimicrobial food packaging materials. Table 4.4 lists different cellulose-based antimicrobial packing.

Table 4.4 Cellulose-Based Antimicrobial Packing

Cellulose material	*Packing product*	*Antimicrobial agent*	*Reference*
Cellulose and polypropylene	Iceberg lettuce	Eugenol (EUG), carvacrol (CAR), and trans-anethole (ANT)	Wieczyńska & Cavoski, 2018
Cellulose and chitosan	Cheese	Monolaurin (ML)	Lotfi et al., 2018
CMC/chitosan (layer by layer)	Strawberry	Chitosan	Yan et al., 2019
PLA/nanocellulose	Ground beef	Mentha piperita EO, Buniumpercicum EO	Talebi et al., 2018
sodium alginate–CMC	Sauced silver carp fillets	*Ziziphora clinopodioides*, apple peel extract, and zinc oxide nanoparticle (ZnO)	Rezaei & Shahbazi, 2018
BC	Processed meat	Nisin	Nguyen et al., 2008
BC	Cheese	Sakacin-A	Rollini et al., 2020
BNC	Ground meat	Lactic acid bacteria (LAB)	Yordshahi et al., 2020

4.2.3 EO-Incorporated Nanocellulose Polymer Films

Natural polymer films have been identified as an alternative to synthetic food packaging materials (Yang et al., 2020). Cellulose, one of the known natural polymers along with antimicrobial EO, has been recognized as an important material for food packaging applications. The incorporation of EOs into the cellulose and nanocellulose matrix can impart many functional features such as antimicrobial activity, hydrophobicity, and barrier properties (Gómez-Estaca et al., 2010; Silva et al., 2019a). Moreover, EOs also contain essential minerals, antioxidants, vitamins, and others (Raghav et al., 2016). EOs can be encapsulated in cellulose nanofibers. The high encapsulation efficiency of cellulose nanofibers with sweet orange EO was reported by (Junior et al., 2018). Vegetable oil incorporated cellulose nanofiber coating exhibits improved shelf life of blackberries (Silva et al., 2020). Table 4.5 shows different nanocellulose films having EO as an antimicrobial component.

Table 4.5 Nanocellulose Films/Coatings Having EO as an Antimicrobial Component

Polymer matrix-nanocellulose films	*EO*	*Pathogen affected*	*Reference*
Nanocrystalline cellulose/starch	Cinnamon EO	*Bacillus subtilis, Staphylococcus aereus, Escherichia coli.*	Syafiq et al., 2020
CNC/Polylactic acid	Oregano EO	*Listeria monocytogenes*	Salmieri et al., 2014b
CNC	Thyme EO	*Listeria innocua*	Criado et al., 2019
Nanocellulose fiber/chitosan	*Origanum vulgare ssp.* gracile EO	*E. coli* *B. cereus*	Jahed et al., 2017
Cellulose nanofiber/sodium caseinate	Cinnamon EO	*S. aureus* *E. Coli* *S. typhimurium* *P. aeruginosa*	Ranjbaryan et al., 2019
Microfibrillated cellulose/soy protein	Clove EO	*Bacillus cereus, Escherichia coli, Salmonella enteritidis and Staphylococcus aureus*	Ortiz et al., 2018
Cellulose nanofiber/lemon waste powder nanocomposite	SavoryEO	*L. monocytogenes* *S. aureus* *S. enterica* *P. aeruginosa* *E. coli*	Soofi et al., 2021

(Continued)

Table 4.5 (Continued)

Polymer matrix-nanocellulose films	*EO*	*Pathogen affected*	*Reference*
2, 2, 6, 6-tetramethylpiperidine-1-oxyl (TEMPO)-oxidized cellulose nanofiber/β cyclodextrin	Carvacrol	*Bacillus subtilis*	Saini et al., 2017
Cellulose nanofibrils	Soybean EO	*E. coli*	Valencia et al., 2019
Nanocellulose fibrils	Miswak root extract	*S. aureus* *E. coli*	Ahmadi et al., 2019

4.2.4 Mechanism of Antimicrobial Activity of EO

The incorporation of EOs into the different packing materials and coatings could enhance their antimicrobial properties. EOs are the final products of secondary metabolism in plants (Sambasivaraju & Za, 2016). EOs penetrate through the lipids present in the bacterial cells, causing the destruction of cell structure and leakage of cell fluid, resulting in the death of the organism (Chaiwarit et al., 2018). Valencia et al. (2019) described that the antimicrobial activity of EOs arises from the lipophilicity which described the ability of EOs to pass through the liposome bilayer, and it will destroy intercellular components by its permeability through the cell membrane. Gómez-Estaca et al. (2010) suggested that the hydrophobic nature of EOs results in the passage of the EO through the cell membrane and mitochondria and disturbs the cell structure, thus destroying the foodborne pathogen. The low pH value contributes to the hydrophobicity of EO. Also, the increase in hydrophobicity of the film surface embedded with oils caused an increase in the water-resistance capacity, which is an unavoidable parameter in active packing (Valencia et al., 2019).

Soofi et al. (2021) describe various possible mechanisms for the antibacterial effect of EO-incorporated films. They were disruption of phospholipid bilayer of cell membranes, the formation of hydroperoxidase from fatty acids by oxidation, and the breakdown of the essential lipids, proteins, and bacterial genetic material. The phenolic components are responsible for the antimicrobial activity of EOs (Soofi et al., 2021). The phenolic components of EOs easily attack the phospholipids, which form the cell membrane of bacterial cells, or pathogens, which cause the increase in permeability into the cell walls and results in the leaking of intercellular fluid causes the desolation of bacteria (Abdollahi et al., 2019).

4.2.5 Nanocellulose as a Super Barrier

The fragile nature and postharvest diseases reduce the shelf-life time of fruits. Nanocellulose-active agent conjugated edible coating/film can improve the shelf life of fruits and vegetables. R. Syafiq et al. (2021) reported that the water uptake capacity of sugar palm starch/sugar palm CNC films was reduced by the addition of cinnamon EO. Vartiainen et al. (2016) developed multilayer packaging consisting of cellulose nanofiber, aluminum oxide, and polyglycolic acid that showed promising oxygen barrier performance which could be useful for fresh-food packaging applications. Ranjbaryan et al. (2019) observed that the sodium caseinate films incorporated with cinnamon EO and 2.5% cellulose nanofiber showed antioxidant activity at 66.04%. They also reported that the water vapor permeability of the film was decreased from 10.24×10^{-9} to 9.47×10^{-9} g/m.h.pa at 5% CNF loading. Pacaphol et al. (2019) observed the retention of color, moisture content, and chlorophyll and a decrease in aerobic respiration rate in spinach leaves coated with an edible nanocellulose suspension (0.3–0.5% w/v). The quality retention of uncoated (control) and coated spinach leaves on the second and third days is shown in Figure 4.2.

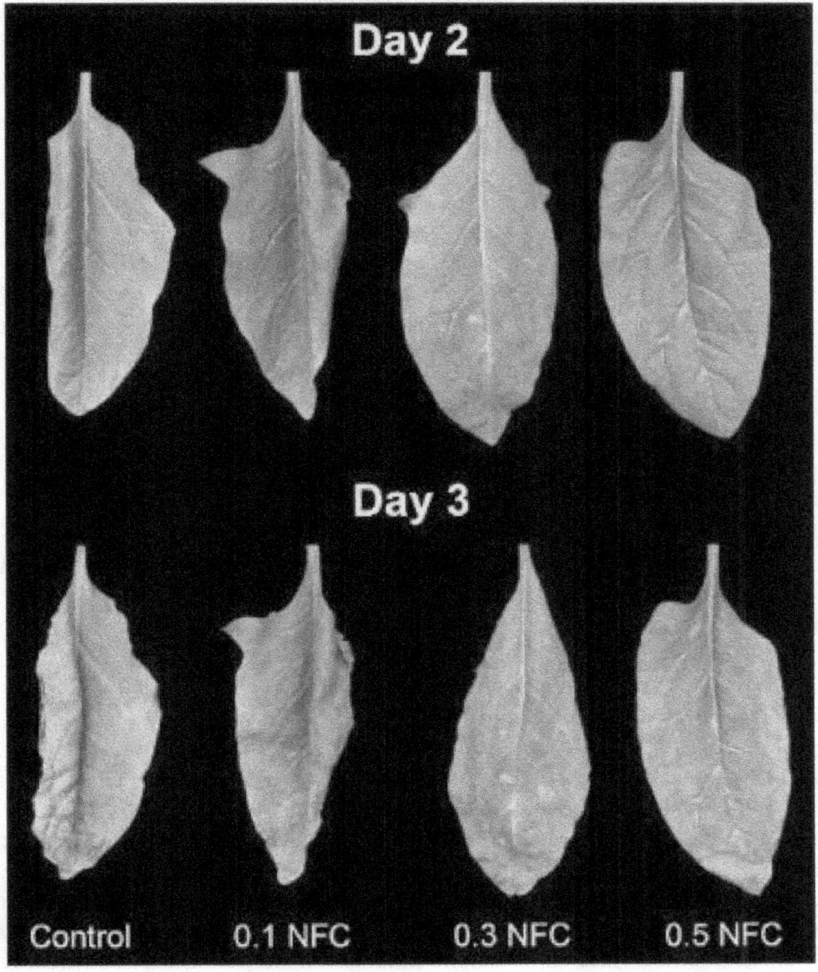

Figure 4.2 Quality retention of uncoated (control) and coated spinach leaves on the second and third day (Pacaphol et al., 2019).

Source: Reproduced with thanks from Elsevier, License Number: 5107611469506.

4.2.6 Cellulose-Based Sensors

Food wastage occurs mainly due to alteration, wastage due to microbial activity, and shelf-life expiration (Motelica et al., 2020). Also, the demand for high-quality foods is increasing, which has made researchers look into smart packing technologies (Smolander, 2003). Smart packing technology includes a sensor or indicator that provides information about the food's freshness or quality, traceability, tamper indication, or safety. Thus, food packing integrated with sensors offers protection for perishable food items, and it helps ensure the quality and integrity of the packaged food items. Consequently, smart or intelligent packing provides us clean, quality foods from manufacturers to the customer (Hogan & Kerry, 2008; Kerry et al., 2006; Kuswandi, 2017; Vanderroost et al., 2014). For instance, Kuswandi et al. (2012) developed a novel curcumin-based freshness sensor to determine volatile amines. BC membrane integrated with curcumin changes color from yellow to orange and then to reddish orange during spoilage. An advantage of this biosensor membrane is that the materials used are edible and well suitable for food packaging applications. Similarly, biosensors based on bromophenol blue integrated with BC membrane have also been reported to exhibit as an on-pack color indicator to detect guava ripeness. The membrane changes color with the decrease in pH as the volatile organic acids are released into the package during the ripening process of guava. Thus, the indicator changes color from blue to green to show the over-ripeness of guava. Results also revealed that the color indicator could be used to detect the change of other parameters such as soluble solids contents, texture, softness, and sensory evaluation of guava stored at 4°C or 28°C (Kuswandi et al., 2013b). A membrane composed of BC and methyl red was also reported for the detection of ripening in strawberries (Kuswandi et al., 2013a).

4.3 Conclusion

Cellulose and its derivatives have been used in a wide range of industries, particularly for food packaging applications. The fundamental features that draw everyone to this green source are its renewability and biodegradability. Because cellulose is hydrophilic packing material, adding hydrophobic oil to it improves its active packing efficiency. Fillers, when used in small amounts, can improve the characteristics of films. The development of nanocellulose-based packaging materials is a hot topic since they possess good barrier properties; enhanced shelf life, transparency, and glossy appearance; edible sources; and so on. The most pressing need for postharvest technologies is the preservation of fruits, vegetables, and meats for long periods without deterioration. The nanocellulose-based edible films are a good choice for improving the shelf life of fruits and vegetables. These films may also help to keep the fruits and vegetables fresh by reducing transpiration and respiration loss. As a result, they could easily cut postharvest losses and provide greater food security. The use of cellulose-based edible films in smart and intelligent food packaging is advancing fast. Also, the studies on the antimicrobial activity of edible cellulose films are promising. Thus, this natural covering provides a viable future alternative for synthetic plastics for food quality control.

References

Abdellatif, E., Kałużna, M., Janse, J. D., Sobiczewski, P., Helali, F., Lamichhane, J. R., & Rhouma, A. (2017). Phenotypic and genetic characterization of pseudomonas syringae strains associated with the recent citrus bacterial blast and bacterial black pit epidemics in Tunisia. *Plant Pathology*, *66*(7), 1081–1093.

Abdollahi, M., Damirchi, S., Shafafi, M., Rezaei, M., & Ariaii, P. (2019). Carboxymethyl cellulose-agar biocomposite film activated with summer savory essential oil as an antimicrobial agent. *International Journal of Biological Macromolecules*, *126*, 561–568. https://doi.org/10.1016/j.ijbiomac.2018.12.115

Acosta, S., Chiralt, A., Santamarina, P., Rosello, J., González-Martínez, C., & Cháfer, M. (2016). Antifungal films based on starch-gelatin blend, containing essential oils. *Food Hydrocolloids*, *61*, 233–240. https://doi.org/10.1016/j.foodhyd.2016.05.008

Aghdam, M. S., Sevillano, L., Flores, F. B., & Bodbodak, S. (2013). Heat shock proteins as biochemical markers for postharvest chilling stress in fruits and vegetables. *Scientia Horticulturae*, *160*, 54–64. https://doi.org/10.1016/j.scienta.2013.05.020

Ahmadi, R., Ghanbarzadeh, B., Ayaseh, A., Kafil, H. S., Özyurt, H., Katourani, A., & Ostadrahimi, A. (2019). The antimicrobial bio-nanocomposite containing non-hydrolyzed cellulose nanofiber (CNF) and Miswak (*Salvadora persica L.*) extract. *Carbohydrate Polymers*, *214*, 15–25. https://doi.org/10.1016/j.carbpol.2019.03.010

Ali, J. B., Danladi, A., Bukhari, M. M., Nyakuma, B. B., Mamza, P., Mohamad, Z. B., Musa, A. B., & Inuwa, I. M. (2020). Extraction and characterization of cellulose nanofibres and cellulose nanocrystals from sammaz-14 maize cobs. *Journal of Natural Fibers*, *19*(7), 2756–2771. https://doi.org/10.1080/15440478.2020.1856279

Archbold, D. D., Hamilton-Kemp, T. R., Barth, M. M., & Langlois, B. E. (1997). Identifying natural volatile compounds that control gray mold (*Botrytis cinerea*) during postharvest storage of strawberry, blackberry, and grape. *Journal of Agricultural and Food Chemistry*, *45*(10), 4032–4037.

Aulin, C., Gällstedt, M., & Lindström, T. (2010). Oxygen and oil barrier properties of microfibrillated cellulose films and coatings. *Cellulose*, *17*(3), 559–574. https://doi.org/10.1007/s10570-009-9393-y

Bastos, M. do S. R., Laurentino, L. da S., Canuto, K. M., Mendes, L. G., Martins, C. M., Silva, S. M. F., Furtado, R. F., Kim, S., Biswas, A., & Cheng, H. N. (2016). Physical and mechanical testing of essential oil-embedded cellulose ester films. *Polymer Testing*, *49*, 156–161. https://doi.org/10.1016/j.polymertesting.2015.11.006

Basumatary, I. B., Mukherjee, A., Katiyar, V., & Kumar, S. (2020). Biopolymer-based nanocomposite films and coatings: Recent advances in shelf-life improvement of fruits and vegetables. *Critical Reviews in Food Science and Nutrition*, *62*(7), 1912–1935. https://doi.org/10.1080/10408398.2020.1848789

Biswas, A., Furtado, R. F., Bastos, M. D. S. R., Benevides, S. D., Oliveira, M. D. A., Boddu, V., & Cheng, H. N. (2018). Preparation and characterization of carboxymethyl cellulose films with embedded essential oils. *Journal of Materials Science Research*, *7*(4), 16. https://doi.org/10.5539/jmsr.v7n4p16

Bojanowski, A., Avis, T. J., Pelletier, S., & Tweddell, R. J. (2013). Management of potato dry rot. *Postharvest Biology and Technology*, *84*, 99–109. https://doi.org/10.1016/j.postharvbio.2013.04.008

Bonilla Lagos, M. J. (2013). *Development of bioactive edible films and coatings with antioxidant and antimicrobial properties for food use*. PhD Thesis (pp. 1–397). Universitat Politècnica De València.

Chaiwarit, T., Ruksiriwanich, W., Jantanasakulwong, K., & Jantrawut, P. (2018). Use of orange oil loaded pectin films as antibacterial material for food packaging. *Polymers, 10*(10). https://doi.org/10.3390/polym10101144

Chen, Y. W., Lee, H. V., & Hamid, S. B. A. (2016). Preparation of nanostructured cellulose via Cr(III)- and Mn(II)-transition metal salt catalyzed acid hydrolysis approach. *BioResources, 11*(3), 7224–7241. https://doi.org/10.15376/biores.11.3.7224-7241

Chiabrando, V., Garavaglia, L., & Giacalone, G. (2019). The postharvest quality of fresh sweet cherries and strawberries with an active packaging system. *Foods, 8*(8), 1–12. https://doi.org/10.3390/foods8080335

Criado, P., Fraschini, C., Jamshidian, M., Salmieri, S., Desjardins, N., Sahraoui, A., & Lacroix, M. (2019). Effect of cellulose nanocrystals on thyme essential oil release from alginate beads: Study of antimicrobial activity against Listeria innocua and ground meat shelf life in combination with gamma irradiation. *Cellulose, 26*(9), 5247–5265. https://doi.org/10.1007/s10570-019-02481-2

Dashipour, A., Razavilar, V., Hosseini, H., Shojaee-Aliabadi, S., German, J. B., Ghanati, K., Khakpour, M., & Khaksar, R. (2015). Antioxidant and antimicrobial carboxymethyl cellulose films containing Zataria multiflora essential oil. *International Journal of Biological Macromolecules, 72*, 606–613. https://doi.org/10.1016/j.ijbiomac.2014.09.006

da Silva, E. L. P., de Carvalho, T. C., Ayub, R. A., & de Almeida, M. C. M. (2019a). Nanocellulose coating associated with lemongrass essential oil at postharvest of blackberry fruits. *Preprints*, 2019100131. https://doi.org/10.20944/preprints201910.0131.v1

Dey, D., Dharini, V., Periyar Selvam, S., Rotimi Sadiku, E., Mahesh Kumar, M., Jayaramudu, J., & Nath Gupta, U. (2020). Physical, antifungal, and biodegradable properties of cellulose nanocrystals and chitosan nanoparticles for food packaging application. *Materials Today: Proceedings*. https://doi.org/10.1016/j.matpr.2020.04.885

Dhall, R. K. (2013). Advances in edible coatings for fresh fruits and vegetables: A review. *Critical Reviews in Food Science and Nutrition, 53*(5), 435–450. https://doi.org/10.1080/10408398.2010.541568

do Evangelho, J. A., da Silva Dannenberg, G., Biduski, B., el Halal, S. L. M., Kringel, D. H., Gularte, M. A., Fiorentini, A. M., & da Rosa Zavareze, E. (2019). Antibacterial activity, optical, mechanical, and barrier properties of corn starch films containing orange essential oil. *Carbohydrate Polymers, 222*, 114981. https://doi.org/10.1016/j.carbpol.2019.114981

Edney, K. L., Tan, A. M., & Burchill, R. T. (1977). Susceptibility of apples to infection by *Gloeosporium* album. *Annals of Applied Biology, 86*(1), 129–132. https://doi.org/10.1111/j.1744-7348.1977.tb01823.x

Fakhouri, F. M., Martelli, S. M., Caon, T., Velasco, J. I., & Mei, L. H. I. (2015). Edible films and coatings based on starch/gelatin: Film properties and effect of coatings on quality of refrigerated Red Crimson grapes. *Postharvest Biology and Technology, 109*, 57–64.

Fotie, G., Limbo, S., & Piergiovanni, L. (2020). Manufacturing of food packaging based on nanocellulose: Current advances and challenges. *Nanomaterials, 10*(9), 1726.

Fujisawa, S., Okita, Y., Fukuzumi, H., Saito, T., & Isogai, A. (2011). Preparation and characterization of TEMPO-oxidized cellulose nanofibril films with free carboxyl groups. *Carbohydrate Polymers, 84*(1), 579–583. https://doi.org/10.1016/j.carbpol.2010.12.029

Ghosh, T., Nakano, K., Mulchandani, N., & Katiyar, V. (2021). Curcumin loaded iron functionalized biopolymeric nanofibre reinforced edible nanocoatings for improved shelf life of cut pineapples. *Food Packaging and Shelf Life, 28*, 100658.

Gómez-Estaca, J., López de Lacey, A., López-Caballero, M. E., Gómez-Guillén, M. C., & Montero, P. (2010). Biodegradable gelatin-chitosan films incorporated with essential oils as antimicrobial agents for fish preservation. *Food Microbiology, 27*(7), 889–896. https://doi.org/10.1016/j.fm.2010.05.012

Han, K. S., Lee, S. C., Lee, J. S., & Soh, J. W. (2012). First report of pink mold rot on tomato fruit caused by *Trichothecium roseum* in Korea. *Research in Plant Disease, 18*(4), 396–398.

Han, Y., Yu, M., & Wang, L. (2018). Physical and antimicrobial properties of sodium alginate/carboxymethyl cellulose films incorporated with cinnamon essential oil. *Food Packaging and Shelf Life, 15*, 35–42. https://doi.org/10.1016/j.fpsl.2017.11.001

Hogan, S. A., & Kerry, J. P. (2008). Smart packaging of meat and poultry products. In Joseph, K., & Butler P. (eds.). *Smart packaging technologies for fast moving consumer goods* (pp. 33–59). John Wiley & Sons, Ltd. https://doi.org/10.1002/9780470753699.ch32

Hu, Z., Zhai, R., Li, J., Zhang, Y., & Lin, J. (2017). Preparation and characterization of nanofibrillated cellulose from bamboo fiber via ultrasonication assisted by repulsive effect. *International Journal of Polymer Science*. https://doi.org/10.1155/2017/9850814

Hubbe, M. A., Ferrer, A., Tyagi, P., Yin, Y., Salas, C., Pal, L., & Rojas, O. J. (2017). Nanocellulose in thin films, coatings, and plies for packaging applications: A review. *BioResources, 12*(1), 2143–2233. https://doi.org/10.15376/biores.12.1.2143-2233

Hubbe, M., Rojas, O. J., Lucia, L. A., & Sain, M. (2008). Cellulosic nanocomposites: A review. *Bioresources, 3*(3), 929–980.

Jahed, E., Khaledabad, M. A., Bari, M. R., & Almasi, H. (2017). Eff ect of cellulose and lignocellulose nano fibers on the properties of *Origanum vulgare ssp.* gracile essential oil-loaded chitosan films. *Reactive and Functional Polymers, 117*, 70–80. https://doi.org/10.1016/j.reactfunctpolym.2017.06.008

James, A., Zikankuba, V., & Yildiz, F. (2017). Postharvest management of fruits and vegetable: A potential for reducing poverty, hidden hunger and malnutrition in sub-Sahara Africa. *Cogent Food & Agriculture, 3*(1), 1312052. https://doi.org/10.1080/23311932.2017.1312052

Jantrawut, P., Boonsermsukcharoen, K., Thipnan, K., Chaiwarit, T., Hwang, K. M., & Park, E. S. (2018). Enhancement of antibacterial activity of orange oil in pectin thin film by microemulsion. *Nanomaterials, 8*(7), 1–12. https://doi.org/10.3390/nano8070545

Jipa, I. M., Stoica-Guzun, A., & Stroescu, M. (2012). Controlled release of sorbic acid from bacterial cellulose based mono and multilayer antimicrobial films. *LWT—Food Science and Technology, 47*(2), 400–406. https://doi.org/10.1016/j.lwt.2012.01.039

Johnson, R. K., Zink-Sharp, A., Renneckar, S. H., & Glasser, W. G. (2009). A new bio-based nanocomposite: Fibrillated TEMPO-oxidized celluloses in hydroxypropylcellulose matrix. *Cellulose, 16*(2), 227–238. https://doi.org/10.1007/s10570-008-9269-6

Jung, J., Simonsen, J., Wang, W., & Zhao, Y. (2018). Evaluation of consumer acceptance and quality of thermally and high hydrostatic pressure processed blueberries and cherries subjected to cellulose nanofiber (CNF) incorporated water-resistant coating treatment. *Food and Bioprocess Technology, 11*(7), 1412–1421. https://doi.org/10.1007/s11947-018-2114-5

Junior, H., Souza, B., Victória, R., Fernandes, D. B., Borges, S. V., Henrique, P., Felix, C., Viana, L. C., Maria, A., Lago, T., & Botrel, D. A. (2018). Utility of blended polymeric formulations containing cellulose nanofibrils for encapsulation and controlled release of sweet orange essential oil utility of blended polymeric formulations containing cellulose nanofibrils for encapsulation and controlled R. *Food and Bioprocess Technology*, 1188–1198.

Kaewklin, P., Siripatrawan, U., Suwanagul, A., & Lee, Y. S. (2018). Active packaging from chitosan-titanium dioxide nanocomposite film for prolonging storage life of tomato fruit. *International Journal of Biological Macromolecules, 112*, 523–529. https://doi.org/10.1016/j.ijbiomac.2018.01.124

Kerry, J. P., O'Grady, M. N., & Hogan, S. A. (2006). Past, current and potential utilisation of active and intelligent packaging systems for meat and muscle-based products: A review. *Meat Science*, *74*(1), 113–130. https://doi.org/10.1016/j.meatsci.2006.04.024

Khan, A., Gallah, H., Riedl, B., Bouchard, J., Safrany, A., & Lacroix, M. (2016). Genipin cross-linked antimicrobial nanocomposite films and gamma irradiation to prevent the surface growth of bacteria in fresh meats. *Innovative Food Science and Emerging Technologies*, *35*, 96–102. https://doi.org/10.1016/j.ifset.2016.03.011

Khedher, S. B., Mejdoub-Trabelsi, B., & Tounsi, S. (2021). Biological potential of *Bacillus subtilis* V26 for the control of Fusarium wilt and tuber dry rot on potato caused by *Fusarium* species and the promotion of plant growth. *Biological Control*, *152*, 104444.

Kuswandi, B. (2017). Freshness sensors for food packaging. In *Reference module in food science*. Elsevier. https://doi.org/10.1016/b978-0-08-100596-5.21876-3

Kuswandi, B., Jayus, Larasati, T. S., Abdullah, A., & Heng, L. Y. (2012). Real-time monitoring of shrimp spoilage using on-package sticker sensor based on natural dye of curcumin. *Food Analytical Methods*, *5*(4), 881–889. https://doi.org/10.1007/s12161-011-9326-x

Kuswandi, B., Kinanti, D. P., Jayus, A. A., & Heng, L. Y. (2013a). Simple and low-cost freshness indicator for strawberries packaging. *Acta Manilana*, *61*(June), 147–159.

Kuswandi, B., Maryska, C., Jayus, A. A., & Heng, L. Y. (2013b). Real time on-package freshness indicator for guavas packaging. *Journal of Food Measurement and Characterization*, *7*(1), 29–39. https://doi.org/10.1007/s11694-013-9136-5

Latorre, B. A., Briceño, E. X., & Torres, R. (2011). Increase in *Cladosporium spp*. populations and rot of wine grapes associated with leaf removal. *Crop Protection*, *30*(1), 52–56. https://doi.org/10.1016/j.cropro.2010.08.022

Lavoine, N., Desloges, I., Sillard, C., & Bras, J. (2014). Controlled release and long-term antibacterial activity of chlorhexidine digluconate through the nanoporous network of microfibrillated cellulose. *Cellulose*, *21*(6), 4429–4442. https://doi.org/10.1007/s10570-014-0392-2

Lee, S. Y., Chun, S. J., Kang, I. A., & Park, J. Y. (2009). Preparation of cellulose nanofibrils by high-pressure homogenizer and cellulose-based composite films. *Journal of Industrial and Engineering Chemistry*, *15*(1), 50–55. https://doi.org/10.1016/j.jiec.2008.07.008

Lotfi, M., Tajik, H., Moradi, M., Forough, M., Divsalar, E., & Kuswandi, B. (2018). Nanostructured chitosan/monolaurin film: Preparation, characterization and antimicrobial activity against *Listeria monocytogenes* on ultrafiltered white cheese. *LWT*, *92*, 576–583. https://doi.org/10.1016/j.lwt.2018.03.020

Lu, P., Guo, M., Xu, Z., & Wu, M. (2018). Application of nanofibrillated cellulose on BOPP/LDPE film as oxygen barrier and antimicrobial coating based on cold plasma treatment. *Coatings*, *8*(6). https://doi.org/10.3390/coatings8060207

Mahajan, P. V., Caleb, O. J., Singh, Z., Watkins, C. B., & Geyer, M. (2014). Postharvest treatments of fresh produce. *Philosophical Transactions of the Royal Society A: Mathematical, Physical and Engineering Sciences, 372*. https://doi.org/10.1098/rsta.2013.0309

Maleki, G., Sedaghat, N., Woltering, E. J., Farhoodi, M., & Mohebbi, M. (2018). Chitosan-limonene coating in combination with modified atmosphere packaging preserve postharvest quality of cucumber during storage. *Journal of Food Measurement and Characterization, 12*(3), 1610–1621. https://doi.org/10.1007/s11694-018-9776-6

Mei, J., Wei, D., Disi, J. O., Ding, Y., Liu, Y., & Qian, W. (2012). Screening resistance against *Sclerotinia sclerotiorum* in *Brassica* crops with use of detached stem assay under controlled environment. *European Journal of Plant Pathology, 134*(3), 599–604. https://doi.org/10.1007/s10658-012-0040-3

Mohsenabadi, N., Rajaei, A., Tabatabaei, M., & Mohsenifar, A. (2018). Physical and antimicrobial properties of starch-carboxy methyl cellulose film containing rosemary essential oils encapsulated in chitosan nanogel. *International Journal of Biological Macromolecules, 112*, 148–155. https://doi.org/10.1016/j.ijbiomac.2018.01.034

Motelica, L., Ficai, D., Ficai, A., Oprea, O. C., Kaya, D. A., & Andronescu, E. (2020). Biodegradable antimicrobial food packaging: Trends and perspectives. *Foods, 9*(10), 1–36. https://doi.org/10.3390/foods9101438

Nguyen, V. T., Gidley, M. J., & Dykes, G. A. (2008). Potential of a nisin-containing bacterial cellulose film to inhibit *Listeria monocytogenes* on processed meats. *Food Microbiology, 25*(3), 471–478. https://doi.org/10.1016/j.fm.2008.01.004

Nie, S., Zhang, C., Zhang, Q., Zhang, K., Zhang, Y., Tao, P., & Wang, S. (2018). Enzymatic and cold alkaline pretreatments of sugarcane bagasse pulp to produce cellulose nanofibrils using a mechanical method. *Industrial Crops and Products, 124*, 435–441. https://doi.org/10.1016/j.indcrop.2018.08.033

Noshirvani, N., Ghanbarzadeh, B., Gardrat, C., Rezaei, M. R., Hashemi, M., Le Coz, C., & Coma, V. (2017). Cinnamon and ginger essential oils to improve antifungal, physical and mechanical properties of chitosan-carboxymethyl cellulose films. *Food Hydrocolloids, 70*, 36–45. https://doi.org/10.1016/j.foodhyd.2017.03.015

Orsuwan, A., & Sothornvit, R. (2018). Active banana flour nanocomposite films incorporated with garlic essential oil as multifunctional packaging material for food application. *Food and Bioprocess Technology, 11*(6), 1199–1210. https://doi.org/10.1007/s11947-018-2089-2

Ortiz, C. M., Salgado, P. R., Dufresne, A., & Mauri, A. N. (2018). Microfibrillated cellulose addition improved the physicochemical and bioactive properties of biodegradable films based on soy protein and clove essential oil. *Food Hydrocolloids, 79*, 416–427. https://doi.org/10.1016/j.foodhyd.2018.01.011

Otoni, C. G., Avena-Bustillos, R. J., Azeredo, H. M. C., Lorevice, M. V., Moura, M. R., Mattoso, L. H. C., & McHugh, T. H. (2017). Recent advances on edible films based on fruits and vegetables—A review. *Comprehensive Reviews in Food Science and Food Safety, 16*(5), 1151–1169. https://doi.org/10.1111/1541-4337.12281

Pacaphol, K., Seraypheap, K., & Aht-Ong, D. (2019). Development and application of nanofibrillated cellulose coating for shelf life extension of fresh-cut vegetable during postharvest storage. *Carbohydrate Polymers, 224*, 115167. https://doi.org/10.1016/j.carbpol.2019.115167

Papoutsis, K., Mathioudakis, M. M., Hasperué, J. H., & Ziogas, V. (2019). Non-chemical treatments for preventing the postharvest fungal rotting of citrus caused by *Penicillium digitatum* (green mold) and *Penicillium italicum* (blue mold). *Trends in Food Science & Technology, 86*, 479–491.

Phanthong, P., Reubroycharoen, P., Hao, X., Xu, G., Abudula, A., & Guan, G. (2018). Nanocellulose: Extraction and application. *Carbon Resources Conversion*, *1*(1), 32–43. https://doi.org/10.1016/j.crcon.2018.05.004

Phokum, C., Jitareerat, P., Phochanachai, S., & Cheevadhanarak, S. (2006). Detection and classification of soft rot Erwinia of vegetables in Thailand by DNA polymerase chain reaction. *Acta Horticulturae*, *712*(II), 917–925. https://doi.org/10.17660/actahortic.2006.712.121

Poiini, P., Subramanian, K. S., Janavi, G. J., & Subramanian, J. (2020). Synthesis of nano-film from nanofibrillated cellulose of banana pseudostem (Musa spp.) to extend the shelf life of tomato. *BioResources*, *15*(2), 2882–2905. https://doi.org/10.15376/biores.15.2.2882-2905

Raghav, P. K., Agarwal, N., Saini, M., Vidhyapeeth, J., & Vidhyapeeth, J. (2016). Edible coating of fruits and vegetables: A review. *International Journal of Scientific and Modern Education*, *I*(I), 188–204.

Rajinipriya, M., Nagalakshmaiah, M., Robert, M., & Elkoun, S. (2018). Homogenous and transparent nanocellulosic films from carrot. *Industrial Crops and Products*, *118*, 53–64. https://doi.org/10.1016/j.indcrop.2018.02.076

Ramos, B., Miller, F. A., Brandão, T. R. S., Teixeira, P., & Silva, C. L. M. (2013). Fresh fruits and vegetables—An overview on applied methodologies to improve its quality and safety. *Innovative Food Science and Emerging Technologies*, *20*, 1–15. https://doi.org/10.1016/j.ifset.2013.07.002

Ranjbaryan, S., Pourfathi, B., & Almasi, H. (2019). Reinforcing and release controlling effect of cellulose nanofiber in sodium caseinate films activated by nanoemulsified cinnamon essential oil. *Food Packaging and Shelf Life*, *21*. https://doi.org/10.1016/j.fpsl.2019.100341

Resende, N. S., Gonçalves, G. A. S., Reis, K. C., Tonoli, G. H. D., & Boas, E. V. B. V. (2018). Chitosan/cellulose nanofibril nanocomposite and its effect on quality of coated strawberries. *Journal of Food Quality*. https://doi.org/10.1155/2018/1727426

Rezaei, F., & Shahbazi, Y. (2018). Shelf-life extension and quality attributes of sauced silver carp fillet: A comparison among direct addition, edible coating and biodegradable film. *LWT–Food Science and Technology*, *87*, 122–133. https://doi.org/10.1016/j.lwt.2017.08.068

Rollini, M., Musatti, A., Cavicchioli, D., Bussini, D., Farris, S., Rovera, C., Romano, D., De Benedetti, S., & Barbiroli, A. (2020). From cheese whey permeate to Sakacin-A/bacterial cellulose nanocrystal conjugates for antimicrobial food packaging applications: A circular economy case study. *Scientific Reports*, *10*(1), 1–14. https://doi.org/10.1038/s41598-020-78430-y

Romanazzi, G., Smilanick, J. L., Feliziani, E., & Droby, S. (2016). Integrated management of postharvest gray mold on fruit crops. *Postharvest Biology and Technology*, *113*, 69–76.

Saini, S., Quinot, D., Lavoine, N., Belgacem, M. N., & Bras, J. (2017). β-cyclodextrin-grafted TEMPO-oxidized cellulose nanofibers for sustained release of essential oil. *Journal of Materials Science*, *52*(7), 3849–3861. https://doi.org/10.1007/s10853-016-0644-7

Salmieri, S., Islam, F., Khan, R. A., Hossain, F. M., Ibrahim, H. M. M., Miao, C., Hamad, W. Y., & Lacroix, M. (2014a). Antimicrobial nanocomposite films made of poly(lactic acid)-cellulose nanocrystals (PLA-CNC) in food applications: Part A–effect of nisin release on the inactivation of Listeria monocytogenes in ham. *Cellulose*, *21*(3), 1837–1850. https://doi.org/10.1007/s10570-014-0230-6

Salmieri, S., Islam, F., Khan, R. A., Hossain, F. M., Ibrahim, H. M. M., Miao, C., Hamad, W. Y., & Lacroix, M. (2014b). Antimicrobial nanocomposite films made of poly(lactic acid)—cellulose nanocrystals (PLA-CNC) in food applications: Part B–effect of oregano essential oil release on the inactivation of Listeria monocytogenes in mixed vegetables. *Cellulose, 21*(6), 4271–4285. https://doi.org/10.1007/s10570-014-0406-0

Sambasivaraju, D., & Za, F. (2016). Evaluation of antibacterial activity of Coriandrum sativum (L.) against gram—positive and gram—negative bacteria. *International Journal of Basic and Clinical Pharmacology, 5*(6), 2653–2656.

Sánchez-González, L., Cháfer, M., Chiralt, A., & González-Martínez, C. (2010). Physical properties of edible chitosan films containing bergamot essential oil and their inhibitory action on *Penicillium italicum*. *Carbohydrate Polymers, 82*(2), 277–283. https://doi.org/10.1016/j.carbpol.2010.04.047

Sánchez-González, L., Pastor, C., Vargas, M., Chiralt, A., Gonzalez-Martinez, C., & Chafer, M. (2011). Effect of hydroxypropylmethylcellulose and chitosan coatings with and without bergamot essential oil on quality and safety of cold-stored grapes. *Postharvest Biology and Technology, 60*(1), 57–63. https://doi.org/10.1016/j.postharvbio.2010.11.004

Sanzani, S. M., Nigro, F., Mari, M., & Ippolito, A. (2009). Symposium paper (pest management: diseases) innovations in the control of postharvest diseases of fresh fruits and vegetables. *Arab Journal of Plant Protection, 27*(2), 240–244.

Sebti, I., Ham-Pichavant, F., & Coma, V. (2002). Edible bioactive fatty acid-cellulosic derivative composites used in food-packaging applications. *Journal of Agricultural and Food Chemistry, 50*(15), 4290–4294. https://doi.org/10.1021/jf0115488

Shahbazi, Y. (2018). Application of carboxymethyl cellulose and chitosan coatings containing Mentha spicata essential oil in fresh strawberries. *International Journal of Biological Macromolecules, 112*, 264–272. https://doi.org/10.1016/j.ijbiomac.2018.01.186

Shahmohammadi Jebel, F., & Almasi, H. (2016). Morphological, physical, antimicrobial and release properties of ZnO nanoparticles-loaded bacterial cellulose films. *Carbohydrate Polymers, 149*, 8–19. https://doi.org/10.1016/j.carbpol.2016.04.089

Shankar, S., & Rhim, J. W. (2018). Antimicrobial wrapping paper coated with a ternary blend of carbohydrates (alginate, carboxymethyl cellulose, carrageenan) and grapefruit seed extract. *Carbohydrate Polymers, 196*, 92–101. https://doi.org/10.1016/j.carbpol.2018.04.128

Sibomana, M. S., Ziena, L. W., Schmidt, S., & Workneh, T. S. (2017). Influence of transportation conditions and postharvest disinfection treatments on microbiological quality of fresh market tomatoes (cv. Nemo-Netta) in a South African supply chain. *Journal of Food Protection, 80*(2), 345–354. https://doi.org/10.4315/0362-028X.JFP-16-229

Silva, E. L. P, de Carvalho, T. C., Antonio Ayub, R., & Menezes de Almeida, M. C. (2020). Blackberry extend shelf life by nanocellulose and vegetable oil coating. *Horticulture International Journal, 4*(2), 54–60. https://doi.org/10.15406/hij.2020.04.00158

Silva, F., Caldera, F., Trotta, F., Nerín, C., & Domingues, F. C. (2019b). Encapsulation of coriander essential oil in cyclodextrin nanosponges: A new strategy to promote its use in controlled-release active packaging. *Innovative Food Science and Emerging Technologies, 56*(May), 102177. https://doi.org/10.1016/j.ifset.2019.102177

Simsek, M., Eke, B., & Demir, H. (2020). Characterization of carboxymethyl cellulose-based antimicrobial films incorporated with plant essential oils. *International Journal of Biological Macromolecules, 163*, 2172–2179. https://doi.org/10.1016/j.ijbiomac.2020.09.075

Smolander, M. (2003). The use of freshness indicators in packaging. In *Novel food packaging techniques*. Woodhead Publishing Limited. https://doi.org/10.1533/9781855737020.1.127

Soofi, M., Alizadeh, A., Hamishehkar, H., Almasi, H., & Roufegarinejad, L. (2021). Preparation of nanobiocomposite film based on lemon waste containing cellulose nanofiber and savory essential oil: A new biodegradable active packaging system. *International Journal of Biological Macromolecules*, *169*, 352–361. https://doi.org/10.1016/j.ijbiomac.2020.12.114

Suput, D., Lazic, V., Popovic, S., & Hromis, N. (2015). Edible films and coatings: Sources, properties and application. *Food and Feed Research*, *42*(1), 11–22. https://doi.org/10.5937/ffr1501011s

Syafiq, R., Sapuan, S. M., & Zuhri, M. R. M. (2021). Antimicrobial activity, physical, mechanical and barrier properties of sugar palm based nanocellulose/starch biocomposite films incorporated with cinnamon essential oil. *Journal of Materials Research and Technology*, *11*, 144–157. https://doi.org/10.1016/j.jmrt.2020.12.091

Syafiq, R., Sapuan, S. M., Zuhri, M. Y. M., Ilyas, R. A., Nazrin, A., Sherwani, S. F. K., & Khalina, A. (2020). Antimicrobial activities of starch-based biopolymers and biocomposites incorporated with plant essential oils: A review. *Polymers*, *12*(10), 1–26. https://doi.org/10.3390/polym12102403

Syahida, N., Fitry, I., Zuriyati, A., & Hanani, N. (2020). Effects of palm wax on the physical, mechanical and water barrier properties of fish gelatin films for food packaging application. *Food Packaging and Shelf Life*, *23*, 100437. https://doi.org/10.1016/j.fpsl.2019.100437

Tahir, H. E., Xiaobo, Z., Mahunu, G. K., Arslan, M., Abdalhai, M., & Zhihua, L. (2019). Recent developments in gum edible coating applications for fruits and vegetables preservation: A review. *Carbohydrate Polymers*, *224*, 115141.

Talebi, F., Misaghi, A., Khanjari, A., Kamkar, A., Gandomi, H., & Rezaeigolestani, M. (2018). Incorporation of spice essential oils into poly-lactic acid film matrix with the aim of extending microbiological and sensorial shelf life of ground beef. *LWT*, *96*, 482–490. https://doi.org/10.1016/j.lwt.2018.05.067

Talibi, I., Askarne, L., Boubaker, H., Boudyach, E. H., Msanda, F., Saadi, B., & Ait Ben Aoumar, A. (2012). Antifungal activity of Moroccan medicinal plants against citrus sour rot agent *Geotrichum candidum*. *Letters in Applied Microbiology*, *55*(2), 155–161. https://doi.org/10.1111/j.1472-765X.2012.03273.x

Tavassoli-Kafrani, E., Gamage, M. V., Dumée, L. F., Kong, L., & Zhao, S. (2020). Edible films and coatings for shelf life extension of mango: A review. *Critical Reviews in Food Science and Nutrition*, *62*(9), 2432–2459. https://doi.org/10.1080/10408398.2020.1853038

Thomas, S. K., Begum, P. S., Midhun Dominic, C. D., Salim, N. V., Hameed, N., Rangappa, S. M., Siengchin, S., & Parameswaranpillai, J. (2021a). Isolation and characterization of cellulose nanowhiskers from Acacia caesia plant. *Journal of Applied Polymer Science*, *138*(15), 50213.

Thomas, S. K., Parameswaranpillai, J., Krishnasamy, S., Begam, P. S., Nandi, D., Siengchin, S., George, J. J., Hameed, N., Salim, N. V., & Sienkiewicz, N. (2021b). A comprehensive review on cellulose, chitin, and starch as fillers in natural rubber biocomposites. *Carbohydrate Polymer Technologies and Applications*, 100095.

Thornton, C. R., Slaughter, D. C., & Davis, R. M. (2010). Detection of the sour-rot pathogen Geotrichum candidum in tomato fruit and juice by using a highly specific monoclonal antibody-based ELISA. *International Journal of Food Microbiology*, *143*(3), 166–172. https://doi.org/10.1016/j.ijfoodmicro.2010.08.012

Vahling-Armstrong, C., Dung, J. K. S., Humann, J. L., & Schroeder, B. K. (2016). Effects of postharvest onion curing parameters on bulb rot caused by *Pantoea agglomerans, Pantoea ananatis* and *Pantoea allii* in storage. *Plant Pathology, 65*(4), 536–544. https://doi.org/10.1111/ppa.12438

Valencia, L., Nomena, E. M., Mathew, A. P., & Velikov, K. P. (2019). Biobased cellulose nanofibril-oil composite films for active edible barriers. *ACS Applied Materials and Interfaces, 11*(17), 16040–16047. https://doi.org/10.1021/acsami.9b02649

Vanderroost, M., Ragaert, P., Devlieghere, F., & De Meulenaer, B. (2014). Intelligent food packaging: The next generation. *Trends in Food Science and Technology, 39*(1), 47–62. https://doi.org/10.1016/j.tifs.2014.06.009

Varghese, S. A., Siengchin, S., & Parameswaranpillai, J. (2020). Essential oils as antimicrobial agents in biopolymer-based food packaging: A comprehensive review. *Food Bioscience*, 100785.

Vartiainen, J., Shen, Y., Kaljunen, T., Malm, T., Vähä-Nissi, M., Putkonen, M., & Harlin, A. (2016). Bio-based multilayer barrier films by extrusion, dispersion coating and atomic layer deposition. *Journal of Applied Polymer Science, 133*(2), 1–6. https://doi.org/10.1002/app.42260

Vilela, C., Moreirinha, C., Domingues, E. M., Figueiredo, F. M. L., Almeida, A., & Freire, C. S. R. (2019). Antimicrobial and conductive nanocellulose-based films for active and intelligent food packaging. *Nanomaterials, 9*(7), 1–16. https://doi.org/10.3390/nano9070980

Wang, L., Liu, F., Jiang, Y., Chai, Z., Li, P., Cheng, Y., Jing, H., & Leng, X. (2011). Synergistic antimicrobial activities of natural essential oils with chitosan films. *Journal of Agricultural and Food Chemistry, 59*(23), 12411–12419. https://doi.org/10.1021/jf203165k

Wieczyńska, J., & Cavoski, I. (2018). Antimicrobial, antioxidant and sensory features of eugenol, carvacrol and trans-anethole in active packaging for organic ready-to-eat iceberg lettuce. *Food Chemistry, 259*, 251–260. https://doi.org/10.1016/j.foodchem.2018.03.137

Yadollahi, A., Arzani, K., & Khoshghalb, H. (2010). The role of nanotechnology in horticultural crops postharvest management. *Acta Horticulturae, 875*, 49–56. https://doi.org/10.17660/ActaHortic.2010.875.4

Yan, J., Luo, Z., Ban, Z., Lu, H., Li, D., Yang, D., Aghdam, M. S., & Li, L. (2019). The effect of the layer-by-layer (LBL) edible coating on strawberry quality and metabolites during storage. *Postharvest Biology and Technology, 147*, 29–38. https://doi.org/10.1016/j.postharvbio.2018.09.002

Yang, W., Fortunati, E., Dominici, F., Giovanale, G., Mazzaglia, A., Balestra, G. M., Kenny, J. M., & Puglia, D. (2016). Synergic effect of cellulose and lignin nanostructures in PLA based systems for food antibacterial packaging. *European Polymer Journal, 79*, 1–12. https://doi.org/10.1016/j.eurpolymj.2016.04.003

Yang, Y., Liu, H., Wu, M., Ma, J., & Lu, P. (2020). Bio-based antimicrobial packaging from sugarcane bagasse nanocellulose/nisin hybrid films. *International Journal of Biological Macromolecules, 161*, 627–635. https://doi.org/10.1016/j.ijbiomac.2020.06.081

Yordshahi, A. S., Moradi, M., Tajik, H., & Molaei, R. (2020). Design and preparation of antimicrobial meat wrapping nanopaper with bacterial cellulose and postbiotics of lactic acid bacteria. *International Journal of Food Microbiology, 321*, 108561. https://doi.org/10.1016/j.ijfoodmicro.2020.108561

Zhang, C., Li, W., Zhu, B., Chen, H., Chi, H., Li, L., Qin, Y., & Xue, J. (2018). The quality evaluation of postharvest strawberries stored in nano-Ag packages at refrigeration temperature. *Polymers, 10*(8). https://doi.org/10.3390/polym10080894

Chapter 5

Applications of Inorganic/ Metallic Nanoparticles in Postharvest Management and the Mechanisms Involved

Kalyan Barman, Swati Sharma, Shatakashi Mishra and Jagdish Singh

5.1 Introduction

The current world is grappling with a booming population of 7.8 billion and enormous food loss. Over 57.1 million tonnes of food are wasted in the U.S. Similarly, in the U.K., 8.3 million tonnes of food end up as waste. The losses in developing economies like India have been estimated to be 5–18% in fresh horticultural commodities, which can be even higher in some actual situations. It is vital to minimize postharvest loss and food waste to help reduce hunger and malnutrition. Moreover, it will enhance profitability by preventing the wastage of resources and inputs used in crop cultivation. It is, therefore, required to employ the current postharvest management technologies in the supply chain to minimize food loss and waste.

The use of nanotechnology in postharvest management can assist the long-term goals of developing sustainable cost-effective technologies for reducing food loss and waste. The nanoscale materials often unlock promising opportunities for enhancing the texture and quality of fresh and processed food products and developing nano-sensors or nano-packaging, nano-antimicrobials, and others. The United States Food and Drug Administration (U.S. FDA) and European Commission (EC) are the major regulatory bodies associated with food nanotechnology. Several workers consider it a potent

DOI: 10.1201/9781003142287-5

upcoming technology to improve storage life, safety and food security (Mousavi and Rezaei, 2011). The enhanced characteristics like barrier and optical (brightness, clarity) properties of nanomaterials are the crux of their broad applications in the preservation of fruit, vegetables, flowers and other food products. Furthermore, they also find applications in unraveling different mechanisms that trigger various chemical changes in stored and processed foods, which may aid in developing shelf-life extension strategies.

5.2 Nanotechnology

Nanotechnology is the structural modification of matter from macro to nano domain (10^{-9} m). It was first cited by Richard Feynman in his famous 1959 lecture named "There Is Plenty of Room at the Bottom". It mainly involves utilizing nanoparticles, with a size ranging between 1 and 100 nm. Many nanoparticles, for example, ferritin, are an intricate part of biological machinery while several others serve as active centers in numerous enzymatic reactions. Most of the nanostructures found in nature are the result of super-saturation triggered either by inorganic pathways or biological mechanisms aided by microbes. Nanoparticles consist of discrete units of nanocrystals and other minerals. The distinctive size of the nanoparticle offers a large surface area-to-mass ratio (Algar, 2018). Owing to large surface area and sensitivity to pH, these particles often form agglomerates. Progressive analytical tools like atomic force microscopy and nano-secondary ion mass spectrometry can be employed to identify and investigate these nanostructures. Nanoparticles occur in a variety of dimensions and geometry. It can exist as a quantum dot (one-dimensional), a carbon nanotube (two-dimensional) or a gold nanoparticle (three-dimensional).

It also has the potential for enhancing the storage life of horticultural produce. Fresh horticultural produce is highly perishable at ambient temperatures. Thus, maintaining freshness, quality and marketability of fresh fruit, vegetables and flowers is a challenge. The possible nanotechnology application in horticulture and postharvest quality maintenance of fruit and vegetables is being studied by researchers worldwide. The nanoscale conversion of any macro structure significantly affects its thermodynamic, solubility and magnetic properties, which is the driving force behind its application in preserving fresh produce and improving shelf life. The immense potential of nano-emulsions and coating formulations can be utilized to extend fruit and vegetable storage life. Moreover, it can be used for various applications, namely, coating, packaging, traceability, quality assessment, safety determination and antimicrobial effects (Chen and Yada, 2011). It opens up opportunities for maintaining the quality of horticultural produce. The different approaches for synthesizing nanomaterials are the self-assembly approach, scanning tunneling microscopy and others (Figure 5.1; Nickols-Richardson, 2007). However, the eco-friendly synthesis, disposal, degradation pattern and rates and nontoxicity of nanoparticles must be ensured before their use. This chapter presents a brief overview of the possible applications of inorganic nanoparticles in postharvest management, the mechanisms involved and the future scope.

5.2.1 Synthesis of Inorganic Nanoparticles

The process for synthesis of metallic nanoparticles can be divided into 'top-down approach' and 'bottom-up approach' (Figure 5.1). The top-down approach is the process

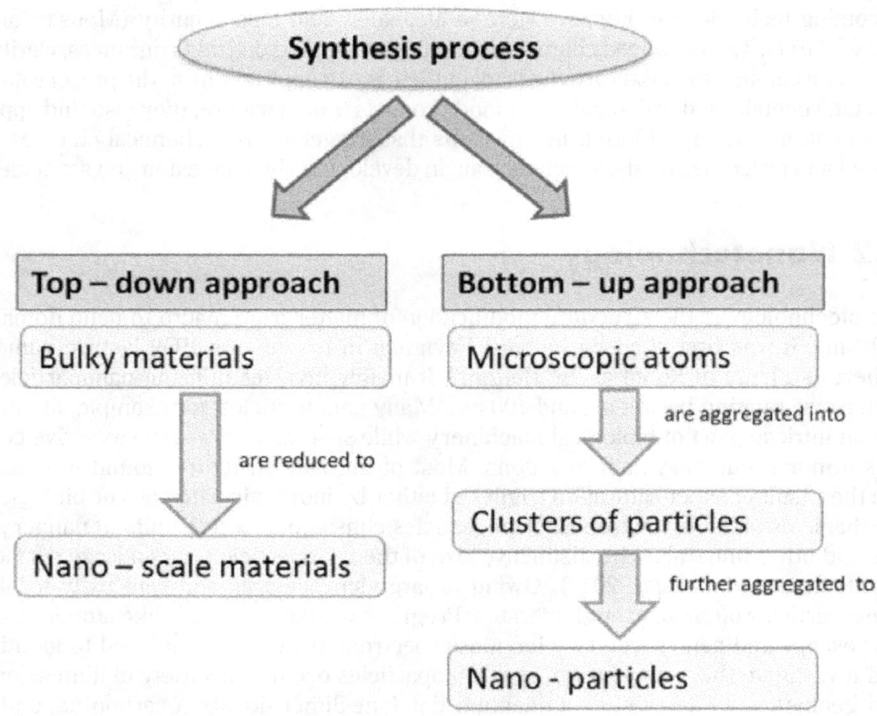

Figure 5.1 Approaches for the synthesis of nanomaterials.

in which the nanomaterial is fabricated from material of relatively higher dimensions by degradation processes to reduce its size. The methods usually employed for attaining the micro-fine nanoscale material are milling, laser ablation, thermal degradation and homogenization, among others. The most prevalent size-reducing technologies are milling, microfluidization and electrospraying, among others (Huang, 2012). The bottom-up approach involves assembling atomic or molecular entities together to form aggregates or clusters and, consequently, nanoparticles, for example, the micelle formation process. The methods employed for bottom-up synthesis are pyrolysis, chemical vapor deposition and sol-gel, among others.

5.2.2 Mechanism of Action of Nanoparticles

Nanoparticles, when incorporated into a polymer matrix, form covalent bonds and tend to alter the barrier property, forcing the molecules to follow a tortuous path. The antimicrobial mechanism is mainly imparted by altering the cell's permeability and inducing intracellular leakage. Nanoparticles like silver often disrupt adenosine triphosphate (ATP) synthesis and the DNA replication process. These particles when incorporated in edible coatings help in lowering the metabolic rate of fruit and vegetables, which may ultimately be helpful for enhancing shelf life and reducing food loss and waste.

5.3 Metal-Based Nanoparticles in Postharvest Management

Metallic nanoparticles like silver (Ag), copper (Cu), zinc (Zn), titanium dioxide (TiO_2), gold and others are extensively used in medical and biological sciences. However, the application of nanoparticles in postharvest technology can be narrowed down to Ag, Zn, Cu and TiO_2. Iron oxides (oxides of iron, CAS Reg. No. 1332-37-2) have been approved for application in food packaging. Oxides of zinc and titanium have also been approved by European Food Safety Authority (EFSA) as GRAS chemicals (He et al., 2019). The properties of nanoparticles often are significantly different than the same particles at macro levels and might be useful for different applications. The antimicrobial action of silver nanoparticles (10 ± 5 nm) against several fungi (Abdelmalek et al., 2016) and bacteria (Tan et al., 2013) makes it a promising component of edible films and coatings. The photocatalytic efficiency of nano-TiO_2 aids in the oxidation of ethylene into H_2O and CO_2 and, hence, can be incorporated into the packaging of fresh fruits and vegetables (Han and Nie, 2004). The applications of metallic nanoparticles in postharvest management are detailed next.

5.3.1 Nano-Based Edible Films and Coatings

Edible coatings are the application of layers of edible matrices as a thin coating on the commodity or as a preformed film more specifically known as an edible film. An edible coating is usually applied by a dipping, spraying or brushing method whereas an edible film is converted into sheets before wrapping around the commodity. Nanoparticle-infused edible coatings have reportedly increased the efficiency of the coatings/films by safeguarding produce against mechanical, chemical and pathogenic deterioration. The chitosan matrix in combination with silicon nanoparticles was reported to improve the storability of jujube (Yu et al., 2012). Silver nanoparticles, owing to their low toxicity, have been used in several edible matrices like chitosan, sodium alginate and others in crops like mango (Chowdappa et al., 2014), mushroom (Jiang et al., 2013) and strawberry (Moussa et al., 2013). Zinc-oxide nanoparticles dispersed in chitosan have been investigated on crops like oranges (Wu et al., 2021). Table 5.1

Table 5.1 Application of Nanomaterial-Based Films and Coatings in Different Fruits and Vegetables

Nanoparticle	*Matrix*	*Crop/ Commodity*	*Findings*	*References*
Silver (Ag) coating	Polyvinyl-pyrrolidone	Asparagus	Silver nano-coated asparagus spears were stored for 25 days at 2°C and 20 days at 10°C	An et al., 2008
	–	Cabbage	Significant reduction was observed in enzymatic browning in white cabbage.	Khan et al., 2015

(Continued)

Table 5.1 (Continued)

Nanoparticle	*Matrix*	*Crop/ Commodity*	*Findings*	*References*
	Montmorillonite (MMT)	Fresh fruit salad	Ag-MMT (20 mg) enhanced shelf life.	Costa et al., 2011
	Chitosan	Mango	Silver–chitosan composite (0.5% and 1%) coating lowered anthracnose incidence by 45.7% and 71.3%.	Chowdappa et al., 2014
	Sodium alginate	Mushroom	Alginate-based nano-coating preserved the quality of mushrooms stored at 4 ± 1°C for 15 days.	Jiang et al., 2013
	Chitosan	Strawberry	Fruit coated with Ag-chitosan showed inhibition of fungal growth during storage.	Moussa et al., 2013
Zinc oxide (ZnO)	Pressurized Argon (Ar)	Orange	Treated fruit showed lower mass loss.	Wu et al., 2021
	Chitosan	Fresh-cut kiwifruit	Lowered ethylene evolution and better fruit texture retention were observed when combined treatment of ultrasound and nano-ZnO.	Meng et al., 2014
	Citric acid	Mango	Fruit treated with ZnO nanoparticles and citric acid showed higher firmness and lower loss.	Jiang et al., 2012a
	Chitosan	Mandarin	Chitosan in combination with nano-ZnO-coated mandarin fruit showed lower weight loss.	Jiang et al., 2012b
	Carrageenan	Mango	Nano ZnO (0.5% and 1%) in combination with carrageenan showed antimicrobial properties against *Escherichia coli*.	Meindrawan et al., 2018

Nanoparticle	Matrix	Crop/ Commodity	Findings	References
	Sodium alginate	Strawberry	Treated fruit showed better retention of sensory attributes.	Emamifar and Bavaisi, 2020
		Vegetable dishes	Nano-ZnO-treated vegetable dishes showed antimicrobial effects without loss of texture.	Yu et al., 2014
Titanium dioxide (TiO_2)	Soy protein	Grape	Grapes coated with nano-TiO_2 showed an increase in total soluble solids and titratable acidity.	Hoseiniyan et al., 2020
	Zein + TiO_2 nanofiber	Cherry tomato	The treatment was used as an ethylene absorber to enhance storage.	Böhmer-Maas et al., 2020
	Alginate + Aloe vera gel	Tomato	The combined treatment of calcium alginate, *Aloe vera* and nano-TiO_2 lowered spoilage.	Salama and Aziz, 2020
	Chitosan	Water chestnut	Extension in shelf life was observed.	Xihong et al., 2008

presents the application of different metallic nano-based coatings/films in fresh fruit and vegetables. The major purpose of nano-based edible coatings and films is to aid the generation of a modified atmosphere around the produce by regulating the diffusion of moisture and respiratory gases. This helps in preserving the quality and sensory attributes of fresh produce and extending shelf life. In addition, the nanoparticle-infused coating and films exhibit an enhanced action against pathogenic microbes, particularly *E. coli* and *Pseudomonas aeruginosa* (Andrade et al., 2014).

5.3.2 Nano-Packaging

Nano-packaging employs the incorporation of nano-sized particles into packaging material to enhance its barrier attributes. Several inorganic nanoparticle-based alternative packaging materials are available in which the nanoparticle can be either placed as a sachet and a pad inside the package or can be incorporated into the packaging material. Du Pont's 'Light Stabilizer 210' is a plastic additive containing nano-titanium dioxide, which aids in slowing ultraviolet (UV)–mediated damage of packed food (Sorrentino et al., 2007). Several inorganic metals like silver, copper, zinc, gold and titanium are restructured to nanoscale and assembled into the polymer matrix (Table 5.2). TiO_2- and ZnO-blended polyethylene terephthalate and polybutylene succinate retard

Table 5.2 Antimicrobial Effects of Nanoparticles Used in Packaging

Nanoparticles	*Matrix*	*Tested against microorganism*	*Observations*	*References*
Silver	Chitosan + Alginate	Gram-positive and gram-negative	Antimicrobial property without any toxicity toward mammalian cells was observed.	Travan et al., 2009
	Chitosan	*S. aureus, Klebsiella pneumonia* and *E. coli*	Bacteriostatic effect was recorded.	Pinto et al., 2012
	Agar-agar	*E. coli, S. aureus* and *Candida albicans*	Maximum antimicrobial action was observed against *Candida albicans.*	Ghosh et al., 2010
Silver (pads)	Cellulose	Fresh-cut melon	Silver nanoparticles impregnated pads significantly maintained lower yeast count during storage.	Fernandez et al., 2010
Silver + Zinc oxide	LDPE	Yeast, molds, total aerobic bacteria	Least degradation of organoleptic quality was noted up to 28 days of storage.	Emamifar et al., 2010
Silver	LDPE	Total aerobic bacteria	Significant reduction in total bacterial and mold counts was noted in barberry.	Motlagh et al., 2013
Silver	EVOH	*Salmonella spp., L. monocytogenes*	Reduction in bacterial count was observed.	Martínez-Abad et al., 2012
Silver	Polystyrene	*S. aureus, Bacillus subtilis, Enterococcus faecalis, Pseudomonas aeruginosa, E. coli, Candida albian, A. niger*	Polystyrene and silver nanocomposite exhibited antimicrobial properties. It was noted that it can be used for food packaging.	Youssef and Abdel-Aziz, 2013

Nanoparticles	Matrix	Tested against microorganism	Observations	References
Zinc oxide	Polyvinyl chloride	*E. coli* and *S. aureus*	Microbe inactivation was noted. It can be potentially used for active food packaging.	Li et al., 2010
	Calcium alginate	*E. coli*	Biocidal action against *E. coli* was observed.	Bajpai et al., 2012
Copper oxide + Zinc oxide nanoparticle	Chitosan film	*Campylobacter, S. aureus, Salmonella*	Extension in storage period of treated packaged guava was observed. CuO nanoparticles showed effective antimicrobial properties.	Kalia et al., 2021
Titanium dioxide	High-density polyethylene film	*E. coli* and *S. aureus*	Maximum bactericidal property was noted in HDPE film coated with 0.6% nano-titanium dioxide.	Saraschandraa et al., 2013
TiO_2	LDPE		Extension in shelf life of strawberries packed in LDPE incorporated with nano-TiO_2 was observed.	Li et al., 2017
Silver + Titanium dioxide	Polyethylene	*Penicillium*	Carrot stored in nano-Ag (1%) + TiO_2 (0.1%) blended container packaging showed less decay during storage.	Metak and Ajaal, 2013
Zirconium molybdate	Chitosan Alginate	Several microbes	Significant reduction in microbial growth in treated mango juice (zirconium molybdate in chitosan and alginate polysaccharide) was observed.	Aly et al., 2021

the growth of *E. coli* and *Staphylococcus aureus* by enhancing thermal and antibacterial properties of the polymer (Threepopnatkul et al., 2014). Similarly, silver nanoparticles incorporated in poly(acrylic acid) film proved effective against gram-negative and gram-positive bacteria (Bardajee et al., 2012). Qiu-Hui (2006) observed that nano-packed green tea retained higher functional properties over normal packaging. The strawberries packed in low-density polyethylene (LDPE) reinforced with silver nanoparticles retained better sensory and functional properties compared to those stored in normal LDPE bags (Motlagh et al., 2020; Yang et al., 2010). Zhang et al. (2019) observed that nano-packaging presented a net beneficial gain and a lower carbon footprint over conventional packaging materials.

5.3.2.1 Nanoparticles as Nano-Antimicrobials and Nano-Biocomposites

Nanocomposites are simply polymers supplemented with multiphasic nanoparticles (up to 5% by weight) to improve the innate properties of the polymer. Furthermore, incorporating nano-antimicrobials into films is helpful in repressing microbial growth. The most commonly used polymer matrices for manufacturing nanocomposites are cellulose (Lloret et al., 2012), polyhydroxy butyrate (Sarfraz et al., 2021), LDPE (Emamifar et al., 2010) and polyethylene (Metak and Ajaal, 2013).

5.3.2.1.1 Copper/Copper Oxide

Copper oxide (CuO)–based nanoparticles possess semiconducting properties that can be employed in fabricating antifouling and antimicrobial packaging (Palza et al., 2015). The use of copper in food packaging can be attributed to its antimicrobial property, economy over silver nanoparticles and comparative easy blending with polymers. The denaturation of microbial DNA initiated by copper nanoparticles is the underlying mechanism behind its antimicrobial property. Copper nanoparticles are known to disrupt the cell membrane protein by lipid peroxidation that consequently leads to the death of fungi (Ingle et al., 2014). The investigations by Shankar and Rhim (2014) to determine the effect of copper salt against *Listeria monocytogenes* and *E. coli* by Llorens et al. (2012) in pineapple and melon juice further exhibit its antibacterial and antifungal property.

5.3.2.1.2 Silver

Silver is the most popular noble metal that is immensely utilized in formulating antimicrobial film due to its potential anti-inflammatory and antimicrobial properties (<35 ppb). Silver particles bind to the receptors on bacterial enzymes, react with thiol group of numerous important enzymes (Sánchez-Valdes et al., 2009) and hinder bacterial transcription and respiration. A multilayered silver nanoparticle–based polyethylene composite film, chitosan-based nano-silver film and silver- and zinc-oxide nanoparticle–based hybrid film (Li et al., 2020) have shown biocidal activity.

5.3.2.1.3 Zinc Oxide

Zinc nanoparticle–coated polyvinyl chloride film exhibited biocidal activity against *E. coli* and *Staphylococcus aureus* and aids enhancement in the tensile strength of polymer (Li et al., 2009). A poly(butylene adipate-co-terephthalate) and ZnO-based film was reported to be effective against *S. aureus*, *Pseudomonas aeruginosa* and *Bacillus subtilis* (Seray et al., 2021). Zn nanoparticle–based coatings tend to upgrade the oxygen barrier and mechanical properties of a polymer (Lepot et al.,

2011; Anžlovar et al., 2012). The ZnO nanoparticles bind themselves to the bacterial wall through electrostatic forces. The irregular geometry of the surface of Zn^{2+} ion causes abrasion on the bacterial cell wall (Saliani et al., 2016). In addition, the generation of free radicals like OH– and H_2O_2, among others, results in enhanced antimicrobial action.

5.3.2.1.4 Titanium Dioxide

Titanium dioxide is the most commonly used compound in nanotechnology-based composite films. The photocatalytic property of TiO_2 imparts strong oxidative power to it, which is subsequently responsible for its antimicrobial action against a wide range of spoilage-causing organisms. The ability of TiO_2 nanoparticles to photodegrade ethylene gas led to the fabrication of TiO_2-coated plastic film in fruit packaging that improved the storability of the fresh fruit (Maneerat and Hayata, 2006). The application of TiO_2-coated polypropylene film on fresh-cut lettuce, which was further irradiated with UV-A rays, exhibited a significant decrease in *E.coli* cell count (Chawengkijwanich and Hayata, 2008). Bodaghi et al. (2013) observed that pears packed in TiO_2/LDPE-based nanocomposite material and stored under illuminated fluorescence exhibited a decrease in bacterial and yeast cell count over control fruit. Similarly, coating or packaging material containing nano-TiO_2 showed a reduction in spoilage, fruit softening and lower ethylene evolution during storage of Chinese water chestnuts and strawberries (Xihong et al., 2008; Luo et al., 2013).

5.3.2.1.5 Gold

The application of gold nanoparticles in postharvest management is in an early stage. It presents beneficial characteristics like chemical stability and conspicuous surface chemistry, which enable easy microbial interaction with the nanoparticle's surface. So far, limited work utilizing gold nanoparticles in nano-packaging has been conducted. Pagno et al. (2015) investigated the development of biofilms based on a quinoa starch matrix dispensed with gold nanoparticles.

5.3.3 Nanosensors/Nanobiosensors

Nanosensors are devices that utilize nanoparticles that may act as indicators against food adulteration, pathogen detection, food spoilage and/or pesticide residues, thereby enabling an awareness for consumers and the promotion of food safety. The optical and magnetic properties of nanoparticles are utilized in the development of nanosensors. They precisely detect and quantify the matter with the help of indicators that emit a signal, a transducer that transmits a signal and a converter that finally changes the signal into analytical form. The sensitivity of devices toward environmental fluctuations improve reliability. Food-based sensors can be electrochemical or optical in nature. Gold (Au) nanoparticle–based sensor electrodeposited on graphene ribbons has been fabricated to detect tert-butylhydroquinone in edible oil and bakery products (Delfino et al., 2020). Similarly, TiO_2/Pd for pesticide detection in potatoes, onions and cabbage (Naser-Sadrabadi et al., 2020) and CeO_2 for mycotoxin detection in milk (Goud et al., 2020) have been developed into food-based nanosensors. Recently, a hybrid Ag/Au film with a sensing ability for glucose and H_2O_2 has been formulated by Scroccarello et al. (2021). The major applications of nanosensors in postharvest are associated with processed food and intelligent/smart packaging.

The applications of nanosensors in postharvest management can be listed as follows:

- Nanosensors embedded in food packaging aid in monitoring potential storage life and food quality.
- Several nanosensors are employed in electronic tongue/nose for detecting aroma volatiles, which may aid in determination of freshness of food (Garcia et al., 2006), for example, 'release on command' packaging innovation controlled by a nano-bioswitch (Yadollahi et al., 2009).
- Fluorescent nanoparticles can be used detecting pathogens and toxins in food (Burris and Stewart, 2012).
- Nanoparticles can be combined with intelligent/smart packaging to ensure easy tracking and enhanced awareness and information.
- The gas emission during the ripening process can be tracked (Pimtong-Ngam et al., 2007).

5.3.4 Nanoencapsulation/Nanoemulsions

The encapsulation of functionally important nutraceuticals like vitamins, carotenes and essential oils at the nanoscale in an ultra-thin membrane is termed a nano-capsule. The capsule is generally downsized to ≤1000 nm and contains active ingredients inside the core. It is a potent technology for assisting in the efficacious delivery of flavor, oils, probiotics and vitamins that are highly prone to oxidative, thermo- and photodegradation (Thies, 2007). The nano-capsules provide stability to chemicals against fluctuating environments and ensure the sustained release of flavor or nutrients. It improves the bioavailability of nutrients. Spice-based essential oils from crops like oregano and cassia have been encapsulated using cornstarch (Parris et al., 2005). Metallic nanoparticles are often amalgamated with an edible matrix. An ultra-thin calcium-alginate cross-linked membrane was developed in which calcium nanoparticles were embedded in the matrix (Kahner et al., 2010).

5.3.5 Postharvest Management of Flowers

Flowers are very delicate and prone to rapid senescence associated the postharvest loss. Floriculture is a multibillion-dollar sunrise industry, which relies on uniform and quality produce (Naing and Kim, 2020). Nanotechnology offers promising ways to improve the shelf life of flowers and floral products. The major factors that deteriorate the postharvest quality of flowers are ethylene-induced senescence and microbial attack on the wounded/cut flower stems (Naing et al., 2017; Park et al., 2017). Stem plugging due to microbial growth impedes the water absorption process of harvested cut-flower stems. Silver nanoparticles have been used commonly in improving the vase life of flowers. Nano-silver, when applied at the rate of 40 mg/l in *Lisianthus*, reduced ethylene evolution and microbial growth. Similarly, in rose cultivar 'Red Ribbon', nano-silver (15 mg/l) + sucrose (5%) reduced the decay percentage (Hatami et al., 2013). Similar observations were also noted in carnations (Naing et al., 2017; Park et al., 2017; Liu et al., 2008; Lin et al., 2019), freesia (Hajizadeh, 2016), gerberas (Liu et al., 2008; Solgi et al., 2009; Nazari and Saba, 2017), gladiolus (Maity et al., 2019), alstroemeria (Langroudi et al., 2020) and lilies (Nemati et al., 2013, 2014; Vinodh et al.,

2013). The mechanism of slowing down the ethylene evolution lies behind the ability of silver nanoparticles to negatively regulate the activity of two major genes, ACC synthase 1 and ACC oxidase 1, responsible for the synthesis of ethylene. The nanoparticles initiate peroxidation and ablate the bacterial cell wall, resulting in the inhibition of microbial growth. This helps in the steady uptake of vase solutions and maintaining turgidity in the stem and other floral parts.

5.4 Risks

The use of nano-based technologies in postharvest management has recorded a continual rise in the recent past. The applications of nano-based emulsions or sensors are mentioned in nano-packaging of fruit. However, there are certain apprehensions associated with the potent risks that need elucidation and safety assessment before commercialization of any such technology at a massive level on a case-by-case basis. Some claims of the migration of silver and copper nanoparticles from packaging material into the food matrix from surrounding have been made (Cushen et al., 2014; Jain et al., 2018). There is a need to investigate the toxicological and any adverse effects of nanoparticles before they can be employed convincingly for their benefits to ensure food safety. The regulatory mechanisms should be stringently followed and upgraded from time to time to keep account of the related advancements and prevent any associated perils. Many have been categorized as GRAS chemicals, but the unique nature of the nanoparticles, particularly behavior in biological systems and food matrices, needs to be studied in detail. A well-formulated policy encompassing the environmental, safety and health concerns shall definitely boost its way for commercial applications in food commodities.

5.5 Conclusion

Nanotechnology has the potential to provide several advantages in the agriculture, horticulture and food industries. Ranging from advanced packaging systems, nanosensors, nano-coatings and nano-food supplements, nanotechnology will surely be instrumental in transforming the food industry. However, the incomplete risk assessment and known and unknown fears are holding up its wide commercial use. The migration and possible health hazards of nanoparticles need to be researched and food safety be guaranteed before incorporating nanotechnology in the postharvest supply chain. The efficacy of applications of nanotechnology, namely, nano-coatings, packaging and their standardization, need to be assessed and validated. Moreover, safety evaluation, action mechanism, characterization, biodegradability, environmental effects, validation and long-term toxicological studies of the metallic nanoparticles, their proper disposal and any possible health and environmental hazards should be looked into carefully. Stringent care should be taken to ensure that environmental and health safety is not compromised on a case-by-case basis.

References

Abdelmalek, G. A. M. and Salaheldin, T. A. (2016). Silver nanoparticles as a potent fungicide for citrus phytopathogenic fungi. *Journal of Nanomedicine Research*, **3**(5), 00065.

Algar, W. R., Jeen, T., Massey, M., Peveler, W. J. and Asselin, J. (2018). Small surface, big effects and big challenges: Toward understanding enzymatic activity at the inorganic nanoparticle—substrate interface. *Langmuir*, **35**(22), 7067–7091.

Aly, A. A., Ali, I. M., Khalil, M., Hameed, A. M., Alrefaei, A. F., Alessa, H., Alfi, A. A., Hassan, M. A. A., Abo El-Naga, M. Y., Hegazy, A. A., Rabie, M. M. and Ammar, M. S. (2021). Chemical, microbial and biological studies on fresh mango juice in presence of nanoparticles of zirconium molybdate embedded chitosan and alginate. *Arabian Journal of Chemistry*, **14**(4), 103066.

An, J., Zhang, M., Wang, S. and Tang, J. (2008). Physical, chemical and microbiological changes in stored green asparagus spears as affected by coating of silver nanoparticles-PVP. *LWT–Food Science and Technology*, **41**(6), 1100–1107.

Andrade, P. F., de Faria, A. F., da Silva, D. S., Bonacin, J. A. and do CarmoGonçalves, M. (2014). Structural and morphological investigations of β-cyclodextrin-coated silver nanoparticles. *Colloids and Surfaces B: Biointerfaces*, **118**, 289–297.

Anžlovar, A., Crnjak Orel, Z. and Kogej, K. (2012). Polyol-mediated synthesis of zinc oxide nanorods and nanocomposites with poly (methyl methacrylate). *Journal of Nanomaterials*, **2012**, 1–9.

Bajpai, S. K., Chand, N. and Chaurasia, V. (2012). Nano zinc oxide-loaded calcium alginate films with potential antibacterial properties. *Food and Bioprocess Technology*, **5**(5), 1871–1881.

Bardajee, G. R., Hooshyar, Z. and Rezanezhad, H. (2012). A novel and green biomaterial based silver nanocomposite hydrogel: Synthesis, characterization and antibacterial effect. *Journal of Inorganic Biochemistry*, **117**, 367–373.

Bodaghi, H., Mostofi, Y., Oromiehie, A., Zamani, Z., Ghanbarzadeh, B., Costa, C., Conte, A. and Del Nobile, M. A. (2013). Evaluation of the photocatalytic antimicrobial effects of a TiO_2 nanocomposite food packaging film by in vitro and in vivo tests. *LWT–Food Science and Technology*, **50**(2), 702–706.

Böhmer-Maas, B. W., Fonseca, L. M., Otero, D. M., da Rosa Zavareze, E. and Zambiazi, R. C. (2020). Photocatalytic zein-TiO_2 nanofibers as ethylene absorbers for storage of cherry tomatoes. *Food Packaging and Shelf Life*, **24**, 100508.

Burris, K. P. and Stewart Jr., C. N. (2012). Fluorescent nanoparticles: Sensing pathogens and toxins in foods and crops. *Trends in Food Science and Technology*, **28**(2), 143–152.

Chawengkijwanich, C. and Hayata, Y. (2008). Development of TiO_2 powder-coated food packaging film and its ability to inactivate *Escherichia coli* in vitro and in actual tests. *International Journal of Food Microbiology*, **123**(3), 288–292.

Chen, H. and Yada, R. (2011). Nanotechnologies in agriculture: New tools for sustainable development. *Trends in Food Science and Technology*, **22**(11), 585–594.

Chowdappa, P., Gowda, S., Chethana, C. S. and Madhura, S. (2014). Antifungal activity of chitosan-silver nanoparticle composite against *Colletotrichum gloeosporioides* associated with mango anthracnose. *African Journal of Microbiology Research*, **8**(17), 1803–1812.

Costa, C., Conte, A., Buonocore, G. G. and Del Nobile, M. A. (2011). Antimicrobial silver-montmorillonite nanoparticles to prolong the shelf life of fresh fruit salad. *International Journal of Food Microbiology*, **148**(3), 164–167.

Cushen, M., Kerry, J., Morris, M., Cruz-Romero, M. and Cummins, E. (2014). Evaluation and simulation of silver and copper nanoparticle migration from polyethylene nanocomposites to food and an associated exposure assessment. *Journal of Agricultural and Food Chemistry*, **62**(6), 1403–1411.

Delfino, J. R., da Silva, J. L., Marques, A. L. and Stradiotto, N. R. (2020). Antioxidants detection in aviation biokerosene by high-performance liquid chromatography using gold nanoparticles anchored in reduced graphene oxide. *Fuel*, **260**, 116315.

Emamifar, A. and Bavaisi, S. (2020). Nanocomposite coating based on sodium alginate and nano-ZnO for extending the storage life of fresh strawberries (*Fragaria* × *ananassa* Duch.). *Journal of Food Measurement and Characterization*, **14**(2), 1012–1024.

Emamifar, A., Kadivar, M., Shahedi, M. and Soleimanian-Zad, S. (2010). Evaluation of nanocomposite packaging containing Ag and ZnO on shelf life of fresh orange juice. *Innovative Food Science and Emerging Technologies*, **11**(4), 742–748.

Fernandez, A., Picouet, P. and Lloret, E. (2010). Cellulose-silver nanoparticle hybrid materials to control spoilage-related microflora in absorbent pads located in trays of fresh-cut melon. *International Journal of Food Microbiology*, **142**(1–2), 222–228.

Garcia, M., Aleixandre, M., Gutiérrez, J. and Horrillo, M. C. (2006). Electronic nose for wine discrimination. *Sensors and Actuators B: Chemical*, **113**(2), 911–916.

Ghosh, S., Kaushik, R., Nagalakshmi, K., Hoti, S. L., Menezes, G. A., Harish, B. N. and Vasan, H. N. (2010). Antimicrobial activity of highly stable silver nanoparticles embedded in agar—agar matrix as a thin film. *Carbohydrate Research*, **345**(15), 2220–2227.

Goud, K. Y., Reddy, K. K., Satyanarayana, M., Kummari, S. and Gobi, K. V. (2020). A review on recent developments in optical and electrochemical aptamer-based assays for mycotoxins using advanced nanomaterials. *Microchimica Acta*, **187**(1), 1–32.

Hajizadeh, H. S. (2016). The study of freesia (*Freesia spp.*) cut flowers quality in relation with nano silver in preservative solutions. *Acta Horticulturae*, **1131**, 1–10.

Han, Y. S. and Nie, L. H. (2004). The mechanism of protecting fresh and preparation of nano TiO_2 thin film. *Journal of Zhuzhou Institute of Technology*, **18**, 148–150.

Hatami, M., Hatamzadeh, A., Ghasemnezhad, M. and Ghorbanpour, M. (2013). The comparison of antimicrobial effects of silver nanoparticles (SNP) and silver nitrate ($AgNO_3$) to extend the vase life of 'Red Ribbon' cut rose flowers. *Trakia Journal of Sciences*, **2**, 144–151.

He, X., Deng, H. and Hwang, H. M. (2019). The current application of nanotechnology in food and agriculture. *Journal of Food and Drug Analysis*, **27**(1), 1–21.

Hoseiniyan, F., Amiri, S., Rezazadeh Bari, M., Rezazad Bari, L. and Dodangeh, S. (2020). Effect of soy protein isolate and TiO_2 edible coating on quality and shelf-life of grapes varieties Hosseini and Ghezel Ozom. *Food Science and Technology*, **17**(100), 29–41.

Huang, Q. (Ed.). (2012). *Nanotechnology in the food, beverage and nutraceutical industries*. Elsevier, pp. 4–5.

Ingle, A. P., Duran, N. and Rai, M. (2014). Bioactivity, mechanism of action, and cytotoxicity of copper-based nanoparticles: A review. *Applied Microbiology and Biotechnology*, **98**(3), 1001–1009.

Jain, A., Ranjan, S., Dasgupta, N. and Ramalingam, C. (2018). Nanomaterials in food and agriculture: An overview on their safety concerns and regulatory issues. *Critical Reviews in Food Science and Nutrition*, **58**(2), 297–317.

Jiang, M., Xiao-Jun, H. U., Jiao, M. S., Huang, S. J., Qiu-Ping, L. U. and Qiao-Lin, F. U. (2012a). Preservative effect of citric acid and nano-ZnO treatment on storage of mango at room temperature. *Northern Horticulture*, **11**, 172–176.

Jiang, M., Ye, X. L. and Qiu, Q. (2012b). Study on the preservation effect of chitosan and nano-ZnO compound coating on Shatang mandarin. *Science and Technology of Food Industry*, **33**, 348–351.

Jiang, T., Feng, L. and Wang, Y. (2013). Effect of alginate/nano-Ag coating on microbial and physicochemical characteristics of shiitake mushroom (*Lentinus edodes*) during cold storage. *Food Chemistry*, **141**(2), 954–960.

Kahner, R., Waerder, B., Arslan, H. K. and Rehage, H. (2010). New types of self-organizing interfacial alginate membranes. *Colloid and Polymer Science*, **288**(4), 461–468.

Kalia, A., Kaur, M., Shami, A., Jawandha, S. K., Alghuthaymi, M. A., Thakur, A. and Abd-Elsalam, K. A. (2021). Nettle-leaf extract derived ZnO/CuO nanoparticle-biopolymer-based antioxidant and antimicrobial nanocomposite packaging films and their impact on extending the post-harvest shelf life of guava fruit. *Biomolecules*, **11**(2), 224.

Khan, A. U., Wei, Y., Ahmad, A., Khan, Z. U. H., Tahir, K., Khan, S. U., Muhammad, N., Khan, F. U. and Yuan, Q. (2015). Enzymatic browning reduction in white cabbage, potent antibacterial and antioxidant activities of biogenic silver nanoparticles. *Journal of Molecular Liquids*, **215**, 39–46.

Langroudi, M. E., Hashemabadi, D., KalateJari, S. and Asadpour, L. (2020). Effects of silver nanoparticles, chemical treatments and herbal essential oils on the vase life of cut alstroemeria (Alstroemeria 'Summer Sky') flowers. *The Journal of Horticultural Science and Biotechnology*, **95**(2), 175–182.

Lepot, N., Van Bael, M. K., Van den Rul, H., D'Haen, J., Peeters, R., Franco, D. and Mullens, J. (2011). Influence of incorporation of ZnO nanoparticles and biaxial orientation on mechanical and oxygen barrier properties of polypropylene films for food packaging applications. *Journal of Applied Polymer Science*, **120**(3), 1616–1623.

Li, D., Ye, Q., Jiang, L. and Luo, Z. (2017). Effects of nano-TiO_2-LDPE packaging on postharvest quality and antioxidant capacity of strawberry (*Fragaria ananassa* Duch.) stored at refrigeration temperature. *Journal of the Science of Food and Agriculture*, **97**(4), 1116–1123.

Li, W., Huang, Z., Cai, R., Yang, W., He, H. and Wang, Y. (2020). Rational design of Ag/ZnO hybrid nanoparticles on sericin/agarose composite film for enhanced antimicrobial applications. *International Journal of Molecular Sciences*, **22**(1), 105.

Li, X., Xing, Y., Jiang, Y., Ding, Y. and Li, W. (2009). Antimicrobial activities of ZnO powder-coated PVC film to inactivate food pathogens. *International Journal of Food Science & Technology*, **44**, 2161–2168.

Li, X. H., Xing, Y. G., Li, W. L., Jiang, Y. H. and Ding, Y. L. (2010). Antibacterial and physical properties of poly (vinyl chloride)-based film coated with ZnO nanoparticles. *Food Science and Technology International*, **16**(3), 225–232.

Lin, X., Li, H., Lin, S., Xu, M., Liu, J., Li, Y. and He, S. (2019). Improving the postharvest performance of cut spray 'Prince' carnations by vase treatments with nano-silver and sucrose. *The Journal of Horticultural Science and Biotechnology*, **94**(4), 513–521.

Liu, J., Zhang, Z., Joyce, D. C., He, S., Cao, J. and Lv, P. (2008). Effects of postharvest nano-silver treatments on cut-flowers. *Acta Horticulturae*, **847**, 245–250.

Llorens, A., Lloret, E., Picouet, P. and Fernandez, A. (2012). Study of the antifungal potential of novel cellulose/copper composites as absorbent materials for fruit juices. *International Journal of Food Microbiology*, **158**(2), 113–119.

Lloret, E., Picouet, P. and Fernández, A. (2012). Matrix effects on the antimicrobial capacity of silver based nanocomposite absorbing materials. *LWT–Food Science and Technology*, **49**(2), 333–338.

Luo, Z. S., Ye, Q. Y. and Li, D. D. (2013). Influence of nano-TiO_2 modified LDPE film packaging on quality of strawberry. *Modern Food Science and Technology*, **29**(10), 2340–2344.

Maity, T. R., Samanta, A., Saha, B. and Datta, S. (2019). Evaluation of *Piper betle* mediated silver nanoparticle in post-harvest physiology in relation to vase life of cut spike of Gladiolus. *Bulletin of the National Research Centre*, **43**(1), 1–11.

Maneerat, C. and Hayata, Y. (2006). Antifungal activity of TiO_2 photocatalysis against *Penicillium expansum* in vitro and in fruit tests. *International Journal of Food Microbiology*, **107**(2), 99–103.

Martínez-Abad, A., Lagaron, J. M. and Ocio, M. J. (2012). Development and characterization of silver-based antimicrobial ethylene—vinyl alcohol copolymer (EVOH) films for food-packaging applications. *Journal of Agricultural and Food Chemistry*, **60**(21), 5350–5359.

Meindrawan, B., Suyatma, N. E., Wardana, A. A. and Pamela, V. Y. (2018). Nanocomposite coating based on carrageenan and ZnO nanoparticles to maintain the storage quality of mango. *Food Packaging and Shelf Life*, **18**, 140–146.

Meng, X., Zhang, M. and Adhikari, B. (2014). The effects of ultrasound treatment and nano-zinc oxide coating on the physiological activities of fresh-cut kiwifruit. *Food and Bioprocess Technology*, **7**(1), 126–132.

Metak, A. M. and Ajaal, T. T. (2013). Investigation on polymer based nano-silver as food packaging materials. *International Journal of Biological, Biomolecular, Agricultural, Food and Biotechnological Engineering*, **7**, 772–777.

Motlagh, N. V., Aliabadi, M., Rahmani, E. and Ghorbanpour, S. (2020). The effect of nano-silver packaging on quality maintenance of fresh strawberry. *International Journal of Nutrition and Food Engineering*, **14**(9), 123–128.

Motlagh, N. V., Hamed Mosavian, M. T. and Mortazavi, S. A. (2013). Effect of polyethylene packaging modified with silver particles on the microbial, sensory and appearance of dried barberry. *Packaging Technology and Science*, **26**(1), 39–49.

Mousavi, S. R. and Rezaei, M. (2011). Nanotechnology in agriculture and food production. *Journal of Applied Environmental and Biological Sciences*, **1**(10), 414–419.

Moussa, S. H., Tayel, A. A., Alsohim, A. S. and Abdallah, R. R. (2013). Botryticidal activity of nanosized silver-chitosan composite and its application for the control of gray mold in strawberry. *Journal of Food Science*, **78**(10), M1589–M1594.

Naing, A. H. and Kim, C. K. (2020). Application of nano-silver particles to control the postharvest biology of cut flowers: A review. *Scientia Horticulturae*, **270**, 109463.

Naing, A. H., Win, N. M., Han, J. S., Lim, K. B. and Kim, C. K. (2017). Role of nano-silver and the bacterial strain *Enterobacter cloacae* in increasing vase life of cut carnation 'Omea'. *Frontiers in Plant Science*, **8**, 1590.

Naser-Sadrabadi, A., Zare, H. R. and Benvidi, A. (2020). Photochemical deposition of palladium nanoparticles on TiO_2 nanoparticles and their application for electrocatalytic measurement of nitrate ions in potato, onion and cabbage using bipolar electrochemical method. *Measurement*, **166**, 108222.

Nazari, F. and Saba, M. K. (2017). Combination effect of 1-methylcyclopropene (1-MCP) with ajowan essential oil and silver nanoparticles on postharvest life of gerbera (*Gerbera jamesonii*) cut flowers. *HortScience*, **52**(11), 1550–1555.

Nemati, S. H., Esfandiyari, B., Tehranifar, A., Rezaei, A. and Ashrafi, S. J. (2014). Effect of nano-silver particles on postharvest life of *Lilium orientalis* cv. 'Shocking'. *International Journal of Postharvest Technology and Innovation*, **4**(1), 46–53.

Nemati, S. H., Tehranifar, A., Esfandiari, B. and Rezaei, A. (2013). Improvement of vase life and postharvest factors of *Lilium orientalis* 'Bouquet' by silver nano particles. *Notulae Scientia Biologicae*, **5**(4), 490–493.

Nickols-Richardson, S. M. (2007). Nanotechnology: Implications for food and nutrition professionals. *Journal of American Dietetic Association*, **107**(9), 1494–1497.

Pagno, C. H., Costa, T. M., de Menezes, E. W., Benvenutti, E. V., Hertz, P. F., Matte, C. R. and Flôres, S. H. (2015). Development of active biofilms of quinoa (*Chenopodium *quinoa* W.) starch containing gold nanoparticles and evaluation of antimicrobial activity. *Food Chemistry*, **173**, 755–762.

Palza, H., Quijada, R. and Delgado, K. (2015). Antimicrobial polymer composites with copper micro-and nanoparticles: Effect of particle size and polymer matrix. *Journal of Bioactive and Compatible Polymers*, **30**(4), 366–380.

Park, D. Y., Naing, A. H., Ai, T. N., Han, J. S., Kang, I. K. and Kim, C. K. (2017). Synergistic effect of nano-sliver with sucrose on extending vase life of the carnation cv. Edun. *Frontiers in Plant Science*, **8**, 1601.

Parris, N., Cooke, P. H. and Hicks, K. B. (2005). Encapsulation of essential oils in zein nanospherical particles. *Journal of Agricultural and Food Chemistry*, **53**(12), 4788–4792.

Pimtong-Ngam, Y., Jiemsirilers, S. and Supothina, S. (2007). Preparation of tungsten oxide—tin oxide nanocomposites and their ethylene sensing characteristics. *Sensors and Actuators A: Physical*, **139**(1–2), 7–11.

Pinto, R. J., Fernandes, S. C., Freire, C. S., Sadocco, P., Causio, J., Neto, C. P. and Trindade, T. (2012). Antibacterial activity of optically transparent nanocomposite films based on chitosan or its derivatives and silver nanoparticles. *Carbohydrate Research*, **348**, 77–83.

Qiu-Hui, H. (2006). Effect of a new fashion nano-packing on preservation quality of green tea. *Food Science*, **4**, 61.

Salama, H. E. and Aziz, M. S. A. (2020). Optimized alginate and aloe vera gel edible coating reinforced with nano TiO_2 for the shelf-life extension of tomatoes. *International Journal of Biological Macromolecules*, **165**, 2693–2701.

Saliani, M., Jalal, R. and Goharshadi, E. K. (2016). Mechanism of oxidative stress involved in the toxicity of ZnO nanoparticles against eukaryotic cells. *Nanomedicine Journal*, **3**(1), 1–14.

Sánchez-Valdes, S., Ortega-Ortiz, H., Ramos-de Valle, L. F., Medellín-Rodríguez, F. J. and Guedea-Miranda, R. (2009). Mechanical and antimicrobial properties of multilayer films with a polyethylene/silver nanocomposite layer. *Journal of Applied Polymer Science*, **111**(2), 953–962.

Saraschandraa, N., Pavithrab, M. and Sivakumar, A. (2013). Antimicrobial applications of TiO_2 coated modified polyethylene (HDPE) films. *Archives of Applied Science Research*, **5**(1), 189–194.

Sarfraz, J., Gulin-Sarfraz, T., Nilsen-Nygaard, J. and Pettersen, M. K. (2021). Nanocomposites for food packaging applications: An overview. *Nanomaterials*, **11**, 10. https://doi.org/nano11010010.

Scroccarello, A., Della Pelle, F., Ferraro, G., Fratini, E., Tempera, F., Dainese, E. and Compagnone, D. (2021). Plasmonic active film integrating gold/silver nanostructures for H_2O_2 readout. *Talanta*, **222**, 121682.

Seray, M., Skender, A. and Hadj-Hamou, A. S. (2021). Kinetics and mechanisms of Zn^{2+} release from antimicrobial food packaging based on poly (butylene adipate-co-terephthalate) and zinc oxide nanoparticles. *Polymer Bulletin*, **78**(2), 1021–1040.

Shankar, S. and Rhim, J. W. (2014). Effect of copper salts and reducing agents on characteristics and antimicrobial activity of copper nanoparticles. *Materials Letters*, **132**, 307–311.

Solgi, M., Kafi, M., Taghavi, T. S. and Naderi, R. (2009). Essential oils and silver nanoparticles (SNP) as novel agents to extend vase-life of gerbera (*Gerbera jamesonii* cv. 'Dune') flowers. *Postharvest Biology and Technology*, **53**(3), 155–158.

Sorrentino, A., Gorrasi, G. and Vittoria, V. (2007). Potential perspectives of bio-nanocomposites for food packaging applications. *Trends in Food Science and Technology*, **18**(2), 84–95.

Tan, H., Ma, R., Lin, C., Liu, Z. and Tang, T. (2013). Quaternized chitosan as an antimicrobial agent: Antimicrobial activity, mechanism of action and biomedical applications in orthopedics. *International Journal of Molecular Sciences*, **14**(1), 1854–1869.

Thies, C. (2007). Microencapsulation of flavors by complex coacervation. In Lakkis, J. M. (ed.), *Encapsulation and controlled release technologies in food systems*. Blackwell, pp. 149–169.

Threepopnatkul, P., Wongnarat, C., Intolo, W., Suato, S. and Kulsetthanchalee, C. (2014). Effect of TiO_2 and ZnO on thin film properties of PET/PBS blend for food packaging applications. *Energy Procedia*, **56**, 102–111.

Travan, A., Pelillo, C., Donati, I., Marsich, E., Benincasa, M., Scarpa, T., Semeraro, S., Turco, G., Gennaro, R. and Paoletti, S. (2009). Non-cytotoxic silver nanoparticle-polysaccharide nanocomposites with antimicrobial activity. *Biomacromolecules*, **10**(6), 1429–1435.

Vinodh, S., Kannan, M. and Jawaharlal, M. (2013). Effect of nanosilver and sucrose on post harvest quality of cut Asiatic Lilium cv. Tresor. *The Bioscan*, **8**(3), 901–904.

Wu, D., Zhang, M., Xu, B. and Guo, Z. (2021). Fresh-cut orange preservation based on nano-zinc oxide combined with pressurized argon treatment. *LWT–Food Science and Technology*, **135**, 110036.

Xihong, L., Songshan, Q., Yunfeng, H., Yage, X. and Jing, Z. (2008). Study on the multiple coating material of chitosan and nano-titania in the preservation of fresh-cut chinese water chestnut. *Food and Fermentation Industries*, **1**.

Yadollahi, A., Arzani, K. and Khoshghalb, H. (2009). The role of nanotechnology in horticultural crops postharvest management. *Southeast Asia Symposium on Quality and Safety of Fresh and Fresh-Cut Produce*, **875**, 9–56.

Yang, F. M., Li, H. M., Li, F., Xin, Z. H., Zhao, L. Y., Zheng, Y. H. and Hu, Q. H. (2010). Effect of nano-packing on preservation quality of fresh strawberry (*Fragaria ananassa* Duch. cv. Fengxiang) during storage at 4°C. *Journal of Food Science*, **75**(3), C236–C240.

Youssef, A. M. and Abdel-Aziz, M. S. (2013). Preparation of polystyrene nanocomposites based on silver nanoparticles using marine bacterium for packaging. *Polymer-Plastics Technology and Engineering*, **52**(6), 607–613.

Yu, N., Zhang, M., Islam, M. N., Lu, L., Liu, Q. and Cheng, X. (2014). Combined sterilizing effects of nano-ZnO and ultraviolet on convenient vegetable dishes. *LWT–Food Science and Technology*, **61**(2), 638–643.

Yu, Y., Zhang, S., Ren, Y., Li, H., Zhang, X. and Di, J. (2012). Jujube preservation using chitosan film with nano-silicon dioxide. *Journal of Food Engineering*, **113**(3), 408–414.

Zhang, B. Y., Tong, Y., Singh, S., Cai, H. and Huang, J. Y. (2019). Assessment of carbon footprint of nano-packaging considering potential food waste reduction due to shelf life extension. *Resources, Conservation and Recycling*, **149**, 322–331.

Chapter 6

Mechanism of Silver Nanoparticle-Based Postharvest Technologies

Udari Wijesinghe, Upekshya Welikala, and Gobika Thiripuranathar

List of Abbreviations

ACC	Aminocyclopropane-1-carboxylic acid
ACS	Acetyl–CoA synthetase
Ag	Silver
Ag^+	Silver ion
AgNP	Silver nanoparticle
Au	Gold
BEO	Bergamot essential oil
EVOH	Ethylene-vinyl alcohol
H_2O_2	Hydrogen peroxide
LDPE	Low-density polyethylene
LSPR	Localized surface plasmon resonance
MNP	Metal nanoparticle
MMT	Montmorillonite
NP	Nanoparticle
PLA	Polylactic acid
PVA	Polyvinyl alcohol
PVC	Polyvinyl chloride
ROS	Reactive oxygen species
Si	Silicon
SERS	Surface-enhanced Raman spectroscopy
TiO_2	Titanium dioxide

 DOI: 10.1201/9781003142287-6

TSS	Total soluble solid
TTI	Time–temperature indicator
UV	Ultraviolet
ZnO	Zinc oxide

6.1 Introduction

In recent decades, changes in modern consumers' lifestyles have led to an increased demand for fresh horticulture products (i.e., fruits, vegetables, flowers, and ornamental plants; Mihindukulasuriya and Lim 2014). Consumers' interest has surged for fresh produce in microbiologically safe, fresh and fresh-like, nutritious, shelf-stable, and convenient products developed using environmentally friendly technologies. Horticultural products have a characteristic perishable nature, and it is estimated that about 20–30% of fresh produce deteriorates at different stages (Liu et al. 2020). Furthermore, 40% of productivity is required to enhance agricultural products to fulfill the needs of the increasing population of the world by 2050 (Restuccia et al. 2010; Ijaz et al. 2020). To meet these requirements, fresh produce needs to be adequately stored or consumed immediately after harvesting. Therefore, postharvest treatments need to be utilized to reduce fresh produce losses by improving quality and increasing shelf life (Liu et al. 2020; Mihindukulasuriya and Lim 2014; Restuccia et al. 2010). Thus, different physical, chemical, and gaseous preservation methods have been invented to reduce crop deprivation (Liu et al. 2020; Llorens et al. 2012). However, these practices do not always guarantee product integrity, which forces producers to apply several treatments during the postharvest management of crops to initiate profit-oriented postharvest technologies. At the same time, there is a lack of regulation on quality control, mechanical properties, storage conditions, product handling, and packaging that indiscriminately use convectional remedies (Liu et al. 2020; Restuccia et al. 2010; Kumar et al. 2017; De Azeredo 2013).

Nanotechnology is a multidisciplinary approach that plays an important role in the development of new materials and devices with unique physiochemical properties resulting in novel applications in diverse fields (Carbone et al. 2016). Nanotechnology has recently evolved as a promising tool in food and horticultural science, providing new insights into postharvest technologies to overcome postharvest losses (Carbone et al. 2016; Ijaz et al. 2020; Restuccia et al. 2010). Metal and metal-oxide NPs such as Ag, Au, ZnO, TiO_2, and copper oxide have been widely studied for their ability to minimize diseases, improve the quality of food, extend the shelf life, and prevent or delay spoilage (Llorens et al. 2012; De Azeredo 2013; Deshmukh et al. 2019). Among all the metallic NPs (MNPs), AgNPs are widely studied due to their unique optical, catalytic, sensing, biological, electrochemical, and antimicrobial properties (De Azeredo 2013; Deshmukh et al. 2019). Additionally, AgNPs exhibit low volatility and high stability at high temperatures. Having a high surface area-to-volume ratio, AgNPs are found to be more effective as germicides than the other Ag forms (Ag nitrate and Ag thiosulfate) in postharvest management to control various foodborne infections (Jung and Zhao 2016; Llorens et al. 2012; Carbone et al. 2016; Sardella et al. 2019). AgNPs with different size, shape, structural stability, and target affinity provide fascinating biocidal properties to eradicate a wide range of gram-negative (*Acinetobacter, Escherichia, Pseudomonas, Salmonella,* and *Vibrio*), gram-positive bacteria (*Bacillus, Clostridium,*

Enterococcus, Listeria, Staphylococcus, and *Streptococcus*; Mohammed Fayaz et al. 2009), fungi (*Aspergillus niger, Alternaria* spp., *Botrytis cinerea, Fusarium oxysporum, Penicillium* spp., and *Rhizoctonia solani*), and viruses, leading to an upsurge in their applications in food science and horticulture (Llorens et al. 2012; De Azeredo 2013; Deshmukh et al. 2019; Sardella et al. 2019). Moreover, antibacterial, antifungal, antiviral, antioxidant, anti-odor, and anti-browning activities enable scientists to design and incorporate AgNPs inventively in the production of novel food products. Depending on the hosting material, such as polymers and stabilizing agents, these activities can significantly get improved (Llorens et al. 2012; Ijaz et al. 2020; De Azeredo 2013). This chapter overviews the latest studies on AgNP-mediated postharvest crop remedies for fruits, vegetables, cut flowers, and ornamental plants. Furthermore, the major focus is on those related to factors such as crop processing and packaging, the development of AgNP-doped nanosensors, and biocompatible nano-laminated coatings to increase shelf life, sensory and quality attributes, and toxicological and ecotoxicological risks.

6.2 Application of AgNPs in Postharvest Management

Postharvest deprivation of horticultural crops is not discretely affected by postharvest activities, however, influenced by pre-harvest activities during growth and development (Kumar et al. 2020; Silberbauer and Schmid 2017). Improper handling, temperature, exposure to sunlight, and insufficient humidity can quicken the deterioration and contamination of food products and withering of cut flowers and ornamental plants, which ultimately decrease the shelf life of produce crops (Kumar et al. 2020). Therefore, AgNP-mediated preservation techniques were introduced to prevent or slow spoilage/withering; sustain nutritional value, appearance, texture, edibility, and flavor; and prolong the shelf life (Kumar et al. 2020; Silberbauer and Schmid 2017). Additionally, Ag nanofillers exhibit blockage of ethylene, antimicrobial, and anti-browning, antioxidant properties and diminish the rate of water loss, transpiration and respiration rates, delay tissue softening, preservation of TSSs content, vitamin C content, and freshness (Gao et al. 2017; Ijaz et al. 2020; Hoffmanna et al. 2019). Moreover, AgNPs deliver active biocidal properties to food or the space around the food, thereby improving food quality by extending shelf life and reducing the risk of pathogens (Khan et al. 2016; Honarvar et al. 2016; Carbone et al. 2016; Ayhan 2017). Furthermore, AgNP-incorporated additives are more stable than organic and inorganic antimicrobial materials (Silberbauer and Schmid 2017). The bactericidal effect of AgNPs has already been well established (Mohammed Fayaz et al. 2009). However, the exact antibacterial mechanisms associated with AgNPs have not yet been fully elucidated, although scientists have identified three probable mechanisms (Durán et al. 2016; Marambio-Jones and Hoek 2010; Deshmukh et al. 2019): (1) AgNP penetration: AgNPs can penetrate bacterial cells and interact with the content present in the cytoplasm, especially with sulfur-containing proteins and phosphorus in DNA, disrupting adenosine triphosphate (ATP) production, hinder the permeation of phosphate and protons across the membrane, affect respiratory chain enzymes activities, and DNA replication (Durán et al. 2016; Deshmukh et al. 2019; Kumar et al. 2020); (2) release of Ag^+ from AgNPs: Even though Ag^+ and AgNPs can penetrate the cell membrane, Ag^+ is highly reactive when compared to AgNPs. The primary functional mechanism of lysis is due to the formation of free radicals such as H_2O_2 and superoxide, which induce oxidative stress, impair cellular signaling, and depletes the antioxidant enzymes, resulting in changes

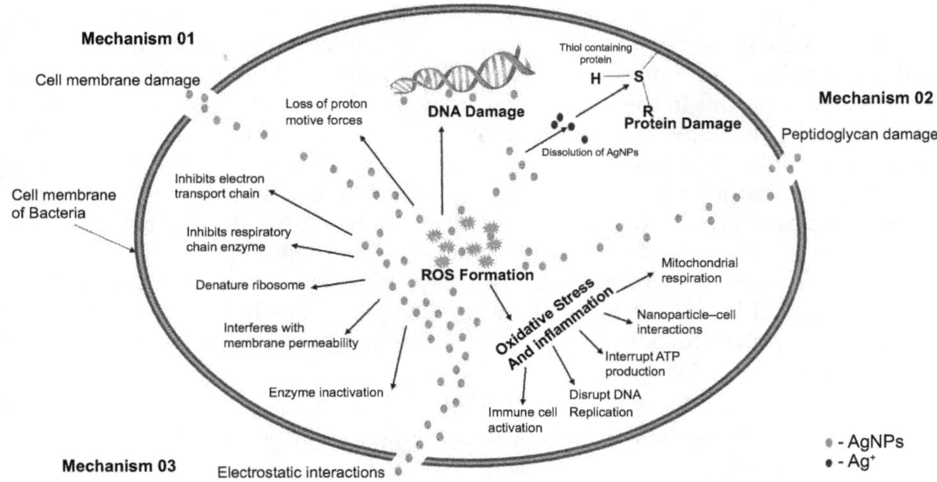

Figure 6.1 Mechanism of antimicrobial activity of AgNPs.

to the cell membrane's permeability, causing cell membrane damage and ultimately cause the cellular contents leakage. Under high levels of stress conditions such as UV light and DNA damage, ROS generation from NPs increases and induces cell damage or death rapidly (Durán et al. 2016; Deshmukh et al. 2019; Kumar et al. 2020; Marambio-Jones and Hoek 2010); (3) direct damage to the cell membrane: Binding to the surface of the cell membrane can drastically interfere with its functions such as permeability, prolong the lag phase of the growth cycle, and increase the generation time of microorganisms and respiration rate, causing the death of the bacterial cells (Durán et al. 2016; Deshmukh et al. 2019; Kumar et al. 2020; Marambio-Jones and Hoek 2010).

The capacity of AgNPs to penetrate the cell membranes and their antimicrobial activity mainly depend on their size and shape (Kumar et al. 2020; Pal et al. 2007; Deshmukh et al. 2019). A recent size-dependent study of NPs <20 nm has demonstrated their excellent binding capability with sulfur and phosphorous functionalized biomolecules of bacteria with maximum permeability of the membrane (Deshmukh et al. 2019; Kumar et al. 2020). Furthermore, AgNPs of <10 nm create pores on the cell wall of bacteria and cause the cytoplasmic leak to govern cell death without interacting with the intracellular and extracellular macromolecules (Deshmukh et al. 2019; Kumar et al. 2020; Pal et al. 2007). At the same time, truncated triangular-shaped AgNPs exhibited higher antibacterial activity against *Escherichia coli* than spherical and rod-shaped NPs (Deshmukh et al. 2019; Kumar et al. 2020). Figure 6.1 illustrates the impact of AgNPs on bacterial cell, either one or a combination of these effects can lead to bacterial death (Panyala et al. 2008; Marambio-Jones and Hoek 2010).

6.2.1 Control of Postharvest Biology of Cut Flowers and Ornamental Plants

The floriculture industry is one of the most profitable sectors in horticulture. Besides, fresh vegetables and fruits, cut flowers, plants, and edible flowers of ornamental plants also have a limited shelf life (Da Silva 2003; Mlcek and Rop 2011). The vase life is primarily controlled by the synthesis of ethylene in the petals and the proliferation of microorganisms (*Pseudomonas fulva*, *Escherichia coli*, and *Enterobacter cloacae*) in the cut stems (Naing and Kim 2020; Hassan et al. 2014). Moreover, the diameter and length of florets and stems changes in fresh weight, petal color, cultivation methods, harvest time, transport condition, and postharvest handling are also identified as factors that affect the quality and longevity of cut flowers and plants (Da Silva 2003; Vehniwal and Abbey 2019). AgNPs applied to the vase solution were found to be useful as an ethylene inhibitor, antimicrobial, antioxidant agent, and stomatal aperture regulator (Rashidiani et al. 2020; Byczyńska 2017; Naing and Kim 2020; Lü et al. 2010). Several studies have been carried out on the use of AgNPs to extend the vase life of cut flowers and ornamental plants as presented in Table 6.1.

Table 6.1 Effect of AgNPs on Postharvest Longevity and Quality of Cut Flowers and Ornamental Plants

Species	*Impact*	*Reference*
Carnation (*Dianthus caryophyllus*)	Reduced ethylene production and bacterial growth and improved antioxidant ability, anthocyanin content, and oxidative stress	Hashemabadi 2014; Rashidiani et al. 2020
Chrysanthemum (*Chrysanthemum morifolium L.*)	• Increased the antibacterial ability, survival, and succulence of cut flowers; reduced stem bacteria colonies and weight loss	Rashidiani et al. 2020
Gerbera (*Gerbera jamesonii*)	• Increased relative fresh weight and water and solution uptake	Solgi et al. 2009
Rose (*Rosa hybrida*)	• Reduced bacterial contamination and improved water balance and vase life	Amingad et al. 2017
Herbaceous Peony (*Paeonia lactiflora Pall.*)	• Increased water uptake and inhibited microbial growth.	Zhao et al. 2018
Gladiolus (*Gladiolus hybridus*)	• Enhanced floret opening rate, daily ornamental value, and water uptake and reduced bacterial colonization, biofilm formation, water loss, and bacterial growth	Li et al. 2017
Tuberose (*Polianthus tuberosa*)	• Increased relative fresh weight, relative water uptake, stem diameter, and chlorophyll content	Bahremand et al. 2014
Gardenias (*Gardenia jasminoides*)	• Enhanced antibacterial properties, water relations, and relative fresh weight	Lin et al. 2019
Alstroemeria (Alstroemeria)	• Decreased bent neck and improved chlorophyll content	Alimoradi et al. 2013

Species	*Impact*	*Reference*
Aspidistra (Aspidistra elatior)	• Increased greenness index of the laminas, fresh weight, and relative solution uptake	Byczyńska 2017
Freesia (Freesia)	• Improved number of opened florets	Hajizadeh 2015
Lily (*Lilium orientalis*)	• Increased solution uptake, initial fresh weight, and decreased bacterial count	Nemati et al. 2013
Tulip (*Tulipa gesneriana*)	• Increased final fresh weight, final stem length, and chlorophyll content	Byczyńska 2017
Chrysanthemum cv. Puma (*Dendranthema grandiflora Tzeleu*)	• Lowered toxicity for flower stalks and the environment and promoted inflorescence opening and increased vase life	Carrillo-López, Morgado-González and Morgado-González 2016
Morus alba (*Morus alba L.*)	• Enhanced the activity of enzymatic and nonenzymatic antioxidants and reduced the accumulation of free radicals and ROS	Das and Mandal 2020
Brassica 'Pusa Jai Kisan' (*Brassica juncea*)	• Increased fresh weight and water uptake	Mehta et al. 2016

Ethylene is a main controlling factor that determines postharvest longevity, changes the cell structure, and increases the ROS content, leading to several physiological changes like diminishing antioxidant enzyme activities and chlorophyll content (Rashidiani et al. 2020; Park et al. 2017; Solgi et al. 2009). Ethylene production is regulated by the genes that encode the enzymes ACS, ACC synthase, and cysteine proteinase (Huang et al. 2007; Vehniwal and Abbey 2019). Consequently, a reduction in ethylene content by AgNPs *via* the suppression of ACS and ACC was reported in carnations, lilies, and rose flowers (Hashemabadi 2014; Rashidiani et al. 2020; Nemati et al. 2013; Hassan et al. 2014). Moreover, significant ethylene inhibition was observed when the vase solution was treated with AgNPs and sucrose, as sucrose can suppress senescence-related genes in carnations and rose cultivar, 'First Red' (Park et al. 2017; Hoeberichts et al. 2007). The incorporation of AgNPs in a vase solution inhibited microbial growth and decreased malondialdehyde content, thereby enhancing the vase life of gerberas (Solgi et al. 2009). Even though AgNPs reduce ethylene production and extended the longevity of cut flowers by suppressing ethylene biosynthesis genes, the possible physiological mechanisms have not yet been fully explored.

Water uptake is also considered another crucial factor in enhancing the vase lifetime as it is required for flower-bud opening and water-balance maintenance in the cut stems (Naing and Kim 2020). Mechanical damage during the harvesting of cut flowers and plants can cause damage in the xylem, leading to microbial contamination, bacterial accumulation, and air embolism in the vascular system (Naing and

Kim 2020; Carrillo-López et al. 2016). The deposition of microbes in the xylem vessel lumen can hinder water uptake, nutrient uptake, and transport, causing water imbalance in the cut flowers and plants (He et al. 2018; Rashidiani et al. 2020; Park et al. 2017; Carrillo-López et al. 2016). Therefore, AgNPs having fascinating bactericidal, fungicidal, and viricidal properties, reduce the cytoplasmic membrane thickness, loosen the cell wall, and cause damage to microbial cells (Carrillo-López et al. 2016; Park et al. 2017).

Furthermore, AgNPs are known to preserve fresh weight and membranes stability, retain chlorophyll content, and reduce weight loss, H_2O_2 content, transpiration rate, and ROS content in gerberas, roses, carnations, freesia, alstroemeria, chrysanthemums, lilies, tuberoses, and tulips (Rashidiani et al. 2020; Byczyńska 2017; Lü et al. 2010; Solgi et al. 2009). Moreover, stem hydraulic conductance suppression, stomatal opening reduction, and transpiration were indicated in the presence of AgNPs (Hassan et al. 2014). It was reported that even at low concentrations, AgNPs increased the vase life, and decreased weight loss, unlike high concentration of AgNPs, which causes wilting due to low water uptake through the stem (Carrillo-López et al. 2016; Byczyńska 2017). The use of AgNPs increased water absorption and fresh weight, by reducing lipid peroxidation in combination with humic acid, extended the vase life in tuberous flower alstroemeria (Carrillo-López et al. 2016). In addition, AgNPs have also been shown to favor inflorescences opening at low concentrations, which is environmentally friendly, decreases the bent neck, and increases the anthocyanin content. Furthermore, it has been reported that AgNPs and Ag^+ mobility in the flower stems is minimal, and therefore, the accumulation of Ag^+ in the vascular system is low (Carrillo-López et al. 2016; Byczyńska 2017).

6.2.2 Edible Coatings

The traditional coatings are nonbiodegradable, carcinogenic, non-eco-friendly and reduce the overall crop quality (Jung and Zhao 2016; Silberbauer and Schmid 2017). Edible coatings or films are prepared from a thin layer of natural polymers, such as polysaccharides, proteins, and lipids, which can be directly consumed. These are inexpensive, biodegradable, and harmless to the environment and are nontoxic if consumed even at sufficient levels (Jung and Zhao 2016; Dhall 2013; Ayhan 2017). Executing different methodologies such as spraying, dipping, brushing, and panning followed by drying can be applied on the food in different ways as shown in Figure 6.2: (1) direct application of liquid solution; (2) coating using molten compounds (Jung and Zhao 2016; Ncama et al. 2018).

Due to biopolymers' hydrophilicity and poor mechanical properties, AgNPs are biocompatible and safe for public health and have been incorporated into the food coating material (Ayhan 2017; Fernández et al. 2010; Gao et al. 2017). AgNP-formulated edible coatings upgrade the barrier performance (O_2, CO_2, moisture); retard respiration, water loss, solute diffusion, microbial contamination, browning, oxidation rate, metabolic rate, and improve physicochemical characteristics, such as strength, mechanical properties, texture, shine or gloss; and suppress decay of the coated products (Ncama et al. 2018; Rai et al. 2019; Yadollahi et al. 2009; Dhall 2013; Flores-López et al. 2016). Recent works on incorporating Ag nanoemulsions into different coating matrices, and their effects on postharvest products are mentioned in Table 6.2.

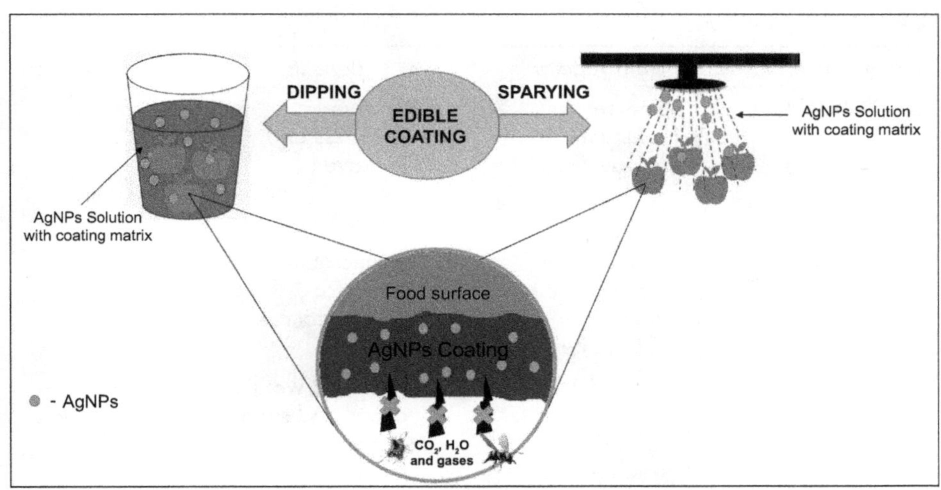

Figure 6.2 The schematic representation of AgNPs coating system applied on a sample fruit.

Table 6.2 AgNP-Incorporated Edible Coatings to Preserve the Postharvest Quality of Fresh Products

Coating material	*Fruit/vegetable*	*Benefits*	*References*
AgNPs–calcium alginate with MMT	Carrots (*Daucus carota subsp. sativus*)	Enhanced shelf life, controlled the microbial growth of *Escherichia coli*, *Staphylococcus aureus*, maintained the sensory quality and reduced decay rates.	Ayhan 2017
AgNP–Cellulose	Fresh-cut melon (*Cucumis melo' Santa Claus*)	Controlled spoilage and enhanced shelf life	Fernández et al. 2010
AgNPs	Cherry tomatoes (*Solanum lycopersicum var.cerasiforme*)	Preserved TSS, vitamin C content, and appearance; reduced the rate of decay and weight loss and resisted external contaminants	Gao et al. 2017

(Continued)

Table 6.2 (Continued)

Coating material	*Fruit/vegetable*	*Benefits*	*References*
AgNPs–MMT	Tomatoes (*Solanum lycopersicum*)	Preserved freshness and improved sensory qualities and shelf life	Ijaz et al. 2020
AgNPs–Agar	Apples (*Pyrus malus*), lime and oranges (*Citrussinensis*)	Antimicrobial agent for *Staphylococcus aureus*	Gudadhe et al. 2013
AgNPs–alginic acid	Carrots (*Daucus carota subsp. sativus*) and pears (*Pyrus*)	Controlled weight decay and sugar content	Mohammed Fayaz et al. 2009
AgNPs–polyvinyl pyrrolidone	Asparagus (*Asparagus officinalis L*)	Reduced weight loss, ascorbic acid content, color change; inhibited the increase of tissue firmness and exhibited antifungal activity against yeast and molds	An et al. 2008
AgNPs–Chitosan and glycol	Red grapes (*Vitis vinifera*)	Surface protection, antibacterial agent for *Escherichia coli* and *Staphylococcus aureus*, and reduced weight loss.	Mohammed Fayaz et al. 2009
AgNPs–Chitosan	Mango (*Mangifera indica*)	Reduced postharvest losses, antifungal agents against *Colletotrichum gloeosporioides*, yeast, and molds.	Chowdappa et al. 2014; Carbone et al. 2016

Coating material	*Fruit/vegetable*	*Benefits*	*References*
AgNPs	Citrus fruit (*Aurantioideae Citrus*)	Exhibited antifungal activity toward *Penicillium digitatum* and *Penicillium italicum*	Chen et al. 2019
AgNPs–sodium alginate	Fresh fruits and vegetables	Showed antibacterial activity for *Escherichia coli, Staphylococcus aureus*, and surface sterilization	Mohammed Fayaz et al. 2009
Alginate–AgNPs	Shiitake mushroom (*Lentinus edodes*)	Inhibited mushroom weight loss, softening, browning, lowered mesophilic, psychrophilic, pseudomonad, yeasts, and mold microbial count	Jiang et al. 2013

6.2.3 Active Food Packaging

Active packaging is focused on protecting food products from microbial contamination and deterioration (Rai et al. 2019). Hosting of AgNPs into the polymer matrix can be achieved through different strategies like coating, absorbing, and directly incorporating an antimicrobial pad/sachet into the synthesizing process (Llorens et al. 2012). Various types of matrices are used to host the AgNPs, namely, non-degradable (polyethylene, PVC, LDPE, PVA, EVOH) and biodegradable polymers (cellulose, starch, chitosan, alginate, and agarose; Carbone et al. 2016; De Paiva et al. 2008). Doping AgNPs into polymer matrix enhanced the shelf life and properties of packaging materials, such as polymer flexibility, mechanical strength, resistance to heat, reduced weight, moisture/humidity barrier, gas barrier properties against CO_2, moisture, O_2, and volatiles present inside the food (Amini and Azadfallah 2016; Carbone et al. 2016; Ray and Okamoto 2003; De Paiva et al. 2008). Furthermore, it offers surplus function besides protecting by releasing/absorbing food additives like minerals, antioxidants, preservatives, flavorings, probiotics, and vitamins into the food (Honarvar et al. 2016). Packaging materials containing AgNPs could inhibit the production of free radicals, regulate the air components of the storage micro-environment, remove ethylene, inhibit microbial growth, and promote product quality and shelf life (Sharma et al. 2019; Rai et al. 2019; Llorens et al. 2012). The developed

AgNP-based films completely inhibit the growth of *Escherichia coli*, *Staphylococcus aureus*, *Bacillus subtilis*, *Enterococcus faecalis*, *Staphylococcus epidermidis*, *Salmonella typhomurim*, and *Candida albican* (Carbone et al. 2016). Furthermore, AgNPs can also inhibit the growth of fungi and their aflatoxin production by disturbing cellular functions, which cause deformation in fungal hyphae of *Aspergillus niger*, *Alternaria alternata*, *Alternaria brassicicola*, *Alternaria solani*, *Botrytis cinerea*, *Fusarium oxysporum*, *Penicillium* spp., *Scleronitia homoeocarpa*, *Rhizoctonia solani*, and *Colletotrichum* spp. (Sardella et al. 2019; Li et al. 2017). Noteworthy antifungal activity of AgNP-incorporated pullulan against *Aspergillus niger* was also reported to reduce postharvest losses of one highly perishable vegetable, onions (Ruchika 2015). A combination of AgNPs and antifungal agents like fluconazole can increase the antifungal effectiveness of AgNPs, causing a decrease in spore number, abnormality, and hypertrophy of *Sclerotium cepivorum* (Ruchika 2015). In addition, AgNPs demonstrated anti-browning activity on enzymatic browning in cabbage (*Brassica oleracea L. capitata*) by preventing polyphenol oxidase activity (Khan et al. 2016). According to Jiang et al. (2013) and Akther et al. (2019), AgNPs behave as an ethylene absorber, which reduces the ethylene content of the storage environment and delays the ripening process of fruits and vegetables. AgNPs containing Polylactic acid (PLA), BEO, and TiO_2NPs displayed slower ripening rates with extended shelf lives when used on fresh mangoes (Li et al. 2017). Ag-zeolite (AgNPs/SiO_2) composite–based antimicrobial polymers are stable and release Ag^+ into stored foods, resulting in continuous antimicrobial activity (Becaro et al. 2016). Thus, nanopolymer composite–based AgNPs may find broad application in antimicrobial food packages and improve the postharvest shelf life of food. A summary of the selected publications on active packaging systems are stated in Table 6.3.

Table 6.3 AgNP-Incorporated Active Food Packaging Materials

Polymer matrix	*Tested food*	*Tested microorganisms*	*References*
AgNPs–Cellulose	Fresh-cut melon	Antimicrobial activity against total mesophilic aerobic bacteria, psychotropic bacteria, yeasts, and molds	Fernández et al. 2010
AgNPs–MMT–Alginate	Fresh-cut melon	Improved sensory qualities, shelf life to about 11 days, and inhibited mesophilic and psychotropic bacteria	Danza et al. 2017
AgNPs–LDPE–ZnONPs	Orange juice	Reduced *Lactobacillus plantarum*, yeast, mold, total aerobic bacterial count, and enhanced shelf life up to 112 days	Polet et al. 2020; Emamifar et al. 2011
AgNPs–LDPE	Barberry	Antibacterial activity against total aerobic bacteria	Carbone et al. 2016
AgNPs–LDPE–TiO_2NPs–Kaolin	Strawberry	Enhanced shelf life	Meyer et al. 2010

Polymer matrix	*Tested food*	*Tested microorganisms*	*References*
AgNPs–EVOH	Apple	Inhibited *Salmonella spp.* and *Listeria monocytogenes*	Carbone et al. 2016
AgNPs–LDPE–TiO_2NPs	Apple, carrot	Inhibited *Penicillium, Lactobacillus, Escherichia coli, Staphylococcus aureus*, and *Listeria monocytogenes*	Metak et al. 2015
AgNPs–Sodium alginate	Pears, carrots	Inhibited *Escherichia coli* and *Staphylococcus aureus*	Fayaz et al. 2009
AgNPs–Calcium alginate–MMT	Fresh-cut carrots	Enhanced shelf life	Costa et al. 2012
AgNPs–TiO_2NPs	Chinese bayberry	Inhibited fruit respiration and decay	Wang et al. 2010
AgNPs–SiO_2–TiO_2	Fresh cut carrot	Reduced antimicrobial activity	Becaro et al. 2016
AgNPs–Chitosan	Fresh-cut melon	Decreased the rate of respiration and production of ethylene and maintained sensory qualities up to 13 days	Kumar et al. 2020
AgNPs–PLA–BEO–TiO_2NPs	Mango	Increased physicochemical, sensory quality, delayed ripening, and retarded the microorganism growth on mango pericarp	Li et al. 2017
AgNPs–Sodium alginate	Carrot, pears	Exhibited antibacterial activity against *Escherichia coli* and *Staphylococcus aureus* and minimized weight loss	Fayaz et al. 2009
AgNPs–Polyethylene–TiO_2NPs	Rice	Inhibited *Aspergillus flavus* and reduced the mildew of rice during storage, enhanced storage quality and shelf life	Li et al. 2017
AgNPs–Cellulose	–	Reduced citrate production and *Escherichia coli* growth	Tankhiwale and Bajpai 2012

(Continued)

Table 6.3 (Continued)

Polymer matrix	*Tested food*	*Tested microorganisms*	*References*
AgNPs–Gelatin	–	Exhibited strong antibacterial activity against foodborne pathogens *Salmonella typhimurium* and *Bacillus cereus*	Kanmani and Rhim 2014
AgNPs–Carboxy-methylcellulose	–	Lowered water vapor permeability and improved quality and shelf life of the product and inhibited *Enterococcus faecalis* and *Escherichia coli*	De Siqueira et al. 2014
AgNPs–Hydroxypropyl methylcellulose	–	Increased good mechanical and barrier properties and tensile strength of the film, decreased water vapor values and bacterial count, and improved product quality and shelf life	De Moura et al. 2012
AgNPs–PVA–Bacterial cellulose nanocrystals	–	Exhibited higher oxygen barrier capacity with slightly improved water vapor permeability and antimicrobial activities against *Escherichia coli*	Wang et al. 2019
AgNPs–LDPE	–	Exhibited antimicrobial activity against *Escherichia coli* and *Staphylococcus aureus*	Jokar et al. 2012

6.2.4 Intelligent Food Packaging

Traditional packaging systems act as a passive barrier against air, dust, and moisture. Their impotence toward food safety and bioterrorism has led to the development of intelligent packaging systems that sense and monitor any biological or chemical changes in packaged food or the surrounding environment to amplify product shelf life (Mihindukulasuriya and Lim 2014; Realini and Marcos 2014). To compensate for this process, nanosensors have been developed with MNPs in conjunction with polymers to detect pathogenic bacteria, food contaminating toxins, adulterants, vitamins, dyes, fertilizers, pesticides, taste and smell, and transit processes in smart packaging (Yadollahi et al. 2009; Manjunatha et al. 2016). Additionally, smart packaging ensures the integrity of the food package and the authenticity of the food product to consumers' fresh-food perceptions. Long-term storage of postharvest crops leads to the degradation of organic compounds, vitamins, and other antioxidant components (Kumar et al. 2017). The quality of the stored food can be measured by identifying the level of glucose, sucrose, fructose, D-sorbitol, L-malic acid, citric acid, succinic acid, H_2O_2, L-glutamic acid, and alcohol (Xu et al. 2017). Several recent reports indicated the use of AgNP-doped smart packaging devices like barcodes, time-temperature indicators, ripeness indicators, gas indicators, and biosensors to sense the quality of food components to detect toxins and food pathogens in the packaging systems (Pathakoti et al. 2017; Kumar et al. 2020). Metal-oxide NPs are widely used as gas sensors to detect H_2O_2, acetone, ethanol, and others. It is reported that tin oxide–based gas sensors decorated with AgNPs improve sensing properties by facilitating the fast electron transfer and catalyzing the oxidation of reducing gas molecules (Xu et al. 2017). Furthermore, AgNP-embedded ZnO nanorods enhanced ethanol's gas-sensor performance almost three times when compared with traditionally doped ZnO or pure ZnO (Xiang et al. 2011). This sensor shows long-term stability and selectivity with no apparent degradation after being exposed to ethanol vapor of 30 ppm for 100 d (Xiang et al. 2011). An ultrasensitive, selective, and fast-response ethanol sensor based on Ag-TiO_2 core–shell NPs has also been reported (Zhu et al. 2014). Furthermore, TiO_2 coated on the Ag nanocore increased ethanol's sensory responses at 0.09 ppm lower detection limit with high selectivity at room temperature. Cyclic voltammetry experiments depict that AgNPs facilitate H_2O_2 reduction, and many researchers have developed H_2O_2 sensors as it is an essential mediator in food.

The fabrication of an amperometric glucose biosensor based on Ag nanowires and chitosan-glucose oxidase film revealed excellent selectivity, low detection limit, fast response, good stability, good biocompatibility, and high mechanical strength, and no interference from various amino acid and acetic acids (Rhim et al. 2013). Amperometric glucose biosensors have excellent performance in detecting food adulterants, which is attributed to high conductivity, large surface-to-volume ratio, and good biocompatibility (Rhim et al. 2013).

In addition to AgNPs, Ag-Au nanocomposites are also used to detect the quality of semi-perishable horticultural crops during storage, pesticides (organophosphates), and fertilizers (melamine) using colorimetric, fluorescence, and UV-visible spectroscopy (Zhao et al. 2013; Mihindukulasuriya and Lim 2014). Colorimetric sensors based on LSPR of PVA-capped AgNPs have been developed to evaluate the catalytic ability of AgNPs. MNPs facilitate different colors depending on size, shape, inter-particle distance, and the local environment of the noble MNPs. In the presence of heat or light, AgNPs undergo degradation while decomposing adulterants, which causes an alteration to the AgNPs' original color due to a remarkable change in the LSPR absorbance strength measured by UV–visible (UV-Vis) spectroscopy. Thus, colorimetric sensors have potential applications in medical, food science, and environmental monitoring as renewable, simplified, and low-cost sensors (Filippo et al. 2009).

Moreover, yellow AgNPs have been used to develop a visual sensor to identify healthy onions from spoiled onions. Yellow AgNPs display selectivity and specificity toward volatile sulfur compounds released from spoiled onions, which could be determined by using UV-Vis spectroscopy and colorimetric analysis (Sachdev et al. 2016). Colorimetric sensors are used to detect spoilage in fruits, vegetables, and hydrogen sulfide levels; besides, AgNPs can also be used as an anti-odorant agent (Welch et al. 2005). These types of gas nanosensors are essential and have practical advantages over the traditional-doping methods to sense the postharvest quality of crops.

AgNPs are used to prepare flexible transparent conductive films, display devices, touch sensors, solar cells, organic photovoltaic devices, and light-emitting diodes due to their low sheet resistance and flexibility (Menamparambath et al. 2015). AgNPs are utilized to prepare the nano-barcodes in the food product in the form of strips or labels to determine the food adulterants and toxins. NP-based invisible and sophisticated encoding systems and nanodisc codes have been used to verify food products' originality (Bhattacharya et al. 2006; Birtwell et al. 2008). Invisible nano-barcodes gauge brand protection, product authenticity, and improved food safety (Jaiswal et al. 2019). Time and temperature control are essential as fresh crops generally perish due to storage longer than shelf life, storage time and temperature, and exposure to O_2 and moisture (Mihindukulasuriya and Lim 2014). Currently, AgNP-mediated food-freshness quality indicators, temperature indicators, and TTIs have been developed for suppliers or consumers to identify the freshness of crop products, whether the packaging has been breached, or whether the product has been kept at the appropriate temperatures throughout the supply chain (Kumar et al. 2017; Tan et al. 2012). AgNPs, having high thermal conductivity, increase the sensitivity and change the color of polydiacetylene/AgNPs embedded in carboxymethylcellulose TTI films, with the change in temperature over time (Saenjaiban et al. 2020). Ag-Au nanodiscs and Ag-coated Si nanowires have been developed to confirm product authenticity, stability, and selectivity (An et al. 2008; Shi et al. 2018). The optimized Ag-coated Si nanowire arrays exhibit great potential for molecular sensing at lower detection concentrations of molecules related to food-contaminating toxins and *Bacillus anthracis* spores (An et al. 2008). AgNPs and Ag^+ can also diffuse from packages to the food surface, and therefore, an analysis of Ag^+ is essential. In recent years, cytosine–cytosine pairs are most widely used due to their high sensitivity and selectivity toward Ag^+ but unable in situ detection of DNAzyme in complex systems. Therefore, SERS nanosensors were developed based on 2-mercaptoisonicotinic acid-functionalized AuNPs to analyze AgNPs and Ag^+. This sensor exhibits a detection limit of 25 nM, and it shows higher sensitivity and selectivity toward the detection of Ag^+ in an aqueous solution against other metal ions (Kumar et al. 2017; Tan et al. 2012).

6.3 Toxicity and Safety Regulatory Issues of AgNPs

6.3.1 Evidence of Toxicity for Mammalian Cells

Enhancing food shelf life by maintaining its quality, freshness, and safety and reducing postharvest losses could be achieved by adding antimicrobial additives with the advent of AgNPs (Cushen et al. 2014; Becaro et al. 2016). Despite the capability of NPs to restructure the food industry, such as the incorporation of AgNPs, to develop smart and active packaging (which improves food safety and quality, storage, transportation, etc.), it is also essential to consider consumer safety, environmental, ethical policy, and regulations (Šimon et al. 2008).

NPs can enter the human body through different pathways, such as inhalation (lung), ingestion (oral), and dermal (skin). Recycling and disposing of MNPs contained in packaging, cut flowers, and ornamental plants enable respiratory and dermal exposure (Ferdous and Nemmar 2020). The potential risk of MNPs migrating from food packaging to postharvest products allows oral exposure, and it is exceptionally undesirable when it comes to food packaging safety (Becaro et al. 2016). Exposure to AgNPs induces local effects, such as inflammation and oxidative stress, and these can penetrate through the gastric mucosal barrier, skin barrier, and lung epithelial barrier, causing translocation of AgNPs into blood and lymphatics (Ferdous and Nemmar 2020; Korani et al. 2015). These accumulate in secondary organs such as the liver, brain, lungs, kidneys, and spleen (Šimon et al. 2008; Oberdörster et al. 2005; Korani et al. 2015).

Migration of AgNPs from packaging to food occurs between very small NPs with an order of 1-nm radius has been reported (Šimon et al. 2008). Relatively low dynamic viscosity polymer matrices that do not interact with NPs, such as Ag with polyolefins, could cause migration. Thus, migration is a function of the size of AgNPs and polymer material, and Cushen et al. (2014) proved that there is no significant effect of time and temperature on the migration of AgNPs to the products. A lack of migration data causes considerable restrictions to determine the risk of consumption of postharvest products packed in AgNP-composed materials and the availability of analytical methods to detect and quantify the NPs are also limited (e.g., AAS, ICP-MS; Šimon et al. 2008; Becaro et al. 2016). In the presence of food matrices and lower concentrations of AgNPs, currently available methods also fail to determine the size distribution, shape, and form of the AgNPs (Šimon et al. 2008). According to US safety guidelines and the European Food Safety Authority (2005), the upper limit of Ag^+ in food matrices is 0.05 mg of Ag per kg and 0.05 mg per L in water (Becaro et al. 2016). The upper limit for plants depends on the total reduction ability for Ag's reduction potential, which is about 0.35 wt.% Ag by dry plant weight (Haverkamp and Marshall 2009). Increment in these values will cause significant harm to humans and other living forms.

A lack of knowledge on the effects of NPs on human health and the environment is a significant concern. Nanomaterials and their respective bulk materials have different physicochemical and biological properties, and hence, the impact of nanomaterials on human health and the environment is much more severe when compared to the bulk material (Oberdörster et al. 2005; Šimon et al. 2008). It is shown that NPs harm the biological systems as they can easily penetrate cellular barriers and cause oxidative damage to the cell by inducing radical oxygen generation (Šimon et al. 2008; Hashemabadi 2014). With the advent of nanotechnology, additional health risks have been initiated due to the high toxicity of NPs when compared with bulk materials, due to the increased surface area per unit mass leading to the adsorption of large biological molecules (Honarvar et al. 2016; Šimon et al. 2008). AgNPs diffuse through the chronic membrane and interact with the embryo tissues can cause the accumulation of AgNPs inside the nuclei, which could damage the DNA and thereby fetal growth damages (Li et al. 2010; Ribeiro et al. 2014). The health effects of AgNPs on different secondary organs are summarized in Table 6.4.

Table 6.4 Health Effects of AgNPs on Different Secondary Organs

Organ	*Health effects*	*Reference*
Kidney	Mitochondrial damage and glomerular shrinkage	AshaRani et al. 2009; Ferdous and Nemmar 2020
Lungs	Decrease lung function	AshaRani et al. 2009
Brain (via blood-brain barrier)	Neurodegeneration	Ferdous and Nemmar 2020
Heart	Exacerbates cardiac-ischemic reperfusion injury and DNA damage	Korani et al. 2015; Ferdous and Nemmar 2020
Spleen	Splenic congestion	Korani et al. 2015
Liver	Hepatocellular damage	AshaRani et al. 2009
Thymus	Apoptosis	Ferdous and Nemmar 2020
Skin	Agyria	Ferdous and Nemmar 2020
Stomach	Damage to gastrointestinal linings	Hoffmanna et al. 2019
Testis (via blood-testis barrier)	Azoospermia	Haldar et al. 2013
Systemic circulation	Prothrombotic effect	Ferdous and Nemmar 2020

6.3.2 Ecotoxicological Risks

The environmental effects of AgNPs have been demonstrated through experiments with *Daphina magna*, *Daphina pulex*, and *Ceriodaphnia dubai*, which are some of the standard test models used for ecotoxicological analysis, approved by the US Environmental Protection Agency, the Organisation for Economic Co-operation and Development, and the American Society for Testing and Materials (Li et al. 2010). The toxicity of AgNPs is a function of coating material, shape, crystallinity, and hydrodynamic diameter (size; Poynton et al. 2012; Marambio-Jones and Hoek 2010). The dissolution of AgNPs and the release Ag^+ depend on the pH, temperature, the concentration of NPs, and the coating material (Ribeiro et al. 2014). Furthermore, the toxicity of AgNPs is due to the release of Ag^+, and Li et al. (2010) state that the LC_{50} for Ag^+ is 2 μg per L in 48 h, and therefore, AgNPs are considered to be one of the most toxic NPs (Ribeiro et al. 2014). Dissolved Ag^+ constrains sodium influx in organisms, leads to mortality, and causes embryonic problems (Li et al. 2010). Also, polymer stability is an important factor that determines the leaching of Ag^+ and AgNPs to be zero or minimal. Even though different coatings have been used to increase the strength of AgNPs, determining the toxicity of coating materials is essential.

The leaching of Ag^+ from AgNP-incorporated packaging materials into the environment causes the accumulation of Ag^+ in water streams, aquifers, soil, and sediments (Oberdörster et al. 2005). Plants and aquatic lives can uptake free Ag^+ and

AgNPs, leading to translocation and accumulation and ultimately entering the food chain cause health effects to humans and other living forms (Yan and Chen 2019). Therefore, both Ag^+ and AgNPs can be toxic to plants, humans, and the environment (Yan and Chen 2019).

6.3.3 Phytotoxicity

Evaluating the phytotoxicity of Ag^+ and AgNPs toward plants is carried out by measuring parameters of growth, seed germination, biomass, and leaf surface area (Dietz and Herth 2011). The accumulation of Ag^+ and AgNPs in plants can affect nutrient intercellular transport and further causes a reduction in seed germination, root growth, and biomass, damages to plant hormones, and plant morphology (Yan and Chen 2019; Dietz and Herth 2011). This reduces water and nutrient uptake, inhibiting plant growth by affecting chlorophyll synthesis and photosynthesis and, ultimately, decreasing crop harvest (Alimoradi et al. 2013). Li et al. (2010) state that Ag^+ acts as photosynthesis inhibitor in the green algae *Chlamydomonas reinhardtii*. All these effects depend on the size and concentration of AgNPs used and the type of plant investigated (Yan and Chen 2019). As AgNPs and Ag^+ are released from coating and packaging materials, appropriate steps should be taken to reduce or prevent this. Yan and Chen (2019) have listed the impacts of AgNPs of different sizes and doses using various models. The entry of Ag^+ into the water stream can cause water pollution, leading to toxicity and undesirable conditions for aquatic plants and animals (Oberdörster et al. 2005).

6.4 Conclusion

This chapter highlights the application of AgNPs in postharvest management technologies to extend the crop shelf life by improving performance properties, quality, and safety of fruits, vegetables, and floricultural crops. Furthermore, AgNPs have quite beneficial effects on physicochemical and physiological quality, processing, and industrial feasibility in contrast with traditional management techniques, which will ultimately benefit both the producers and consumers. AgNPs are fascinate biocidal agents over a broad spectrum of microorganisms, such as *Escherichia coli*, *Staphylococcus aureus*, *Bacillus subtilis*, *Enterobacter cloacae*, *Aspergillus niger*, and *Pseudomonas* spp., among others. Hence, AgNP-embedded packaging and coatings matrices can be used effectively to improve the preservation qualities of valuable floricultural and horticultural crops over extended storage. Moreover, the large potential of nanosensors to detect the trace numbers of contaminants is also being explored. Besides, it was identified that AgNPs could reduce ethylene production by lowering the ethylene biosynthesis and thereby increasing vase life and acting as an antioxidant, anti-ripening agent, gas barrier, and additive to maintain the original taste and flavor. Overall, studies confirm that AgNPs are a promising tool in postharvest management, although there are still risks associated with the toxicity of AgNPs that need to be fully understood. Nonetheless, Ag nanocomposites appear to provide superior economic and environmental benefits to the medical, agricultural, food, and horticultural industries.

References

Akther, Tahira, Vabeiryureilai Mathipi, Nachimuthu Senthil Kumar, MubarakAli Davoodbasha, and Hemalatha Srinivasan. 2019. Fungal-mediated synthesis of pharmaceutically active silver nanoparticles and anticancer property against A549 cells through apoptosis. *Environmental Science and Pollution Research* 26 (13):13649–13657.

Alimoradi, Meysam, Mehrdad Jafararpoor, and Ahmadreza Golparvar. 2013. Improving the keeping quality and vase life of cut Alstroemeria flowers by post-harvest nano silver treatments. *International Journal of Agriculture and Crop Sciences (IJACS)* 6 (11):632–635.

Amingad, Varun, KN Sreenivas, B Fakrudin, GK Seetharamu, TH Shankarappa, and R Venugopalan. 2017. Comparison of silver nanoparticles and other metal nanoparticles on postharvest attributes and bacterial load in cut roses var. Taj Mahal. *International Journal of Pure & Applied Bioscience* 5:579–584.

Amini, Elahe, and Mohammad Azadfallah. 2016. Mohammad Layeghi & Reza Talaei-Hassanloui. *Cellulose* 23:557–570.

An, Jianshen, Min Zhang, Shaojin Wang, and Juming Tang. 2008. Physical, chemical and microbiological changes in stored green asparagus spears as affected by coating of silver nanoparticles-PVP. *LWT–Food Science and Technology* 41 (6):1100–1107.

AshaRani, PV, Grace Low Kah Mun, Manoor Prakash Hande, and Suresh Valiyaveettil. 2009. Cytotoxicity and genotoxicity of silver nanoparticles in human cells. *ACS Nano* 3 (2):279–290.

Ayhan, Zehra. 2017. Packaging and preservation methods of minimally processed produce. *Minimally Processed Refrigerated Fruits and Vegetables*:239–268.

Bahremand, Sonia, Jamshid Razmjoo, and Homaun Farahmand. 2014. Effects of nano-silver and sucrose applications on cut flower longevity and quality of tuberose (Polianthus tuberosa). *International Journal of Horticultural Science and Technology* 1 (1):67–77.

Becaro, Aline A, Fernanda C Puti, Alan R Panosso, Juliana C. Gern, Humberto M. Brandão, Daniel S. Correa, and Marcos D. Ferreira. 2016. Postharvest quality of fresh-cut carrots packaged in plastic films containing silver nanoparticles. *Food and Bioprocess Technology* 9 (4):637–649.

Bhattacharya, Arunabh, Jameela Banu, Mizanur Rahman, Jennifer Causey, and Gabriel Fernandes. 2006. Biological effects of conjugated linoleic acids in health and disease. *The Journal of Nutritional Biochemistry* 17 (12):789–810.

Birtwell, SW, GS Galitonov, H Morgan, and NI Zheludev. 2008. Superimposed nano-structured diffraction gratings as high capacity barcodes for biological and chemical applications. *Optics Communications* 281 (7):1789–1795.

Byczyńska, Andżelika. 2017. Nano-silver as a potential biostimulant for plant—a review. *World Scientific News* 86 (3):180–192.

Carbone, Marilena, Domenica Tommasa Donia, Gianfranco Sabbatella, and Riccarda Antiochia. 2016. Silver nanoparticles in polymeric matrices for fresh food packaging. *Journal of King Saud University-Science* 28 (4):273–279.

Carrillo-López, Luis M, Antonio Morgado-González, and Aurora Morgado-González. 2016. Biosynthesized silver nanoparticles used in preservative solutions for Chrysanthemum cv. Puma. *Journal of Nanomaterials* 1769250:1–10.

Chen, Jinyin, Yuting Shen, Chuying Chen, and Chunpeng Wan. 2019. Inhibition of key citrus postharvest fungal strains by plant extracts in vitro and in vivo: A review. *Plants* 8 (2):26.

Chowdappa, P, Shivakumar Gowda, CS Chethana, and S Madhura. 2014. Antifungal activity of chitosan-silver nanoparticle composite against Colletotrichum gloeosporioides associated with mango anthracnose. *African Journal of Microbiology Research* 8 (17):1803–1812.

Costa, C, Amalia Conte, GG Buonocore, M Lavorgna, and Matteo Alessandro Del Nobile. 2012. Calcium-alginate coating loaded with silver-montmorillonite nanoparticles to prolong the shelf-life of fresh-cut carrots. *Food Research International* 48 (1):164–169.

Cushen, M, J Kerry, M Morris, M Cruz-Romero, and E Cummins. 2014. Evaluation and simulation of silver and copper nanoparticle migration from polyethylene nanocomposites to food and an associated exposure assessment. *Journal of Agricultural and Food Chemistry* 62 (6):1403–1411.

Danza, Alessandra, Amalia Conte, Marcella Mastromatteo, and Matteo Alessandro Del Nobile. 2017. Preservation techniques to extent the shelf life of fresh-cut melon. *Journal of Food Processing and Preservation* 41 (3):e12906.

Das, Dipayan, and Palash Mandal. 2020. Use of biogenic silver nanoparticles in enhancing shelf life of Morus alba L. at post harvest stage. *Scientific Reports* 10 (1):1–18.

De Azeredo, HMC. 2013. Antimicrobial nanostructures in food packaging. *Trends in Food Science & Technology* 30 (1):56–69.

De Moura, Marcia R, Luiz HC Mattoso, and Valtencir Zucolotto. 2012. Development of cellulose-based bactericidal nanocomposites containing silver nanoparticles and their use as active food packaging. *Journal of Food Engineering* 109 (3):520–524.

De Paiva, Lucilene Betega, Ana Rita Morales, and Francisco R Valenzuela Díaz. 2008. Organoclays: Properties, preparation and applications. *Applied Clay Science* 42 (1–2):8–24.

De Siqueira, Alexandre Fioravante, Wagner Massayuki Nakasuga, Aylton Pagamisse, Carlos Alberto Tello Saenz, and Aldo Eloizo Job. 2014. An automatic method for segmentation of fission tracks in epidote crystal photomicrographs. *Computers & Geosciences* 69:55–61.

Deshmukh, SP, SM Patil, SB Mullani, and SD Delekar. 2019. Silver nanoparticles as an effective disinfectant: A review. *Materials Science and Engineering: C* 97:954–965.

Dhall, RK. 2013. Advances in edible coatings for fresh fruits and vegetables: A review. *Critical Reviews in Food Science and Nutrition* 53 (5):435–450.

Dietz, Karl-Josef, and Simone Herth. 2011. Plant nanotoxicology. *Trends in Plant Science* 16 (11):582–589.

Durán, Nelson, Marcela Durán, Marcelo Bispo De Jesus, Amedea B Seabra, Wagner J Fávaro, and Gerson Nakazato. 2016. Silver nanoparticles: A new view on mechanistic aspects on antimicrobial activity. *Nanomedicine: Nanotechnology, Biology and Medicine* 12 (3):789–799.

Emamifar, Aryou, Mahdi Kadivar, Mohammad Shahedi, and Sabihe Soleimanian-Zad. 2011. Effect of nanocomposite packaging containing Ag and ZnO on inactivation of Lactobacillus plantarum in orange juice. *Food Control* 22 (3–4):408–413.

European Food Safety Authority. 2005. Opinion of the Scientific Committee on a request from EFSA related to a harmonised approach for risk assessment of substances which are both genotoxic and carcinogenic. *EFSA Journal* 3 (10):282.

Fayaz, A Mohammed, K Balaji, PT Kalaichelvan, and R Venkatesan. 2009. Fungal based synthesis of silver nanoparticles—an effect of temperature on the size of particles. *Colloids and Surfaces B: Biointerfaces* 74 (1):123–126.

Ferdous, Zannatul, and Abderrahim Nemmar. 2020. Health impact of silver nanoparticles: A review of the biodistribution and toxicity following various routes of exposure. *International Journal of Molecular Sciences* 21 (7):2375.

Fernández, Avelina, Pierre Picouet, and Elsa Lloret. 2010. Cellulose-silver nanoparticle hybrid materials to control spoilage-related microflora in absorbent pads located in trays of fresh-cut melon. *International Journal of Food Microbiology* 142 (1–2):222–228.

Filippo, E, A Serra, and D Manno. 2009. Poly (vinyl alcohol) capped silver nanoparticles as localized surface plasmon resonance-based hydrogen peroxide sensor. *Sensors and Actuators B: Chemical* 138 (2):625–630.

Flores-López, María L, Miguel A Cerqueira, Diana Jasso de Rodríguez, and António A Vicente. 2016. Perspectives on utilization of edible coatings and nano-laminate coatings for extension of postharvest storage of fruits and vegetables. *Food Engineering Reviews* 8 (3):292–305.

Gao, Li, Qinqin Li, Yinghu Zhao, Haifang Wang, Yaqing Liu, Youyi Sun, Fang Wang, Wanli Jia, and Xiaodong Hou. 2017. Silver nanoparticles biologically synthesised using tea leaf extracts and their use for extension of fruit shelf life. *IET Nanobiotechnology* 11 (6):637–643.

Gudadhe, Janhavi A, Alka Yadav, Aniket Gade, Priscyla D Marcato, Nelson Durán, and Mahendra Rai. 2013. Preparation of an agar-silver nanoparticles (A-AgNp) film for increasing the shelf-life of fruits. *IET Nanobiotechnology* 8 (4):190–195.

Hajizadeh, HS. 2015. The study of freesia (*Freesia spp.*) cut flowers quality in relation with nano silver in preservative solutions. *III International Conference on Quality Management in Supply Chains of Ornamentals* 1131:1–10.

Haldar, Koyel Mallick, Basudeb Haldar, and Goutam Chandra. 2013. Fabrication, characterization and mosquito larvicidal bioassay of silver nanoparticles synthesized from aqueous fruit extract of putranjiva, Drypetes roxburghii (Wall.). *Parasitology Research* 112 (4):1451–1459.

Hashemabadi, Davood. 2014. The role of silver nano-particles and silver thiosulfate on the longevity of cut carnation (Dianthus caryophyllus) flowers. *Journal of Environmental Biology* 35 (4):661.

Hassan, FAS, EF Ali, and B El-Deeb. 2014. Improvement of postharvest quality of cut rose cv.'First Red' by biologically synthesized silver nanoparticles. *Scientia Horticulturae* 179:340–348.

Haverkamp, RG, and AT Marshall. 2009. The mechanism of metal nanoparticle formation in plants: Limits on accumulation. *Journal of Nanoparticle Research* 11 (6):1453–1463.

He, Yijia, Lichao Qian, Xu Liu, Ruirui Hu, Meirong Huang, Yule Liu, Guoqiang Chen, Dusan Losic, and Hongwei Zhu. 2018. Graphene oxide as an antimicrobial agent can extend the vase life of cut flowers. *Nano Research* 11 (11):6010–6022.

Hoeberichts, Frank A, Wouter G Van Doorn, Oscar Vorst, Robert D Hall, and Monique F Van Wordragen. 2007. Sucrose prevents up-regulation of senescence-associated genes in carnation petals. *Journal of Experimental Botany* 58 (11):2873–2885.

Hoffmanna, Tuany Gabriela, Daniel Amaral Petersa, Betina Louise Angiolettia, Sávio Leandro Bertolia, Leonardo Vieira Péresb, Mercedes Gabriela Ratto Reiterc, and Carolina Krebs de Souzaa. 2019. Potentials nanocomposites in food packaging. *Chemical Engineering* 75:253–258.

Honarvar, Zohreh, Zahra Hadian, and Morteza Mashayekh. 2016. Nanocomposites in food packaging applications and their risk assessment for health. *Electronic Physician* 8 (6):2531.

Huang, Jiale, Qingbiao Li, Daohua Sun, Yinghua Lu, Yuanbo Su, Xin Yang, Huixuan Wang, Yuanpeng Wang, Wenyao Shao, Ning He, Jinqing Hong, and Cuixue Chen. 2007. Biosynthesis of silver and gold nanoparticles by novel sundried *Cinnamomum camphora* leaf. *Nanotechnology* 18 (10):105104.

Ijaz, Mohsin, Maria Zafar, Sumera Afsheen, and Tahir Iqbal. 2020. A review on Ag-nanostructures for enhancement in shelf time of fruits. *Journal of Inorganic and Organometallic Polymers and Materials* 30:1475–1482.

Jaiswal, Lily, Shiv Shankar, and Jong-Whan Rhim. 2019. Applications of nanotechnology in food microbiology. *Methods in Microbiology* 46:43–60.

Jiang, Tianjia, Lifang Feng, and Yanbo Wang. 2013. Effect of alginate/nano-Ag coating on microbial and physicochemical characteristics of shiitake mushroom (*Lentinus edodes*) during cold storage. *Food Chemistry* 141 (2):954–960.

Jokar, Maryam, Russly Abdul Rahman, Nor Azowa Ibrahim, Luqman Chuah Abdullah, and Chin Ping Tan. 2012. Melt production and antimicrobial efficiency of low-density polyethylene (LDPE)-silver nanocomposite film. *Food and Bioprocess Technology* 5 (2):719–728.

Jung, J, and Y Zhao. 2016. Antimicrobial packaging for fresh and minimally processed fruits and vegetables. *Antimicrobial Food Packaging*:243–256.

Kanmani, Paulraj, and Jong-Whan Rhim. 2014. Nano and nanocomposite antimicrobial materials for food packaging applications. *Future Medicine*:34–49.

Khan, Arif Ullah, Yun Wei, Aftab Ahmad, Zia Ul Haq Khan, Kamran Tahir, Shahab Ullah Khan, Nawshad Muhammad, Faheem Ullah Khan, and Qipeng Yuan. 2016. Enzymatic browning reduction in white cabbage, potent antibacterial and antioxidant activities of biogenic silver nanoparticles. *Journal of Molecular Liquids* 215:39–46.

Korani, Mitra, Elham Ghazizadeh, Shahla Korani, Zahra Hami, and Afshin Mohammadi-Bardbori. 2015. Effects of silver nanoparticles on human health. *European Journal of Nanomedicine* 7 (1):51–62.

Kumar, P, P Mahajan, R Kaur, and S Gautam. 2020. Nanotechnology and its challenges in the food sector: A review. *Materials Today Chemistry* 17:100332.

Kumar, Santosh, Avik Mukherjee, and Joydeep Dutta. 2020. Chitosan based nanocomposite films and coatings: Emerging antimicrobial food packaging alternatives. *Trends in Food Science & Technology* 97:196–209.

Kumar, Vineet, Praveen Guleria, and Surinder Kumar Mehta. 2017. Nanosensors for food quality and safety assessment. *Environmental Chemistry Letters* 15 (2): 165–177.

Li, Hongbo, Hongmei Li, Jiping Liu, Zhihong Luo, Daryl Joyce, and Shenggen He. 2017. Nano-silver treatments reduced bacterial colonization and biofilm formation at the stem-ends of cut gladiolus 'Eerde'spikes. *Postharvest Biology and Technology* 123:102–111.

Li, Ting, Brian Albee, Matti Alemayehu, Rocio Diaz, Leigha Ingham, Shawn Kamal, Maritza Rodriguez, and Sandra Whaley Bishnoi. 2010. Comparative toxicity study of Ag, Au, and Ag-Au bimetallic nanoparticles on Daphnia magna. *Analytical and Bioanalytical Chemistry* 398 (2):689–700.

Lin, Shuqin, Hongmei Li, Xijin Xian, Xiaohui Lin, Zhenpei Pang, Jiping Liu, and Shenggen He. 2019. Nano-silver pretreatment delays wilting of cut gardenia foliage by inhibiting bacterial xylem blockage. *Scientia Horticulturae* 246:791–796.

Liu, Hongyu, Huan Zhang, Jie Wang, and Junfu Wei. 2020. Effect of temperature on the size of biosynthesized silver nanoparticle: Deep insight into microscopic kinetics analysis. *Arabian Journal of Chemistry* 13 (1):1011–1019.

Llorens, Amparo, Elsa Lloret, Pierre A Picouet, Raul Trbojevich, and Avelina Fernandez. 2012. Metallic-based micro and nanocomposites in food contact materials and active food packaging. *Trends in Food Science & Technology* 24 (1):19–29.

Lü, Peitao, Shenggen He, Hongmei Li, Jinping Cao, and Hui-lian Xu. 2010. Effects of nano-silver treatment on vase life of cut rose cv. Movie Star flowers. *Journal of Food, Agriculture and Environment* 8 (2):1118–1122.

Manjunatha, SB, DP Biradar, and Yallappa R Aladakatti. 2016. Nanotechnology and its applications in agriculture: A review. *Journal of Farm Sciences* 29 (1):1–13.

Marambio-Jones, Catalina, and Eric MV Hoek. 2010. A review of the antibacterial effects of silver nanomaterials and potential implications for human health and the environment. *Journal of Nanoparticle Research* 12 (5):1531–1551.

Mehta, CM, Rashmi Srivastava, Sandeep Arora, and AK Sharma. 2016. Impact assessment of silver nanoparticles on plant growth and soil bacterial diversity. *3 Biotech* 6 (2):254.

Menamparambath, Mini Mol, C Muhammed Ajmal, Kwang Hee Kim, Daejin Yang, Jongwook Roh, Hyeon Cheol Park, Chan Kwak, Jae-Young Choi, and Seunghyun Baik. 2015. Silver nanowires decorated with silver nanoparticles for low-haze flexible transparent conductive films. *Scientific Reports* 5 (1):1–9.

Metak, Amal M, Farhad Nabhani, and Stephen N Connolly. 2015. Migration of engineered nanoparticles from packaging into food products. *LWT–Food Science and Technology* 64 (2):781–787.

Meyer, Joel N, Christopher A Lord, Xinyu Y Yang, Elena A. Turner, Appala R. Badireddy, Stella M. Marinakos, Ashutosh Chilkoti, Mark R. Wiesner, and Melanie Auffan. 2010. Intracellular uptake and associated toxicity of silver nanoparticles in *Caenorhabditis elegans*. *Aquatic Toxicology* 100 (2):140–150.

Mihindukulasuriya, SDF, and L-T Lim. 2014. Nanotechnology development in food packaging: A review. *Trends in Food Science & Technology* 40 (2):149–167.

Mlcek, Jiri, and Otakar Rop. 2011. Fresh edible flowers of ornamental plants—A new source of nutraceutical foods. *Trends in Food Science & Technology* 22 (10): 561–569.

Mohammed Fayaz, A, K Balaji, M Girilal, PT Kalaichelvan, and R Venkatesan. 2009. Mycobased synthesis of silver nanoparticles and their incorporation into sodium alginate films for vegetable and fruit preservation. *Journal of Agricultural and Food Chemistry* 57 (14):6246–6252.

Naing, Aung Htay, and Chang Kil Kim. 2020. Application of nano-silver particles to control the postharvest biology of cut flowers: A review. *Scientia Horticulturae* 270:109463.

Ncama, Khayelihle, Lembe Samukelo Magwaza, Asanda Mditshwa, and Samson Zeray Tesfay. 2018. Plant-based edible coatings for managing postharvest quality of fresh horticultural produce: A review. *Food Packaging and Shelf Life* 16:157–167.

Nemati, Seyed Hossein, Ali Tehranifar, Behnam Esfandiari, and Azar Rezaei. 2013. Improvement of vase life and postharvest factors of lilium orientalis â€˜ Bouquetâ€™ by silver nano particles. *Notulae Scientia Biologicae* 5 (4):490–493.

Oberdörster, Günter, Eva Oberdörster, and Jan Oberdörster. 2005. Nanotoxicology: An emerging discipline evolving from studies of ultrafine particles. *Environmental Health Perspectives* 113 (7):823–839.

Pal, Sukdeb, Yu Kyung Tak, and Joon Myong Song. 2007. Does the antibacterial activity of silver nanoparticles depend on the shape of the nanoparticle? A study of the gram-negative bacterium *Escherichia coli*. *Applied and Environmental Microbiology* 73 (6):1712–1720.

Panyala, Nagender Reddy, Eladia María Peña-Méndez, and Josef Havel. 2008. Silver or silver nanoparticles: A hazardous threat to the environment and human health? *Journal of Applied Biomedicine* 6 (3).

Park, Chang Min, Jiyong Heo, and Yeomin Yoon. 2017. Oxidative degradation of bisphenol A and 17α-ethinyl estradiol by Fenton-like activity of silver nanoparticles in aqueous solution. *Chemosphere* 168:617–622.

Pathakoti, Kavitha, Manjunath Manubolu, and Huey-Min Hwang. 2017. Nanostructures: Current uses and future applications in food science. *Journal of Food and Drug Analysis* 25 (2):245–253.

Polet, Madeleine, Laurie Laloux, Sébastien Cambier, Johanna Ziebel, Arno C Gutleb, and Yves-Jacques Schneider. 2020. Soluble silver ions from silver nanoparticles induce a polarised secretion of interleukin-8 in differentiated Caco-2 cells. *Toxicology Letters* 325:14–24.

Poynton, Helen C, James M Lazorchak, Christopher A Impellitteri, Bonnie J. Blalock, Kim Rogers, H. Joel Allen, Alexandre Loguinov, J. Lee Heckman, and Shekar Govindasmawy. 2012. Toxicogenomic responses of nanotoxicity in Daphnia magna exposed to silver nitrate and coated silver nanoparticles. *Environmental Science & Technology* 46 (11):6288–6296.

Rai, Mahendra, Avinash P Ingle, Indarchand Gupta, Raksha Pandit, Priti Paralikar, Aniket Gade, Marco V. Chaud, and Carolina Alves dos Santos. 2019. Smart nanopackaging for the enhancement of food shelf life. *Environmental Chemistry Letters* 17 (1):277–290.

Rashidiani, Nahid, Farzad Nazari, Taimoor Javadi, and Saadi Samadi. 2020. Comparative postharvest responses of carnation and chrysanthemum to synthesized silver nanoparticles (AgNPs). *Advances in Horticultural Science* 34 (2):133–145.

Ray, Suprakas Sinha, and Masami Okamoto. 2003. Polymer/layered silicate nanocomposites: A review from preparation to processing. *Progress in Polymer Science* 28 (11):1539–1641.

Realini, Carolina E, and Begonya Marcos. 2014. Active and intelligent packaging systems for a modern society. *Meat Science* 98 (3):404–419.

Restuccia, Donatella, U Gianfranco Spizzirri, Ortensia I Parisi, Giuseppe Cirillo, Manuela Curcio, Francesca Iemma, Francesco Puoci, Giuliana Vinci, and Nevio Picci. 2010. New EU regulation aspects and global market of active and intelligent packaging for food industry applications. *Food Control* 21 (11):1425–1435.

Rhim, JW, LF Wang, and SI Hong. 2013. Preparation and characterization of agar/silver nanoparticles composite films with antimicrobial activity. *Food Hydrocolloids* 33 (2):327–335.

Ribeiro, Fabianne, Julián Alberto Gallego-Urrea, Kerstin Jurkschat, Alison Crossley, Martin Hassellöv, Cameron Taylor, Amadeu M.V.M. Soares, and Susana Loureiro. 2014. Silver nanoparticles and silver nitrate induce high toxicity to Pseudokirchneriella subcapitata, Daphnia magna and Danio rerio. *Science of the Total Environment* 466:232–241.

Ruchika, Ashok Kumar. 2015. Performance analysis of zinc oxide based alcohol sensors. *International Journal of Applied Sciences and Engineering Research* 4 (4):427–436.

Sachdev, Divya, Vinay Kumar, Priyanka H Maheshwari, Renu Pasricha, and Neeraj Baghel. 2016. Silver based nanomaterial, as a selective colorimetric sensor for visual detection of post harvest spoilage in onion. *Sensors and Actuators B: Chemical* 228:471–479.

Saenjaiban, Aphisit, Teeranuch Singtisan, Panuwat Suppakul, Kittisak Jantanasakulwong, Winita Punyodom, and Pornchai Rachtanapun. 2020. Novel color change film as a time—temperature indicator using polydiacetylene/silver nanoparticles embedded in carboxymethyl cellulose. *Polymers* 12 (10):2306.

Sardella, Davide, Ruben Gatt, and Vasilis P Valdramidis. 2019. Metal nanoparticles for controlling fungal proliferation: Quantitative analysis and applications. *Current Opinion in Food Science* 30:49–59.

Sharma, Poorva, VP Shehin, Navpreet Kaur, and Pratibha Vyas. 2019. Application of edible coatings on fresh and minimally processed vegetables: A review. *International Journal of Vegetable Science* 25 (3):295–314.

Shi, Chong, Yuanyue Wu, Donglu Fang, Fei Pei, Alfred Mugambi Mariga, Wenjian Yang, and Qiuhui Hu. 2018. Effect of nanocomposite packaging on postharvest senescence of Flammulina velutipes. *Food Chemistry* 246:414–421.

Silberbauer, Alina, and Markus Schmid. 2017. Packaging concepts for ready-to-eat food: Recent progress. *Journal of Packaging Technology and Research* 1 (3): 113–126.

Šimon, Peter, Qasim Chaudhry, and Dušan Bakoš. 2008. Migration of engineered nanoparticles from polymer packaging to food–a physicochemical view. *Journal of Food & Nutrition Research* 47 (3).

Solgi, Mousa, Mohsen Kafi, Toktam Sadat Taghavi, and Roohangiz Naderi. 2009. Essential oils and silver nanoparticles (SNP) as novel agents to extend vase-life of gerbera (*Gerbera jamesonii* cv. 'Dune') flowers. *Postharvest Biology and Technology* 53 (3):155–158.

Tan, Hairen, Rudi Santbergen, Arno HM Smets, and Miro Zeman. 2012. Plasmonic light trapping in thin-film silicon solar cells with improved self-assembled silver nanoparticles. *Nano Letters* 12 (8):4070–4076.

Tankhiwale, R., and S. K. Bajpai. 2012. Preparation, characterization and antibacterial applications of ZnO-nanoparticles coated polyethylene films for food packaging. *Colloids and Surfaces B: Biointerfaces* 90:16–20.

Teixeira da Silva, JA. 2003. The cut flower: Postharvest considerations. *Journal of Biological Sciences* 3 (4):406–442.

Vehniwal, SS, and L Abbey. 2019. Cut flower vase life—influential factors, metabolism and organic formulation. *Horticulture International Journal* 3 (6):275–281.

Wang, Kaituo, Shifeng Cao, Peng Jin, Huaijing Rui, and Yonghua Zheng. 2010. Effect of hot air treatment on postharvest mould decay in Chinese bayberry fruit and the possible mechanisms. *International Journal of Food Microbiology* 141 (1–2):11–16.

Wang, Xin, Ting Hou, Haiyang Lin, Wenxin Lv, Haiyin Li, and Feng Li. 2019. In situ template generation of silver nanoparticles as amplification tags for ultrasensitive surface plasmon resonance biosensing of microRNA. *Biosensors and Bioelectronics* 137:82–87.

Welch, CM, CE Banks, AO Simm, and RG Compton. 2005. Silver nanoparticle assemblies supported on glassy-carbon electrodes for the electro-analytical detection of hydrogen peroxide. *Analytical and Bioanalytical Chemistry* 382 (1):12–21.

Xiang, Dong-Xi, Qian Chen, Lin Pang, and Cong-Long Zheng. 2011. Inhibitory effects of silver nanoparticles on H1N1 influenza A virus in vitro. *Journal of Virological Methods* 178 (1–2):137–142.

Xu, QingBo, LiJing Xie, Helena Diao, Fang Li, YanYan Zhang, FeiYa Fu, and Xiang-Dong Liu. 2017. Antibacterial cotton fabric with enhanced durability prepared using silver nanoparticles and carboxymethyl chitosan. *Carbohydrate Polymers* 177:187–193.

Yadollahi, A, K Arzani, and H Khoshghalb. 2009. The role of nanotechnology in horticultural crops postharvest management. Paper read at Southeast Asia Symposium on Quality and Safety of Fresh and Fresh-Cut Produce 875.

Yan, An, and Zhong Chen. 2019. Impacts of silver nanoparticles on plants: A focus on the phytotoxicity and underlying mechanism. *International Journal of Molecular Sciences* 20 (5):1003.

Zhao, Daqiu, Menglin Cheng, Wenhui Tang, Ding Liu, Siyu Zhou, Jiasong Meng, and Jun Tao. 2018. Nano-silver modifies the vase life of cut herbaceous peony (Paeonia lactiflora Pall.) flowers. *Protoplasma* 255 (4):1001–1013.

Zhao, Wei, Yongfang Zhang, Bin Du, Dong Wei, Qin Wei, and Yanfang Zhao. 2013. Enhancement effect of silver nanoparticles on fermentative biohydrogen production using mixed bacteria. *Bioresource Technology* 142:240–245.

Zhu, Zhen, Cheng-Tse Kao, and Ren-Jang Wu. 2014. A highly sensitive ethanol sensor based on Ag@ TiO_2 nanoparticles at room temperature. *Applied Surface Science* 320:348–355.

Chapter 7

Nanoclay-Based Methods for the Postharvest Processing of Horticultural Produce

Gabriela E. Viacava and María R. Ansorena

List of Abbreviations

ALK-HNT	Alkaline-modified halloysite nanotube
CPI	Croaker protein isolate
DPBD	Trans,trans-1,4-diphenyl-1,3-dibutadiene
EC	Edible coating
EU	European Union
FDA	U.S. Food and Drug Administration
HNT	Halloysite nanotube
LDPE	Low-density polyethylene
LLDPE	Linear low-density polyethylene
MMT	Montmorillonite
OMMT	Organically modified montmorillonite
OP	Oxygen permeability
OPEO	Orange peel essential oil
OTR	Oxygen transmission rate
PCN	Polymer clay nanocomposite
PE	Polyethylene
PET	Polyethylene terephthalate
PLA	Polylactic acid
PMMT	Protonated montmorillonite
PP	Polypropylene

DOI: 10.1201/9781003142287-7

PVC	Polyvinyl chloride
RH	Relative humidity
SPS	Sweet potato starch
TEO	Thyme essential oil
WVP	Water vapor permeability
WVTR	Water vapor transmission rate

7.1 Introduction

Fruits and vegetables are perishable foods with a limited shelf life mainly because of their continuous metabolic activity after harvest, among which respiration, transpiration, and ethylene production are the most important. The minimal processing of these foods promotes additional physical disorders, physiological aging, biochemical changes, and microbial growth, which accelerate produce decay and reduce its shelf life compared with intact fruits and vegetables. Maintaining the postharvest quality of fresh and minimally processed produce has long been a challenging task. Although adequate management of temperature during processing, storage, transport, and marketing is the most important parameter to minimize the quality loss of horticultural products, other technologies are also of high importance. For instance, food packaging technology can help minimize the metabolic activity of the produce and prevent or inhibit microbial growth, thereby extending products' shelf life (Artés and Allende, 2014). Novel approaches in this area include the incorporation of nanomaterials into food packaging systems, and several advanced materials with improved functional properties have already been developed over the past two decades through nanoscience and nanotechnology (He et al., 2019).

Nanotechnology is a highly interdisciplinary field that involves the application of materials that have a size or structure of 1–100 nm (Mihindukulasuriya and Lim, 2014; Bumbudsanpharoke and Ko, 2015). These nanomaterials can be combined with polymers to produce nanocomposites with enhanced thermal and mechanical resistance and barrier properties compared to the neat polymer. These improved properties of nanocomposites have been very attractive in the area of food packaging. Nanotechnology can also improve the performance of active packaging systems, allowing better protection of the food (Mihindukulasuriya and Lim, 2014). In general, active packaging involves the use of substances that release desirable compounds (such as antioxidants, antimicrobial agents, and flavors) or remove detrimental factors (such as ethylene, oxygen, or water vapor). In this scenario, nanomaterials can act as the active compound itself, as a delivering carrier of the active substance, protecting, stabilizing, and controlling its release or as a scavenging system of undesirable factors (Echegoyen, 2015).

Several nanomaterials can be used to improve materials for food packaging applications. Among them, clay nanoparticles were the first to be applied mainly because of their wide availability, natural occurrence, low cost, and low environmental impact (Viacava et al., 2017). Besides, clay minerals enjoy 'generally regarded as safe' status according to the FDA and also the European Food Safety Authority (Ibarguren et al., 2014). Incorporating nanoclays into polymer-based packaging materials has led to the development of PCNs with improved mechanical and physical properties. For instance, when clay nanoparticles are well dispersed in the polymer matrix, they provide substantial improvements in the gas barrier properties of the nanocomposite since they increase the tortuosity of the diffusive route, hindering the penetrant to diffuse through the matrix (Choudalakis and Gotsis, 2009; Azeredo, 2013). Nanoclays can also be used in active packaging systems as delivering carriers of active compounds (Meira et al., 2014) or as scavengers of undesired substances, such as ethylene (Guo et al., 2018).

The incorporation of nanoclay minerals into different materials for food packaging applications can result in the exposure of consumers to nanoparticles. For instance, skin contact or absorption of the nanoparticle can take place when manipulating the food package. Moreover, accidental ingestion of nanoclays can occur if they interact with food components and migrate into the food product. Although nanoclays are considered safe food additives (Ibarguren et al., 2014), some toxic effects have been observed in different experimental models (Houtman et al., 2014; Maisanaba et al., 2015). Therefore, migration tests and toxicity and safety assessments are needed.

This chapter provides a contemporary summary of cutting-edge developments in nanoclay packaging technology applied to horticultural produce, with a special focus on enhancing barrier and mechanical properties and active packaging. Recent advances in the synthesis and applications of PCNs as novel food packaging materials are also discussed. Risk assessment and safety concerns are also highlighted.

7.2 Composition and Structure of Clay Nanoparticles

Nanoclays are nanoparticles of layered mineral silicates organized in layered structural units that can form complex crystallites by stacking these layers (Ray and Okamoto, 2003). An individual layer unit is composed of octahedral and/or tetrahedral sheets (Uddin, 2017). In the octahedral sheets, the octahedrons are centered by aluminum or magnesium, which coordinates six oxygen atoms, some of them also bond to a hydrogen atom, located at the corners. In the tetrahedral sheets, the center of each tetrahedron is occupied by a silicon atom, and the four corners consist of oxygen atoms. Each tetrahedron is linked to an adjacent tetrahedron through three corners, while the fourth corner is attached by a covalent bond to an adjacent octahedral sheet (Jawaid et al., 2016). These sheet arrangements impact various defining aspects of nanoclays.

Nanoclay materials present three major sheet arrangements: 1:1, 2:1, and 2:1:1. In 1:1, an octahedral sheet is linked to a tetrahedral sheet; in 2:1 structures, each octahedral sheet is between two tetrahedral sheets; and in 2:1:1 structure, the octahedral sheet is connected to an octahedral and two tetrahedral sheets (Majeed et al., 2013; Barton and Karathanasis, 2016).

Clays can be divided, based on the structure, into several subtypes such as bentonite, laponite, smectite, chlorite, kaolinite, illite, and halloysite, among others (Nazir et al., 2016). MMT nanoclay (a 2:1 layered silicate belonging to the smectite clay subtype) is the most commonly used clay for obtaining polymer/clay nanocomposites with enhanced mechanical properties for food packaging applications (Choudalakis and Gotsis, 2009; Yusoh et al., 2017; Irshidat and Al-Saleh, 2018). Halloysite nanoclay (1:1 nanotube) belongs to the kaolin clay group and has been also incorporated in the formulation of polymer-based packaging to enhance its mechanical and thermal properties. Additionally, halloysite nanoclays present a tubular structure that allows the incorporation of guest molecules into their cavities (Lvov and Abdullayev, 2013). Thus, this last nanoclay is also used in active packaging systems to carry biologically active molecules or capture undesirable substances, such as ethylene (Gaaz et al., 2017; Lazzara et al., 2018).

7.3 Synthesis of PCNs

In PCN synthesis, the crucial challenges that need to be achieved are chemical compatibility between the polymer matrix and the nanoclay and reaching a homogeneous dispersion of the nanoclay within the polymer matrix (Dasgupta and Ranjan, 2018). Both intercorrelated features have a significant influence on the polymer/nanoclay composite morphology and determine their final bulk properties (Shah et al., 2015; Thakur and Kessler, 2015).

Surface functionalization enhances the properties and the characteristics of nanoparticles through surface modification, improving the interfacial interactions between the nanoclay structures and the polymer matrix (Zhang and Park, 2018, Liu et al., 2014). Three different combinations of polymer matrix–nanoclay structures can be achieved (immiscible, intercalated, exfoliated) depending on the synthesis approach (Guo et al., 2018; Figure 7.1).

In the immiscible structure, it is not possible for the polymer to diffuse between the clay layers, and the polymers are separated from the clay layers. Normally, an improvement in some properties like modulus is observed, but a deficiency in some other properties like strength and toughness is observed as well. In the intercalated nanocomposites, clay layers maintain the well-ordered multi-structure of alternating polymeric and clay layers. This alternation leads to an enhancement of the mechanical and thermal properties (Karamane et al., 2017; Ozkose et al., 2017; Saad et al., 2017).

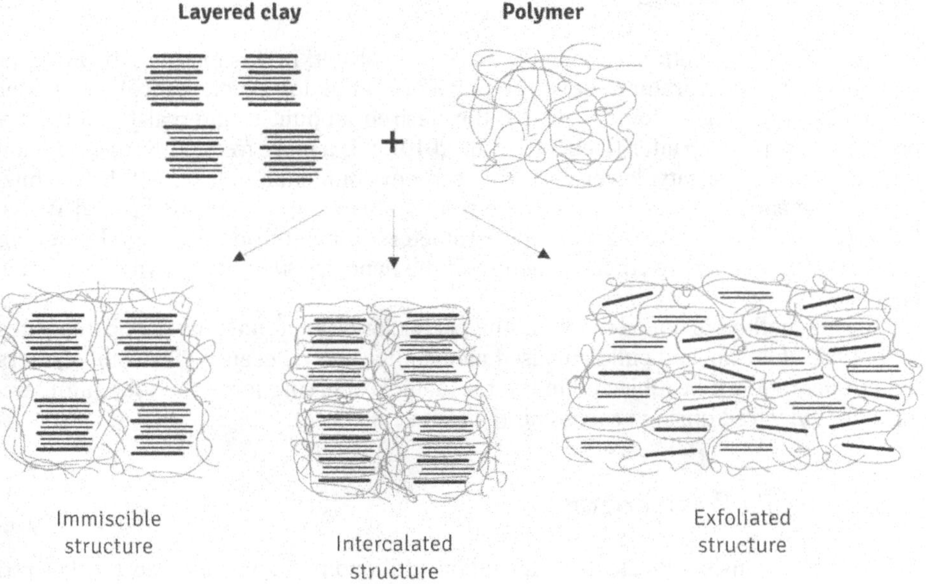

Figure 7.1 Immiscible, intercalated, and exfoliated polymer–clay structures.

However, in the exfoliated structure, clay layers are fully separated by the polymer, leading to improved mechanical properties (Cherifi et al., 2018).

Different synthesis approaches have been used for the preparation of PCNs including the solution-blending, melt-blending, and in situ polymerization method. The solution-blending technique leads to uniform nanoclay dispersions in the polymer matrix, when compared to the melt-blending method, mainly due to its low viscosity and high stirring speed. Melt blending is considered eco-friendly and with high economic potential (Valapa et al., 2017). By comparison, the in situ polymerization method provides uniform nanoclay dispersions and can be easily modified by changing the polymerization conditions (Vo et al., 2016).

7.3.1 Solution Blending

Solution blending is a technique in which the clay is pre-swell or exfoliated in a suitable solvent in which the polymer is soluble. Usually used solvents are water, acetone, chloroform, or toluene (Beyer, 2002). Swelling of the clay layers occurs firstly when the clay is added to the polymer solution. Then, the polymer chains intercalate and displace the solvent within the interlayers of the clay. The PCNs are formed upon solvent removal, although large quantities of volatile solvent are necessary for this approach (Gurses, 2015, Fischer et al., 1999).

Although the solution blending method leads to uniform nanoclay dispersions, the disadvantages of this method are the requirement of suitable polymer/clay/solvent combinations and the high costs related to solvents disposal, including their impact on the environment (Ray and Bousmina, 2005).

7.3.2 Melt Blending

The melt-blending method does not require any solvent. In this method, the polymer is melted (at a temperature above the polymers' softening point, statically or under shear interlayers) and combined with the desired amount of intercalated nanoclay particles using an extruder (Debnath et al., 2010). It is a more frequently used method because of its simplicity, lower cost, reduced environmental impact, and better mixing performance when compared to the solution-blending method (Ports and Weiss, 2010; Albdiry et al., 2013). Important parameters to be considered, include melting conditions as well as the chemical nature of the nanoclay filler and polymers (Dennis et al., 2001).

This method is not adequate in the particular case of most bio-nanocomposites (e.g., PLA, chitosan, and pullulan-based nanocomposites), because these biopolymers tend to degrade when subjected to the mechanical shearing forces and the high temperatures involved during processing (Unalan et al., 2014).

7.3.3 In Situ Polymerization

This method consists in the polymerization of monomer species when the layered materials are present (Abedi and Abdouss, 2014; Asensio et al., 2018, Huang et al., 2006). One of the main advantages of the in situ polymerization process is the versatile molecular designs of the polymer matrix by flexible tuning of the matrix composition and structure (Salmi et al., 2013). The principal drawback of this method consists in its unsuitableness for bio-macromolecules (proteins and polysaccharides) which are needed for bio-nanocomposite generation.

7.4 Nanoclay Applications in the Postharvest Processing of Horticultural Produce

Nanoclays have recently been studied to improve the performance of fruits and vegetables packaging systems because of their wide availability, relatively low cost and simple processability, and significant packaging properties improvement. Main research trends are related to (1) nanocomposites (the presence of nanoclays in polymer-based packaging materials can considerably enhance the packaging properties [mainly mechanical and physical] and thus improve the shelf life of packaged foods) and (2) active packaging (nanoclays are used as part of an active packaging system that delivers functional compounds that positively interact with the food; Guo et al., 2018).

Although many studies were conducted to develop nanoclay-based systems for potential use in food packaging, most of them focused their research on the characterization of the improved properties of the packaging, and only few have tested their effects on real food systems. Moreover, applications for packaging fruits and vegetables are scarce. This section provides a thorough review of current developments in nanoclay packaging technology being effectively applied to horticultural produce, with a particular emphasis on the barrier and mechanical properties and active packaging.

7.4.1 Organic Polymers Reinforced with Nanoclays

Organic polymeric materials (i.e., PE, PP, or PVC) are generally used for packaging fruits and vegetables since they present several advantages, such as good barrier properties, low weight, ease of processing, and low cost. However, in order to improve barrier and mechanical properties, polymer blends or complex multilayer films are generally used. Unfortunately, their higher production costs limit the economically viable utilization of these films (Echegoyen, 2015). Furthermore, these multilayered structures are generally not recyclable. Therefore, new production directions are related for developing packaging solutions that not only improve the mechanical and barrier properties of the polymers but also target sustainable development. In this context, mixing plastics with functional fillers such as clay nanoparticles can offer the desired functionalities for packaging perishable fruits and vegetables while increasing their recyclability (Sarfraz et al., 2021). As mentioned earlier, the inclusion of nanoclays into the structure of polymer materials increases the tensile strength of the packaging by acting as a reinforcement agent that fills gaps in the polymer matrix. Additionally, the barrier properties of nanoclay polymer–based packaging are also improved thanks to the increased tortuosity of the diffusion or permeation path for gases. In this sense, gas molecules cannot follow a straight route perpendicular to the film surface but must diffuse around the nanoparticles, resulting in a longer mean path for gas diffusion through the film (Echegoyen, 2015). Table 7.1 summarizes the most relevant studies that have investigated the performance of these packaging materials using real food systems.

Table 7.1 Polymer Nanoclay Composites Applied to Horticultural Produce

Nanoclay type	*Matrix*	*Food product*	*References*
Montmorillonite	Polyethylene + nano-Ag + nano-TiO_2	Kiwifruit	Hu et al., 2011
Kaolin	Polyethylene + nano-Ag + anatase TiO_2 + rutile TiO_2	Chinese jujube	Li et al., 2009
Kaolin	Polyethylene + nano-Ag + anatase TiO_2 + rutile TiO_2	Strawberry	Yang et al., 2010
Cloisite® 20A	Polyethylene	Peach	Ebrahimi et al., 2018
Montmorillonite	Alginate + acerola puree	Acerola fruit	Azeredo et al., 2012
NA	Chitosan	Lemon fruit	Taghinezhad and Ebadollahi, 2017
NA	Chitosan	Thomson oranges	Khoshtaghaza and Taghinezhad, 2016
Organo-clay montmorillonite	Whitemouth croaker protein isolate	Papaya slices	Cortez-Vega et al., 2014
Montmorillonite	Carnauba wax + orange peel essential oil	Blood oranges	Nasirifar et al., 2018
Modified Na-bentonite	Carnauba wax	'Valencia' orange fruit	Motamedi et al., 2018

Note: NA denotes not available.

Hu et al. (2011) developed a novel nanocomposite-based packaging blending PE with nano-Ag, nano-TiO_2, and MMT by extrusion, obtaining good dispersivity of nanoparticles in the matrix and good adhesion between fillers and PE matrix. However, they also observed a slight agglomeration of nanoparticles that probably took place when making the nanocomposite. The nanocomposite film presented 32.86% and 19.54% lower WVP and OP, respectively, compared to the pure PE film, and 31.69% higher longitudinal strength. Authors associated these results with the introduction of the exfoliated MMT in the polymer matrix. Packing of kiwifruit in the nanocomposite film proved to be an efficient technology for inhibiting ethylene production, extending organoleptic characteristics (firmness and color), reducing degradation of nutritional components (ascorbic acid and polyphenol compounds), and preventing physiologic changes (lipid peroxidation, polyphenol oxidase, peroxidase activities) of the fruit during long cold storage (4°C, 42 days). However, the improved postharvest quality of the nanocomposite-packed kiwifruit was associated with not only the presence of the MMT in the PE matrix but also the content of nano-Ag and nano-TiO_2. Similarly, Li et al. (2009) and Yang et al. (2010) also found that a nano-packaging material synthesized by blending PE with nano-powder (nano-Ag, kaolin, anatase TiO_2, rutile TiO_2) improved the preservation quality and extended the shelf life of Chinese jujube and strawberry fruits, respectively, during refrigerated storage.

More recently, Ebrahimi et al. (2018) investigated the effects of PE-based films reinforced with cloisite® 20A (an ion-exchange natural MMT) by the melt-blending method, on quality parameters of mature peach fruit stored for 42 days at 2° C. Mechanical characterization of the nanocomposite film evidenced a good distribution and exfoliation of the nanoclay in the PE matrix. Regarding barrier properties, a reduction in O_2, CO_2, and water vapor permeabilities of 51.48%, 42.88%, and 40.32%, respectively, were observed compared to those of the neat PE films. When used to pack peach fruit, the clay-reinforced PE film properly suited fruit respiration, prevented physiologic changes, delayed ripening, and inhibited bacterial growth, thus extending the shelf life of peaches. In particular, a significant reduction of 42%, 28%, and 15% was observed in the weight loss, softening, and internal browning, respectively, of nanocomposite-packed peaches, in comparison to the unpacked fruit. Moreover, less pH, total soluble solids, and titratable acidity changes were found in cloisite–PE-packed peaches when compared to the fruit packed in the neat PE film or unpacked. In addition, the nanocomposite film reduced the activity of enzymes associated with oxidation processes (polyphenol oxidase) but increased the activity of those related to defense mechanisms of the fruit (peroxidase and catalase). Accordingly, the authors indicated that packaging peaches in the cloisite-reinforced PE film could be an industrial and cost-effective method for delaying ripening and prolonging the postharvest quality of the fruit.

7.4.2 ECs Reinforced with Nanoclays

ECs are edible formulations applied as a thin layer (normally <100 µm) over a food material with the objective of extending its shelf life by the control of moisture and gas exchange, lipid diffusion, and aroma and flavor losses. They also offer protection from physical damage caused by mechanical impact. Additionally, ECs are environmentally friendly since they are biodegradable and can reduce the use of petroleum source–derived polymeric films (Ghoshal, 2018). Most ECs used for packaging fruits and vegetables are fabricated from biopolymers like lipids, polysaccharides, and proteins. However, their inherent water sensitivity, inferior barrier properties, relatively low stiffness, and strength have restrained their potential and feasible application. Incorporation and well dispersion of clay minerals into biopolymer-based EC formulations can significantly enhance its physical and mechanical properties (Table 7.1).

Azeredo et al. (2012) used MMT to reinforce ECs based on alginate-acerola puree and evaluated the ability of the nanocomposite coating to enhance acceptance and stability of acerola fruits packed in PET trays and stored for 7 days at 6° C. The coating was effective to extend fruit stability by decreasing weight loss, ascorbic acid loss, and decay incidence. Moreover, MMT-coated acerolas presented color properties very similar to those of fresh fruits, maintaining their visual acceptance. Overall, incorporating MMT into the EC improved its water vapor barrier properties and probably reduced its OP, helping reduce the respiration rate and weight loss of coated acerola fruit during storage and thus to delay the ripening process.

Taghinezhad and Ebadollahi (2017) investigated the effect of a chitosan-clay nanocomposite coating on the chemical and mechanical properties of lemon fruit during cold storage (63 days at 8° C ± 2° C and 75–80% RH). According to their results, coated samples showed the lowest values of total soluble solids and punch force, and the highest values of titratable acidity, firmness, and peel shear forces. Furthermore, weight loss during storage was lower in coated samples in comparison to uncoated ones. The authors suggested that the chitosan-clay coating may have lowered the fruit respiration rate and moisture loss, thus extending the shelf life and preserving the quality of the lemons during storage. Similarly, Khoshtaghaza and Taghinezhad (2016) also found that chitosan-clay coatings prevent quality loss in Thomson oranges stored for 3 months at 6°C and 85–90% RH, as evidenced by the higher pH, color retention, peel moisture, and firmness found in the nanocomposite-coated fruits in comparison with the other samples (coated with fogger wax or uncoated).

Cortez-Vega et al. (2014) evaluated the effectiveness of ECs formulated with Whitemouth CPI with and without organo-clay MMT to increase the shelf life of minimally processed papaya slices stored for 12 days at 5°C. According to their results, although both coatings (pure CPI coating and CPI coating with MMT) were effective to reduce the harmful effects of minimal processing on papaya, fruits coated with CPI containing MMT showed lower mass loss, lower microbial growth (particularly psychrotrophic microorganisms, and molds and yeasts), and a smaller decrease of firmness, lightness, and pH during storage. Consequently, the addition of MMT to the CPI allowed obtaining an EC with greater potential to increase the shelf life of minimally processed papaya without affecting its sensorial attributes.

In another study, Nasirifar et al. (2018) assessed the effect of different lipid-based coatings containing MMT (carnauba wax, carnauba wax incorporated with OPEO, (1%), carnauba wax with MMT [2%], and carnauba wax with OPEO [0.5%] and MMT [1%]) on the physicochemical properties and shelf life of blood oranges stored at 7°C and 85% RH for 100 days. According to their results, both coatings formulated with MMT were the most appropriate to preserve the fruits' freshness and extend its shelf life. In particular, blood oranges coated with these formulations presented higher antioxidant activity and vitamin C content, higher levels of titratable acidity, lower firmness loss, and better visual appearance in terms of brightness, in comparison with the other samples. These results were attributed to the excellent oxygen and water vapor barrier properties of the reinforced coatings with MMT.

Similarly, Motamedi et al. (2018) also evaluated the effect of carnauba wax–nanoclay coatings on the postharvest quality of 'Valencia' orange fruit stored under simulated processing and marketing environmental conditions. In order to make the nanoclay surface well suited for hydrophobe wax, the authors used a modified Na-bentonite nanoclay. Different concentrations of the nanoclay on the carnauba wax coating were assessed (0.0, 0.5, and 1.0 wt%), and its efficiency was compared against the uncoated fruit and two commercial waxes. Noteworthy are that results indicated that the synthesized nanoclay wax coatings were more efficient to prevent the postharvest deterioration of orange fruit than the most commonly used commercial waxes. In particular, fruits coated with the 1% nanoclay wax formulation achieved the highest sensory scores, since the panelists positively highlighted their color, juiciness, aroma, and flavor attributes after storage time. Moreover, although the three synthesized nanoclay wax coatings reduced the fruit weight loss more than the commercial wax treatments, the best result was achieved with the 1% nanoclay wax coating. Authors suggested that a higher concentration of nanoclay in the coating formulation enhanced its water vapor barrier properties, leading to a greater reduction in fruit water loss. Additionally, fruits coated with the 1% nanoclay wax emulsions also presented the highest antioxidant activity, results associated with a reduced OP of the nanoclay wax EC, which, in turn, retards oxidative processes and preserves the antioxidant activity of the treated fruits. The authors concluded that the designed nanoclay carnauba wax coatings could be a potential and cost-effective alternative to commercial waxes.

7.4.3 Nanoclays in Active Packaging Systems

Active packaging systems are those designed to interact with the packaged food or the environment surrounding the food, by releasing desired substances or by removing detrimental compounds, in order to improve the shelf life or the sensory characteristics of the food (Sarfraz et al., 2021). Several studies have evaluated the performance of different nanoclays as part of active packaging systems for fruits and vegetables with antimicrobial properties or as ethylene scavengers (Table 7.2).

Table 7.2 Nanoclays in Active Packaging Systems

Nanoclay type	*Active agent/property*	*Food product*	*References*
Halloysite nanotubes	Thyme essential oil as antimicrobial agent	Cherry tomato	Lee et al., 2017
Montmorillonite	Silver nanoparticles as antimicrobial agent	Fresh fruit salad	Costa et al., 2011
Organically-modified montmorillonite	Carvacrol and thymol as antimicrobial agents	Strawberry	Campos-Requena et al., 2015
Montmorillonite	Thyme essential oil as antimicrobial agent	Baby spinach leaves	Issa et al., 2017
Montmorillonite	Silver nanoparticles as antimicrobial agent	Fresh-cut carrots	Costa et al., 2012
Modified clay with salicylate anion (LDH/Na^+ salicylate)	Salicylate anion as antimicrobial agent	Fresh apricots	Gorrasi and Bugatti, 2016
Raw and alkali-treated halloysite nanotubes	Ethylene scavenging	NA	Gaikwad et al., 2018
Halloysite nanotubes	Ethylene scavenging	Banana, tomato, and strawberry	Tas et al., 2017
Alkaline modified halloysite nanotubes	Ethylene scavenging	Cherry tomatoes	Boonsiriwit et al., 2020
NA	Ethylene scavenging	Banana	Santosa and Widodo, 2013
NA	Ethylene scavenging	Banana	Chamara et al., 2000
Protonated montmorillonite	Ethylene scavenging	Blueberry fruit	Álvarez-Hernández et al., 2019
Sepiolite	Ethylene scavenging	'Mirlo naranja' apricots	Álvarez-Hernández et al., 2020

Note: NA denotes not available.

7.4.3.1. Nanoclays in Antimicrobial Packaging Systems

Recently, nanoclays have been successfully employed as delivering carriers of antimicrobial compounds of lipophilic nature, such as essential oils and their active components (De Silva et al., 2015; Gorrasi, 2015; Lee and Park, 2015; Biddeci et al., 2016). However, most of these studies are focused on the characterization of the functional properties of the capsules and only few have tested their real effects on food products. In this last case, Lee et al. (2017) took advantage of the tubular structure of HNTs and developed an antimicrobial capsule by incorporating TEO into HNTs, which was further applied to cherry tomatoes. In this research, the HNTs were alkaline-modified to increase the pore volume of the HNTs and thus improve its TEO-loading capacity. The authors observed the ALK-HNTs capsules achieved a sustained release of the essential oil for more than 21 days when they were stored at 4 and 25°C inside a PP container. *In vitro* studies determined that this sustained TEO release presented antimicrobial activity against *E. coli* O157:H7 in the vapor phase. For *in vivo* studies, cherry tomatoes were inoculated with *E. coli* O157:H7, packaged in sterilized propylene containers and exposed to the TEO–ALK–HNT capsules at 4 and 25°C for up to 5 days. These samples presented a significant reduction in the number of *E. coli* O157:H7, total mesophilic aerobic bacteria, and molds and yeasts during storage, achieving different reductions depending on the exposure dose.

Costa et al. (2011) evaluated the effectiveness of silver-MMT (Ag-MMT) nanoparticles as an active antimicrobial system to assure safety and increase the shelf life of fresh fruit salad (a mixture of sliced kiwifruit and pineapple). The authors exploited the fact that MMT, used as silver-supporting material, established weak electrostatic interactions with silver ions, favoring its controlled release and antimicrobial action. In their investigation, different concentrations of active Ag-MMT nanoparticles (10, 15, and 20 mg) were left on the bottom of PP boxes containing the fruits. Microbiological and organoleptic qualities were assessed during 3 weeks of storage at 5°C. The authors found that the Ag-MMT nanoparticles were effective in reducing spoilage microorganism growth, particularly at the highest assayed concentration. Additionally, sensory parameters, particularly the firmness attribute, were the determinants of a product's shelf life, and only samples stored with 20 mg of Ag-MMT nanoparticles showed an increase in shelf life of about 5 days compared to the other samples.

The potential of polymer/clay or biopolymer/clay nanocomposites for active food packaging has also been studied. In this sense, Campos-Requena et al. (2015) synthesized an antimicrobial film based on LDPE/organically modified MMT (OMMT) containing a mixture of carvacrol and thymol by the melt-blending method and evaluated its *in vivo* antimicrobial activity against *Botrytis cinerea* in stored inoculated strawberries (18 ± 2°C, and 60 ± 5% RH, 5 days). The nanocomposite film (50 mm in diameter) was attached to the internal lid of a 340-cm^3 PET container, and three or four inoculated fruits were placed in each container. An improvement in the mechanical, thermal, and rheological properties of the film was achieved with the developed nanocomposite, with respect to the pure LDPE film. Moreover, the authors observed that the release rate of the essential oils was significantly reduced thanks to the intercalated structure of the nanocomposite. The films containing an equal mixture of carvacrol and thymol effectively inhibited *B. cinerea* growth in stored inoculated strawberries, without significant changes in organoleptic attributes, demonstrating the potential of the LDPE/OMMT nanocomposite film containing essential oils as an antimicrobial active packaging.

In another study, an active biodegradable nanocomposite film was developed by incorporating TEO, as natural antimicrobial agent, and MMT nanofillers into SPS (Issa et al., 2017). The performance of this film in the microbiological quality and the shelf life of baby spinach leaves inoculated with two pathogens (*Escherichia coli* and *Salmonella typhimurium*) and stored at 4°C for 8 days was also studied. The authors observed that the baby spinach samples wrapped in the films showed lower counts of bacterial, yeast, and mold populations in comparison with the unwrapped samples. These results were associated with the antimicrobial properties of the thyme oil and the reduced OP that the incorporation of MMT provides to the starch film, which limits the growth of aerobic microorganisms on wrapped samples. Moreover, as the storage time increased, the antimicrobial effectiveness of the film also increased, since the growth of inoculated pathogens was reduced by the end of the storage period. This effect was attributed to the contribution of MMT nanoclay, which can provide a controlled and sustained release of antimicrobial TEO to the surface of the wrapped vegetable.

Costa et al. (2012) evaluated the effect of an active alginate coating loaded with Ag-MMT nanoparticles on the shelf life of fresh-cut carrots. The incorporation of Ag-MMT into the alginate coating controlled the microbial growth of carrots better than the coating treatment without nanoparticles. In particular, minimally processed carrots coated with alginate/Ag-MMT did not overcome the maximum legal microbiological concentration imposed for fresh-cut vegetables during the entire storage period. Additionally, sensory attributes of nano-coated samples were better preserved. Noteworthy is that the incorporation of Ag-MMT in the alginate coating increased the shelf life of fresh-cut carrots to 70 days, against 10 and 4 days for pure alginate-coated samples and uncoated ones, respectively. Although in this study MMT was used mainly as support material for the antimicrobial Ag nanoparticles, improved barrier properties of the alginate biopolymer could also be achieved, aiding in preservation (Incoronato et al., 2010). Similarly, Gorrasi and Bugatti (2016) developed an EC based on pectin and a modified clay with salicylate anion as an antimicrobial agent, and demonstrated the efficiency of the formulation in extending the shelf life of fresh apricots.

7.4.3.2. Nanoclays in Ethylene-Scavenging Systems

Although ethylene (C_2H_4) gas has some positive effects on the maturation of fruits and vegetables, ethylene scavenging is a desired feature in the packaging of those produces sensitive to this plant hormone. Natural clay minerals having a porous structure (such as halloysite nanotube, Japanese Oya stone, cristobalite, bentonite) can act as ethylene scavengers and, therefore, be incorporated in active packaging in the form of ethylene-permeable sachets or nanocomposite films (Gaikwad et al., 2020).

HNTs were recently evaluated as substitutes for traditional ethylene scavengers taking advantage of their hollow tubular structure. For instance, Gaikwad et al. (2018) investigated the performance of raw and alkali-treated HNTs as ethylene scavengers to maintain the postharvest quality of fresh fruits and vegetables. Results showed a high ethylene adsorption capacity of alkali-treated HNTs (1 g of alkali-treated HNTs scavenged 49 µL of ethylene gas within 24 h), which was attributed to its improved pore structure and increased pore size.

Tas et al. (2017) developed an ethylene-scavenging film by melting HNTsand PE solids using a twin-screw extruder followed by blown-film extrusion. They found that films prepared with 1% HNTs presented a homogeneous distribution of HNTs, while higher HNT loadings diminished the dispersion quality due to the aggregation of nanoparticles. Additionally, films prepared with 1% HNT/PE presented better oxygen and water vapor barrier properties, compared to PE films without HNT. In particular, a 22% and 32% decrease in OTR and WVTR, respectively, were achieved when compared to the neat PE performance. In this sense, authors indicated that HNTs introduced a tortuous path in the PE matrix for the diffusion of oxygen and water vapor. Conversely, higher HNTs loadings did not improve barrier properties probably because agglomeration of HNTs formed empty spaces in the film that increased the permeation of gases, surpassing the barrier effect. However, films containing 5% HNT presented 20% more ethylene-adsorption capacity compared to neat PE films. It seems that the hollow structure of HNTs along with their large aspect ratio allowed the adsorption of ethylene gas in large amounts. However, the authors tested the developed films for packaging perishable products stored at ambient temperature and found that 5% HNT/PE films slowed down the ripening process of bananas, retained the firmness of tomatoes, and reduced the weight loss of strawberries.

Boonsiriwit et al. (2020) studied the suitability of incorporating ALK-HNTs into LDPE for developing active packaging films with ethylene-scavenging activity. Nanocomposite films were prepared by extrusion with different concentrations of ALK-HNTs (1, 3, and 5%) and were characterized in terms of thermal, morphological, and mechanical properties, as well as ethylene adsorption capacity. The authors also monitored the quality of cherry tomatoes packaged in the LDPE film containing 3% of ALK-HNTs and stored during 21 days at 8° C. They found that higher ALK-HNTs concentrations in the LDPE film formed large aggregates of the nanoclay, reducing its well dispersion and decreasing the film mechanical properties. However, the ethylene adsorption capacity of the film was significantly higher at increased ALK-HNT loadings. Additionally, cherry tomatoes packaged in the 3% ALK-HNTs/LDPE film presented lower losses of firmness and weight compared to samples packaged in the control LDPE film. The authors associated these results with the decreased presence of ethylene gas in the headspace of tomatoes packaged in the nanocomposite film. It is known that ethylene promotes the respiration of the cherry tomatoes, with a consequent increase in moisture evaporation and further weight loss, withering, and deterioration (Dhital et al., 2018).

Clays have also been used as a $KMnO_4$ support to extend postharvest life of fresh produce. $KMnO_4$ is the best-known ethylene scavenger since it oxidizes C_2H_4 to CO_2 and H_2O. In order to favor the interaction and reaction between ethylene gas and $KMnO_4$, this last one is generally supported on microporous mineral particles, like natural and/or modified clays (Álvarez-Hernández et al., 2018). For example, Santosa and Widodo (2013) used a natural clay found in Indonesia as support material for $KMnO_4$, and studied its performance to delay banana ripening. In this study, bananas were placed in transparent plastic bags (six bananas per bag) and a cloth sachet containing $KMnO_4$ supported in clay powder was added inside each plastic bag. Then, the bags were arranged inside cardboard containers and stored at room temperature (27–30° C) for 18 days. They found that using 30 g of scavenger powder per 6 units of packaged fruits was an effective method to delay the ripening processes and extend storage of bananas for 18 days. Chamara et al. (2000) also studied the efficacy of $KMnO_4$ supported on clay bricks in combination with modified atmosphere to extend the storage life of bananas and observed that this treatment significantly lowered in-package ethylene and carbon dioxide contents and minimized changes in firmness and total soluble solids content, compared with control fruits packed without C_2H_4 scavenger.

More recently, Álvarez-Hernández et al. (2019) developed a $KMnO_4$ support material using protonated MMT (PMMT) and evaluated its performance on the postharvest quality of blueberry fruit stored during 46 days at 2 and 10° C. The combined effect of the ethylene scrubber and modified atmosphere packaging technology significantly increased the shelf life of blueberry fruit from 7 to 21 days at 10° C and from 21 to 46 days at 2° C when compared to the fruit stored under air conditions. Although similar positive effects were obtained using a commercial $KMnO_4$ support material (based on zeolite), the significantly less quantity of material used with the developed C_2H_4 scrubber (sixfold lower) reduced the costs associated with the application of this effective packaging technology. The same authors developed another $KMnO_4$-based ethylene scavenger using sepiolite clay as support material and evaluated its effects on the postharvest quality of 'Mirlo naranja' apricots stored under modified atmosphere packaging at 2° C and under air packaging conditions at 15° C (Álvarez-Hernández et al., 2020). Sepiolite is a nontoxic, inexpensive, and eco-friendly fibrous clay, which possesses microporous channels in the fiber direction that confers it high C_2H_4 adsorption properties (Alver and Sakizci, 2012). The novel C_2H_4 scavenger developed by Álvarez-Hernández et al. (2020) provided a more favorable in-package gaseous composition for apricot storage than the commercially used scavenger ($KMnO_4$ supported on zeolite). Moreover, 50% less quantity of the developed scavenger was needed compared with the commercial one. However, the authors emphasized that further studies were necessary to evaluate the performance of the developed novel scavenger when the fruits were stored at not recommended conditions of temperature, as well as to determine the costs associated with its manufacturing.

7.5 Commercially Available Products Containing Nanoclays for Food Packaging

There are several nanoclay products already available in the market and specially designed for food and beverage packaging purposes. Most of these products exploit the high-barrier properties of PCNs for applications in modified atmosphere packaged foods (meats, cheese, confectionery, cereals, boil-in-the-bag products), in extrusion coatings combined with paperboard for fruit juice and dairy products, and in co-extrusion processes for the fabrication of beer and carbonated drinks bottles (Unalan et al., 2014). For instance, Nanocor™ (USA) possesses more than 40 patents for the manufacture and commercialization of a wide variety of polymer nanocomposites incorporating clays; other registered similar products include Durethan® (previously owned by Bayer, now from LANXESS, Germany), Aegis™ (Honeywell polymers, Korea), Imperm® (ColorMatrix Corp., USA), NanoTuff™ (Nylon Corporation of America, USA), and NanoSeal™ (NanoPack Inc., USA) (Echegoyen, 2015). Although not particularly designed for that purpose, some of these products can be potentially applied for packaging horticultural produce. For example, Honeywell has developed a polyamide-based nanocomposite under the Aegis® trade name, which incorporates clay particles as passive barrier against oxygen. In particular, Aegis® HFX is a nylon resin formulated for use in co-injection PE bottles that provides glass-like clarity and recyclability and satisfies shelf-life requirements for hot-fill applications including fruit juices. Thanks to an alliance between Mitsubishi Gas Chemical Company Inc. and Nanocor™, a nanocomposite film obtained by melt compounding Nanocor's nanoclay additives with Mitsubishi's MXD 6-nylon has been developed under the name of "M9" for use in barrier multilayer films and thermoformed containers. Another hybrid film, called Durethan® KU2-2601, has been developed via the in situ polymerization strategy and is composed of polyamide and layered silicates. This nanocomposite presents enhanced gloss and stiffness and the embedded nanoclay prevents the penetration of gases into the film and escaping of moisture. It is intended for use in applications where conventional polyamide is more permeable and too expensive, for example, paperboard juice containers. In fact, it can protect highly oxygen-sensitive package food products, such as orange juice, at a much lower cost (Ranjan et al., 2014).

The use of PCNs in food packaging is still a developing area of research and novel low-cost materials that improve the packaging properties using relatively simple manufacturing processes are continuously being designed.

7.6 Toxicity and Risk Assessment

The toxicity evaluation of nanoparticles is a topic of growing interest mainly because they are included in different food applications. Particularly, toxicity information about nanoclays is still scarce mainly in relation to the specific interaction of these nanoparticles with food and environment. Particularly, in food packaging applications, the principal risk is the migration of nanoclay particles to the food product and the subsequent ingestion by the consumer. In this regard, European regulations (European Commission, 2011) have established limits in nanoparticle migration from polymer composites into food/food simulants based on standard migration tests assessed by the European Union (EU) and the FDA.

Regulatory compliance in the particular case of nanoclays for food applications takes into account two important issues: The first is the size of the nanoclay particles. According to this perspective, applications of PCNs containing intercalated and/or exfoliated nanoplatelets must be specifically authorized. The other aspect is that the surfactants used for organic modification of nanoclays need to be authorized either by the EU Regulation No 10/2011 or by the FDA (Störmer et al., 2017). However, the EU regulation established a limit of 10 mg dm^{-2} to the total permitted migration from a nanocomposite. PCN-migration studies are few in number and ambiguous. Until now, evaluations about the migration of nanoclay particles into vegetables using biodegradable clay nanocomposite packaging aroused promising results establishing that the overall migration was within the limits established by the European directives (European Commission, 2007), thus indicating that these materials are suitable for food packaging applications (Avella et al., 2005). Another study (Mauricio-Iglesias et al., 2010) showed that Uvitex OB, an authorized additive for food contact use in Europe, (2002/72/EEC) has a very low release from MMT/wheat-gluten film, compared to a loss up to 60% in LLDPE films. Additionally, the MMT migration of the wheat gluten–based films, using aluminum and silicon as markers, was low and within the limits set by the European Food Safety Authority for maximum aluminum dietary intake. Another paper demonstrated that in both additives, triclosan and DPBD, diffusion was lower in poly(amide)/clay nanocomposite than in neat poly(amide) (de Abreu et al., 2010; Šimon et al., 2008). However, the work of Mauricio-Iglesias et al. (2010) showed that the migration of nanoclay components to food simulants, when the MMT/wheat gluten\ was subjected to a high-pressure treatment, was higher probably due to structural and chemical changes during the processing of the film. In the same sense, nanoparticles could come in contact with food as a consequence of mechanical impact.

In this sense, it is important the evaluation of migration studies of nanoparticles during mechanical and thermal treatments and after and before flexing the packaging material (Bandyopadhyay and Ray, 2019).

While previous studies demonstrated that PCNs may slow down the migration of potentially harmful additives into foods, the health risk associated with the ingestion of food containing nanoclay particles that comes from the packaging is not yet fully understood. Current toxicological research is fragmentary and incomplete. What is really known is that when the particles are smaller, the faster and easier are absorbed generating cell and tissue damage (Han et al., 2011). Although toxicity evaluations of nanoclay particles have increased during the last years, results are not clear since some studies describe the toxic effects of nanoclays and, by comparison, others indicate the absence of toxicity (Brandelli, 2018). In general, *in vitro* studies reveal cytotoxicity in nanoclays depending on the concentration and the type of clay. However, *in vivo* studies in rodents indicate no systemic toxicity even at doses as high as 5 g/kg (Maisanaba et al., 2015). Based on these results, a systematic evaluation of toxicological effects is needed.

In conclusion, data regarding toxicological effects of clay nanoparticles is still scarce and poor. The development of new strategies is needed in order to detect and categorize nanoclays and other nanoparticles in complex food matrices. A case-by-case approach is needed depending on the type of clay and taking into account that each clay has its own toxicological profile. Although clay nanocomposites may represent the next revolution for food packaging applications, more toxicological information is needed to protect consumers from any potential hazards associated with the use of PCNs.

7.7 Conclusion

The safety of fruits and vegetables and the preservation of food quality depends on the continuous development of innovative food packaging materials. In this regard, the incorporation of nanoclay particles in the formulation of food packaging materials constitutes one of the most promising classes of materials mainly due to the enhancements in the mechanical and barrier properties of the packaging. This, in turn, helps preserve the quality attributes of foods with a considerable improvement in shelf life. However, migration tests and short- and long-term toxicity studies are still needed in order to guarantee the consumer's health.

References

Abedi, S. and Abdouss, M. 2014. A review of clay-supported Ziegler-Natta catalysts for production of polyolefin/clay nanocomposites through in situ polymerization. *Applied Catalysis A: General* 475, 386–409.

Albdiry, M., Yousif, B., Ku, H. and Lau, K. 2013. A critical review on the manufacturing processes in relation to the properties of nanoclay/polymer composites. *Journal of Composite Materials* 47, 1093–1115.

Álvarez-Hernández, M.H., Artés-Hernández, F., Ávalos-Belmontes, F., Castillo-Campohermoso, M.A., Contreras-Esquivel, J.C., Ventura-Sobrevilla, J.M. and Martínez-Hernández, G.B. 2018. Current scenario of adsorbent materials used in ethylene scavenging systems to extend fruit and vegetable postharvest life. *Food and Bioprocess Technology* 11, 511–525.

Álvarez-Hernández, M.H., Martínez-Hernández, G.B., Avalos-Belmontes, F., Rodríguez-Hernández, A.M., Castillo-Campohermoso, M.A. and Artés-Hernández, F. 2019. An innovative ethylene scrubber made of potassium permanganate loaded on a protonated montmorillonite: A case study on blueberries. *Food and Bioprocess Technology* 12, 524–538.

Álvarez-Hernández, M.H., Martínez-Hernández, G.B., Avalos-Belmontes, F., Miranda-Molina, F.D. and Artés-Hernández, F. 2020. Postharvest quality retention of apricots by using a novel sepiolite-loaded potassium permanganate ethylene scavenger. *Postharvest Biology and Technology* 160, 111061.

Alver, B.E. and Sakizci, M. 2012. Ethylene adsorption on acid-treated clay minerals. *Adsorption Science and Technology* 30, 265–273.

Artés, F. and Allende, A. 2014. Minimal processing of fresh fruit, vegetables, and juices. pp. 583–597. *In*: Da-Wen Sun (ed.). *Emerging Technologies for Food Processing* (2nd Edition). Academic Press, London.

Asensio, M., Herrero, M., Núñez, K., Gallego, R., Merino, J.C. and Pastor, J.M. 2018. In situ polymerization of isotactic polypropylene sepiolite nanocomposites and its copolymers by metallocene catalysis. *European Polymer Journal* 100, 278–289.

Avella., M., De Vlieger, J.J., Errico, M.E., Fischer, S., Vacca, P. and Volpe, M.G. 2005. Biodegradable starch/clay nanocomposite films for food packaging applications. *Food Chemistry* 93, 467–474.

Azeredo, H.M.C., Miranda, K.W.E., Ribeiro, H.L., Rosa, M.F. and Nascimento, D.M. 2012. Nanoreinforced alginate-acerola puree coatings on acerola fruits. *Journal of Food Engineering* 113, 505–510.

Azeredo, H. 2013. Antimicrobial nanostructures in food packaging. *Trends in Food Science and Technology* 30, 56–69.

Bandyopadhyay, J. and Ray, S.S. 2019. Are nanoclay-containing polymer composites safe for food packaging applications?—An overview. *Journal of Applied Polymer Science* 136, 47214.

Barton, C.D. and Karathanasis, A.D. 2016. Clay minerals. pp. 187–192. *In*: R. Lal (ed.). *Encyclopedia of Soil Science*. Marcel Dekker, New York.

Beyer, G. 2002. Nanocomposites: A new class of flame retardants for polymers. *Plastics, Additives and Compounding* 4, 22–28.

Biddeci, G., Cavallaro, G., Di Blasi, F., Lazzara, G., Massaro, M., Milioto, S., Parisi, F., Riela, S. and Spinelli, G. 2016. Halloysite nanotubes loaded with peppermint essential oil as filler for functional biopolymer film. *Carbohydrate Polymers* 152, 548–557.

Boonsiriwit, A., Xiao, Y., Joung, J., Kim, M., Singh, S. and Lee, Y.D. 2020. Alkaline halloysite nanotubes/low density polyethylene nanocomposite films with increased ethylene absorption capacity: Applications in Cherry tomato packaging. *Food Packaging and Shelf Life* 25, 100533.

Brandelli, A. 2018. Toxicity and safety evaluation of nanoclays. pp. 57–76. *In*: M. Rai and J.K. Biswas (eds.). *Nanomaterials: Ecotoxicity, Safety, and Public Perception*. Springer Nature, Switzerland, AG. https://doi.org/10.1007/978-3-030-05144-0_4.

Bumbudsanpharoke, N. and Ko, S. 2015. Nano-food packaging: An overview of market, migration research, and safety regulations. *Journal of Food Science* 80, 910–923.

Campos-Requena, V.H., Rivas, B.L., Pérez, M.A., Figueroa, C.R. and Sanfuentes, E.A. 2015. The synergistic antimicrobial effect of carvacrol and thymol in clay/polymer nanocomposite films over strawberry gray mold. *LWT–Food Science and Technology* 64, 390–396.

Chamara, D., Illeperuma, K. and Galappatty, P. 2000. Effect of modified atmosphere and ethylene absorbers on extension of storage life of 'Kolikuttu' banana at ambient temperature. *Fruits* 55, 381–388.

Cherifi, Z., Boukoussa, B., Zaoui, A., Belbachir, M. and Meghabar, R. 2018. Structural, morphological and thermal properties of nanocomposites poly (GMA)/clay prepared by ultrasound and in-situ polymerization. *Ultrasonics Sonochemistry* 48, 188–198.

Choudalakis, G. and Gotsis, A. 2009. Permeability of polymer/clay nanocomposites: A review. *European Polymer Journal* 45, 967–984.

Cortez-Vega, W.R., Pizato, S., Andreghetto de Souza, J.S. and Prentice, C. 2014. Using edible coatings from Whitemouth croaker (*Micropogonias furnieri*) protein isolate and organo-clay nanocomposite for improve the conservation properties of fresh-cut 'Formosa' papaya. *Innovative Food Science and Emerging Technologies* 22, 197–202.

Costa, C., Conte, A., Buonocore, G.G. and Del Nobile, M.A. 2011. Antimicrobial silver-montmorillonite nanoparticles to prolong the shelf life of fresh fruit salad. *International Journal of Food Microbiology* 148, 164–167.

Costa, C., Conte, A., Buonocore, G.G., Lavorgna, M. and Del Nobile, M.A. 2012. Calcium-alginate coating loaded with silver-montmorillonite nanoparticles to prolong the shelf-life of fresh-cut carrots. *Food Research International* 48, 164–169.

Dasgupta, N. and Ranjan, S. 2018. Nanotechnology in food packaging. pp. 129–150. *In*: N. Dasgupta and S. Ranjan (eds.). *An Introduction to Food Grade Nanoemulsions. Environmental Chemistry for a Sustainable World*. Springer, Singapore.

de Abreu, D.A.P., Cruz, J.M., Angulo, I. and Losada, P.P. 2010. Mass transport studies of different additives in polyamide and exfoliated nanocomposite polyamide films for food industry. *Packaging Technology and Science* 23, 59–68.

De Silva, R.T., Pasbakhsh, P., Lee, S.M. and Kit, A.Y. 2015. ZnO deposited/encapsulated halloysite-poly(lactic acid) (PLA) nanocomposites for high performance packaging films with improved mechanical and antimicrobial properties. *Applied Clay Science* 111, 10–20.

Debnath, D., Dhibar, A.K. and Khatua, B. 2010. Studies on the morphology and properties of PMMA-organoclay nanocomposites with reference to the manufacturing techniques. *Polymer-Plastics Technology and Engineering* 49, 1087–1094.

Dennis, H.R., Hunter, D.L., Chang, D., Kim, S., White, J.L., Cho, J.W. and Paul, D.R. 2001. Effect of melt processing conditions on the extent of exfoliation in organoclay-based nanocomposites. *Polymer* 42, 9513–9522.

Dhital, R., Mora, N.B., Watson, D.G., Kohli, P. and Choudhary, R. 2018. Efficacy of limonene nano coatings on post-harvest shelf life of strawberries. *LWT–Food Science and Technology* 97, 124–134.

Ebrahimi, H., Abedi, B., Bodaghi, H., Davarynejad, G., Haratizadeh, H. and Conte, A. 2018. Investigation of developed clay-nanocomposite packaging film on quality of peach fruit (*Prunus persica* Cv. Alberta) during cold storage. *Journal of Food Processing and Preservation* 42, e13466.

Echegoyen, Y. 2015. Nano-developments for food packaging and labeling applications. pp. 141–166. *In*: Rai, M., Ribeiro, C., Mattoso, L., Duran, N. (eds). *Nanotechnologies in Food and Agriculture*. Springer International Publishing Switzerland, Cham. Switzerland. https://doi.org/10.1007/978-3-319-14024-7_7.

European Commission. 2007. Directive 2007/19/EC. *Official Journal of the European Union*, L 91/1717-36.2

European Commission. 2011. Commission Regulation (EC) No. 10/2011 on plastic materials and articles intended to come into contact with food. *Official Journal of the European Union*, 12, 1–89.

Fischer, H., Gielgens, L. and Koster, T. 1999. Nanocomposites from polymers and layered minerals. *Acta Polymerica* 50, 122–126.

Gaaz, T., Sulong, A., Kadhum, A., Al-Amiery, A., Nassir, M. and Jaaz, A. 2017. The impact of halloysite on the thermo-mechanical properties of polymer composites. *Molecules* 22, 838.

Gaikwad, K.K., Singh, S. and Lee, Y.S. 2018. High adsorption of ethylene by alkali-treated halloysite nanotubes for food-packaging applications. *Environmental Chemistry Letters* 16, 1055–1062.

Gaikwad, K.K., Singh, S. and Negi, Y.S. 2020. Ethylene scavengers for active packaging of fresh food produce. *Environmental Chemistry Letters* 18, 269–284.

Ghoshal, G. 2018. Recent trends in active, smart, and intelligent packaging for food products. pp. 343–374. *In*: A.M. Grumezescu and A.M. Holban (eds.). *Food Packaging and Preservation. Handbook of Food Bioengineering*, Volume 9. Academic Press, London.

Gorrasi, G. 2015. Dispersion of halloysite loaded with natural antimicrobials into pectins: Characterization and controlled release analysis. *Carbohydrate Polymers* 127, 47–53.

Gorrasi, G. and Bugatti, V. 2016. Edible bio-nano-hybrid coatings for food protection based on pectins and LDH-salycilate: Preparation and analysis of physical properties. *LWT–Food Science and Technology* 69, 139–145.

Guo, F., Aryana, S., Han, Y. and Jiao, Y. 2018. A review of the synthesis and applications of polymer-nanoclay composites. *Applied Science* 8, e1696.

Gurses, A. 2015. *Introduction to Polymer-Clay Nanocomposites*. Pan Stanford, Singapore.

Han, W., Yu, Y.J., Li, N.T. and Wang, L.B. 2011. Application and safety assessment for nano-composite materials in food packaging. *Chinese Science Bulletin* 56, 1216–1225.

He, X., Deng, H. and Hwang, H. 2019. The current application of nanotechnology in food and agriculture. *Journal of Food and Drug Analysis* 27, 1–21.

Houtman, J., Maisanaba, S., Puerto, M., Gutiérrez-Praena, D., Jordá, M., Aucejo, S. and Jos, A. 2014. Toxicity assessment of organomodified clays used in food contact materials on human target cell lines. *Applied Clay Science* 90, 150–158.

Hu, Q., Fang, Y., Yang, Y., Ma, N. and Zhao, L. 2011. Effect of nanocomposite-based packaging on postharvest quality of ethylene-treated kiwifruit (*Actinidia deliciosa*) during cold storage. *Food Research International* 44, 1589–1596.

Huang, Y., Yang, K. and Dong, J.Y. 2006. Copolymerization of ethylene and 10-undecen-1-ol using a montmorillonite-intercalated metallocene catalyst: Synthesis of polyethylene/montmorillonite nanocomposites with enhanced structural stability. *Macromolecular Rapid Communications* 27, 1278–1283.

Ibarguren, M., López, D.J. and Escribá, P.V. 2014. The effect of natural and synthetic fatty acids on membrane structure, microdomain organization, cellular functions and human health. *Biochimica et Biophysica Acta* 1838, 1518–1528.

Incoronato, A.L., Buonocore, G.G., Conte, A., Lavorgna, M. and Del Nobile, M.A. 2010. Active systems based on silver-montmorillonite nanoparticles embedded into bio-based polymer matrices for packaging applications. *Journal of Food Protection* 73, 2256–2262.

Irshidat, M.R. and Al-Saleh, M.H. 2018. Thermal performance and fire resistance of nanoclay modified cementitious materials. *Construction and Building Materials* 159, 213–219.

Issa, A., Ibrahim, S.A. and Tahergorabi, R. 2017. Impact of sweet potato starch-based nanocomposite films activated with thyme essential oil on the shelf-life of baby spinach leaves. *Foods* 6, 43.

Jawaid, M., Qaiss, A.K. and Bouhfid, R. 2016. *Nanoclay Reinforced Polymer Composites: Nanocomposites and Bionanocomposites*. Springer, Singapore.

Karamane, M., Raihane, M., Tasdelen, M.A., Uyar, T., Lahcini, M., Ilsouk, M. and Yagci, Y. 2017. Preparation of fluorinated methacrylate/clay nanocomposite via in-situ polymerization: Characterization, structure, and properties. *Journal of Polymer Science Part A: Polymer Chemistry* 55, 411–418.

Khoshtaghaza, M.H. and Taghinezhad, E. 2016. Investigation effect of particle nano coating on storage quality properties of Thomson Orange. *Iranian Journal of Food Science and Technology* 61, 121–133.

Lazzara, G., Cavallaro, G., Panchal, A., Fakhrullin, R., Stavitskaya, A., Vinokurov, V. and Lvov, Y. 2018. An assembly of organic-inorganic composites using halloysite clay nanotubes. *Current Opinion in Colloid and Interface Science* 35, 42–50.

Lee, M.H. and Park, H.J. 2015. Preparation of halloysite nanotubes coated with Eudragit for a controlled release of thyme essential oil. *Journal of Applied Polymer Science* 132, 42771.

Lee, M.H., Seo, H.S. and Park, H.J. 2017. Thyme oil encapsulated in halloysite nanotubes for antimicrobial packaging system. *Journal of Food Science* 82, 922–932.

Li, H., Li, F., Wang, L., Sheng, J., Xin, Z., Zhao, L., Xiao, H., Zheng, Y. and Hu, Q. 2009. Effect of nano-packing on preservation quality of Chinese jujube (*Ziziphus jujuba Mill.* var. *inermis (Bunge) Rehd*). *Food Chemistry* 114, 547–552.

Liu, M., Jia, Z., Jia, D. and Zhou, C. 2014. Recent advance in research on halloysite nanotubes-polymer nanocomposite. *Progress in Polymer Science* 39, 1498–1525.

Lvov, Y. and Abdullayev, E. 2013. Functional polymer—clay nanotube composites with sustained release of chemical agents. *Progress in Polymer Science* 38, 1690–1719.

Maisanaba, S., Pichardo, S., Puerto, M., Gutierrez-Praena, D., Camean, A.M. and Jos, A. 2015. Toxicological evaluation of clay minerals and derived nanocomposites: A review. *Environmental Research* 138, 233–254.

Majeed, K., Jawaid, M., Hassan, A., Abu Bakar, A., Abdul Khalil, H.P.S., Salema, A.A. and Inuwa, I. 2013. Potential materials for food packaging from nanoclay/natural fibres filled hybrid composites. *Materials and Design* 46, 391–410.

Mauricio-Iglesias, M., Peyron, S., Guillard, V. and Gontard, N.J. 2010. Wheat gluten nanocomposite films as food-contact materials: Migration tests and impact of a novel food stabilization technology (high pressure). *Applied Polymer Science* 116, 2526–2535.

Meira, S.M.M., Zehetmeyer, G., Jardim, A.I., Scheibel, J.M., Oliveira, R.V.B. and Brandelli, A. 2014. Polypropylene/montmorillonite nanocomposites containing nisin as antimicrobial food packaging. *Food Bioprocess Technology* 7, 3349–3357.

Mihindukulasuriya, S.D.F. and Lim, L.T. 2014. Nanotechnology development in food packaging: A review. *Trends in Food Science and Technology* 40, 149–167.

Motamedi, E., Nasiri, J., Malidarreh, T.R., Kalantari, S., Naghavi, M.R. and Safari, M. 2018. Performance of carnauba wax-nanoclay emulsion coatings on postharvest quality of 'Valencia' orange fruit. *Scientia Horticulturae* 240, 170–178.

Nasirifar, S.Z., Maghsoudlou, Y. and Oliyaei, N. 2018. Effect of active lipid-based coating incorporated with nanoclay and orange peel essential oil on physicochemical properties of *Citrus sinensis*. *Food Science and Nutrition* 6, 1508–1518.

Nazir, M.S., Kassim, M.H.M., Mohapatra, L., Gilani, M.A., Raza, M.R. and Majeed, K. 2016. Characteristic properties of nanoclays and characterization of nanoparticulates and nanocomposites. pp. 35–55. *In*: M. Jawaid, A. Qaiss and R. Bouhfid (eds.). *Nanoclay Reinforced Polymer Composites. Nanocomposites and Bionanocomposites*. Springer, Singapore.

Ozkose, U.U., Altinkok, C., Yilmaz, O., Alpturk, O. and Tasdelen, M.A. 2017. In-situ preparation of poly(2-ethyl-2-oxazoline)/clay nanocomposites via living cationic ring-opening polymerization. *European Polymeric Journal* 88, 586–593.

Ports, B.F. and Weiss, R. 2010. One-step melt extrusion process for preparing polyolefin/clay nanocomposites using natural montmorillonite. *Industrial & Engineering Chemistry Research* 49, 11896–11905.

Ranjan, S., Dasgupta, N., Chakraborty, A.R., Samuel, S.M., Ramalingam, C., Shanker, R. and Kumar, A. 2014. Nanoscience and nanotechnologies in food industries: Opportunities and research trends. *Journal of Nanoparticle Research* 16, 2464.

Ray, S.S. and Okamoto, M. 2003. Polymer/layered silicate nanocomposites: A review from preparation to processing. *Progress in Polymer Science* 28, 1539–15641.

Ray, S.S. and Bousmina, M. 2005. Biodegradable polymers and their layered silicate nanocomposites: In greening the 21st century materials world. *Progress in Materials Science* 50, 962–1079.

Saad, A., Jlassi, K., Omastová, M. and Chehimi, M.M. 2017. Chapter 6: Clay/conductive polymer nanocomposites. pp. 199–237. *In:* K. Jlassi, M.M. Chehimi and S. Thomas (eds.). *Clay-Polymer Nanocomposites.* Elsevier, Cambridge, MA.

Salmi, Z., Benzarti, K. and Chehimi, M.M. 2013. Diazonium cation-exchanged clay: An efficient, unfrequented route for making clay/polymer nanocomposites. *Langmuir* 9, 13323–13328.

Santosa, E., and Widodo, W.D. 2013. The use of clay as potassium permanganate carrier to delay the ripening of Raja Bulu Banana. *Journal Hortikultura Indonesia* 1, 88–95.

Sarfraz, J., Gulin-Sarfraz, T., Nilsen-Nygaard, J. and Pettersen, M.K. 2021. Nanocomposites for food packaging applications: An overview. *Nanomaterials* 11, 10.

Shah, R., Kausar, A., Muhammad, B. and Shah, S. 2015. Progression from graphene and graphene oxide to high performance polymer-based nanocomposite: A review. *Polymer* 69, 369–383.

Šimon, P., Chaudhry, Q. and Bakoš, D. 2008. Migration of engineered nanoparticles from polymer packaging to food—A physicochemical view. *Journal of Food and Nutrition Research* 47, 105–113.

Störmer, A., Bott, J., Kemmer, D. and Franz, R. 2017. Critical review of the migration potential of nanoparticles in food contact plastics. *Trends in Food Science & Technology* 63, 39–50.

Taghinezhad, E. and Ebadollahi, A. 2017. Potential application of chitosan-clay coating on some quality properties of lemon during storage. *Agricultural Engineering International: CIGR Journal* 19, 189–194.

Tas, C.E., Hendessi, S., Baysal, M., Unal, S., Cebeci, F.C., Menceloglu, Y.Z. and Unal, H. 2017. Halloysite nanotubes/polyethylene nanocomposites for active food packaging materials with ethylene scavenging and gas barrier properties. *Food and Bioprocess Technology* 10, 789–798.

Thakur, V.K. and Kessler, M.R. 2015. Self-healing polymer nanocomposite materials: A review. *Polymer* 69, 369–383.

Uddin, M.K. 2017. A review on the adsorption of heavy metals by clay minerals, with special focus on the past decade. *Chemical Engineering Journal* 308, 438–462.

Unalan, I.U., Cerri, G., Marcuzzo, E., Cozzolino, C.A. and Farris, S. 2014. Nanocomposite films and coatings using inorganic nanobuilding blocks (NBB): Current applications and future opportunities in the food packaging sector. *RSC Advances* 4, 29393–29428.

Valapa, R.B., Loganathan, S., Pugazhenthi, G., Thomas, S. and Varghese, T.O. 2017. Chapter 2: An overview of polymer—clay nanocomposites. pp. 29–81. *In:* K. Jlassi, M.M. Chehimi and S. Thomas (eds.). *Clay-Polymer Nanocomposites.* Elsevier, Cambridge, MA.

Viacava, G.E., Ayala-Zavala, J.F., Vázquez, J. and Ansorena, M.R. 2017. Sustainability challenges involved in use of nanotechnology in agro-food sector. pp. 343–368.

In: R. Bhat and V. Gomez-Lopez (eds.). *Sustainability Challenges in the Agro-Food Sector.* Wiley-Blackwell, West Sussex.

Vo, V.S., Mahouche-Chergui, S., Babinot, J., Nguyen, V.H., Naili, S. and Carbonnier, B. 2016. Photo-induced SI-ATRP for the synthesis of photoclickable intercalated clay nanofillers. *RSC Advances* 6, 89322–89327.

Yang, F.M., Li, H.M., Li, F., Xin, Z.H., Zhao, L.Y., Zheng, Y.H. and Hu, Q.H. 2010. Effect of nano-packing on preservation quality of fresh strawberry (*Fragaria ananassa* Duch. cv Fengxiang) during storage at 4°C. *Journal of Food Science* 75, C236–C240.

Yusoh, K., Kumaran, S.V. and Ismail, F.S. 2017. Surface modification of nanoclay for the synthesis of polycaprolactone (PCL)-clay nanocomposite. Proceedings of the MATEC Web of Conferences, Penang, 6–7.

Zhang, Y. and Park, S.J. 2018. In situ shear-induced mercapto group-activated graphite nanoplatelets for fabricating mechanically strong and thermally conductive elastomer composites for thermal management applications. *Composites Part A: Applied Science and Manufacturing* 112, 40–48.

Chapter 8

Hybrid Nanomaterials for Postharvest Storage

C. Vibha, Jyotishkumar Parameswaranpillai, K. Senthilkumar, Suchart Siengchin, Sabarish Radoor, G.L. Praveen, Midhun Dominic C.D., M. Chandrasekar, and T. Senthil Muthu Kumar

8.1 Introduction

A thorough knowledge of horticultural commodities (i.e., fruits, vegetables, etc.) is necessary for reducing postharvest wastage. Excellent technology and expertise are being introduced through research and development in postharvest storage. Although techniques and technologies are available, several regions are not utilizing them due to their improper usage and higher cost. Consequently, significant losses (both in quality and quantity) in horticultural commodities would occur. For instance, in developing countries, the losses of tomatoes, grapes, and avocados were up to 50%, 27%, and 43%, respectively, thus the efforts to further improve and enhance the proper use of postharvest handling techniques in developing countries [1].

Postharvest diseases may arise at any point of postharvest handling, that is, from harvest to consumption. Diseases caused by plant-pathogenic bacteria or fungi in fruits and vegetables can lead to severe losses during transportation and storage. The high demand for food production proportionally increases pesticide usage in the environment, resulting in the incidence of harmful chemicals in water, soil, and air. Pesticides contain toxic chemicals intended to eliminate various pests, fungi, weeds, and bacteria [2, 3]. The excessive and prolonged use of pesticides may result in the accumulation of harmful chemicals on the surface of fruits and vegetables. This may adversely affect the biota [4].

Nanomaterials and nanotechnology can be effectively utilized for developing environmentally friendly strategies and products for managing and controlling

DOI: 10.1201/9781003142287-8

bacterial diseases. Nowadays, the use of nanotechnology is being utilized by many researchers in different scientific fields, including agriculture. The huge surface area and high functionality of nanomaterials have made them the best choice for sustainable horticulture applications, especially in the postharvest disease management of fruits and vegetables from the scientific level to the industry [5–8]. The application of nanotechnology in postharvest handling includes novel inventions of packaging films, controlling postharvest diseases, improving the appearance and properties of the packaging materials, and developing smart/intelligent packaging [9–13]. In addition, nanotechnology is employed to produce antifungal agents in various merchandise, including fruits and vegetables. For instance, Figure 8.1 demonstrates the application of nanomaterials in postharvest disease control [11].

This chapter discusses how to decrease postharvest losses (both qualitative and quantitative) by employing nanohybrid composites. Nanohybrid composites include those based on (1) titanium dioxide (TiO_2), (2) zinc oxide (ZnO), (3) copper oxide (CuO), (4) chitosan, and (5) cellulose. Using nanohybrid nanocomposites could help increase food availability. For instance, Figure 8.2 explains the application of hybrid nanocomposites in grapefruits.

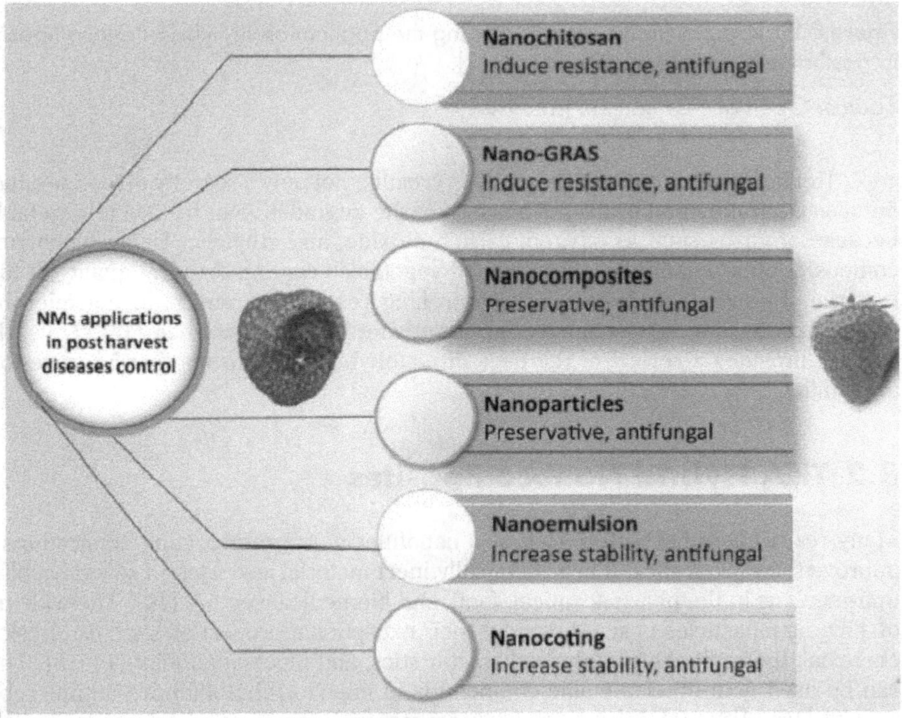

Figure 8.1 Application of nanomaterials in the management of postharvest diseases [11].

Source: Reproduced with thanks from Elsevier, License number: 5107350031442.
Note: NM: nanomaterials, GRAS: generally recognized as safe.

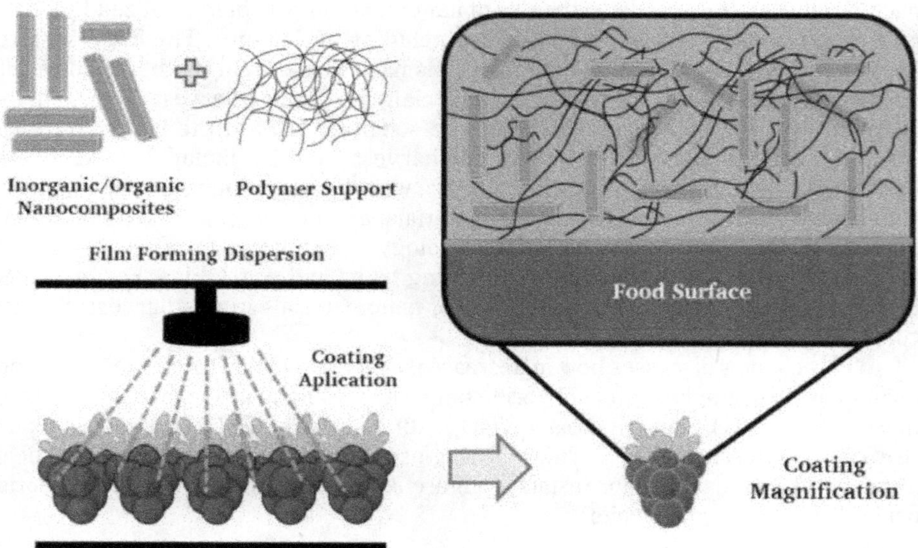

Figure 8.2 Representative image showing the application of hybrid nanocomposites in postharvest disease management [12].

Source: Open access content from MDPI.

Horticultural commodity wastage is creating not only a scarcity of food products but also environmental problems because of the degradation of fruits and vegetables because of gases such as oxygen, carbon dioxide, and ethylene. Using biopolymer composite films/coatings over fruits and vegetables can improve the quality of food materials by maintaining firmness, controlling respiration rate, reducing microbial growth, controlling color change, and improving barrier resistance [13]. Thus, this chapter provides a review of the uses of nanohybrid composites that improve food availability.

8.2 TiO_2 Hybrid Nanocomposites

Many researchers have used TiO_2 as a nanofiller in polymer coating applications to improve their properties. It is a chemically inert material also used in several applications such as in the pharmaceutical, food, and biomedical sectors [14]. The wide use of TiO_2 nanoparticles is supported by their exceptional properties such as physical, chemical, thermal, photocatalytic characteristics, and biocompatibility [14, 15]. They can be used with other metallic compounds to improve photoinduced antimicrobial activity [16]. Although it has many advantages, the main drawback is its agglomeration. For instance, Anaya-Esparza *et al.*, [17] reported that the interaction of biopolymers (e.g., chitosan, cellulose, starch, etc.) with TiO_2 minimizes the spontaneous agglomeration by improving the functional properties of the composite.

TiO_2-based hybrid composites can help preserve the quality of fruits by enhancing their shelf life, for instance, kiwifruit [18], Chinese jujube [19], strawberries

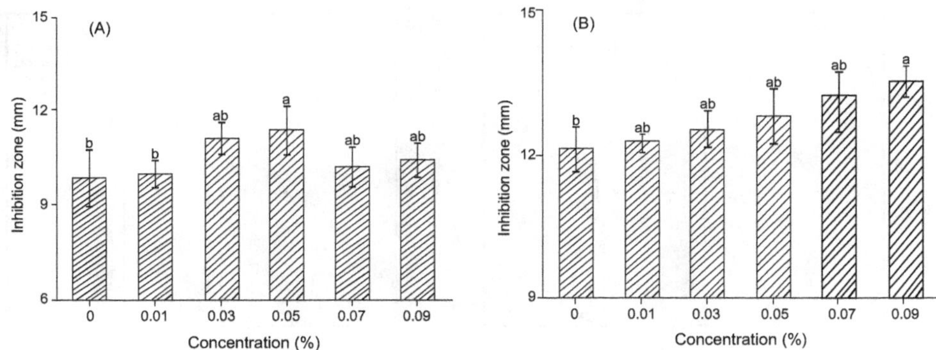

Figure 8.3 Inhibition zone of (A) *E. coli* and (B) *S. aureus* vs. TiO_2 concentration formed for chitosan/TiO_2 nanocomposite [23].

Source: Open access content from MDPI.

[20], and mangoes [21]. These advantageous properties are due to the higher barrier properties of hybrid composite material. Moreover, the customized atmosphere inside hybrid packaging had a vital role in improving postharvest storage [22, 23]. For instance, reactive oxygen species (ROS) generated by TiO_2 were effective in killing bacteria and improving the shelf life of fruits and vegetables. Figure 8.3. shows the inhibition zone of *E. coli* and *S. aureus* versus TiO_2 concentration formed for a chitosan/TiO_2 nanocomposite [23].

8.3 ZnO Nanoparticles

Among a wide range of nanomaterials, ZnO nanoparticles have gained greater attention due to their low cost, high thermochemical stability, and antifungal and antibacterial activity [24–26]. Also, ZnO was recognized as safe (GRAS) by U.S. Food and Drug Administration (FDA) according to Regulation 21 CFR 182.899; thus, they can be used as a food additive to inhibit microbial growth [25].

Sogvar *et al.* [26] observed that the strawberry fruit dipped in a distilled water solution containing nano-ZnO at concentrations of 0.03%, 0.07%, and 0.5% was effective in maintaining the quality of strawberry by showing better firmness, lower weight loss, increased vitamin C and total soluble solids (TSS), increased antimicrobial activity, and enhanced shelf life of the fruit. Lakshmi *et al.* [27] investigated the effect of a chitosan/green-synthesized ZnO nanoparticle (from spinach leaves) coating on the characteristics like firmness, color, pH, weight loss, microbial population, and TSS of fig fruit (*Ficus carica L.*). They proved that the coating of chitosan/ZnO nanoparticles extended the shelf life of fig fruits by maintaining firmness, inhibiting moisture loss and microbial growth, lowering color change (indicating delayed ripening), and increasing TSS with respect to storage time.

Li *et al.* [28] studied the effect of a coating of ZnO nanoparticles on polyvinyl chloride (PVC) films. The researchers observed that the PVC film coated with nano ZnO prolongs the shelf life of fresh-cut Fuji apples by increasing TSS, lowering polyphenol

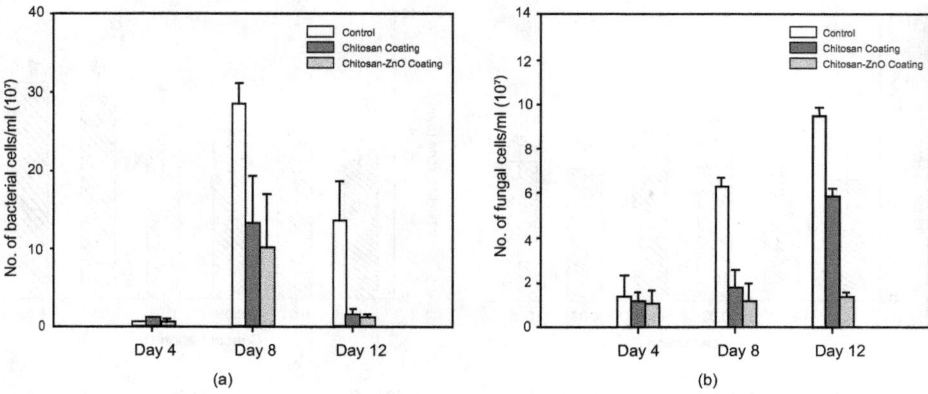

Figure 8.4 Antimicrobial response on okra coated with LDPE, LDEP/chitosan, and LDPE/chitosan–nano-ZnO: (a) bacterial and (b) fungal cells (colony forming units/ml of the sample) incubated for 4, 8, and 12 days.

Source: Open access content from MDPI [30].

oxidase and pyrogallol peroxidase activity, and reducing fruit decay. Emamifar *et al.* [29] investigated and demonstrated that the incorporation of nano-ZnO into a sodium alginate (SA) matrix improved its antimicrobial properties and extended the shelf life of fresh strawberries for up to 20 days. The weight loss (after 20 days of storage) lowered from 13% to 11%, 7%, and 5% with the addition of 0.25, 0.75, and 1.75 g/L of nano-ZnO, respectively, in the SA matrix. The firmness value, which is the measure of penetration force required to puncture a half-cut fruit, also increased from 1.23N to 1.91N, 2.55N, and 3.18N, respectively, with the addition of nano-ZnO at the weights mentioned earlier. Peroxidase (POD) activity decreased with nano-ZnO while the superoxide dismutase (SOD) activity increased, which is believed to show less tissue damage on the surface of the strawberry fruit. Al-Naamani *et al.* [30] explored the effect of chitosan and chitosan-ZnO nanocomposite coatings on low-density polyethylene (LDPE) packaging films on the shelf life of okra (ladyfingers). The study demonstrated that both the chitosan coating and the chitosan-ZnO nanocomposite coating retarded the fungal and microbial growth in okra samples incubated for 4, 8, and 12 days. Among the coating methods, the chitosan-ZnO coating was more effective in reducing the bacterial and fungal activity than the chitosan coating and control sample (without any coating), as shown in Figure 8.4.

8.4 CuO Nanoparticles

CuO nanoparticles are known for their exceptional antimicrobial properties [31]. It has been incorporated in packaging films to improve the shelf life of fresh-cut peppers [31], grapes [32], and guava [33], among others. Eshghi *et al.* [34] developed copper-integrated-nanochitosan hybrid composite suspensions and investigated the physicochemical property and bioactive components of strawberries. The nanocomposite coating considerably suppressed the fungal decay, polyphenol oxidase

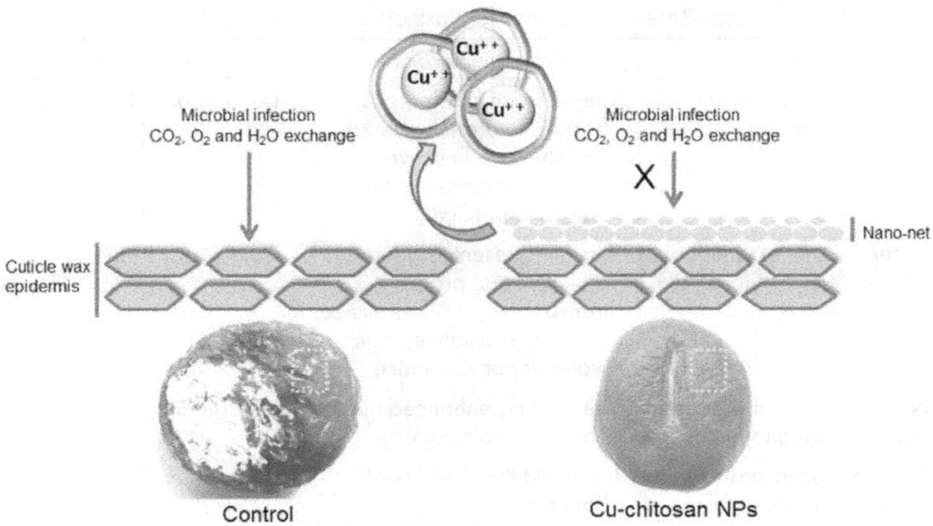

Figure 8.5 Photograph showing untreated and treated tomatoes after 21 days [36].

(PPO), and POD activity of the treated strawberries with respect to the untreated strawberries. Bonilla-Bird *et al.* [35] employed two-photon microscopy to study the penetration of copper particles into plant tissues such as the periderm, cortex, perimedulla, and medulla of sweet potatoes. They dipped the sweet potato for 30 min in CuO nanoparticle, bulk CuO, and $CuCl_2$ at 25, 75, and 125 mg/L. The researchers reported that CuO nanoparticles, even at higher concentrations (125 mg/L), were hardly found in the perimedulla and medulla, which makes them a potential alternative for preserving sweet potatoes for a longer duration after their harvest. Meena *et al.* [36] created a coating of Cu-chitosan nanoparticles by dipping tomatoes in a Cu-chitosan nanoparticle/water solution and observed an extended shelf life for the treated tomato due to the formation of nanonet over the tomato surface as shown in Figure 8.5.

8.5 Chitosan-Based Nanohybrids

Chitosan had attained great attention as a polysaccharide coating material due to its excellent antimicrobial activity (presence of the positively charged amino groups) and its outstanding film-forming ability (protects from mechanical injury in postharvest fruits) [37, 38]. However, chitosan possesses poor moisture barrier properties, limiting its efficiency in controlling moisture transfer. In view of that, there have been several studies carried out to enhance the functionality of chitosan-based coatings by integrating other functional materials, such as cellulose nanocrystals, essential oils, and others, into the chitosan coating matrix [39, 40]. Some chitosan-based composites for postharvest storage applications are provided in Table 8.1.

Table 8.1 Chitosan-Based Composites for Postharvest Storage Application

Modifications of chitosan	*Properties evaluated*	*References*
Cellulose nanocrystal and chitosan coating	The moisture barrier, gas barrier, and antibacterial activity were improved with the incorporation of cellulose nanocrystal (CNC).	Deng *et al.*, 2017 [39]
Citrus limonia essential oil integrated in chitosan film	This film presented good antimicrobial property, transparency, UV resistance, moisture resistance, high water vapor resistance.	Filho *et al.*, 2020 [40]
Nanoemulsion of mandarin essential oil and chitosan.	The coating enhanced the antimicrobial property.	Donsì *et al.*, 2015 [41]
Chitosan incorporated with chitosan nanoparticles	Improved the shelf life of bananas	Lustriane *et al.*, 2019 [42]
Hybrid nanocomposite film containing, chitosan, polyethylene glycol, gelatin, and silver nanoparticles	Improved the shelf life of red grapes	Kumar *et al.*, 2018 [43]
Clay-chitosan nanocomposite	Effective control of *Penicillium digitatum*.	Youssef *et al.*, 2020 [44]
Clay-chitosan nanocomposite	Improved the quality of lemon fruit.	Taghinezhad *et al.*, 2017 [45]
Chitosan/montmorillonite coating	Improved water vapor and oxygen barrier properties. Improved the shelf life of tangerine fruits.	Xu *et al.*, 2018 [46]
Chitosan/ silica nanocomposite	Effective control of gray mold on grapes	Youssef *et al.*, 2019 [47]
Chitosan-silver nanoparticles	Effective control of *Colletotrichum gloeosporioides* (anthracnose) in mangoes	Chowdappa *et al.*, 2014 [48]
Chitosan + pepper tree essential oil	Improved the quality of avocado fruits	Chávez-Magdaleno *et al.*, 2018 [49]

8.6 Cellulose-Based Nanohybrids

Cellulose is a nontoxic, abundant, cheap, renewable, and low-cost material. Therefore, there has been an upsurge in research in the area of cellulose-based composites [50]. Cellulose's small size, crystallinity, good thermo-mechanical properties, and biodegradation are causes for using it in packaging applications [51]. The consumption of blackberries

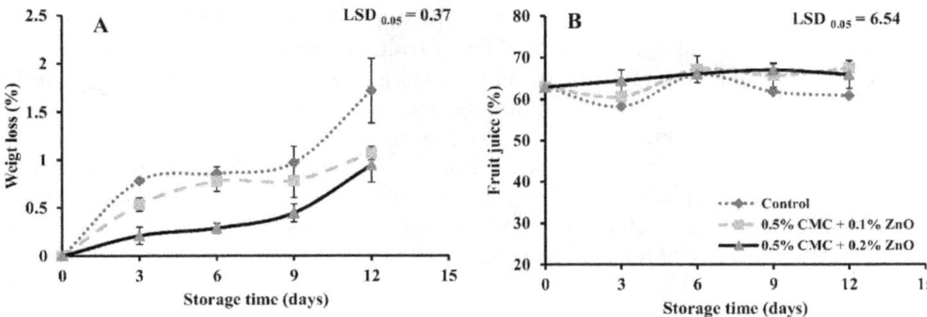

Figure 8.6 Change in weight loss (%) and fruit juice (%) with respect to storage time for the control sample and nano-ZnO-modified CMC samples [53].

Source: Reproduced with thanks from Elsevier, License Number: 5107531506457.

is higher due to their higher nutritional value. But blackberry fruit has a shorter shelf life due to its fragility and postharvesting disease. Silva *et al.* [52] developed lemongrass essential oil–integrated nanofibrillated cellulose coatings to extend the shelf life of blackberry fruits during the period of postharvest storage. The coating helped maintain the fruit's quality for 6 days and did not show any changes in soluble solids, and the visual appearance was also not changed for 6 days. Thus, the coating was very effective in extending the shelf life of blackberry fruits. Saba and Amini [53] applied nano-ZnO/carboxymethyl cellulose coating on pomegranate arils to improve their quality. The results reported that the combination of nano-ZnO/carboxymethyl cellulose (CMC) effectively improved the storage life of pomegranates. Furthermore, the coating reduced the weight loss, improved the juice content, and maintained the overall quality of the pomegranates during storage. Figure 8.6 shows the change in weight loss (%) and fruit juice (%) with respect to storage time for the control sample and nano-ZnO-modified CMC samples.

In another study, Resende *et al.*, [54] evaluated the quality of strawberries using chitosan/cellulose nanofibril-based coating. The coating was developed by varying the formulations between the chitosan, and cellulose nanofibril. The result analysis revealed that the concentration of cellulose nanofibril did not change the film color; however, it affected water permeability and thickness. Pacaphol *et al.* [55] studied the shelf life of vegetables (spinach) using nanofibrillated cellulose coating during postharvesting. Result analysis revealed that the coated leaves showed a significant improvement in moisture level, color, and appearance of the leaves when compared to the uncoated leaves. Thus, the researchers obtained the enhanced storage capacity of spinach using the nanocellulose coat. Widodo *et al.* [56] attempted to develop a nano-ZnO/cellulose coating and investigated its application on salak fruit by evaluating the parameters of weight loss and total acid. The weight loss and the drop in total acid were minimal for the fruits coated with the nano-ZnO coating during storage.

8.7 Conclusion

Maintaining the quality of horticultural commodities (e.g., vegetables, fruits) is important. Apart from the use of nanohybrid composites, many postharvest techniques are available to handle the horticultural commodities because these are highly recommended to maintain a healthy diet. The application of chemical fungicides is another practice for extending the shelf life of fruits/vegetables. Some of postharvest techniques are removing ethylene during the storage period; controlling atmospheric conditions such as temperature, humidity, air, and others; and using nano-based packaging systems. Furthermore, during the marketing of horticultural commodities, proper handling and protection are needed. When horticultural commodities are transported from one place to another place, refrigerated transport is essential. Otherwise, it will increase the deterioration and losses. Thus, a significant effort is needed to maintain the quality of horticultural commodities.

References

[1] Yahia, E.M., editor., 2019. Postharvest technology of perishable horticultural commodities. Woodhead Publishing.

[2] Schirra, M., D'Aquino, S., Cabras, P. and Angioni, A., 2011. Control of postharvest diseases of fruit by heat and fungicides: efficacy, residue levels, and residue persistence. A review. *Journal of Agricultural and Food Chemistry*, 59(16), pp. 8531–8542.

[3] Ruffo Roberto, S., Youssef, K., Hashim, A.F. and Ippolito, A., 2019. Nanomaterials as alternative control means against postharvest diseases in fruit crops. *Nanomaterials*, 9(12), p. 1752.

[4] Bhilwadikar, T., Pounraj, S., Manivannan, S., Rastogi, N.K. and Negi, P.S., 2019. Decontamination of microorganisms and pesticides from fresh fruits and vegetables: a comprehensive review from common household processes to modern techniques. *Comprehensive Reviews in Food Science and Food Safety*, 18(4), pp. 1003–1038.

[5] Singh, A., Singh, N.B., Hussain, I., Singh, H., Yadav, V. and Singh, S.C., 2016. Green synthesis of nano zinc oxide and evaluation of its impact on germination and metabolic activity of Solanum lycopersicum. *Journal of Biotechnology*, 233, pp. 84–94.

[6] Pomastowski, P., Król-Górniak, A., Railean-Plugaru, V. and Buszewski, B., 2020. Zinc oxide nanocomposites—extracellular synthesis, physicochemical characterization and antibacterial potential. *Materials*, 13(19), p. 4347.

[7] Mittal, A.K., Chisti, Y. and Banerjee, U.C., 2013. Synthesis of metallic nanoparticles using plant extracts. *Biotechnology Advances*, 31(2), pp. 346–356.

[8] Abdallah, Y., Liu, M., Ogunyemi, S.O., Ahmed, T., Fouad, H., Abdelazez, A., Yan, C., Yang, Y., Chen, J. and Li, B., 2020. Bioinspired green synthesis of chitosan and zinc oxide nanoparticles with strong antibacterial activity against rice pathogen Xanthomonas oryzae pv. oryzae. *Molecules*, 25(20), p. 4795.

[9] Anaya-Esparza, L.M., Villagrán-de la Mora, Z., Rodríguez-Barajas, N., Ruvalcaba-Gómez, J.M., Iñiguez-Muñoz, L.E., Maytorena-Verdugo, C.I., Montalvo-González, E. and Pérez-Larios, A., 2021. Polysaccharide-based packaging functionalized with inorganic nanoparticles for food preservation. *Polysaccharides*, 2(2), pp. 400–428.

[10] Khan, I., Saeed, K. and Khan, I., 2019. Nanoparticles: properties, applications and toxicities. *Arabian Journal of Chemistry*, 12(7), pp. 908–931.

[11] Alghuthaymi, M., Abd-Elsalam, K.A., Paraliker, P. and Rai, M., 2020. Mono and hybrid nanomaterials: novel strategies to manage postharvest diseases. *Multifunctional Hybrid Nanomaterials for Sustainable Agri-Food and Ecosystems*, pp. 287–317.

[12] Zambrano-Zaragoza, M.L., González-Reza, R., Mendoza-Muñoz, N., Miranda-Linares, V., Bernal-Couoh, T.F., Mendoza-Elvira, S. and Quintanar-Guerrero, D., 2018. Nanosystems in edible coatings: a novel strategy for food preservation. *International Journal of Molecular Sciences*, 19(3), p. 705.
[13] Jafarzadeh, S., Nafchi, A.M., Salehabadi, A., Oladzad-Abbasabadi, N. and Jafari, S.M., 2021. Application of bio-nanocomposite films and edible coatings for extending the shelf life of fresh fruits and vegetables. *Advances in Colloid and Interface Science*, p. 102405.
[14] Anaya-Esparza, L.M., Villagrán-de la Mora, Z., Ruvalcaba-Gómez, J.M., Romero-Toledo, R., Sandoval-Contreras, T., Aguilera-Aguirre, S., Montalvo-González, E. and Pérez-Larios, A., 2020. Use of titanium dioxide (TiO_2) nanoparticles as reinforcement agent of polysaccharide-based materials. *Processes*, 8(11), p. 1395.
[15] Anaya-Esparza, L.M., Villagrán-de la Mora, Z., Rodríguez-Barajas, N., Ruvalcaba-Gómez, J.M., Iñiguez-Muñoz, L.E., Maytorena-Verdugo, C.I., Montalvo-González, E. and Pérez-Larios, A., 2021. Polysaccharide-based packaging functionalized with inorganic nanoparticles for food preservation. *Polysaccharides*, 2(2), pp. 400–428.
[16] Janczarek, M., Endo, M., Zhang, D., Wang, K. and Kowalska, E., 2018. Enhanced photocatalytic and antimicrobial performance of cuprous oxide/titania: the effect of titania matrix. *Materials*, 11(11), p. 2069.
[17] Anaya-Esparza, L.M., Ruvalcaba-Gómez, J.M., Maytorena-Verdugo, C.I., González-Silva, N., Romero-Toledo, R., Aguilera-Aguirre, S. and Pérez-Larios, A., 2020. Chitosan-TiO_2: a versatile hybrid composite. *Materials*, 13(4), 811.
[18] Hu, Q., Fang, Y., Yang, Y., Ma, N. and Zhao, L., 2011. Effect of nanocomposite-based packaging on postharvest quality of ethylene-treated kiwifruit (*Actinidia deliciosa*) during cold storage. *Food Research International*, 44(6), pp. 1589–1596.
[19] Li, H., Li, F., Wang, L., Sheng, J., Xin, Z., Zhao, L., Xiao, H., Zheng, Y. and Hu, Q., 2009. Effect of nano-packing on preservation quality of Chinese jujube (*Ziziphus jujuba Mill.* var. *inermis (Bunge) Rehd*). *Food Chemistry*, 114(2), 547–552.
[20] Yang, F.M., Li, H.M., Li, F., Xin, Z.H., Zhao, L.Y., Zheng, Y.H. and Hu, Q.H., 2010. Effect of nano-packing on preservation quality of fresh strawberry (*Fragaria ananassa* Duch. cv Fengxiang) during storage at 4°C. *Journal of Food Science*, 75(3), pp. C236–C240.
[21] Chi, H., Song, S., Luo, M., Zhang, C., Li, W., Li, L. and Qin, Y., 2019. Effect of PLA nanocomposite films containing bergamot essential oil, TiO_2 nanoparticles, and Ag nanoparticles on shelf life of mangoes. *Scientia Horticulturae*, 249, pp. 192–198.
[22] Yildirim, S., Röcker, B., Pettersen, M.K., Nilsen-Nygaard, J., Ayhan, Z., Rutkaite, R., Radusin, T., Suminska, P., Marcos, B. and Coma, V., 2018. Active packaging applications for food. *Comprehensive Reviews in Food Science and Food Safety*, 17(1), pp. 165–199.
[23] Xing, Y., Li, X., Guo, X., Li, W., Chen, J., Liu, Q., Xu, Q., Wang, Q., Yang, H., Shui, Y. and Bi, X., 2020. Effects of different TiO_2 nanoparticles concentrations on the physical and antibacterial activities of chitosan-based coating film. *Nanomaterials*, 10(7), p. 1365.
[24] Fakhri, L.A., Ghanbarzadeh, B., Dehghannya, J., Abbasi, F. and Ranjbar, H., 2018. Optimization of mechanical and color properties of polystyrene/nanoclay/nano ZnO based nanocomposite packaging sheet using response surface methodology. *Food Packaging and Shelf Life*, 17, 11–24.

[25] Nafady, N.A., Alamri, S.A., Hassan, E.A., Hashem, M., Mostafa, Y.S. and Abo-Elyousr, K.A., 2019. Application of ZnO-nanoparticles to manage Rhizopus soft rot of sweet potato and prolong shelf-life. *Folia Horticulturae*, 31(2), 319–329.

[26] Sogvar, O.B., Saba, M.K., Emamifar, A. and Hallaj, R., 2016. Influence of nano-ZnO on microbial growth, bioactive content and postharvest quality of strawberries during storage. *Innovative Food Science & Emerging Technologies*, 35, pp. 168–176.

[27] Lakshmi, S.J., Roopa Bai, R.S., Sharanagouda, H., Ramachandra, C.T., Nadagouda, S. and Nidoni, U., 2018. Effect of biosynthesized zinc oxide nanoparticles coating on quality parameters of fig (*Ficus carica L.*) fruit. *Journal of Pharmacognosy and Phytochemistry*, 7(3), pp. 10–14.

[28] Li, X., Li, W., Jiang, Y., Ding, Y., Yun, J., Tang, Y. and Zhang, P., 2011. Effect of nano-ZnO-coated active packaging on quality of fresh-cut 'Fuji'apple. *International Journal of Food Science & Technology*, 46(9), pp. 1947–1955.

[29] Emamifar, A. and Bavaisi, S., 2020. Nanocomposite coating based on sodium alginate and nano-ZnO for extending the storage life of fresh strawberries (*Fragaria× ananassa* Duch.). *Journal of Food Measurement and Characterization*, pp. 1–3.

[30] Al-Naamani, L., Dutta, J. and Dobretsov, S., 2018. Nanocomposite zinc oxide-chitosan coatings on polyethylene films for extending storage life of okra (*Abelmoschus esculentus*). Nanomaterials, 8(7), p. 479.

[31] Saravanakumar, K., Sathiyaseelan, A., Mariadoss, A.V.A., Xiaowen, H. and Wang, M.H., 2020. Physical and bioactivities of biopolymeric films incorporated with cellulose, sodium alginate and copper oxide nanoparticles for food packaging application. *International Journal of Biological Macromolecules*, 153, pp. 207–214.

[32] Loyola, N. and Arriola, M., 2017. Increasing the shelf life of post-harvest table grapes (Vitis vinífera cv. Thompson Seedless) using different packaging material with copper nanoparticles to change the atmosphere. *International Journal of Agriculture and Natural Resources*, 44(1), pp. 54–63.

[33] Kalia, A., Kaur, M., Shami, A., Jawandha, S.K., Alghuthaymi, M.A., Thakur, A. and Abd-Elsalam, K.A., 2021. Nettle-leaf extract derived ZnO/CuO nanoparticle-biopolymer-based antioxidant and antimicrobial nanocomposite packaging films and their impact on extending the post-harvest shelf life of guava fruit. *Biomolecules*, 11(2), p. 224.

[34] Eshghi, S., Hashemi, M., MohAmmadi, A., Badii, F., Mohammadhoseini, Z. and Ahmadi, K., 2014. Effect of nanochitosan-based coating with and without copper loaded on physicochemical and bioactive components of fresh strawberry fruit (*Fragaria x ananassa* Duchesne) during storage. *Food and Bioprocess Technology*, 7(8), pp. 2397–2409.

[35] Bonilla-Bird, N.J., Paez, A., Reyes, A., Hernandez-Viezcas, J.A., Li, C., Peralta-Videa, J.R. and Gardea-Torresdey, J.L., 2018. Two-photon microscopy and spectroscopy studies to determine the mechanism of copper oxide nanoparticle uptake by sweetpotato roots during postharvest treatment. *Environmental Science & Technology*, 52(17), 9954–9963.

[36] Meena, M., Pilania, S., Pal, A., Mandhania, S., Bhushan, B., Kumar, S., Gohari, G. and Saharan, V., 2020. Cu-chitosan nano-net improves keeping quality of tomato by modulating physio-biochemical responses. *Scientific Reports*, 10(1), pp. 1–11.

[37] Li, J. and Zhuang, S., 2020. Antibacterial activity of chitosan and its derivatives and their interaction mechanism with bacteria: current state and perspectives. *European Polymer Journal*, p. 109984.

[38] Duan, C., Meng, X., Meng, J., Khan, M.I.H., Dai, L., Khan, A., An, X., Zhang, J., Huq, T. and Ni, Y., 2019. Chitosan as a preservative for fruits and vegetables: a review on chemistry and antimicrobial properties. *Journal of Bioresources and Bioproducts*, 4(1), pp. 11–21.

[39] Deng, Z., Jung, J., Simonsen, J., Wang, Y. and Zhao, Y., 2017. Cellulose nanocrystal reinforced chitosan coatings for improving the storability of postharvest pears under both ambient and cold storages. *Journal of Food Science*, 82(2), 453–462.

[40] de Oliveira Filho, J.G., de Deus, I.P., Valadares, A.C., Fernandes, C.C., Estevam, E.B. and Egea, M.B., 2020. Chitosan film with citrus limonia essential oil: physical and morphological properties and antibacterial activity. *Colloids and Interfaces*, 4(2), p. 18.

[41] Donsì, F., Marchese, E., Maresca, P., Pataro, G., Vu, K.D., Salmieri, S., Lacroix, M. and Ferrari, G., 2015. Green beans preservation by combination of a modified chitosan based-coating containing nanoemulsion of mandarin essential oil with high pressure or pulsed light processing. *Postharvest Biology and Technology*, 106, pp. 21–32.

[42] Lustriane, C., Dwivany, F.M., Suendo, V. and Reza, M., 2018. Effect of chitosan and chitosan-nanoparticles on post harvest quality of banana fruits. *Journal of Plant Biotechnology*, 45(1), 36–44.

[43] Kumar, S., Shukla, A., Baul, P.P., Mitra, A. and Halder, D., 2018. Biodegradable hybrid nanocomposites of chitosan/gelatin and silver nanoparticles for active food packaging applications. *Food Packaging and Shelf Life*, 16, pp. 178–184.

[44] Youssef, K. and Hashim, A.F., 2020. Inhibitory effect of clay/chitosan nanocomposite against *Penicillium digitatum* on citrus and its possible mode of action. *Jordan Journal of Biological Sciences*, 13(3).

[45] Taghinezhad, E. and Ebadollahi, A., 2017. Potential application of chitosan-clay coating on some quality properties of agricultural product during storage. *Agricultural Engineering International: CIGR Journal*, 19(3), pp. 189–194.

[46] Xu, D., Qin, H. and Ren, D., 2018. Prolonged preservation of tangerine fruits using chitosan/montmorillonite composite coating. *Postharvest Biology and Technology*, 143, pp. 50–57.

[47] Youssef, K., de Oliveira, A.G., Tischer, C.A., Hussain, I. and Roberto, S.R., 2019. Synergistic effect of a novel chitosan/silica nanocomposites-based formulation against gray mold of table grapes and its possible mode of action. *International Journal of Biological Macromolecules*, 141, pp. 247–258.

[48] Chowdappa, P., Gowda, S., Chethana, C.S. and Madhura, S., 2014. Antifungal activity of chitosan-silver nanoparticle composite against *Colletotrichum gloeosporioides* associated with mango anthracnose. *African Journal of Microbiology Research*, 8(17), pp. 1803–1812.

[49] Chávez-Magdaleno, M.E., González-Estrada, R.R., Ramos-Guerrero, A., Plascencia-Jatomea, M. and Gutiérrez-Martínez, P., 2018. Effect of pepper tree (*Schinus molle*) essential oil-loaded chitosan bio-nanocomposites on post-harvest control of *Colletotrichum gloeosporioides* and quality evaluations in avocado (*Persea americana*) cv. Hass. *Food Science and Biotechnology*, 27(6), pp. 1871–1875.

[50] Thomas, S.K., Parameswaranpillai, J., Krishnasamy, S., Begam, P.S., Nandi, D., Siengchin, S., George, J.J., Hameed, N., Salim, N.V. and Sienkiewicz, N., 2021. A comprehensive review on cellulose, chitin, and starch as fillers in natural rubber biocomposites. *Carbohydrate Polymer Technologies and Applications*, p. 100095.

[51] Qasim, U., Osman, A.I., Ala'a, H., Farrell, C., Al-Abri, M., Ali, M., Vo, D.V., Jamil, F. and Rooney, D.W., 2020. Renewable cellulosic nanocomposites for food packaging to avoid fossil fuel plastic pollution: a review. *Environmental Chemistry Letters*, pp. 1–29.

[52] Silva, E.L.P.D., Carvalho, T.C.D., Ayub, R.A. and Almeida, M.C.M.D., 2019. Nanocellulose coating associated with lemongrass essential oil at postharvest of blackberry fruits. *Preprints*, p. 2019100131. doi: 10.20944/preprints 201910.0131.v1.

[53] Saba, M.K. and Amini, R., 2017. Nano-ZnO/carboxymethyl cellulose-based active coating impact on ready-to-use pomegranate during cold storage. *Food Chemistry*, 232, pp. 721–726.

[54] Resende, N.S., Gonçalves, G.A., Reis, K.C., Tonoli, G.H. and Boas, E.V., 2018. Chitosan/cellulose nanofibril nanocomposite and its effect on quality of coated strawberries. *Journal of Food Quality* (Vol. 2018) Article ID 1727426. https://doi.org/10.1155/2018/1727426.

[55] Pacaphol, K., Seraypheap, K. and Aht-Ong, D., 2019. Development and application of nanofibrillated cellulose coating for shelf life extension of fresh-cut vegetable during postharvest storage. *Carbohydrate Polymers*, 224, p. 115167.

[56] Widodo, A.K., Pridiana, D.B., Arrizqi, N., Fajrina, M.N., Setyawan, D. and Laksono, D.A., 2020. Combining cellulose and ZnO for extend the shelf life of fruit. In *Materials Science Forum* (Vol. 990, pp. 267–271). Trans Tech Publications Ltd.

Chapter 9

Engineering the Nano-Based Packaging Materials for the Postharvest Storage

Bharat Bhushan, Manoj Kumar Mahawar, Kirti Jalgaonkar, Satish Kumar, Shrikrishna Nishani, Archana Mahapatra, and Sharmila Patil

List of Abbreviations

CMC	Carboxymethyl cellulose
CO_2	Carbon dioxide
iPP	Isotactic polypropylene
LDPE	Low-density polyethylene
MA	Modified atmosphere
NP	Inorganic nanoparticles
O_2	Oxygen
PET	Polyethylene terephthalate
PP	Polypropylene
PS	Polystyrene
PVC	Polyvinyl chloride
RFID	Radiofrequency identification
SLN	Solid lipid nanoparticle
TTI	Time–temperature indicator
UV	Ultraviolet

DOI: 10.1201/9781003142287-9

9.1 Introduction

Horticultural products and their processing steps like handling, storage, and transportation generate a lot of waste. The availability of facilities for postharvest management depends on the economic status of countries. A poorly developed or developing country has enormous wastage from perishables, which are estimated at around 20–30% in developing countries, so even if we succeed to reduce this amount or utilize it for by-products or its value addition, huge savings can be obtained. The identification of critical factors for appropriate postharvest management like temperature and humidity regime at each step after fruit harvest is important in order to apply the appropriate postharvest technologies. In this way, we can minimize the unwanted reactions of fruits and vegetables and prolong their shelf life (Kader, 2004). The postharvest management of perishables will not only ensure food and nutritional security but also reduce intensive farming for enhanced production. In this way, it will also reduce the burden of cutting trees for more cultivable land and minimize the environmental impacts of food production (Kiaya, 2014).

Fruits and vegetables have inherent barriers to evading climatic stress and biotic stress. The natural state of defense and their response is altered by the gaseous and water exchange from the surface in the local environment. The high-respiring and ethylene-responsive fruit have shorter shelf lives than the fruit coated with wax or packaged with synthetic covering. This coating or covering protects the fruit from pathogenic microbes and makes them safe to eat. The wax coating or packaging films are synthetic and may have some harmful effects when consumed with unpeeled fruit.

Food packaging is required to facilitate easy handling, maintain hygiene, damageless transport, and storage, as well as preservation of foods (Hoseinnejad et al., 2018; Jafari et al., 2015). Packaging materials have a great impact on the quality, shelf life, and safety of foods. The materials used in packaging, their physicochemical characteristics, and their interaction with food can decide the acceptability and marketability of packaged products (Domínguez et al., 2018). Traditionally, materials like wood, paper, paper board, and glass were used for food packaging. Thereafter, iron, steel, and aluminum were used as trendy materials. Generally, inactive materials that prevent moisture and O_2 penetration have been used for packaging and storage. These inactive materials include non-degradable plastics like LDPE, PET, PP, PS, and PVC. These materials have taken a place everywhere from home to industries over the last 60 years. The preference for plastics over glass, steel, or iron is attributed to their cost efficiency, low density, strength, and multipurpose characteristics (Hahladakis et al., 2018).

The packaging materials of fresh produce, such as polyethylene and PP films, create a MA for better shelf life but have some functional limitations. It has been proven that with a 10°C change in the storage temperature, the respiration rate of any live commodity increases from twofold to threefold. However, the film permeability to O_2 and CO_2 does not increase to such an extent (Exama et al., 1993). Therefore, once a cool, acclimatized MA-packaged food product is transferred to a higher temperature, it may result in O_2 depletion and hypoxic stress in the produce. Another drawback is associated with the water barrier properties of tough and rigid packaging materials. As a result, the water vapor accumulates inside the MA packages after attaining saturation. The water vapor and resultant water activity favor microbial contamination, posing a challenge to the safety of food products. Another drawback is the possible leaching of monomeric constituents from the packaging material into the foods, which is a primary concern with food safety (Dehnad et al., 2014; Hashemi Tabatabaei et al., 2018). The probable movement of certain small molecules, such as residual monomers and plasticizers, from packaging materials structure into the foods in contact with them as a serious potential pollutant is under consideration (Li et al., 2017; Lin et al., 2017). The greenhouse gas emissions from packaging waste reaching landfills provide their own share of 22% toward total pollution of greenhouse gases. The demand for safe food products coupled with environmental concerns prompt the development of biodegradable packaging. To ensure sustainable production and consumption environmentally cleaner materials are required to replace non-degradable plastics. Keeping in view of the previously mentioned drawbacks of plastic or other non-degradable packaging, usage of these materials could not be counted as the best solution to the problem of food and environmental security.

Now, with the advancement of packaging materials and fabrication technology, various strategies evolved to create active packaging that provides functional properties in packaging films to extend shelf life and improve safety. These include O_2 scavengers, moisture absorbers, ethylene scavengers, and CO_2 emitters. Each and every component acts in its own way to increase the shelf life of food material either through lowering respiration or microbial activity. The packaging material acts as a thermal and mechanical barrier whose functionality can be enhanced by incorporating an O_2 scavenger into the matrix. These O_2 scavengers reduce the oxidation reactions inside the food and increase shelf life. Simultaneously, two or more technologies can also be attempted in a single combinatorial approach. There are numerous types of active packaging ingredients that include preservatives, antimicrobials, and flavors to maintain or improve food quality. The sachets of these are placed inside the packaging to monitor and enhance the shelf life. Smart packaging includes polymer films that give off visual signals at different stages of contamination, physicochemical or textural change in the packaged produce. These components of intelligent packaging can also be included within modern degradable packaging.

Nanotechnology has entered into the arena of the food sector to take the challenge of food spoilage and to increase the durability of the freshness and health of these products while reducing the environmental impact. In recent years, organic and nonorganic coatings have been fabricated through nanotechnology. Nanotechnology has encompassed leaps and bounds in food freshness and helped in the development of smart and cost-effective polymer-based packaging materials. Its biofunctional propensities have shown its potential as the best functional packaging material. This is due to accruable perfections in properties such as enhancements in strength, reduced gas diffusion, and increment in water imperviousness. These nano-coatings have come up as a great interest to the scientific society to study the complex interaction between the food packaging material and food. Its application has surpassed the restrictions of traditional packaging and considers monitoring of stored food material and better package biodegradability. Most of the commonly known metal-based nanomaterials are O_2 scavengers. The nanoparticles of organic material like chitosan and inorganics like silica, titanium, and their compounds have been applied as coatings, especially in fruits with a short storage life.

The physicochemical properties of intelligent polymers or films are significantly different than those of naïve or unmodified biopolymers. The native properties of a biopolymer can be fabricated by chemical modifications, types of composite materials, and porosity. In recent times, a number of studies have reported on the fabrication of food-integrated packaging films with multiple advantages. A thoughtful process with a proof of concept is required to design the composite packaging materials, which can bear all types of stresses and pass all the stages of food processing (Rhim et al., 2013; Youssef and El-Sayed, 2018).

Packaging systems are generally employed to contain the product and function for its protection, manipulation, distribution, transportation, and identification of the particular commodity throughout its supply chain, that is, from material collection to consumers (Kuswandi, 2016). Innovative packaging techniques not only guarantee preservation during transit and effective distribution but also expedite consumer communication. Nanotechnology has benefited society and has emerged with vast potential in various areas of research and our routine daily living. Over the years, nanotechnology has emerged as a frontier and potential technique to augment the quality attributes of different food products. The technological improvements in the sphere of the food packaging sector in recent years have occurred because of nanotechnology and have considerably addressed the food quality, safety, and stability concerns. The nanotechnological advancements in food packaging enable consumers to understand the real-time quality of the product. Using this technique, the controlled release of preservatives/antimicrobials to extend the shelf life of the product has also been explored. Along with food preservation, the nanomaterial-based packaging contains superior barrier and mechanical characteristics and has gained welcoming attention from the marketing and consumer fronts. The commercialization, successful implementation, and responses to nanotechnological applications are governed by the consumer's outlook and authorization of recently introduced technologies (Gupta et al., 2011; Kim et al., 2014). On the other side, the recent advancement in and upliftment of this domain have also raised a few ethical, environmental, and safety concerns. The policies and regulations regarding the incorporation of nanoparticles in food packaging are also being under consideration for review (Sharma et al., 2017). A pictorial representation in the form of a flow chart and a picture showing different nanotechnological applications in the food packaging sector is shown in Figure 9.1 and Figure 9.2, respectively.

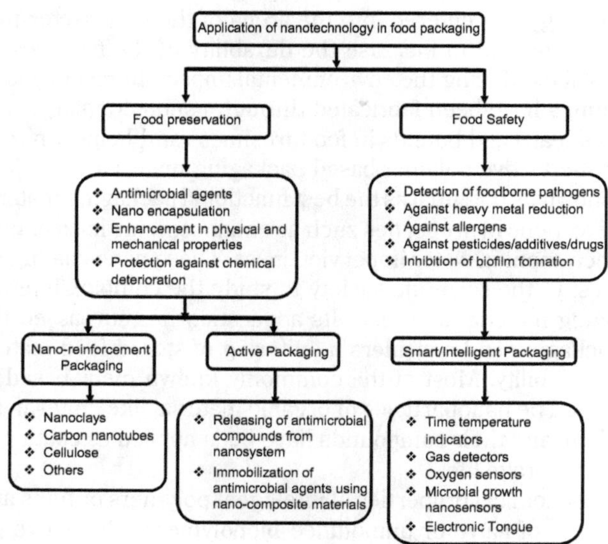

Figure 9.1 Flow chart showing the summarized applications of nanotechnology in food packaging.

Source: Bajpai et al. (2018); Pereda et al. (2019).

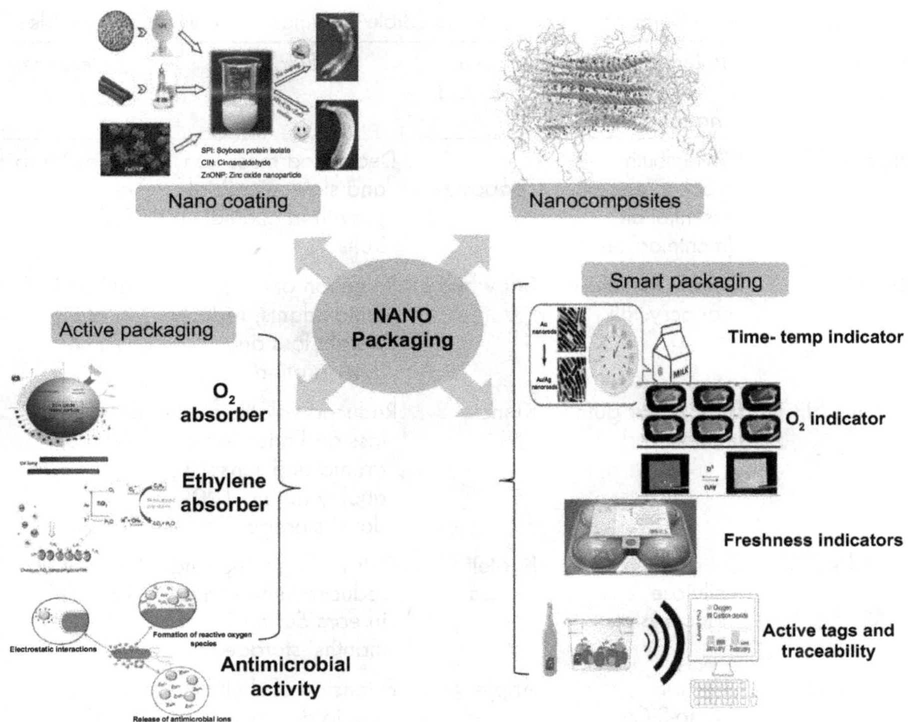

Figure 9.2 Pictorial presentation of different domains of nano-packaging.

9.2 Nanotechnology in Edible Coating

Food coatings are referred to as the film of edible material placed between food components to retard the mass transfer of moisture, lipid, and gas (Guilbert et al., 1997). The technique involves direct application either by the addition of a liquid film-forming solution or by molten compounds to the food product (Baldwin et al., 1996). Such coatings have an extensive application on different food products, including fruits, vegetables, meats, chocolate, cheese, candies, bakery products, and others (Costa et al., 2018; Zambrano-Zaragoza et al., 2018). Nanometric particles exhibit the immense potential to be incorporated into polymer composites, such as edible coatings, to aid in the preservation of food products. Nanomaterials modify the optical, mechanical, thermal, and rheological properties due to their high surface activity. Furthermore, the controlled release of active compounds, antioxidants, nutraceuticals, antimicrobial and anti-browning agents to augment the shelf life of food products can also be achieved, which can address the food safety issues due to pathogenic fungal, yeast, and bacterial growth during the storage of the food products. Nanomaterials in various forms, such as nanoemulsion, SLNs, nanostructured lipid carriers, NPs, nanotubes, and nanofibers, can be incorporated into the polymeric matrix. A brief explanation of the applications of different nanosystems to enhance the functionality of edible coatings is provided in Table 9.1.

Table 9.1 Studies on Different Nanosystems as Edible Coatings for Fruits and Vegetables

Type of nanosystem	*Polymer matrix (active nanomaterial)*	*Food product*	*Findings*	*Reference*
Inorganic NP	Whitemouth croaker/oregano essential oil (montmorillonite)	Cut papaya	Decreased mass loss and slow microbial growth in coated fruits.	Cortez-Vega et al. (2014)
Nanotubes	Gelatin (cellulose nanocrystal)	Strawberry	Retention of antioxidants, reduced weight loss and deterioration	Fakhouri et al. (2014)
Inorganic NP	CMC/guar gum (nano-silver)	Kinnow	Reduced volatile matter loss and retention of aroma and sensory quality during 120 days' storage.	Shah et al. (2016)
Nanotubes	Chitosan (cellulose nanocrystal)	Bartlett pears	Delayed ripening and reduced senescence in ears during 3 months' storage.	Deng et al. (2017)
Inorganic NP	Chitosan (nano-SiOx)	Apple	Extension of shelf life due to decreased respiration rate and firmness of coated fruits	Liu et al. (2017)
Pickering emulsion	Chitosan/oleic acid (cellulose nanocrystal)	Bartlette pear	Improved water vapor resistance and stability of the films resulted in enhanced shelf life	Deng et al. (2018)
SLN	Xanthan gum (candeuba wax)	Saladette tomato	Improved optical, mechanical, and adhesion properties along with functionality	Miranda-Linares et al. (2018)
Nanoemulsion	Quinoa protein/chitosan (thymol)	Strawberry	Antimicrobial coating resulted in lower fungal and yeast load during storage	Robledo et al. (2018)
Nanoemulsion	Sodium alginate (*Citrus sinensis* essential oil)	Tomato	Antibacterial and antibiofilm properties eradicated both sessile and planktonic forms of *Salmonella* and *Listeria* and enhanced shelf life	Das et al. (2020)

Type of nanosystem	*Polymer matrix (active nanomaterial)*	*Food product*	*Findings*	*Reference*
SLN	Xanthan gum/ stearic acid/ sesame oil (zinc oxide)	Apple and tomato	Antibacterial and low weight loss in coated product improved the storability	Joshy et al. (2020)
Nanoemulsion	Sodium Alginate (Citral)	Pineapple	Antibacterial property of citral led to reduction of *Salmonella enterica* and *Listeria monocytogenes* in the coated fruits	Prakash et al. (2020)
SLN	Xanthan gum/ propyl glycol (beeswax)	Strawberry	Increased shelf life by decreasing decay rates, low fungal growth, weight loss, and physiological damage by the end of storage	Zambrano-Zaragoza (2020)
Nanoemulsion	Chitosan/calcium chloride (tea tree oil)	Red bell pepper	Contamination by foodborne pathogens was controlled and the shelf life of coated food improved up to 18 days at 4°C	Sathiyaseelan et al. (2021)

9.3 Nano-Reinforcement Packaging

The addition of various nanoparticles to polymers has enhanced the performance and has resulted in the evolvement of several polymer nanomaterials for food packaging (Duncan, 2011). Such a type of nano-reinforcement helps fill in gaps in packaging materials, which ultimately increases its gas barrier properties, flexibility, viability, tensile strength, and moisture, as well as its temperature stability, and thus plays a role to enhance the shelf life of packaged foods.

Nanoparticle-reinforced polymers (nanocomposites) typically contain ≤5% w/w nanoparticles with high aspect ratios dramatically reducing the gas transfer rate of O_2, CO_2, and water vapor crossing the packages (Farhoodi, 2015). Even though several nanoparticles have been recognized as potential additives to improve polymer performance, the main focus and attention of the packaging industry have been on layered inorganic solids (clays and silicates), due to their low cost, significant enhancements, ready availability, and comparatively simple processing. The application of nanomaterials as food packaging materials in different fruits and vegetables is presented in Table 9.2.

Table 9.2 Application of Nanomaterials as Food Packaging Materials in Different Fruits and Vegetables (2010–2021)

Nanomaterial	*Application*	*Types of food*	*Effect*	*References*
PVC with ZnO NPs	Antimicrobial	Sliced apples	Significant reduction in the decay rate	Li et al. (2011)
Low density Polyethylene (LDPE) films loaded with Ag and ZnO NPs	Antimicrobial	Orange juice	Shelf life has increased up to 28 days and inactivation of *Lactobacillus plantarun*	Emamifar et al. (2011)
Ag montmorillonite NPs	Antimicrobial	Fresh cut carrots	Inhibited the growth of spoilage microorganisms and shelf-life extension by higher than 60 days under 4 ± 1°C	Costa et al. (2012)
Cellulose Ag nanoparticles (AgNPs)	Antimicrobial	Kiwi and melon juices	99.9% reduction of total viable count of bacteria and yeast	Lloret et al. (2012)
Low-density polyethylene with AgNPs	Antimicrobial	Barberry	2.3 log reduction in molds and 2.84 log reductions of total bacteria	Motlagh et al. (2012)
Polyethylene with Ag, TiO2	Antimicrobial	Fresh apples, carrots, orange juice, white sliced bread, soft cheese, atmosphere packaging, milk powder	Inhibited the growth of *Penicillium* and *Lactobacillus* spp.	Metak and Ajaal (2013)
iPP with $CaCO_3$ nanofiller	Antimicrobial	Apple slice	Increased the shelf life up to 10 days	Volpe et al. (2015)
Polyethylene with Ag and TiO_2 NPs	Antimicrobial	Fresh apples, white sliced bread, fresh carrots, prepacked soft cheese, MAP milk powder and orange juice	Decay percentage was significantly lower than control up to 10 days	Metak (2015)

Source: Modified from Sharma et al. (2017) and Vigneshwaran et al. (2019).

9.4 Active Packaging

Nanomaterials are employed to relation to food or its environment to provide enhanced protection. Nanoparticles supplemented with antimicrobial or other functionalities have the capacity of releasing them into packaging. Such kinds of packaged food exhibit improved taste, freshness, and storage duration.

9.4.1 Antimicrobial Films

The addition of materials possessing antimicrobial nature into food packaging has gained substantial attention among researchers. Such antimicrobial films help in restricting the growth of pathogenic and spoilage microbes. Such films are enviable due to their satisfactory structural integrity and barrier properties provided by the nanomaterials and the antimicrobial nature by the agents infused within the film (Rhim and Ng, 2007).

9.4.2 O_2-Scavenging Film

The presence of O_2 is the prime reason for the direct or indirect spoilage of many food items. For instance, fruit browning and rancid vegetable oils are examples of direct oxidation. To restrict the propagation of aerobic microorganisms, incorporating O_2 scavengers into food packages is essential to maintain very low O_2 concentrations, which, in turn, will help in the shelf-life extension of food commodities (Mihindukulasuriya and Lim, 2014).

9.4.3 UV-Absorbing Films

Nanocrystalline titania (TiO_2) is the most frequently used material for UV-absorbing film. It has been reported that doping TiO_2 with silver greatly improved photocatalytic bacterial inactivation (Reddy et al., 2007).

9.4.4 Freshness and Spoilage Indicators

The metabolism of food spoilage microbes produces gases that can be detected by conducting polymer nanocomposites/metal oxides. On the basis of gas emissions, they can be utilized to identify and quantify microorganisms as well as detect food freshness. Numerous gas sensors (using surface properties of materials) have been developed, which are employed to translate chemical interactions into a response signal. Conducting polymers or electro-active conjugated polymers, that can be synthesized either by chemical or electrochemical oxidation, are very important because of their electrical, electronic, magnetic, and optical properties. Sensors based on conducting polymer nanocomposites consist of conducting particles embedded into an insulating polymer matrix. The resistance changes of the sensors produce a pattern that corresponds to the gas under investigation (Kuswandi, 2016).

9.4.5 O_2 Indicators

During food storage, the presence of O_2 is favorable for the proliferation of aerobic microorganisms. Thus, in order to restrict the presence of O_2, there has been a research trend for the development of sensors to make O_2-free food packaging systems (vacuum/nitrogen packaging). The use of different nanomaterials in the active packaging of various fruits and vegetables is summarized in Table 9.3.

Table 9.3 Application of Nanomaterials in Active Packaging of Various Fruits and Vegetables

Nanomaterials	*Carrier*	*Mode of application*	*Fruits/ vegetables*	*Findings*	*Reference*
Silver-montmorillonite	–	Nanopowder sample kept at the bottom of a PP box	Fresh salad (kiwifruit and pineapple)	• Showed less growth of mesophilic bacteria than the control sample • Improved sensorial qualities during storage • Increased shelf life up to 5 days	Costa et al. (2011)
Copper	–	Spray coating	Parsley	• Increased shelf life • Retained higher ascorbic acid • Reduced respiration rate and weight loss • Inhibited yeast growth	Ouzounidou and Gaitis (2011)
Ag_2O	Polyethylene	Packing bags	Fresh cut apples	• Maintain freshness, delayed browning, reduced loss in weight, protect from microbial spoilage • Stored for 12 days at 5°C	Zhou et al. (2011)
Silver	Alginate	Dip coating and packed in polyethylene bags	Shiitake Mushroom	• Stopped microbial growth • Enhance keeping quality for 16 days under the temperature of 4 ± 1°C.	Jiang et al. (2013)
Silver- agar		Dip coating	Apple and lime	• Protect from microbial attack • Minimum loss in weight after 9 days' storage at room temperature	Gudadhe et al. (2013)
Zinc oxide		Dip coating	Green soybean	• Inhibited growth of bacteria, yeast, coliforms, and molds	Yu et al. (2015)
Silver, oolong tea extract		Dip coating	Cherry tomatoes	• Improved storage life by decreasing loss in weight, TSS, titratable acidity, and ascorbic acid content • Extended storage time up to 15 days	Gao et al. (2017)

Zinc oxide	CMC	Dip coating	Pomegranate arils	• Decreased the growth of yeast and mold • Decreased the physiological loss in weight • Found highest total anthocyanin, ascorbic acid, and antioxidant capacity	Saba and Amini (2017)
Zinc oxide		Dip coating followed by packing in retort pouch and radio frequency heating	Carrot	• Reduced the total colony forming units, loss of firmness, change in color, carotenoids during storage • Increased the shelf life up to 60 days	Xu et al. (2017)
Zinc oxide	Carrageenan	Dip coating	Mangoes	• Reduced rate of transpiration and respiration • Provided antimicrobial properties • Increased storage life	Meindrawan et al. (2018)
Zinc oxide	Chitosan	Surface coating on polyethylene films followed by packing	Okra	• Coated LDPE films retarded microbial and fungal growth and maintained the quality of the packed okra up to 12 days	Al-Naamani et al. (2018)
Silver	Polylactic acid	Packing bags	Strawberry	• Inhibited the growth of microorganisms • Lower weight loss • Prevented loss in ascorbic acid • Slowed the reduction in total phenols, firmness, and antioxidant activity	Zhang et al. (2018)
Silver	Chitosan	Dip coating	Fresh-cut melon	• Controlled the growth of microbial load • Reduced the rate of respiration and ethylene production	Ortiz-Duarte et al. (2019)
Titanium dioxide, Silver	Polylactic acid, bergamot essential oils	Packing bags	Mangoes	• Delayed the ripening process • Slowed the microorganisms' growth • Improved the sensory qualities	Chi et al. (2019)
Zinc oxide	Soybean protein Isolate/ Cinnamaldehyde	Dip coating	Banana	• Maintained nutrient content, delayed ripening, and decreased weight loss, fruit respiration, pectin conversion, and carbohydrate analysis. • Prevented fungal growth during storage	Li et al. (2019)

9.5 Intelligent/Smart Packaging

Nanomaterials are specifically designed for recognizing biochemical/microbial variation in the food, for example, distinguishing specific pathogens/gases responsible for food spoilage (Kuswandi et al., 2011). Sensors can identify particular pathogens rising in the food or the development of particular gases due to spoilage of food. It can also be used as a tracing tool for improving food safety. In the case of smart packaging, in particular, nanomaterials can be employed as reactive agents to notify about the condition of the packaged product. Nanosensors must respond to any outside stimuli variation in order to communicate about the product with the objective to give assurances about its quality and safety. Modern advancements in polymer nanomaterials include O_2 indicators, spoilage indicators, product identification, and traceability (Kuswandi, 2016; Kuswandi and Moradi, 2019). Some companies like British Airways, Nestlé, and Mono Prix Supermarket are working on the smart packaging concept by detecting the change in color by means of chemical sensors (Pehanich, 2006).

9.5.1 TTIs

TTIs are intended for the purpose of continuously observing, recording, reflecting the temperature history, diagnosing storage conditions, and translating food safety along the whole cold chain of the stored food. The importance is more when the storage of food products is done under conditions other than the optimal ones. TTIs ensure the manufacturers/consumers feel confident about the tracing of food products along the supply chain (Pereda et al., 2019).

9.5.2 Gas Indicators

Gas sensors are developed for the identification and quantity estimation of spoilage microbes based on their gas emissions. They are used to indicate the food quality, identify variations inside the packaging atmosphere, monitor the quality and safety of packaged food with regard to specific gas composition inside (MA packaging), and diagnosis any leakage in the package (Pereda et al., 2019).

9.5.3 Nanosensors

Nanosensor-equipped packaging is utilized for tracking the inner/outer environment of food products throughout the supply chain. Monitoring the temperature/humidity over time, levels of O_2 exposure, and microbial contamination and then giving appropriate information about these conditions can be examples of such packaging (Bouwmeester et al., 2009). Nanosensors coupled packaging can sense foul smell coming out of the food due to spoilage, and the package itself changes color to alert. They are also useful in eliminating the need for inaccurate expiration dates as they can provide real-time status of food freshness (Liao et al., 2005).

Nanosensors have the capability to ensure the presence of pathogenic bacteria, food-contaminating toxins, adulterants, vitamins, dyes, fertilizers, pesticides, taste, and smell. Spot indicators based on nanosensors are responsive and can be effortlessly labeled on individual packages for food freshness detection (Realini and Marcos, 2014; Jiang et al., 2015). Generally, nanosensors with distinctive characteristics are improving food security (Kumar et al., 2017).

9.5.4 Nanobiosensors

Nanobiosensors are compact analytical devices that contain a biologically derived sensitized element linked to a physicochemical transducer (Turner, 2000). The progression of nanotechnology in biosensors leads to an effective biosensor with a miniature structure that was lacking in conventional biosensors. They can be meticulously utilized for detecting a variety of fertilizers, insecticides, microorganisms, moisture, and contaminants in food materials. Their portable nature enables them for their applications in the field as well as laboratory. Nanobiosensors are already commercialized for detecting spoilage/contamination, the presence of pathogens, and product tampering and for identifying ingredients, as well as tracking products throughout the process value chain (Nachay, 2007). An ideal nanobiosensor must possess the following characteristics (Rai et al., 2012):

- Must distinguish among analyte and any 'other' material
- Stable under ambient storage environment
- Specific interaction between analytes should be independent of any physical parameters
- Minimum reaction time
- Accurate, precise, reproducible measurement of the responses
- Biocompatible, nontoxic, and nonantigenic nature
- Should be economical, portable, and capable of operating by semiskilled manpower

9.5.5 Nano-Barcodes for Product Authenticity

Sensing is an imperative and essential part of an intelligent packaging system to identify authentication and originality of the food product, and it facilitates inventory control (Farahi et al., 2012). The conventional two-dimensional barcodes are well established. However, owing to the limitation of easy tracing, they are more susceptible to alteration, damage, and distortion. Alternatively, manipulating the nanoparticle-based invisible barcodes is a challenge making them more robust and better than traditional barcodes to verify the originality of food products (Wang et al., 2015; Kumar et al., 2017).

9.5.6 Electronic Noses and Electronic Tongues

The artificial stimulation of properties (odor and taste) of food products with human-like efficiency is helpful in creating food having desired organoleptic quality attributes. Engineered nanosensors (electronic noses [smell and taste] and tongues [taste]) have been incorporated into packages to indicate the color change as a warning to the consumer to check for food spoilage and contamination by pathogens (Kuswandi, 2016). An electronic tongue consisting of an array of nanosensors was incorporated in food packaging. This was extremely sensitive to the gases released by microbes, generating a color change to indicate food deterioration (Joseph and Morrison, 2006). Kumar et al. (2017) reported the use of a variety of nanoparticles for making electronic noses (for smell detection) and tongues (for taste detection).

9.5.7 Active Tags and Traceability

RFID is principally used as active tags in packaging to trace and identify the object automatically using electronic information-based systems. This kind of tagging mechanism is very strong with a longer reading range (≥100 m distance) and can perform under extreme temperatures and pressures (Abad et al., 2007, 2009). Nanotechnology-enabled sensor packaging involves the incorporation of cheaper RFID tags, which are comparatively much smaller and more flexible with the provision of printing on thin labels. This definitely has enhanced the tag's versatility and made it suitable for commercial and inexpensive production. The working principle and different applications concerned with the intelligent packaging of horticultural produce are enlisted in Table 9.4.

Table 9.4 Working Principals and Applications for Intelligent Packaging of Horticultural Produce

Smart devices	*Principle*	*Applications*	*References*
TTI	Mechanical, chemical, enzymatic, or microbiological	Monitored the quality of nonpasteurized fresh juices (apple, orange, carrot, beetroot) using visible color change	Biegańska et al. (2014)
Gas indicator	Redox dyes, pH dyes, or enzymes	On-packaging methyl red–based colorimetric indicator for detecting the aldehyde emission from apples	Kim et al. (2018)
Biosensor	Biological materials including antigens, enzymes, and nucleic acids	For assessment of apple quality (release of volatile ethylene and acetaldehyde)	Weber et al. (2009)
		Ethylene detection in bananas, avocados, apples, pears, oranges	Esser et al. (2012)
		• Spatial imaging with kiwifruit and Asian pears • Time-dependent imaging with apples, Asian pears, kiwifruit, Muscat grapes, carrots, red bell peppers	Vong et al. (2019)

Smart devices	Principle	Applications	References
Electrochemical sensors	Cystamine-PAMAM-modified cold AgNPs/ graphene nanoribbons (GNR)-based/ AuNP-based alginate-plasmonic sensor	Detection and quantification of aflatoxin B_1 aflatoxin B_2 and ochratoxin A in peanut extract and peanuts and corn snacks	Castillo et al. (2015) Yousefi et al. (2019)
		Detection of methyl parathion in fruits and vegetables	Govindasamy et al. (2017)
		Irreversible change in color that indicates variation in temperature in perishable foods	Wang et al. (2017)
RFID tags	Radio waves	Environmental parameters (O_2, CO_2, C_2H_4) inside packaged fruit and vegetables are measured and the recorded information is directed to the central system	Vanderroost et al. (2014)
		Modeled usefulness of the RFID systems in a highly perishable product from different opinions including the distributor, retailer, and consumer for estimation of the date of expiry and residual shelf life of the product.	Grunow and Piramuthu (2013)
		Useful for showing the content status of the sealed container prepared for export without opening it. RFID technology coupled with integrated sensors was utilized to track and monitor shipped perishable products.	Todorovic et al. (2014)
		Monitoring of temperature along with humidity for perishable food distribution during conveyance and storage. They improved the efficiency of the RFID-based traceability by developing a machine learning model for instant identification of the direction of tagged products.	Alfian et al. (2020)

Source: Modified from Alfei et al. (2020) and Firouz et al. (2021).

9.6 Conclusion

Preserving nutritional properties, counteracting deterioration, increasing the attention to the continuous monitoring of quality, and extending the shelf life of the food products have led to manufacturing/emergence of different innovative packaging techniques. In this context, the inception of nanotechnology approaches in the formulation of food packaging has offered ample opportunities in preserving the quality and in shelf-life extension of horticultural produce, especially fruits and vegetables. Such an incorporation of nanomaterials will enhance the existing functions of packaging materials and have a substantial positive impact on the sensory perceptions of the consumers, with a decline in wastage and, consequently, and improvement in the economy. This promising technology will ensure the quality and safety of the foods, which, in turn, will benefit both the producer and the consumer. Limitations like complex nature, manufacturing cost for commercial production, lack of legislation for these technologies in food packaging, and consumer acceptance need to be addressed as well. Also, research and development should focus on minimizing the cost and complicity of these technologies so as to spread their applications in the industry. Simultaneously, it is pertinent to emphasize that more research should consider carrying out short- and long-term toxicity studies to study the effects on the environment and humans in order to guarantee the consumer's health.

References

Abad E, Palacio F, Nuin M, De Zárate AG, Juarros A, Gómez J, Marco S (2009). RFID smart tag for traceability and cold chain monitoring of foods: demonstration in an intercontinental fresh fish logistic chain. Journal of Food Engineering, 93(4): 394–399.

Abad E, Zampolli S, Marco S (2007). Flexible tag microlab development: gas sensors integration in RFID flexible tags for food logistic. Sensors and Actuators B: Chemical, 127(1): 2–7.

Alfei S, Marengo B, Zuccari G (2020). Nanotechnology application in food packaging: a plethora of opportunities versus pending risks assessment and public concerns. Food Research International, 137: 109664.

Alfian G, Syafrudin M, Farooq U, Ma'arif MR, Syaekhoni MA, Fitriyani NL, Lee J, Rhee J (2020). Improving efficiency of RFID-based traceability system for perishable food by utilizing IoT sensors and machine learning model. Food Control, 110: 107016.

Al-Naamani, L, Dutta J, Dobretsov S (2018). Nanocomposite zinc oxide-chitosan coatings on polyethylene films for extending storage life of okra (*Abelmoschus esculentus*). Nanomaterials, 8(7): 479.

Bajpai VK, Kamle M, Shukla S, Mahato DK, Chandra P, Hwang SK, Kumar P, Huh YS, Han YK (2018). Prospects of using nanotechnology for food preservation, safety, and security. Journal of Food and Drug Analysis, 26(4): 1201–1214.

Baldwin EA, Nisperos MO, Chen X, Hagenmaier RD (1996). Improving storage life of cut apples and potato with edible coating. Postharvest Biology and Technology, 9(2): 151–163

Biegańska M, Gwiazdowska D, Kozak W, Kluczyńska K (2014). The use of TTI indicators for quality monitoring of freshly squeezed juices. The International Forum on Agri-Food Logistics, 2.

Bouwmeester H, Dekkers S, Noordam MY, Hagens WI, Bulder AS, De Heer C, Voorde SECG, Wijnhoven SWP, Marvin HJP, Sips AJAM (2009). Review of health safety aspects of nanotechnologies in food production. Regulatory Toxicology and Pharmacology, 53(1): 52–62.

Castillo G, Spinella K, Poturnayova A, Snejdarková M, Mosiello L, Hianik T (2015). Detection of aflatoxin B1 by aptamer-based biosensor using PAMAM dendrimers as immobilization platform. Food Control, 52: 9–18.

Chi H, Song S, Luo M, Zhang C, Li W, Li L, Qin Y (2019). Effect of PLA nanocomposite films containing bergamot essential oil, TiO_2 nanoparticles, and Ag nanoparticles on shelf life of mangoes. Scientia Horticulturae, 249: 192–198.

Cortez-Vega WR, Pizato S, de Souza JTA, Prentice C (2014). Using edible coatings from Whitemouth croaker (*Micropogonias furnieri*) protein isolate and organo-clay nanocomposite for improve the conservation properties of fresh-cut 'Formosa' papaya. Innovative Food Science & Emerging Technologies, 22: 197–202.

Costa C, Conte A, Buonocore GG, Del Nobile MA (2011). Antimicrobial silver-montmorillonite nanoparticles to prolong the shelf life of fresh fruit salad. International Journal of Food Microbiology, 148(3): 164–167.

Costa C, Conte A, Buonocore GG, Lavorgna M, Del NMA (2012). Calcium-alginate coating loaded with silver montmorillonite nanoparticles to prolong the shelf-life of fresh-cut carrots. Food Research International, 48: 164–169.

Costa MJ, Maciel LC, Teixeira JA, Vicente AA, Cerqueira MA (2018). Use of edible films and coatings in cheese preservation: opportunities and challenges. Food Research International, 107: 84–92.

Das S, Vishakha K, Banerjee S, Mondal S, Ganguli A (2020). Sodium alginate-based edible coating containing nanoemulsion of Citrus sinensis essential oil eradicates planktonic and sessile cells of food-borne pathogens and increased quality attributes of tomatoes. International Journal of Biological Macromolecules, 162: 1770–1779.

Dehnad D, Emam-Djomeh Z, Mirzaei H, Jafari SM, Dadashi S (2014). Optimization of physical and mechanical properties for chitosan-nanocellulose biocomposites. Carbohydrate Polymers, 105(1): 222–228.

Deng Z, Jung J, Simonsen J, Wang Y, Zhao Y (2017). Cellulose nanocrystal reinforced chitosan coatings for improving the storability of postharvest pears under both ambient and cold storages. Journal of Food Science, 82(2): 453–462.

Deng Z, Jung J, Simonsen J, Zhao Y (2018). Cellulose nanocrystals Pickering emulsion incorporated chitosan coatings for improving storability of postharvest Bartlett pears (*Pyrus communis*) during long-term cold storage. Food Hydrocolloids, 84: 229–237.

Domínguez R, Barba FJ, Gómez B, Putnik P, Kovačević DB, Pateiro M, Santos EM, Lorenzo JM (2018). Active packaging films with natural antioxidants to be used in meat industry: a review. Food Research International, 113: 93–101.

Duncan TV (2011). Applications of nanotechnology in food packaging and food safety: barrier materials, antimicrobials and sensors. Journal of Colloid and Interface Science, 363: 1–24.

Emamifar A, Kadivar M, Shahedi M, Soleimanian-Zad S (2011). Evaluation of nanocomposites packaging containing Ag and ZnO on shelf life of fresh orange juice. Innovative Food Science & Emerging Technologies, 11: 742–748.

Esser B, Schnorr JM, Swager TM (2012). Selective detection of ethylene gas using carbon nanotube-based devices: utility in determination of fruit ripeness. Angewandte Chemie International Edition, 51(23): 5752–5756.

Exama A, Arul J, Lencki RW, Lee LZ, Toupin C (1993). Suitability of plastic films for modified atmosphere packaging of fruits and vegetables. Journal of Food Science, 58: 1365–1370.

Fakhouri FM, Casari ACA, Mariano M, Yamashita F, Mei LI, Soldi V, Martelli SM (2014). Effect of a gelatin-based edible coating containing cellulose nanocrystals (CNC) on the quality and nutrient retention of fresh strawberries during storage. IOP Conference Series: Materials Science and Engineering, 64(1): 012024.

Farahi RH, Passian A, Tetard L, Thundat T (2012). Critical issues in sensor science to aid food and water safety. ACS Nano, 6: 4548–4556.

Farhoodi M (2015). Nanocomposite materials for food packaging applications: characterization and safety evaluation. Food Engineering Reviews, 8: 35–51.

Firouz MS, Mohi-Alden K, Omid M (2021). A critical review on intelligent and active packaging in the food industry: research and development. Food Research International, 141: 110113.

Gao L, Li Q, Zhao Y, Wang H, Liu Y, Sun Y, Wang F, Jia W, Hou X (2017). Silver nanoparticles biologically synthesised using tea leaf extracts and their use for extension of fruit shelf life. IET Nanobiotechnol, 11(6): 637–643.

Govindasamy M, Mani V, Chen S-M, Chen T-W, Sundramoorthy AK (2017). Methyl parathion detection in vegetables and fruits using silver@ graphene nanoribbons nanocomposite modified screen printed electrode. Scientific Reports, 7: Article e46471.

Grunow M, Piramuthu S (2013). RFID in highly perishable food supply chains–remaining shelf life to supplant expiry date?. International Journal of Production Economics, 146(2): 717–727.

Gudadhe JA, Yadav A, Gade A, Marcato PD, Duran N, Rai M (2013). Preparation of an agar-silver nanoparticles (A-AgNp) film for increasing the shelf-life of fruits. IET Nanobiotechnology, 8(4): 190–195.

Guilbert S, Cuq B, Gontard N (1997). Recent innovations in edible and/or biodegradable packaging materials. Food Additives and Contaminants, 14(6): 741–751.

Gupta N, Fischer ARH, Frewer LJ (2011). Socio-psychological determinants of public acceptance of technologies: a review. Public Understanding of Science, 21: 782–795.

Hahladakis JN, Velis CA, Weber R, Iacovidou E, Purnell P (2018). An overview of chemical additives present in plastics: migration, release, fate and environmental impact during their use, disposal and recycling. Journal of Hazardous Materials, 344: 179–199.

Hashemi Tabatabaei R, Jafari SM, Mirzaei H, Mohammadi Nafchi A, Dehnad D (2018). Preparation and characterization of nano-SiO_2 reinforced gelatin-k-carrageenan biocomposites. International Journal of Biological Macromolecules, 111: 1091–1099.

Hoseinnejad M, Jafari SM, Katouzian I (2018). Inorganic and metal nanoparticles and their antimicrobial activity in food packaging applications. Critical Reviews in Microbiology, 44(2): 161–181.

Jafari SM, Khanzadi, M, Mirzaei H, Dehnad D, Chegini FK, Maghsoudlou Y (2015). Hydrophobicity, thermal and micro-structural properties of whey protein concentrate-pullulan-beeswax films. International Journal of Biological Macromolecules, 80: 506–511.

Jiang T, Feng L, Wang Y (2013). Effect of alginate/nano-Ag coating on microbial and physicochemical characteristics of shiitake mushroom (*Lentinus edodes*) during cold storage. Food Chemistry, 141(2): 954–960.

Jiang X, Valdeperez D, Nazarenus M, Wang Z, Stellacci F, Parak WJ, Pino PD (2015). Future perspectives towards the use of nanomaterials for smart food packaging and quality control. Particle & Particle Systems Characterization, 32: 408–416.

Joseph T, Morrison M (2006). Nanotechnology in agriculture and food. A Nanoforum Report, Institute of Nanotechnology May 2006, www.nanoforum.org.

Joshy KS, Jose J, Li T, Thomas M, Shankregowda AM, Sreekumaran S, Nandakumar K, Thomas S (2020). Application of novel zinc oxide reinforced xanthan gum hybrid system for edible coatings. International Journal of Biological Macromolecules, 151: 806–813.

Kader AA (2004). Increasing food availability by reducing post-harvest losses of fresh produce. International Postharvest Symposium, 682: 2169–2176.

Kiaya V (2014). Post-harvest losses and strategies to reduce them. Technical Paper on Postharvest Losses, Action Contre la Faim.

Kim YH, Yang YJ, Kim JS, Choi DS, Park SH, Jin SY, Park JS (2018). Non-destructive monitoring of apple ripeness using an aldehyde sensitive colorimetric sensor. Food Chemistry, 267: 149–156.

Kim YR, Lee EJ, Park SH, Kwon HJ, An SS, Son SW, Seo YR, Pie JE, Yoon M, Kim JH, Kim MK (2014). Comparative analysis of nanotechnology awareness in consumers and experts in South Korea. International Journal of Nanomedicine, 15: 21–27.

Kumar V, Guleria P, Mehta SK (2017). Nanosensors for food quality and safety assessment. Environmental Chemistry Letters, 15: 165–177.

Kuswandi B (2016). Nanotechnology in food packaging. In: S. Ranjan et al. (eds.), Nanoscience in Food and Agriculture 1. Sustainable Agriculture Reviews, 20: 151–183. https://doi.org/10.1007/978-3-319-39303-2_6

Kuswandi B, Moradi M (2019). Improvement of food packaging based on functional nanomaterial. In S Siddiquee, G Melvin, M Rahman (Eds.), Nanotechnology: Applications in Energy, Drug and Food (pp. 309–344). Springer Nature.

Kuswandi B, Wicaksono Y, Jayus, Abdullah A, Heng LY, Ahmad M (2011). Smart packaging: sensors for monitoring of food quality and safety. Sensing and Instrumentation for Food Quality and Safety, 5: 137–146.

Li J, Sun Q, Sun Y, Chen B, Wu X, Le T (2019). Improvement of banana postharvest quality using a novel soybean protein isolate/cinnamaldehyde/zinc oxide bionanocomposite coating strategy. Scientia Horticulturae, 258: 108786.

Li SQ, Ni HG, Zeng H (2017). PAHs in polystyrene food contact materials: an unintended consequence. Science of the Total Environment, 609: 1126–1131.

Li X, Li W, Jiang Y, Ding Y, Yun I, Tang Y (2011). Effect of nano-ZnO-coated active packaging on quality of fresh-cut 'Fuji' apple. International Journal of Food Science and Technology, 46: 1947–1955.

Liao F, Chen C, Subramanian V (2005). Organic TFTs as gas sensors for electronic nose applications. Sensors and Actuators B: Chemical, 107(2): 849–855.

Lin QB, Song XC, Fang H, Wu YM, Wang ZW (2017). Migration of styrene and ethylbenzene from virgin and recycled expanded polystyrene containers and discrimination of these two kinds of polystyrene by principal component analysis. Food Additives & Contaminants: Part A, 34(1): 126–132.

Liu R, Liu D, Liu Y, Song Y, Wu T, Zhang M (2017). Using soy protein SiOx nanocomposite film coating to extend the shelf life of apple fruit. International Journal of Food Science and Technology, 52(9): 2018–2030.

Lloret E, Picouet P, Fernandez A (2012). Matrix effects on the antimicrobial capacity of silver based nanocomposite absorbing materials. LWT–Food Science and Technology, 49: 333–338.

Meindrawan B, Suyatma NE, Wardana AA, Pamela VY (2018). Nanocomposite coating based on carrageenan and ZnO nanoparticles to maintain the storage quality of mango. Food Packaging and Shelf Life, 18: 140–146.

Metak AM, Ajaal TT (2013). Investigation on polymer based nanosilver as food packaging materials. International Journal of Chemical and Molecular Engineering, 7: 772–777.

Metak MM (2015). Effects of nanocomposite based nano-silver and nano-titanium dioxide on food packaging materials. International Journal of Applied Science and Technology, 5: 26–40.

Mihindukulasuriya SDF, Lim LT (2014). Nanotechnology development in food packaging: a review. Trends in Food Science and Technology, 40: 149–167.

Miranda-Linares V, Escamilla-Rendón P, Real-López AD, González-Reza RM, Zambrano-Zaragoza ML (2018). Solid lipid nanoparticles based edible coating for saladette tomato preservation. Acta Horticulturae, (1194): 305–312.

Motlagh NV, Mosavian MTH, Mortazavi SA (2012). Effect of polyethylene packaging modified with silver particles on the microbial, sensory and appearance of dried barberry. Packaging Technology and Science, 26: 39–49.

Nachay K (2007). Analyzing nanotechnology. Food Technology, 61(1): 34–36.

Ortiz-Duarte G, Perez-Cabrera LE, Artes-Hernandez F, Martınez-Hernandez GB (2019). Ag-chitosan nanocomposites in edible coatings affect the quality of fresh-cut melon. Postharvest Biology and Technology, 147: 174–184.

Ouzounidou G, Gaitis F (2011). The use of nano-technology in shelf-life extension of green vegetables. Journal of Innovation Economics and Management, 8: 163–171.

Pehanich M (2006). Small gains in processing and packaging. Food Processing, 11: 46–48

Pereda M, Marcovich NE, Ansorena MR (2019). Nanotechnology in food packaging applications: barrier materials, antimicrobial agents, sensors, and safety assessment. In MT Martínez et al. (Eds.), Handbook of Ecomaterials. (pp. 2035–2056). Springer International Publishing.

Prakash A, Baskaran R, Vadivel V (2020). Citral nanoemulsion incorporated edible coating to extend the shelf life of fresh cut pineapples. LWT–Food Science and Technology, 118: 108851.

Rai V, Acharya S, Dey N (2012). Implications of nanobiosensors in agriculture. Journal of Biomaterials and Nanobiotechnology, 3: 315–324.

Realini CE, Marcos B (2014). Active and intelligent packaging systems for a modern society. Meat Science, 98: 404–419.

Reddy PM, Venugopal A, Subrahmanyam M (2007). Hydroxyapatite-supported Ag-TiO_2 as *Escherichia coli* disinfection photocatalyst. Water Research, 41: 379–386.

Rhim JW, Ng PKW (2007). Natural biopolymer-based nanocomposite films for packaging applications. Critical Reviews in Food Science and Nutrition, 47: 411–433.

Rhim J-W, Park HM, Ha CS (2013). Bio-nanocomposites for food packaging applications. Progress in Polymer Science, 38(10–11): 1629–1652.

Robledo N, López L, Bunger A, Tapia C, Abugoch L (2018). Effects of antimicrobial edible coating of thymol nanoemulsion/quinoa protein/chitosan on the safety, sensorial properties, and quality of refrigerated strawberries (*Fragaria×ananassa*) under commercial storage environment. Food and Bioprocess Technology, 11(8): 1566–1574.

Saba MK, Amini R (2017). Nano-ZnO/carboxymethyl cellulose-based active coating impact on ready-to-use pomegranate during cold storage. Food Chemistry, 232: 721–726.

Sathiyaseelan A, Saravanakumar K, Mariadoss AVA, Rama Chandran C, Hu X, Oh DH, Wang MH (2021). Chitosan-tea tree oil nanoemulsion and calcium chloride tailored edible coating increase the shelf life of fresh cut red bell pepper. Progress in Organic Coatings, 151: 106010.

Shah SWA, Qaisar M, Jahangir M, Abbasi KS, Khan SU, Ali N, Liaquat M (2016). Influence of CMC-and guar gum-based silver nanoparticle coatings combined with low temperature on major aroma volatile components and the sensory quality of kinnow (*Citrus reticulata*). International Journal of Food Science & Technology, 51(11): 2345–2352.

Sharma C, Dhiman R, Rokana N, Panwar H (2017). Nanotechnology: an untapped resource for food packaging. Frontiers in Microbiology, 8: 1735. doi: 10.3389/fmicb.2017.01735

Todorovic V, Neag M, Lazarevic M (2014). On the usage of RFID tags for tracking and monitoring of shipped perishable goods. Procedia Engineering, 69: 1345–1349.

Turner AP (2000). Biosensors-sense and sensitivity. Science, 290(5495): 1315–1317.

Vanderroost M, Ragaert P, Devlieghere F, De Meulenaer B (2014). Intelligent food packaging: the next generation. Trends in Food Science and Technology. 39: 47–62. https://doi.org/10.1016/j.tifs.2014.06.009

Vigneshwaran N, Kadam DM, Patil S (2019). Nanomaterials for active and smart packaging of food. In R Pudake, N Chauhan, C Kole (Eds.), Nanoscience for Sustainable Agriculture. Springer. https://doi.org/10.1007/978-3-319-97852-9_22

Volpe MG, Stasio MD, Paolucci M, Moccia S (2015). Polymers for food shelf-life extension. In G Cirillo, UG Spizzirri, F Iemma (Eds.), Functional Polymers in Food Science from Technology to Biology Food Packaging (pp. 9–61). John Wiley & Sons, Inc.

Vong K, Eda S, Kadota Y, Nasibullin I, Wakatake T, Yokoshima S, Shirasu K, Tanaka K (2019). An artificial metalloenzyme biosensor can detect ethylene gas in fruits and Arabidopsis leaves. Nature Communications, 10(1): 5746.

Wang YC, Lu L, Gunasekaran S (2015). Gold nanoparticle-based thermal history indicator for monitoring low-temperature storage. Microchimica Acta, 182: 1305–1311.

Wang YC, Lu L, Gunasekaran S (2017). Biopolymer/gold nanoparticles composite plasmonic thermal history indicator to monitor quality and safety of perishable bioproducts. Biosensors & Bioelectronics, 92: 109–116.

Weber W, Luzi S, Karlsson M, Fussenegger M (2009). A novel hybrid dual-channel catalytic-biological sensor system for assessment of fruit quality. Journal of Biotechnology, 139(4): 314–317.

Xu J, Zhang M, Bhandari B, Kachele R (2017). ZnO nanoparticles combined radio frequency heating: a novel method to control microorganism and improve product quality of prepared carrots. Innovative Food Science and Emerging Technologies, 44: 46–53.

Yousefi M, Orojzadeh P, Jafari SM (2019). Nanoencapsulation of food ingredients by dendrimers. In SM Jafari (Ed.), Nanoencapsulation in the Food Industry, Biopolymer Nanostructures for Food Encapsulation Purposes (pp. 607–625). Academic Press, Elsevier.

Youssef AM, El-Sayed SM (2018). Bionanocomposites materials for food packaging applications: concepts and future outlook. Carbohydrate Polymers, 193: 19–27.

Yu N, Zhang M, Islam N, Lu L, Liu Q, Cheng X (2015). Combined sterilizing effects of nano-ZnO and ultraviolet on convenient vegetable dishes. LWT–Food Science and Technology, 61(2): 638–643.

Zambrano-Zaragoza M, González-Reza R, Mendoza-Muñoz N, Miranda-Linares V, Bernal-Couoh T, Mendoza-Elvira S, Quintanar-Guerrero D (2018). Nanosystems in edible coatings: a novel strategy for food preservation. International Journal of Molecular Sciences, 19: 705.

Zambrano-Zaragoza ML, Quintanar-Guerrero D, Del Real A, González-Reza RM, Cornejo-Villegas MA, Gutiérrez-Cortez E (2020). Effect of nano-edible coating based on beeswax solid lipid nanoparticles on Strawberry's preservation. Coatings, 10(3): 253.

Zhang C, Li W, Zhu B, Chen H, Chi H, Li L, Qin Y, Xue J (2018). The quality evaluation of postharvest strawberries stored in nano-Ag packages at refrigeration temperature. Polymers, 10: 894.

Zhou L, Lv S, He G, He Q, Sh BI (2011). Effect of PE/AG_2O nano-packaging on the quality of apple slices. Journal of Food Quality, 34(3): 171–176.

Chapter 10

Advancements in the Essential Oil-Based Packaging Materials

Francisco J. Blancas-Benitez, Luis Guillermo Hernández-Montiel, Jonathan Michel Sanchez-Silva, Cristina Moreno-Hernández, Gutiérrez-Martinez P., Héctor J. Cortés-Rivera, Lizet Aguirre-Güitrón, Surelys Ramos-Bell, and Ramsés R. González-Estrada

List of Abbreviations

(γ)	Interfacial tension
(n)	Refractive index
(ρ)	Density
(η_D)	Oils phase
(η)	Viscosity
AP	Active packaging
CO_2	Carbon dioxide
DNA	Deoxyribonucleic acid
EO	Essential oil
FDA	Food and Drug Administration of the United States
FTIR	Fourier-transform infrared spectroscopy
h	Hour
HLB	Hydrophilic–lipophilic balance
HWD	Hot-water dip
HWR	Hot-water rinse
kGy	Kilograys

DOI: 10.1201/9781003142287-10

NO	Nitric oxide
O_2	Oxygen
PFA	Phytogenic feed additives
ppm	Parts per million
SO_2	Sulfur dioxide

10.1 Introduction

Nowadays, quality losses by microbial spoilage in fruits represent a huge problem for production around the world. Diseases in fruits are mainly produced by fungal infections; however, bacteria are also involved, and their management is related to the extensive application at pre- and postharvest stages of chemical fungicides. Nonetheless, even though they are effective for controlling diseases in several situations, such as the presence of chemical residues on fruits and the adaptation of several pathogens to their presence, consumers' preference to consume fruits free of chemical residues has led to the research of other alternatives. In this sense, the use of biopolymers from different sources used as edible coatings represents a smart choice as a postharvest treatment for preserving fruit quality. Different types of additives can be incorporated into the coating-forming solutions such as EOs, which are recognized as GRAS and effective compounds against several bacteria and fungi. The present chapter summarizes recent investigations related to the use of EOs and edible coatings as alternatives to the conventional management of diseases in fruits.

10.2 Factors Involved in Postharvest Losses in Fruits and Vegetables

Quality in fruits and vegetables is a very complex term based on their characteristics, attributes, and physicochemical properties that can define the commercial or nutritional value of these foods (Barman et al., 2015). The perception of quality can differ greatly between producers, traders, and final consumers of the same agricultural product; however, the appearance, weight, firmness, aroma, flavor, nutrient content, and shelf life are some of the main properties related to the acceptability of fruits and vegetables (Bhowmik and Dris, 2004; Bhargava and Bansal, 2018). However, the deterioration of the fruit quality can occur during the ripening stages and senescence process (Kyriacou and Rouphael, 2017), where the harvesting stage, as well as transport and storage management, plays an important role in accelerating or reducing these processes during postharvest (Ennigrou et al., 2017). The main factors that can affect the quality of fruits and vegetables are the presence of mechanical injuries, physiological disorders, fungal or bacterial diseases, changes in the atmospheric composition of storage, and temperatures and humidity that are too high or low (Figure 10.1; Hodges and Toivonen, 2008). Fruits and vegetables are metabolically active after harvest, which is why temperature changes can have a greater effect on quality parameters (Brasil and Siddiqui, 2018). In a recent study, it was reported that the storage of fruits and vegetables at the optimum temperature of refrigeration can reduce senescence processes, allowing a longer shelf life as well as conservation (Ma et al., 2017). However, if the temperature is too low and not adequate for the fruit/vegetable, it can lead to physiological disorders due to irreversible cell damage, causing wilting, darkening, loss of firmness, aroma, and moisture (Wang, 2016; Li et al., 2018). For

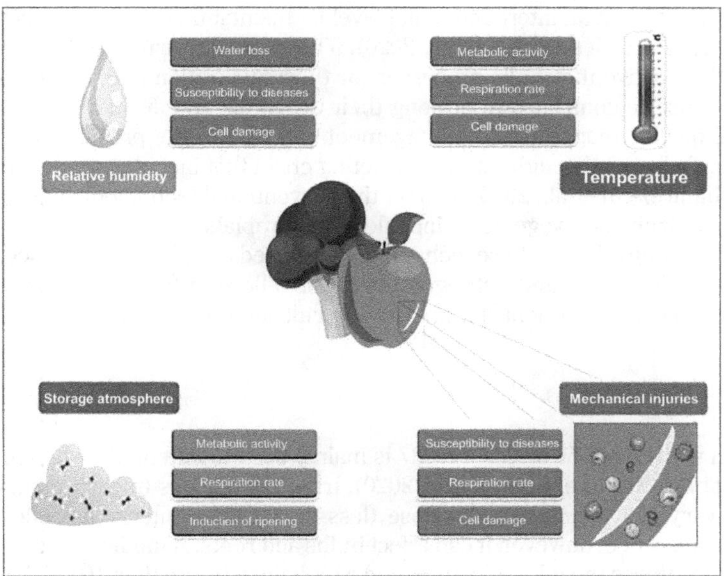

Figure 10.1 Biotic and abiotic factors that affect the quality of fruits and vegetables.

the case of high temperatures, they can cause an increase in metabolic activity and the respiration rates of fruits and vegetables, accelerating changes in color, firmness, weight, aroma, nutritional content, and changes in the cuticle leading to microbial infections (Calín-Sánchez et al., 2020). There is not much information on the effect of storage temperature on the production of volatile organic compounds and their effect on quality (Aubert et al., 2014; Spadafora et al., 2016); the production and presence of compounds such as ethylene, CO_2, O_2 and some inducers of maturation in the storage atmosphere, can play a key role in the metabolic processes of pectin degradation and starch conversion, increasing the release of water affecting the nutritional content (Gong et al., 2019; Jha et al., 2018; Sumonsiri and Barringer, 2013). The control of relative humidity during storage is a determining factor to minimize the exchange of water vapor pressure between the agricultural product and the environment, low relative humidity can facilitate water losses, turgor, and firmness, accelerating color changes and causing wilting, conversely, high relative humidity can favor microbial establishment and fruit decay, affecting shelf life (Nunes, 2008; Oliveira et al., 2016; Zhang and Long, 2017). Fungal diseases have been widely related to postharvest losses, as well as the degradation of the phytochemical and nutritional profile of infected foods, characterized by changes in color, firmness, aroma, and nutritional composition of these foods (Nouri et al., 2020; Ramos-Guerrero et al., 2018; Syafiq et al., 2020).

10.3 Management of Postharvest Diseases to Maintain the Quality of Fruits

The technologies used in the postharvest conservation of fruits and vegetables have allowed the agricultural and industrial sectors to sustain the global demand for food and its distribution at an intercontinental level by maintaining fresh products and their organoleptic properties (Vivek et al., 2020). The harvested fruits and vegetables still maintain their active metabolic processes for their maturation and senescence; therefore, they must be controlled to prolong their useful life and avoid a decrease in their nutritional quality. Inappropriate management can lead to the presence of pathogens and economic losses throughout the marketing chain that includes producers and consumers (Ramírez-Gil et al., 2020). Currently, conventional technologies for preserving the quality of fruits and vegetables include antimicrobials, gas, heat, antioxidants, and irradiation, among others. These technologies are used to reduce senescence, maturity, physiological disorders, and pathogen infections, allowing fresh for consumers. The most common basic treatments for preserving fruit and vegetables are described below.

10.3.1 Irradiation

Irradiation with cobalt-60 or cesium-137 is mainly used to extend the shelf life of fruits and vegetables (Munir and Federighi, 2020). Irradiation doses (measured in kilograys, kGy) can vary. For example, a low dose (less than 1 kGy) interrupts cellular activity delaying senescence; however, it can affect bulbs and roots. A medium dose (between 1 a 10 kGy) decreases the microbial load, and a high dose (more than 10 kGy) completely eliminates microorganisms and pests (Mahajan et al., 2014). Generally, medium or high doses of irradiation are not used in fruits and vegetables because they cause negative effects on the texture and flavor, in addition to accelerating senescence due to the fact that they affect the DNA and proteins of fresh products (Huang et al., 2020). The use of irradiation on food does not cause it to be radioactive, causes reduced chemical changes, and does not alter the nutritional value, keeping fruits and vegetables healthy.

10.3.2 Heat Treatment

At present, postharvest heat treatment is used as an alternative to the use of synthetic pesticides used in the preservation of fruits and vegetables. Common heat treatments are HWD, HWR, hot dry air, and heat from water vapor, among others. The use of heat treatment (for h or days) on fresh produce causes a reduction of cold damage, the preservation of color and flavor, a delay in maturation, the preservation of the structure and composition of epicuticular waxes, a reduction in microbial contaminants, and an increase shelf life, among others (Fallik and Ilic', 2019). Although heat treatment on fruits and vegetables is efficient, it is an option that has not yet been adopted by most producers due to the high energy costs required for its implementation.

10.3.3 Ozone

In many countries, ozone is used as a replacement for the traditional disinfectant agents used on fruits and vegetables. Ozone is a gas of natural origin with reactive oxidant properties that allows it to (1) preserve the shelf life of fresh products and (2) reduce the microbial load on food (Tzortzakis and Chrysargyris, 2017). Ozone has a greater oxidant potential than chlorine, and its antimicrobial action is faster on bacteria and phytopathogenic fungi; on fresh products, it increases; total soluble solids, ascorbic acid, β-carotene, lycopene, flavor, and tissue tenacity, among others (Fundo et al., 2018).

10.3.4 Controlled Atmosphere Storage

In the storage with atmospheres, the levels of CO_2 and O_2 are controlled inside the hermetic warehouses where fruits and vegetables are preserved for their later commercialization (Park et al., 2020). Generally, for the conservation of fresh products, the concentration of CO_2 is higher than that of O_2, modifying in fruits and vegetables their respiration, glycolysis, pigments, fermentation, senescence, ethylene, and shelf life, among others. Despite the beneficial effects of storing fruits and vegetables with controlled atmospheres, CO_2 and O_2 can affect ripening, taste, smell, and fermentation, among others, in addition to being a treatment with a high capital investment cost for construction and maintenance of high-quality warehouses (Ramírez-Acosta et al., 2019).

10.3.5 Ethylene

The exogenous application of ethylene on fruits and vegetables mainly increases color, maturation, and shelf life and maintains the organoleptic properties of the fresh product. However, it can affect color, senescence, and softening, reducing the shelf life of fruits and vegetables (Wei et al., 2020).

10.3.6 Plasma

Plasma is a technique used to reduce pathogens in fruits and vegetables, prolonging their useful life. The cold gas plasma inactivates and erodes microbial spores and causes DNA destruction (Mir et al., 2020).

10.3.7 NO

NO is a free radical gas that acts as a signaling molecule in various physiological processes such as fruit ripening and senescence. The application of endogenous NO prolongs the shelf life of fruits and vegetables during storage, preventing ripening and senescence. In addition, it decreases ethylene biosynthesis and increases oxidative stress (Rabiei et al., 2019).

10.3.8 SO_2

SO_2 is used in fruits and vegetables to reduce the microbial load and preserve the shelf life of food. Its action is antiseptic and inexpensive. However, in incorrect doses, it can affect the texture and color of the fruits (Liu, 2019).

10.4 Food Packaging

Society is currently interested in safe, fresh, and nutritious products; therefore, the food industry must provide a safe and modern packaging system to prevent infectious foodborne diseases (Han, 2014). The contamination of food can occur during production, processing, distribution, and even during preparation (Malhotra et al., 2015). Consequently, packaging plays an important role in improving the shelf life and adding value to food products (Kumar et al., 2018). Packaging is a container for the food and a barrier used to protect the product against external environmental factors and contribute to good distribution and efficient sales (Sharma et al., 2020). The packaging material can be flexible or rigid and is used to package a wide variety of food products (Zahra et al., 2016). The main objectives of packaging are (1) barrier protection, protecting from moisture, O_2, ultraviolet light, temperature, and abiotic and biotic agents, and (2) physical protection, isolating the food product from other things, mechanic shock, and vibration (Khaneghah et al., 2018; Yildirim et al., 2018). The Egyptians were the pioneers in the use of food packaging. The use of tinplate for packaging dates back to the 18th century and the first corrugated cardboard box was produced in England in 1817 (Gupta and Dudeja, 2017). Nowadays, researchers around the world have developed a novel wide variety of smart packing that includes intelligent and AP to provide quality and safe products and reduce environmental pollution related to food packaging waste (Gupta and Dudeja, 2017; Ozdemir and Floros, 2004). Intelligent packaging basically provides information, monitoring conditions of packaged products during transport and storage. Some examples of this type of packaging include indicators of pH, gas leaks, storage time, temperature, freshness, and microbial spoilage (Kruijf et al., 2002). AP is defined as a novel packaging system in which the packaging, the additive, and the product as well as the surrounding atmosphere interact with each other. AP can intentionally incorporate certain additives in the formulation of the packaging material or add active substances within the packaging that interact directly with the product, such as CO_2 scavengers and emitters, O_2 scavengers, ethanol emitters, ethylene scavengers, moisture absorbers, odor and flavor absorber/releaser, and antimicrobials systems (Almasi et al., 2020; Kumar et al., 2018). Scavengers are incorporated to remove undesirable components from the environment inside packaging and emitters/releasers add desired substances (Wyrwa and Barska, 2017). Antimicrobial packaging is an example of releasing system and incorporates into the polymer matrix substances, such as bacteriocins (brevicin, subtilin, lactocin, enisin, etc.), chlorine dioxide, ethanol, organic acids (benzoic, sorbic, propionic), CO_2, enzymes (peroxidase, lysozyme), extracts, nanoparticles (silver nanoparticles), EOs (cumin, oregano, clove, marjoram, rosemary, mustard, etc.), and the use of materials with antimicrobial properties, such as chitosan (Khaneghah et al., 2018). Antimicrobial packaging can be classified into (1) those that have antimicrobial agents added into sachets and are incorporated to the packages, (2) those that adsorb or coat the surfaces of polymer, (3) those that add antimicrobial compounds directly into polymers structure, (4) those that immobilize antimicrobial compounds in the polymers by covalent or ion linkages, and (5) those that use packaging polymers with antimicrobial properties (Appendini and Hotchkiss, 2002). Smart packaging has a wide range of applications in the food industry; its use also has several advantages, such as extending the shelf life, avoiding product loss, and enhancing control of storage conditions.

10.5 EOs: Physicochemical Properties

Plants have the ability to synthesize molecules, especially secondary metabolites, known as PFAs, where EOs represent the largest group of PFAs, and are recognized as safe by the FDA (de Oliveira et al., 2020). EOs are a complex mixture of compounds, mainly alkaloids, flavonoids, isoflavones, monoterpenes, phenolic acids, carotenoids, and aldehydes, where most of them have high lipophilicity, volatility, and oxidative decomposition, and are known for their use as additives in food and various food products (Nazzaro et al., 2017; Seow et al., 2014). EOs can be classified into four groups according to their active compounds: (1) Terpenes are a class of natural hydrocarbons that are derived from the isoprene unit (C_5H_8), and according to the extension of the chain, they have different chemical characteristics and biological properties; 90% of EO are monoterpenes, and they allow a great variety of structures. Some examples of terpenes include p-Cymene, pinene, humulene, limonene, and carvone (Hyldgaard et al., 2012). (2) Terpenoids are compounds related to terpenes and have some rearrangement or functionality with O_2. Some examples are thymol, carvacrol, linalol, geraniol, and menthol; these compounds have more microbial activity, caused by the functional groups. For example, the hydroxyl group of phenolic terpenoids is important in microbial activity since they interact directly with the cell membrane of the microorganism (Veldhuizen et al., 2006). (3) Phenylpropenes constitute a subfamily of organic compounds called phenylpropanoids, which are derived from phenol and correspond to a small part of EOs. Some examples are eugenol, isoeugenol, vanillin, safrole, and cinnamaldehyde (Hyldgaard et al., 2012). Table 10.1 shows some specific compounds with their most important characteristics for the chapter.

Studies show that EOs have powerful antimicrobial activity against a wide variety of microorganisms; thus, they can be incorporated into edible coatings for protecting fruits and vegetables (Burt, 2004; Donsì et al., 2011). However, the use of EOs is limited due to their high volatility and decomposition, as well as their strong taste affects the organoleptic characteristics of the food (Prakash et al., 2018). Therefore, different methods have been developed to avoid these limitations, such as encapsulation in polymeric particles through spray-drying (Wu et al., 2014), coacervation (Kaushik et al., 2015), encapsulation in liposomes by thin film hydration method (Varona et al., 2011), encapsulation in solid lipid nanoparticles by high-pressure homogenization (Weber et al., 2014), and encapsulation by emulsions, that favor the stability of EO, retain its biological activity, and minimize its organoleptic impact on food (Donsì et al., 2011).

Emulsions of the oil-in-water type (o/w) are one of the best methods for preparing polymers with EOs added for their application as edible coatings on fruits, they have a core–layer type structure and it has been shown that this type of encapsulation increases the bioavailability of lipophilic substances such as EOs (Acosta, 2009). When designing an emulsion, the type of emulsion suitable for a given application must be defined, and this is achieved through the hydrophilic/lipophilic properties of the emulsifiers, as explained by the Bancroft rule in conjunction with the HLB parameter. A rule of thumb states that different emulsions have similar characteristics if their emulsifiers have similar HLB numbers. For proper emulsification, the required HLB of an emulsion must match the HLB of the emulsifier mixture used (Kontogeorgis and Kiil, 2016), so it is common to use mixed emulsifiers. The HLB values of mixed surfactants are a key factor in the formation of emulsion droplets, During the formation of a stable o/w emulsion, lipophilic surfactants have a higher affinity for the droplets dispersed in the emulsion than hydrophilic surfactant (Lu et al., 2018), so when designing an emulsion with EOs, the HLB of the surfactants should be between 8–15; the best-known method for estimating HLB is the group contribution methods of Davies and Rideal (Feng et al., 2020; Kontogeorgis and Kiil, 2016).

When using EOs in o/w emulsions, adequate HLB values are needed to maintain the balance of the oil and water phases, in addition, the HLB value has a direct effect on the visual appearance. Lu et al. (2018) prepared nanoemulsions with Citral as

Table 10.1 Physicochemical Properties and Biological Activity of EO Compounds

EO components	*CAS number*	*Vapor pressure (mmHg at 25° C)*	*Refractive index (20°C)*	*Relative density (g/ml)*	*Topological polar surface area (Å²)*	*Plant source*	*Some biological activities*	*References*
Terpenes								
p-Cymene	99–87–6	1.50	1.49	0.86	0	*Cuminum cyminum*	Antifungal, antioxidant	Cristani et al., 2007
Terpenoids								
Carvacrol	499–75–2	N. A	1.522	0.97	20.2	*Thymus maroccanus*	Strong antimicrobial	Cristani et al., 2007
Thymol	89–83–8	0.016	N. A	0.96	20.2	*Thymus vulgaris*	Strong antimicrobial, antiseptic	Cristani et al., 2007
Fhenylpropenes								
Eugenol	97–53–0	0.022	1.544	1.067	29.5	*Eugenia, Caryophyllata*	Antifungal, antibacterial	Hemaiswarya and Doble, 2009
Cinnamaldehyde	104–55–2	0.029	1.621	1.05	17.1	*Cinnamomum, Zeylanicum*	Bactericide, fungicide, insecticide	Ye et al., 2013
Others								
Allicin	539–86–6	N. A	N. A	1.11	61.6	*Allium sativum*	Antifungal, antibacterial	Hyldgaard et al., 2012
Allyl isothiocyanate	57–06–7	3.70	1.53	1.01	44.4	*Brassica nigra*	Antifungal, antibacterial	Hyldgaard et al., 2012

EO and Span 85–Brij 97 as a mixed surfactant, according to their results as the HLB value is increased the emulsion becomes translucent and the encapsulated proportion of citral is significantly increased. In another study, Feng et al. (2020) used d-limonene and El-40 as surfactant, they found an optimal HLB value of 13.5. The formation, stability, and properties of nanoemulsions depend on the physicochemical properties of the oil phase whether it is a component or a mixture of various EOs, for example, polarity, solubility in water, interfacial tension, refractive index, viscosity, density, pressure vapor (volatility), and chemical stability (McClements and Rao, 2011; Tadros et al., 2004). These physicochemical properties can directly influence the formation, stability, and performance of the emulsion, which are summarized in Table 10.2.

Table 10.2 Physicochemical Properties of Oil and Aqueous Phases on the Formation Stability and Performance of Food-Grade Oil-in-Water Emulsion

Physicochemical properties	*Formation*	*Stability*	*Performance*
Viscosity (η) Oil Phase (η_D) Aqueous Phase (η_C)	High-energy methods depend on viscosity ratio (η_D / η_C)	Creaming rate decreases with increasing aqueous phase viscosity (η_C)	Determines formation, texture, and shelf life
Interfacial tension (γ)	Low γ facilitates droplet disruption in high-energy methods. Low γ facilities droplet formation in phase-inversion and spontaneous-emulsification methods.	Low γ promotes droplet coalescence Low γ may lead to poor emulsifier affinity for droplet surfaces	Affects formation and shelf life
Solubility	Water solubility of surfactants and solvents affects droplet formation in solvent displacement and spontaneous emulsification methods	High water solubility of an oil phase promotes Ostwald ripening	Affects shelf life
Polarity	–	Determines partitioning of components between oil and aqueous phases	Affects flavor profiles
Density (ρ)	–	Creaming rate increases with increasing density contrast	Determines shelf life
Refractive index (n)	–	–	Turbidity increases with increasing refractive index contrast

Source: Adapted from McClements and Rao (2011).

One of the main objectives of food scientists working with EO through some encapsulation method is to create emulsions with sufficiently small particles and with long enough kinetic stability for commercial applications, this can be achieved by incorporating substances as stabilizers, such as emulsifiers, texture modifiers, weighting agents, or ripening retarders (McClements and Rao, 2011). An interesting application of encapsulated EOs is the formulation of coatings on fruits and vegetables; they have a high potential to reduce microbial activity and delay the ripening process of the coated produce.

10.6 Edible Films with EOs: Efficacy *In Vitro*

In recent years, there has been an increase in the search for alternative methods for food preservation, highlighting the study of edible films as a protective barrier against dehydration, microbial attack, and other factors that detract from food quality (Fernández Valdés et al., 2015). In addition to film's benefits, an advantage is that active ingredients can be incorporated into these systems, improving the quality of food, as well as its safety (Dhall, 2013). Among the active ingredients that are usually added to these films, compounds with activity against bacteria and fungi stand out, as is the case of EOs, which is important for their phenolic compounds content (Varghese et al., 2020).

It is mentioned that EOs most used for their antifungal properties are those from oregano (*Origanum vulgare* L.), basil (*Ocimum seali* L.), savory (*Satureja hortensis* L.), dill (*Anethum graveolens* L.), laurel (*Lauris nobilis* L.), and thyme (*T. vulgaris* L.; Aguilar-Veloz et al., 2020). In this sense, Hashemi and Mousavi Khaneghah (2017) mention that by using oregano EO as a complementary agent in edible films, it is possible to inhibit *in vitro* growth of *S. Typhimurium*, *P. aeruginosa*, *S. aureus*, and *B. cereus*. However, other oils have been used, for example, turmeric oil was added to edible films to demonstrate its effect on the fungal pathogens *Penicillium sp.* and *Cladosporium sp.* (Campo Velasco et al., 2017). A recent study in which lemon and bergamot oil are added to edible whey protein films demonstrates the antimicrobial potential of these films in controlling the growth of *E. coli*, *S. aureus*, and *A. niger* (Çakmak et al., 2020).

On the other hand, chitosan is a very versatile polysaccharide due to its physicochemical properties, it can be used alone or in combination with additives such as EOs (Saxena et al., 2020). Tea tree EO and its antimicrobial capacity were evaluated on *Listeria monocytogenes* and *Penicillium italicum* in a PDA medium, and according to Sánchez-González et al. (2010), their growth was affected by the microbial interaction with the EOs compared to control plates during the first 3 days of incubation. Edible films were made with amaranth, chitosan, and starch, adding EOs of oregano, cinnamon, and lemongrass in order to inhibit the growth of *Aspergillus niger* and *Penicillium digitatum* (Avila-Sosa et al., 2012). The *in vitro* evaluation by the steam contact technique exhibited higher antifungal activity for the chitosan films (inhibition of *A. niger* with 0.25% EOs of Mexican oregano and cinnamon; inhibition of *P. digitatum* with 0.50% EO). The activity exerted by these oils is mainly attributed to the content of antimicrobial compounds such as carvacrol, eugenol, cinnamaldehyde, and thymol as reported in this study. In another investigation, the films were combined with EO of thyme, cinnamon, and cloves and were evaluated *in vitro* on the mycelial growth of *Rhizopus stolonifer*, *Colletotrichum gloeosporioides*, *Alternaria alternata*, *Fusarium oxysporum* and *Penicillium digitatum*, achieving 100% inhibition (Hernández-López et al., 2018). Similarly, films that had rosemary and ginger oil added to them were formed to evaluate their antimicrobial activity on the pathogens *Enterococcus faecalis*, *Listeria monocytogenes*, *Staphylococcus aureus*, *Escherichia coli*, and *Pseudomonas aeruginosa* (Souza et al., 2019). The results show the summation effect of mechanisms of action since adding EOs enhances the antimicrobial activity of chitosan since a greater reduction of the logarithmic cycle of these fungi was obtained; therefore, these films can be used in the conservation of food. Other polymers have also shown good results in the formation of edible films, such as alginate, that were used to form edible films from nanoemulsions containing EOs (Acevedo-Fani et al., 2015), demonstrating that its antimicrobial effect is increased when thyme oil was added compared to other oils, such as lemongrass or sage. The inoculation of the films with *Escherichia coli* caused up to 4.71 logarithmic reductions after 12 h, and it was shown that the type of oil had a greater influence on the antimicrobial activity than the droplet size of the nanoemulsion. In another study, nanoemulsions with cinnamon oil incorporated were evaluated by Ghani et al. (2018); the authors observed an inhibitory effect on gram-positive and gram-negative bacteria, attributed to the presence of the compound trans-cinnamaldehyde in cinnamon oil. Fasihi et al. (2019) showed good antifungal properties of carbohydrate-based films enriched with cinnamon EO against *Penicillium digitatum*, by the complete inhibition *in vitro* tests. These authors commented on the protection provided by the film to the oil from environmental factors, which increases its antifungal power attributed to the presence of cinnamaldehyde. And it is discussed that the possible mechanisms of action could be the destruction and damage of the cell wall and plasma membrane, the loss of cytoplasm and mitochondria, cells, folding, and inhibitory effects on enzyme systems. Table 10.3 shows a summary of authors who

Table 10.3 Applications of Edible Films Added with EOs (*in vitro*)

Edible films + EO	*Pathogens*	*Reference*
Tea tree EO	*Listeria monocytogenes* *Penicillium italicum*	Sánchez-González et al. (2010)
Oregano EO	*S. Typhimurium* *P. aeruginosa* *S. aureus* *B. cereus*	Hashemi and Mousavi Khaneghah (2017)
Oregano, cinnamon, and lemongrass essential oil	*Aspergillus niger* *Penicillium digitatum*	Avila-Sosa et al. (2012)
Thyme, cinnamon, and clove EO	*Rhizopus stolonifer* *Colletotrichum gloeosporioides* *Alternaria alternata* *Fusarium oxysporum Penicillium digitatum*	Hernández-López et al. (2018)
Cinnamon EO	*Penicillium digitatum*	Fasihi et al. (2019)
Turmeric oil	*Penicillium sp.* *Cladosporium sp.*	Campo Velasco et al. (2017)
Lemon and bergamot EO	*E. coli, S. aureus* *A. niger*	Çakmak et al. (2020)

report the use of edible films added with EOs as an alternative to the use of synthetic fungicides. In general, most of the studies that have evaluated edible films added with EOs *in vitro* conclude that this is a very efficient alternative with which the main objective is achieved, which is to avoid the deterioration of food due to the effect of different pathogens, thanks to the summation of mechanisms of action involved by the compounds integrated into the film.

10.7 Edible Coatings with EOs

The incidence of pathogens on postharvest fruits and vegetables decreases their quality and shelf life. EOs have shown promising results to the control of many postharvest diseases. Antifungal compounds have been identified in some EOs. Some of these EOs are thyme, lemongrass, cinnamon, mint, and clove (Ali et al., 2015; Mohammadi et al., 2015a; Sapper et al., 2019). *In vivo* application of ELs is actually by incorporating them into edible coating, these polymers are used as an encapsulation matrix for their preservation. Materials based on proteins, polysaccharides, and/or lipids are used as edible coatings (Aitboulahsen et al., 2018). The most used are polysaccharides, including gums, starches, pectin, and chitosan. These coatings are considered natural, nontoxic, biodegradable products that have commercial use (Raghav et al., 2016). In this section, the synergy between edible coatings and EOs and their efficiency in controlling the pathogens of different fruits and vegetables are presented.

10.7.1 Gums

Gums are polysaccharide compounds that in contact with water have the ability to form a gel or stabilizer that can function as edible coatings. Arabic gum is one of the most-used polymers for edible coatings on fruits. The high solubility and low viscosity of this gum make it useful as coatings compared with other gums (Raghav et al., 2016). The application of arabic gum as an edible coating has preserved the quality and shelf life of fruits and vegetables such as tomatoes (Ali et al., 2010), strawberries (Tahir et al., 2018), cucumbers (Al-Juhaimi et al., 2012), and mangoes (Daisy et al., 2020). The gelling capacity of the gum helps encapsulate antifungal compounds as EOs. This effect was confirmed on papaya fruit treated with a combination of 10% of gum arabic and 2% of ginger oil. This synergy generated a 30% inhibition of the pathogen *C. gloeosporioides* (Ali et al., 2016). Another study reported more than 70% inhibition of this pathogen by the combination of arabic gum with the application of 0.5% of cinnamon EO (Maqbool et al., 2011). On peach fruits, the development of *B. cinerea* was controlled at 80%; this effect was possible by combining this gum with mint EO (1 mL/L; Alamri et al., 2020). The encapsulating effect of gum arabic with respect to thyme EO was evaluated by FTIR spectroscopy analysis. According to this analysis, the gum showed an efficient encapsulant capacity and did not interact with the thyme oil; in addition, the authors reported the control of *C. gloeosporioides*, attributed to the antifungal effect of the aroma compounds present in thyme EO (Cai et al., 2019).

10.7.2 Starch

Starch-based coatings are obtained from grains or some tubers. The polymers amylopectin and amylose are the constituents of this polysaccharide, specifically the amylose polymer, which gives the starch its ability to form coatings (Raghav et al., 2016). Starch-based coatings are used at the postharvest stage of fruits and vegetables. One of the most used starches is from cassava. In peppers, the application of 4% of starch from cassava and 1000 ppm thyme oil did not allow the proliferation of pathogens of the genera *Alternaria* and *Fusarium*. This effect was attributed by these authors to the antifungal compound carvacrol present in thyme EO (Ordoñez Bolaños et al., 2014). The incorporation of oregano EO at 250 and 500 ppm to starch from cassava in combination with soy protein inhibited *Fusarium solani* approximately 97% (Castillo Silva et al., 2018). In apple fruit, it was possible to control *B. cinerea* up to 75% with applications of starch from cassava and gellan gum in combination with thyme oil (Sapper et al., 2019). In a recent study, thyme EOs with different antifungal compounds, including thymol, carvacrol, and α-terpineol, were identified (Tohidi et al., 2017). Hou et al. (2020) reported that the application of antifungal compounds thymol and carvacrol generated damage to mycelium morphology, ROS accumulation, increased cell permeability, and mitochondrial changes in the pathogen *B. cinerea*.

10.7.3 Chitosan

Chitosan biopolymer is commonly applied as a postharvest agent to the control of pathogenic fungi. The application of this biopolymer reduced the incidence of pathogens by the activation of their three mechanisms of action: the ability to form a coating, antifungal effect, and inductor (Romanazzi et al., 2018). Chitosan is used as a matrix of incorporation of EO. Several *in vivo* studies have confirmed that the synergy between chitosan and EOs is efficient. The combination of this compound increases the antifungal potential and reduces the development of pathogens (El-Mohamedy et al., 2015; Mohammadi et al., 2015b; dos Passos Braga et al., 2019). In cucumber fruit, the individual application of chitosan showed a 38.6% incidence of *Phytophthora drechsleri*, while the combination of chitosan with the EO of cinnamon (*Cinnamomum zeylanicum*) reduced completely the disease incidence after 9 days of refrigerated storage (Mohammadi et al., 2015a). Ali et al. (2015) evaluated the antifungal effect on *Colletotrichum capsici* on bell peppers with the individual and combined application of chitosan and lemongrass EO. With the individual application of chitosan, a disease incidence of 26.7% was obtained, whereas with the use of lemongrass, fruits showed a 40% of disease incidence, while a 46.7% on disease incidence was obtained with the combination of both compounds. In other fruits such as strawberries, avocados, papayas, oranges, and tomatoes, the combination of chitosan and different EOs have been applied showing more than 50% of inhibition of the attacking pathogens (Table 10.4). The antifungal efficiency of chitosan and EOs directly dependent on their synergy potential. FTIR spectroscopy analyses have indicated that this synergy is possible with the union between the NH (chitosan) and OH (EO) functional groups, forming a chemically stable matrix (Shen and Kamdem, 2015). Shao et al. (2015) reported this effect with the application of chitosan and clove oil applied alone or in combination.

Table 10.4 Synergy of Chitosan Coating and EOs to the Control of Postharvest Pathogens

Chitosan	*EO*	*Fruit/ vegetable*	*Pathogen*	*Inhibition (%)*	*References*
Chitosan 0.3%	Cinammon 1.5 g/L	Cucumber	*Phytophthora drechsleri*	100%	Mohammadi et al., 2015a
Chitosan 0.15%	Cinnamon 0.15%	Strawberry	*Botrytis cinerea*	58.3%	Mohammadi et al., 2015b
Chitosan 1%	Lemongrass 1%	Bell pepper	*Colletotrichum capsici*	61.8%	Ali et al., 2015
Chitosan 4 g/L	Lemongrass 4 ml/L	Orange	*Penicillium itallicum*	84.0%	El-Mohamedy et al., 2015
Chitosan 2%	Thyme 0.5%	Strawberry	*Penicillium spp.** *Monilinia spp.** *Botrytis spp.**	87.0%	Martínez et al. (2018)
Chitosan 1%	Thyme 5%	Avocado	*Colletotrichum gloeosporioides*	64.0%	Correa-Pacheco et al. (2017)
Chitosan 0.5 mg/mL	Pepper tree 100 µL	Avocado	*C. gloeosporioides*	100%	Chávez-Magdaleno et al. (2018)
Chitosan 5 mg/mL	Mint 1.25 µL/mL	Papaya	*C. gloeosporioides*	100%	dos Passos Braga et al. (2019)
Chitosan 4 mg/mL	Mint 2.5 µL/mL	Cherry tomato	*Rhizopus stolonifer*	59.4%	Guerra et al. (2015)

**Without pathogen inoculation.*

10.8 Other Coatings with the Incorporation of EOs

The synergy of the different coatings with different EOs continues in evaluation. The antifungal effect of aloe vera gel has been confirmed to control *Aspergillus*, *Penicillium*, and *Fusarium* (Oni et al., 2018). Khaliq et al. (2019) when applied this gel as coating on banana fruit; the results of the investigation showed a 60% inhibition of *Colletotrichum musae*. Pectin-based coatings have also been used as a matrix for EOs. Rodriguez-Garcia et al. (2015) used this coating in synergy with oregano EO in tomato fruit. According to their results, the pathogen *Alternaria alternata* was completely inhibited. A multicomponent coating based on pectin, pomegranate peel fiber, and polyvinyl alcohol was combined with thyme oil for controlling anthracnose in mango fruit. This multicomponent coating and thyme oil controlled the development of the pathogen below 23% (Nandhavathy et al., 2020). *Aureobasudium pullulans* is capable of producing pullulan polysaccharide, which has been used as an edible coating in fruits. Its effect in combination with caraway EO was evaluated on the development of the *Penicillium expansum* in baby carrots. The result obtained was a 10% inhibition.

Pullulan polysaccharide and caraway oil did not show an efficient antifungal effect (Gniewosz et al., 2013). The EO of mint in combination with 4% gelatin coating controlled more than 50% in strawberry fruits without damage from pathogens (*B. cinerea* and *R. stolonifer*) (Aitboulahsen et al., 2018).

Lipid-based coatings such as carnauba wax have also been used as edible coatings. Its application together with the EO from the *Lippia scaberrima* plant did not allow the development of *P. digitatum* in orange fruits (du Plooy et al., 2009). Some compounds such as (d) -limonene, R–(-)–carvone, and 1,8-cineole provide the antifungal potential to this EO (Combrinck et al., 2006).

The interaction between edible coatings and the diverse EOs that exist continues to be studied. Coatings properties, such as water absorption, water vapor permeability, and mechanical properties, can be affected by the incorporation of EOs (Ju et al., 2019). The coatings must act only as the suspension or encapsulant matrix of the EOs, interact as little as possible with these compounds so as not to be affected by their functional properties, and allow the EO to generate its antifungal effect. The investigation of the synergistic effect between these compounds continues in order to get information about their efficacy, as well as their impact, for not only controlling diseases but also maintaining the quality of treated fruits at the postharvest stage.

10.9 Conclusion and Future Perspectives

At present, fruit production is highly affected by phytopathogens, this situation leads to a huge amount of produce losses around the world. The application of chemical fungicides is considered the primary strategy to assure fruit production; however, their uncontrolled use leads to their accumulation not only in the environment but also in fruits and vegetables. The management of diseases at the postharvest stage needs to change to environmentally friendly alternatives. EOs are considered safe compounds, they are effective against a wide range of phytopathogens. Even when they are effective, they need to be protected from the environment, in this sense their incorporation as additives into polymeric matrices represents an alternative to assure their efficacy. Quality is a key factor for a total acceptance by consumers; thus, the selection of the appropriate EO to a specific fruit is crucial so the sensorial compatibility is not altered. However, it is necessary to deepen the knowledge of how these edible coatings can interact with the fruit and how their presence impact in the physicochemical profile of coated fruits.

References

Acevedo-Fani A, Salvia-Trujillo L, Rojas-Graü MA, Martín-Belloso O. 2015. Edible films from essential-oil-loaded nanoemulsions: Physicochemical characterization and antimicrobial properties. Food Hydrocoll;47:168–77.

Acosta E. 2009. Bioavailability of nanoparticles in nutrient and nutraceutical delivery. Curr Opin Colloid Interface Sci;14:3–15.

Aguilar-Veloz LM, Calderón-Santoyo M, Vázquez González Y, Ragazzo-Sánchez JA. 2020. Application of essential oils and polyphenols as natural antimicrobial agents in postharvest treatments: Advances and challenges. Food Sci Nutr;8:2555–68.

Aitboulahsen M, Zantar S, Laglaoui A, Chairi H, Arakrak A, Bakkali M, et al. 2018. Gelatin-based edible coating combined with *Mentha pulegium* essential oil as bioactive packaging for strawberries. J Food Qual;2018:1–7.

Alamri SAM, Hashem M, Alqahtani MSA, Alshehri AMA, Mohamed ZA, Ziedan ESH. 2020. Formulation of mint and thyme essential oils with Arabic gum and Tween to enhance their efficiency in the control of postharvest rots of peach fruit. Can J Plant Pathol;42:330–43.

Ali A, Hei GK, Keat YW. 2016. Efficacy of ginger oil and extract combined with gum arabic on anthracnose and quality of papaya fruit during cold storage. J Food Sci Technol;53:1435–44.

Ali A, Maqbool M, Ramachandran S, Alderson PG. 2010. Gum arabic as a novel edible coating for enhancing shelf-life and improving postharvest quality of tomato (*Solanum lycopersicum* L.) fruit. Postharvest Biol Technol;58:42–7.

Ali A, Noh NM, Mustafa MA. 2015. Antimicrobial activity of chitosan enriched with lemongrass oil against anthracnose of bell pepper. Food Packag Shelf Life;3:56–61.

Al-Juhaimi F, Ghafoor K, Babiker EE. 2012. Effect of gum arabic edible coating on weight loss, firmness and sensory characteristics of cucumber (*cucumis sativus* l.) fruit during storage. Pak J Bot;44:1439–44.

Almasi H, Jahanbakhsh Oskouie M, Saleh A. 2020. A review on techniques utilized for design of controlled release food active packaging. Crit Rev Food Sci Nutr:1–21.

Appendini P, Hotchkiss JH. 2002. Review of antimicrobial food packaging. Innov Food Sci Emerg Technol;3:113–26.

Aubert C, Bony P, Chalot G, Landry P, Lurol S. 2014. Effects of storage temperature, storage duration, and subsequent ripening on the physicochemical characteristics, volatile compounds, and phytochemicals of western red nectarine (*Prunus persica* L. Batsch). J Agric Food Chem;62:4707–24.

Avila-Sosa R, Palou E, Jiménez Munguía MT, Nevárez-Moorillón GV, Navarro Cruz AR, López-Malo A. 2012. Antifungal activity by vapor contact of essential oils added to amaranth, chitosan, or starch edible films. Int J Food Microbiol;153:66–72.

Barman K, Ahmad MDS, Siddiqui MW. 2015. Factors affecting the quality of fruits and vegetables: Recent understandings. In: Siddiqui MW, editor. Postharvest Biology and Technology of Horticultural Crops. Apple Academic Press Inc; pp. 1–55.

Bhargava A, Bansal A. 2018. Fruits and vegetables quality evaluation using computer vision: A review. J King Saud Univ Comput Inf Sci.

Bhowmik PK, Dris R. 2004. Enzymes and quality factors of fruits and vegetables. In: Dris R, Jain SM, editors. Production Practices and Quality Assessment of Food Crops. Springer; pp. 1–25.

Brasil IM, Siddiqui M. 2018. Postharvest quality of fruits and vegetables: An overview. In: Siddiqui MW, editor. Preharvest Modulation of Postharvest Fruit and Vegetable Quality. An Overview. Academic Press; pp. 1–40.

Burt S. 2004. Essential oils: Their antibacterial properties and potential applications in foods-a review. Int J Food Microbiol;94:223–53.

Cai C, Ma R, Duan M, Lu D. 2019. Preparation and antimicrobial activity of thyme essential oil microcapsules prepared with gum arabic. RSC Adv;9:19740–7.

Çakmak H, Özselek Y, Turan OY, Fıratlıgil E, Karbancioğlu-Güler F. 2020. Whey protein isolate edible films incorporated with essential oils: Antimicrobial activity and barrier properties. Polym Degrad Stab;179:109285.

Calín-Sánchez A, Lipan L, Cano-Lamadrid M, Kharaghani A, Masztalerz K, Carbonell-Barrachina AA. 2020. Comparison of traditional and novel drying techniques and its effect on quality of fruits, vegetables and aromatic herbs. Foods;9:1–27.

Campo Velasco JA, Vanegas Mahecha P, Andrade-Mahecha MM. 2017. Aceite esencial de cúrcuma (*Curcuma longa* L.) como agente antifúngico en recubrimientos comestibles aplicados a zapallo (*Cucurbita maxima*) mínimamente procesado. Rev Ciênc Agrár;40:641–54.

Castillo Silva DA, Salazar Anacona K, Mosquera Sánchez SA, Rengifo Canizales E. 2018. Efecto de recubrimientos de almidón modificado de yuca, proteina aislada de soya y aceite esencial de orégano aplicados a la papaya. Rev UDCA Actual Divulg Científica;21:71–80.

Chávez-Magdaleno ME, Luque-Alcaraz AG, Gutíerrez-Martínez P, Cortez-Rocha MO, Burgos-Hernández A, Lizardi-Mendoza J. 2018. Effect of chitosan-pepper tree (*Schinus molle*) essential oil. Rev Mex De Ingeniería Química;17:29–45.

Combrinck S, Bosman AA, Botha BM, Du Plooy W, McCrindle RI, Retief E. 2006. Effect of post-harvest drying on essential oil and glandular trichomes of *Lippia scaberrima* Sond. J Essent Oil Res;18:80–84.

Correa-Pacheco ZN, Bautista-Baños S, Valle-Marquina MÁ, Hernández-López M. 2017. The effect of nanostructured chitosan and chitosan-thyme essential oil coatings on *Colletotrichum gloeosporioides* growth in vitro and on cv hass avocado and fruit quality. J Phytopathol;165:297–305.

Cristani M, D'Arrigo M, Mandalari G, Castelli F, Sarpietro MG, Micieli D. 2007. Interaction of four monoterpenes contained in essential oils with model membranes: Implications for their antibacterial activity. J Agric Food Chem;55:6300–8.

Daisy LL, Nduko JM, Joseph WM. 2020. Effect of edible gum Arabic coating on the shelf life and quality of mangoes (*Mangifera indica*) during storage. J Food Sci Technol;57:79–85.

de Oliveira MS, da Costa WA, Silva SG. 2020. Essential Oils: Bioactive Compounds, New Perspectives and Applications. IntechOpen; pp. 1–222.

Dhall RK. 2013. Advances in edible coatings for fresh fruits and vegetables: A review. Crit Rev Food Sci Nutr;53:435–50.

Donsì F, Annunziata M, Sessa M, Ferrari G. 2011. Nanoencapsulation of essential oils to enhance their antimicrobial activity in foods. LWT Food Sci Technol;44:1908–14.

dos Passos Braga S, Lundgren GA, Macedo SA, Tavares JF, dos Santos Vieira WA, Câmara MPS. 2019. Application of coatings formed by chitosan and Mentha essential oils to control anthracnose caused by Colletotrichum gloesporioides and C. brevisporum in papaya (Carica papaya L.) fruit. Int J Biol Macromol;139:631–9.

du Plooy W, Regnier T, Combrinck S. 2009. Essential oil amended coatings as alternatives to synthetic fungicides in citrus postharvest management. Postharvest Biol Technol;53:117–22.

El-Mohamedy RSR, El-Gamal NG, Bakeer ART. 2015. Application of chitosan and essential oils as alternatives fungicides to control green and blue moulds of citrus fruits. Int J Curr Microbiol Appl Sci;4:629–43.

Ennigrou A, Casabianca H, Laarif A, Hanchi B, Hosni K. 2017. Maturation-related changes in phytochemicals and biological activities of the Brazilian pepper tree (*Schinus terebinthifolius* Raddi) fruits. S Afr J Bot;108:407–15.

Fallik E, Ilic Z. 2019. Control of postharvest decay of fresh produce by heat treatments; the risks and the benefit. In Postharvest Pathology of Fresh Horticultural Produce. CRC Press.

Fasihi H, Noshirvani N, Hashemi M, Fazilati M, Salavati H, Coma V. 2019. Antioxidant and antimicrobial properties of carbohydrate-based films enriched with cinnamon essential oil by pickering emulsion method. Food Packag Shelf Life; 19:147–54.

Feng J, Wang R, Chen Z, Zhang S, Yuan S, Cao H. 2020. Formulation optimization of D-limonene-loaded nanoemulsions as a natural and efficient biopesticide. Colloids Surf A Physicochem Eng Asp:124746.

Fernández Valdés D, Bautista Baños S, Fernández Valdés D, Ocampo Ramírez A, García Pereira A, Falcón Rodríguez A. 2015. Películas y recubrimientos comestibles: Una alternativa favorable en la conservación poscosecha de frutas y hortalizas Eatable films and coverings: A favorable alternative in the postharvesIng. conservation of fruits and vegetables. Rev Ciencias Técnicas Agropecu;24:52–7.

Fundo JF, Miller FA, Tremarin A, Garcia E, Brandão TRS, Silva CLM. 2018. Quality assessment of Cantaloupe melon juice under ozone processing. Innov Food Sci Emerg Technol;47:461–6.

Ghani S, Barzegar H, Noshad M, Hojjati M. 2018. The preparation, characterization and in vitro application evaluation of soluble soybean polysaccharide films incorporated with cinnamon essential oil nanoemulsions. Int J Biol Macromol;112:197–202.

Gniewosz M, Kraśniewska K, Woreta M, Kosakowska O. 2013. Antimicrobial activity of a pullulan-caraway essential oil coating on reduction of food microorganisms and quality in fresh baby carrot. J Food Sci;78:242–8.

Gong D, Bi Y, Li Y, Zong Y, Han Y, Prusky D. 2019. Both *Penicillium expansum* and *Trichothecim roseum* infections promote the ripening of apples and release specific volatile compounds. Front Plant Sci;10:1–14.

Guerra ICD, de Oliveira PDL, de Souza Pontes AL, Lúcio ASSC, Tavares JF, Barbosa-Filho JM. 2015. Coatings comprising chitosan and Mentha piperita L. or Mentha×villosa Huds essential oils to prevent common postharvest mold infections and maintain the quality of cherry tomato fruit. Int J Food Microbiol;214:168–78.

Gupta RK, Dudeja P. 2017. Food packaging. In: Dudeja P, Gupta RK, Minhas AS, editors. Food Safety in the 21st Century. Academic Press; pp. 547–53.

Han JH. 2014. Edible films and coatings: A review. Innov. Food Packag: 213–55.

Hashemi SMB, Mousavi Khaneghah A. 2017. Characterization of novel basil-seed gum active edible films and coatings containing oregano essential oil. Prog Org Coatings;110:35–41.

Hemaiswarya S, Doble M. 2009. Synergistic interaction of eugenol with antibiotics against Gram negative bacteria. Phytomedicine;16:997–1005.

Hernández-López M, Guillén-Sánchez J, Bautista-Baños S, Guillén-Sánchez D. 2018. Evaluación de películas biodegradables en el control de hongos postcosecha de la papaya. Cultiv Trop;39:52–60.

Hodges DM, Toivonen PMA. 2008. Quality of fresh-cut fruits and vegetables as affected by exposure to abiotic stress. Postharvest Biol Technol;48:155–62.

Hou H, Zhang X, Zhao T, Zhou L. 2020. Effects of *Origanum vulgare* essential oil and its two main components, carvacrol and thymol, on the plant pathogen *Botrytis cinerea*. PeerJ;8:1–25.

Huang H, Belwal T, Li L, Wang Y, Aalim H, Luo Z. 2020. Effect of modified atmosphere packaging of different oxygen levels on cooking qualities and phytochemicals of brown rice during accelerated aging storage at 37°C. Food Packag Shelf Life;25:100529.

Hyldgaard M, Mygind T, Meyer RL. 2012. Essential oils in food preservation: Mode of action, synergies, and interactions with food matrix components. Front Microbiol;3:12.

Jha PK, Xanthakis E, Chevallier S, Jury V, Le-Bail A. 2018. Assessment of freeze damage in fruits and vegetables. Food Res Int;121:479–96.

Ju J, Xie Y, Guo Y, Cheng Y, Qian H, Yao W. 2019. Application of edible coating with essential oil in food preservation. Crit Rev Food Sci Nutr;59:2467–80.

Kaushik P, Dowling K, Barrow CJ, Adhikari B. 2015. Microencapsulation of omega-3 fatty acids: A review of microencapsulation and characterization methods. J Funct Foods;19:868–81.

Khaliq G, Abbas HT, Ali I, Waseem M. 2019. Aloe vera gel enriched with garlic essential oil effectively controls anthracnose disease and maintains postharvest quality of banana fruit during storage. Hortic Environ Biotechnol;60:659–69.

Khaneghah AM, Hashemi SMB, Limbo S. 2018. Antimicrobial agents and packaging systems in antimicrobial active food packaging: An overview of approaches and interactions. Food Bioprod Process;111:1–19.

Kontogeorgis GM, Kiil S. 2016. Introduction to Applied Colloid and Surface Chemistry. Wiley Online Library.

Kruijf ND, Beest MV., Rijk R, Sipiläinen T, Paseiro-Losada P, Meulenaer BD. 2002. Active and intelligent packaging: Applications and regulatory aspects. Food Addit Contam;19:144–62.

Kumar KVP, Jessie Suneetha W, Kumari BA. 2018. Active packaging systems in food packaging for enhanced shelf life. Food Packag Shelf Life;7:2044–6.

Kyriacou MC, Rouphael Y. 2017. Scientia horticulturae towards a new definition of quality for fresh fruits and vegetables. Sci Hortic;234.

Li D, Zhu Z, Sun D. 2018. Effects of freezing on cell structure of fresh cellular food materials: A review. Trends Food Sci Technol;75:46–55.

Liu Y-B. 2019. Sulfur dioxide fumigation for postharvest control of mealybugs on harvested table grapes. J Econ Entomol;112:597–602.

Lu W-C, Huang D-W, Wang C-C, Yeh C-H, Tsai J-C, Huang Y-T. 2018. Preparation, characterization, and antimicrobial activity of nanoemulsions incorporating citral essential oil. J Food Drug Anal;26:82–9.

Ma L, Zhang M, Bhandari B, Gao Z. 2017. Recent developments in novel shelf life extension technologies of fresh-cut fruits and vegetables. Trends Food Sci Technol;64:23–34.

Mahajan PV, Caleb OJ, Singh Z, Watkins CB, Geyer M. 2014. Postharvest treatments of fresh produce. Philos Trans R Soc A Math Phys Eng Sci;372:20130309.

Malhotra B, Keshwani A, Kharkwal H. 2015. Antimicrobial food packaging: Potential and pitfalls. Front Microbiol;6:1–9.

Maqbool M, Ali A, Alderson PG, Mohamed MTM, Siddiqui Y, Zahid N. 2011. Postharvest application of gum arabic and essential oils for controlling anthracnose and quality of banana and papaya during cold storage. Postharvest Biol Technol;62:71–6.

Martínez K, Ortiz M, Albis A, Castañeda CGG, Valencia ME, Tovar CDG. 2018. The effect of edible chitosan coatings incorporated with thymus capitatus essential oil on the shelf-life of strawberry (*Fragaria x ananassa*) during cold storage. Biomolecules;8:1–23.

McClements DJ, Rao J. 2011. Food-grade nanoemulsions: Formulation, fabrication, properties, performance, biological fate, and potential toxicity. Crit Rev Food Sci Nutr;51:285–330.

Mir SA, Siddiqui MW, Dar BN, Shah MA, Wani MH, Roohinejad S. 2020. Promising applications of cold plasma for microbial safety, chemical decontamination and quality enhancement in fruits. J Appl Microbiol;129:474–85.

Mohammadi A, Hashemi M, Hosseini SM. 2015a. Chitosan nanoparticles loaded with *Cinnamomum zeylanicum* essential oil enhance the shelf life of cucumber during cold storage. Postharvest Biol Technol;110:203–13.

Mohammadi A, Hashemi M, Hosseini SM. 2015b. The control of Botrytis fruit rot in strawberry using combined treatments of Chitosan with Zataria multiflora or *Cinnamomum zeylanicum* essential oil. J Food Sci Technol;52:7441–8.

Munir MT, Federighi M. 2020. Control of foodborne biological hazards by ionizing radiations. Foods;9:878.

Nandhavathy G, Dharini V, Anand Babu P, Nambiar RB, Periyar Selvam S, Sadiku ER. 2020. Determination of antifungal activities of essential oils incorporated-pomegranate peel fibers reinforced-polyvinyl alcohol biocomposite film against mango postharvest pathogens. Mater Today Proc:1–5.

Nazzaro F, Fratianni F, Coppola R, De Feo V. 2017. Essential oils and antifungal activity. Pharmaceuticals;10:1–20.

Nouri B, Mohtasebi SS, Rafiee S. 2020. Quality detection of pomegranate fruit infected with fungal disease. Int J Food Prop;23:9–21.

Nunes MCN. 2008. Impact of environmental conditions on fruit and vegetable quality. Stewart Postharvest Rev;4:1–14.

Oliveira A, Castro PM, Amaro AL, Sain J, Pintado M. 2016. Optimization of temperature, relative humidity and storage time before and after packaging of baby spinach leaves using response surface methodology. Food Bioprocess Technol;9:2070–2079.

Oni OO, Olalemi AO, Balogun OB. 2018. The effect of aloe vera gel on microorganisms associated with the deterioration of sweet orange fruits (*Citrus sinensis*). J Adv Microbiol;12:1–12.

Ordoñez Bolaños DY, Zuñiga Camacho D, Hoyos Concha JL, Mosquera Sánchez SA, Mosquera Sánchez LP. 2014. Efecto de recubrimiento de almidón de yuca modificado y aceite de tomillo aplicado al pimiento (*Capsicum annuum*)/Effect of coating of starch from cassava modified and thyme oil applied to the pepper (*Capsicum annuum*). Rev Mex Ciencias Agrícolas;5:795.

Ozdemir M, Floros JD. 2004. Active food packaging technologies. Crit Rev Food Sci Nutr;44:185–93.

Park DH, Park JJ, Olawuyi IF, Lee WY. 2020. Quality of white mushroom (*Agaricus bisporus*) under argon-and nitrogen-based controlled atmosphere storage. Sci Hortic (Amsterdam);265:109229.

Prakash A, Baskaran R, Paramasivam N, Vadivel V. 2018. Essential oil based nanoemulsions to improve the microbial quality of minimally processed fruits and vegetables: A review. Food Res Int;111:509–23.

Rabiei V, Kakavand F, Zaare-Nahandi F, Razavi F, Aghdam MS. 2019. Nitric oxide and γ-aminobutyric acid treatments delay senescence of cornelian cherry fruits during postharvest cold storage by enhancing antioxidant system activity. Sci Hortic (Amsterdam);243:268–73.

Raghav PK, Agarwal N, Saini M, Vidhyapeeth J, Vidhyapeeth J. 2016. Edible coating of fruits and vegetables. Int J Sci Mod Educ:188–204.

Ramírez-Acosta S, Arias-Borrego A, Gómez-Ariza JL, García-Barrera T. 2019. Metabolomic study of bioactive compounds in strawberries preserved under controlled atmosphere based on GC-MS and DI-ESI-QqQ-TOF-MS. Phytochem Anal;30:198–207.

Ramírez-Gil JG, López JH, Henao-Rojas JC. 2020. Causes of hass avocado fruit rejection in preharvest, harvest, and packinghouse: Economic losses and associated variables. Agronomy;10:8.

Ramos-Guerrero A, González-Estrada RR, Montalvo-González E, Miranda-Castro SP, Gutiérrez-Martínez P. 2018. Effect of the application of inducers on soursop fruit (*Annona muricata* L.): Postharvest disease control, physiological behaviour and activation of defense systems. Emir J Food Agric;30.

Rodriguez-Garcia I, Cruz-Valenzuela MR, Silva-Espinoza BA, Gonzalez-Aguilar GA, Moctezuma E, Gutierrez-Pacheco MM. 2015. Oregano (*Lippia graveolens*) essential oil added within pectin edible coatings prevents fungal decay and increases the antioxidant capacity of treated tomatoes. J Sci Food Agric;96:3772–8.

Romanazzi G, Feliziani E, Sivakumar D. 2018. Chitosan, a biopolymer with triple action on postharvest decay of fruit and vegetables: Eliciting, antimicrobial and film-forming properties. Front Microbiol;9:1–9.

Sánchez-González L, González-Martínez C, Chiralt A, Cháfer M. 2010. Physical and antimicrobial properties of chitosan-tea tree essential oil composite films. J Food Eng;98:443–52.

Sapper M, Palou L, Pérez-Gago MB, Chiralt A. 2019. Antifungal starch—gellan edible coatings with thyme essential oil for the postharvest preservation of apple and persimmon. Coatings;9:333.

Saxena A, Sharma L, Maity T. 2020. Enrichment of Edible Coatings and Films with Plant Extracts or Essential Oils for the Preservation of Fruits and Vegetables. Elsevier Inc.

Seow YX, Yeo CR, Chung HL, Yuk H-G. 2014. Plant essential oils as active antimicrobial agents. Crit Rev Food Sci Nutr;54:625–44.

Shao X, Cao B, Xu F, Xie S, Yu D, Wang H. 2015. Effect of postharvest application of chitosan combined with clove oil against citrus green mold. Postharvest Biol Technol;99:37–43.

Sharma S, Barkauskaite S, Jaiswal AK, Jaiswal S. 2020. Essential oils as additives in active food packaging. Food Chem:128403.

Shen Z, Kamdem DP. 2015. Development and characterization of biodegradable chitosan films containing two essential oils. Int J Biol Macromol;74:289–96.

Souza VGL, Rodrigues C, Ferreira L, Pires JRA, Duarte MP, Coelhoso I. 2019. In vitro bioactivity of novel chitosan bionanocomposites incorporated with different essential oils. Ind Crops Prod;140:111563.

Spadafora ND, Amaro AL, Pereira MJ, Müller CT, Pintado M, Rogers HJ. 2016. Multi-trait analysis of post-harvest storage in rocket salad (*Diplotaxis tenuifolia*) links sensorial, volatile and nutritional data. Food Chem;211:114–23.

Sumonsiri N, Barringer SA. 2013. Application o f SIFT-MS in monitoring volatile compounds in fruits and vegetables. Curr Anal Chem;9:631–41.

Syafiq R, Sapuan SM, Zuhri MYM, Ilyas RA, Nazrin A, Sherwani SFK. 2020. Antimicrobial activities of starch-based biopolymers and biocomposites incorporated with plant essential oils: A review. Polymers;12:1–26.

Tadros T, Izquierdo P, Esquena J, Solans C. 2004. Formation and stability of nano-emulsions. Adv Colloid Interface Sci;108:303–18.

Tahir HE, Xiaobo Z, Jiyong S, Mahunu GK, Zhai X, Mariod AA. 2018. Quality and postharvest-shelf life of cold-stored strawberry fruit as affected by gum arabic (*Acacia senegal*) edible coating. J Food Biochem;42:1–10.

Tohidi B, Rahimmalek M, Arzani A. 2017. Essential oil composition, total phenolic, flavonoid contents, and antioxidant activity of Thymus species collected from different regions of Iran. Food Chem;220:153–61.

Tzortzakis N, Chrysargyris A. 2017. Postharvest ozone application for the preservation of fruits and vegetables. Food Rev Int;33:270–315.

Varghese SA, Siengchin S, Parameswaranpillai J. 2020. Essential oils as antimicrobial agents in biopolymer-based food packaging—A comprehensive review. Food Biosci;38:100785.

Varona S, Martin A, Cocero MJ. 2011. Liposomal incorporation of lavandin essential oil by a thin-film hydration method and by particles from gas-saturated solutions. Ind Eng Chem Res;50:2088–97.

Veldhuizen EJA, Tjeerdsma-van Bokhoven JLM, Zweijtzer C, Burt SA, Haagsman HP. 2006. Structural requirements for the antimicrobial activity of carvacrol. J Agric Food Chem;54:1874–9.
Vivek K, Singh SS, Ritesh W, Soberly M, Baby Z, Baite H, Mishra S, Pradhan RC. 2020. A review on postharvest management and advances in the minimal processing of fresh-cut fruits and vegetables. J Microbiol Biotechnol Food Sci;2020:1178–87.
Wang, C. Y. (2016). Chilling and freezing injury. The commercial storage of fruits, vegetables, and florist and nursery stocks. Agriculture handbook,;66:62–67.
Weber S, Zimmer A, Pardeike J. 2014. Solid lipid nanoparticles (SLN) and nanostructured lipid carriers (NLC) for pulmonary application: A review of the state of the art. Eur J Pharm Biopharm;86:7–22.
Wei H, Seidi F, Zhang T, Jin Y, Xiao H. 2020. Ethylene scavengers for the preservation of fruits and vegetables: A review. Food Chem:127750.
Wu Y, Zou L, Mao J, Huang J, Liu S. 2014. Stability and encapsulation efficiency of sulforaphane microencapsulated by spray drying. Carbohydr Polym;102:497–503.
Wyrwa J, Barska A. 2017. Innovations in the food packaging market: Active packaging. Eur Food Res Technol;243:1681–92.
Ye H, Shen S, Xu J, Lin S, Yuan Y, Jones GS. 2013. Synergistic interactions of cinnamaldehyde in combination with carvacrol against food-borne bacteria. Food Control;34:619–23.
Yildirim S, Röcker B, Pettersen MK, Nilsen-Nygaard J, Ayhan Z, Rutkaite R. 2018. Active packaging applications for food. Compr Rev Food Sci Food Saf;17:165–99.
Zahra SA, Butt YN, Nasar S, Akram S, Fatima Q, Ikram J. 2016. Food packaging in perspective of microbial activity: A review. J Microbiol Biotechnol Food Sci;6:752–7.
Zhang R, Long J. 2017. Study on drying uniformity of static small-sized drying box for fruits and vegetables. Procedia Eng;205:2615–22.

Chapter 11

Bioactive Nano-Based Packaging for Postharvest Storage of Horticultural Produce

Nazila Oladzadabbasabadi, Abdorreza Mohammadi Nafchi, Fazilah Ariffin, and Karim A.A.

11.1 Introduction

Food packaging is used to preserve food from environmental contaminants and other factors such as shocks, odors, temperature change, dust, physical damage, mechanical damage, compressive forces, microorganisms, light, and humidity (W. Lin et al., 2019; Luís et al., 2019; Youssef & El-Sayed, 2018). It is crucial to guarantee the quality and protection of food while also increasing shelf life and reducing postharvest losses and wastage (Pedreschi et al., 2013; Pessu et al., 2011). Agriculture and the food industry worldwide suffer notably from postharvest losses due to numerous factors from harvest to consumption (Xue Mei et al., 2020). These losses mainly depend on each country's management condition and economic resources (Hengsdijk & de Boer, 2017; Kumar et al., 2006). More than 40% of food losses in developed countries are reported at the retail and consumer levels, especially for the roots and tubers, cereals, fruits and vegetables, oil crops, and pulses. In contrast, more than 40% of the food losses occur in developing countries during the postharvest and processing stages (Gustavsson et al., 2011; Heller & Keoleian, 2015). Furthermore, the highest postharvest losses in vegetable crops and fruits are attributed to degradation caused by microbial contamination after harvest and during transportation and cold storage time (Kasso & Bekele, 2018; Santos et al., 2020). Fruits are extremely susceptible to fungal attacks because of their

DOI: 10.1201/9781003142287-11

higher moisture content, low pH, and nutrient composition. In contrast, vegetables are usually less acidic, and their deterioration occurs mostly by bacteria (Tripathi & Dubey, 2004). Although the extent of postharvest losses due to microorganism attack, mechanical damage, and physical damage are very difficult to define, these losses are substantial. One of the postharvest technologies used before and after harvest is a chemical treatment that increases shelf life and minimizes horticultural produce spoilage (James & Zikankuba, 2017; Smith, 1962). The increased demand for high-quality and nutritious fresh fruits and vegetables has prompted the food industry to create novel and better food-quality methods and also to prolong the shelf life. It is possible to design packaging materials to reduce these microbiological, chemical, and physical events (Ncama et al., 2018; Yousuf et al., 2018). However, the durability, stiffness, and other characteristics, such as thermal stability, gas barrier, low resistance to solvents, and absence of antibacterial properties, make polymers insufficient for packaging applications (Mousavian et al., 2020). The use of the bionanocomposite principle has recently been shown to be an innovative alternative for optimizing the previously mentioned properties efficiently (Jafarzadeh et al., 2021; Youssef & El-Sayed, 2018). Here, nanocomposite synthesis is a promising alternative to conventional chemical methods. The inadequacy of overall properties of polymers packing, including the thermal stability, strength, moisture, oxygen permeability, and antibacterial properties, can effectively be managed in nanocomposites (Jafarzadeh et al., 2020, 2021; Li et al., 2017b)

Advances in smart antimicrobial packaging systems and materials production guarantee the protection of products and their shelf life once they are commercially approved (Kim et al., 2011). Nanotechnology is a potential and emerging field with multiple applications in different areas. The beneficial effects of nanoparticles in agriculture have already been proven (Usman et al., 2020). In recent years, in nanotechnology, much attention has been given to improving the antimicrobial properties of packaging (Cruz-Romero et al., 2013). Active packaging systems with integrated antimicrobial and antioxidant agents provide new possibilities for developing novel packaging materials that preserve food quality and enhance food and product shelf life (Nerín, 2010; Yildirim et al., 2018).

11.2 Factors Influencing Postharvest Loss

Postharvest loss includes quantity and quality distortions of products from harvest to consumption by the consumer (Chegere, 2018; Nourbakhsh et al., 2016). The physical, physiological, mechanical, and environmental factors are responsible for the postharvest loss. Generally, weight reduction, a low value of nutrients, color change, and texture, among other factors, are indications of postharvest loss of products (Abass et al., 2014; Ladaniya, 2008). Temperature variation, sunshine radiation, climatic conditions within the different geographical locations, vibration, and microbial contamination, among others, may affect postharvest products during transportation and storage (Jung et al., 2018; Kasso & Bekele, 2018; Perrone et al., 2020). However, fruits and vegetables continue to survive throughout harvesting. The energy used to carry out such processes is an outcome of the respiratory process, carbon dioxide, heat, and moisture, which are produced by the breakdown of carbohydrate (El-Ramady et al., 2015; Mishra & Gamage, 2007). Furthermore, the transpiration process transfers the water vapor from the plant surface to the air around them. These reactions produce weight loss in products and an initial shrinkage of tissues (Madani et al., 2019). Hence, the regulation and control of these factors are necessary for increasing the shelf life of horticultural produce. Microbial spoilage by fungi and bacteria is responsible for the highest loss after harvesting (Mari et al., 2016; S. Zhang et al., 2019). In addition, fungal damage could be more significant in fruits and vegetables than losses in the field because they occur if the product is not cooled quickly or is not transported and stored under the required conditions (Stathers et al., 2020).

11.3 Packaging of Horticultural Produce

Due to advancements in material science and consumer demand, the food packaging industry continues to change. The packaging helps ensure a uniform and efficient delivery and helps preserve food, food quality, and food safety while extending food's shelf life and consumer convenience (Marsh & Bugusu, 2007). The use of food packaging dates back to the 18th century (Dobrucka & Cierpiszewski, 2014). In the 20th century, many advances in packaging technology emerged, such as biodegradable packaging, edible coatings/films, smart packaging, intelligent packaging, active packaging, and nanomaterial packaging, as described in some very comprehensive reviews (Chen et al., 2020; Emamhadi et al., 2020; J. H. Han, 2014; Siracusa et al., 2008; Yildirim et al., 2018; Yousefi et al., 2019). These technologies and materials can enhance food safety, ensure the quality of food, decrease environmental impacts, extend shelf life, and increase the packaged product's attractiveness to sellers and customers (J.-W. Han et al., 2018). For susceptible fruit, transport is a vital part of the supply chain. Owing to static and dynamic forces and mechanical factors damage can occur during the operation and transport (Dagdelen & Aday, 2021; Hussein et al., 2020). Different friction forces that occur during transport between farms and retailers cause mechanical damage to fresh postharvest products (Hussein et al., 2020). Vibration frequencies increase mechanical damage, which leads to skin discoloration and fruit rejection during quality control. This harm could also lead to the increased risk that the wounded texture may be infected by bacterial and fungal agents, leading to a shorter shelf life (Dagdelen & Aday, 2021). Packaging creates a beneficial environment for the food, which protects it from mechanical shocks, vibrations, microbial contamination, mites, insects, mist, and others. Active packaging is a creative packaging technology that enables the product and its environment to interact to prolong the product's shelf life and ensure its microbial protection while preserving the packaged food quality (Wyrwa & Barska, 2017).

11.4 Potential Bioactive Nano-Based Packaging for Horticultural Produce

Active packaging is one of the most sophisticated techniques currently being used to maintain quality, protect from harmful environmental attacks, and maintain the sensory characteristics of products. Active packaging has a positive effect on the product and environment, enhancing and protecting products' quality longer than traditional packaging (Barbosa-Pereira et al., 2014). Active packaging systems with antimicrobials and antioxidants provide new opportunities to manufacture new packaging materials that are capable of preserving food quality and enhancing food safety and a product's shelf life (Cruz-Romero et al., 2013). There is a controlled release of bioactive substances (antimicrobials and antioxidants previously applied to the packaging) in some active packaging, which prevents the direct addition of active ingredients to the products (Barbosa-Pereira et al., 2014) The field of nanotechnology has gained much attention in recent years in the development of antimicrobial packaging (Cruz-Romero et al., 2013). Recent developments in nanotechnology have led to the manufacture of nanoparticles in various forms and sizes and the generation of nanostructured compounds with antimicrobial effects. Compared to larger particles, the larger area per mass makes nano-sized particles more biologically active (Nile et al., 2020; Oberdörster et al., 2005). Research indicates that the application of nanomaterials such as SiO2, TiO2, clay, and nanocellulose can significantly improve the mechanical and barrier properties of the biopolymers as well as the antioxidant and antimicrobial activity of packaging.

Due to the nanofiller incorporation, the properties of polymer such as the tensile strength, thermal stability, stiffness, toughness, and optical properties could be improved significantly. In the case of bionanocomposite films based on fish gelatin and chitosan nanoparticles, an enhancement of tensile strength, water vapor barrier, and elastic modulus could be observed, which proved that chitosan nanoparticles have the ability to improve the film quality used for packaging. The results of scanning electron microscope (SEM) images showed that nanoparticles were diffused uniformly in the fish gelatin film. Hence, the use of nanotechnology could enhance the application of films for packaging (Hosseini et al., 2015).

11.5 Application

Since the mechanical properties of biodegradable polymers are not strong, nano-sized materials are used to enhance their properties (Choudhury et al., 2019). Additionally, nanoparticles in the coating formulations can have an important role in quality conservation, antimicrobial activity enhancement, and storage-time extension (W. Liu et al., 2020). Nanotechnology applications increase the bioavailability, texture, taste, and quality of foodstuffs, which could be achieved by changing the particle size and surface loading on foodstuffs (Yu et al., 2018). Previous studies have demonstrated that chitosan coatings could effectively reduce mold growth and extend the shelf life of blueberries and strawberries (Duan et al., 2011; Moslehi et al., 2021). Moreover, metal oxide could also be applied as an antioxidant and antimicrobial agent. ZnO has already been identified as an effective antibacterial among metal oxides. ZnO particles are demonstrated to have antimicrobial activity even against high-pressure and temperature-resistant bacteria (Babapour et al., 2021). Kazemi et al. (2020) have described packaging modified with cinnamon essential oil, modified atmosphere containing nitrogen packaging, and the spraying of ZnO nanoparticles. Here, the strong antimicrobial efficiency of nanoparticles was found to perform with the essential oils to enhance the chemical solubility and stability and reduce the quick evaporation and degradation of the active essential oil structure. The findings showed that utilizing the modified packaging could prolong the shelf life of fresh pistachio up to 12 weeks, by retaining other factors, such as taste, odor, and texture unchanged. Besides, from a health perspective, modified packaging was also found to be beneficial due to the inhibition of microbial growth and accumulation of aflatoxin in the pistachio. The shelf life of a modified packaging system could further be enhanced by incorporating the ZnO nanoparticles. In a study, Sogvar et al. (2016) coated strawberries with nano-ZnO solutions for 5 min at 20°C, and the coated samples were found to have lost weight at a slower rate. Moreover, the treatment was found to preserve vitamin C, anthocyanin, phenolic content, and antioxidant activity during storage. The nano-ZnO treatment also reduced microbial activity during the 18 days of storage. Total aerobic mesophilic bacteria in coated samples (5% nano-ZnO) were 3.67 Log CFU g^{-1}, while for the control sample, the same was 4.35 Log CFU g^{-1}. The yeast and mold populations in the control samples were high when compared with the coated fruit. These results suggest that during prolonged storage, nano-ZnO could be a suitable option for maintaining the quality of fresh strawberries and controlling their deterioration. Li et al. (2017) have described film containing polyethylene and Ag/TiO2 to have an application as an antimicrobial packaging film for rice. The findings revealed that the antimicrobial packaging had a very positive impact on the antimildew and physicochemical properties when compared with polyethylene packaging. The small quantity of silver migration from antimicrobial packaging films could remarkably also inhibit *A. flavus* and helped with rice mildew reduction throughout the storage time. A lower microbial population for the packaging was recorded due to the Ag/TiO2 presence in the film. Finally, rice packed with antimicrobial nano-Ag packaging was demonstrated to have improved quality as well as pasting and texture properties.

In another study, mango fruit was described to get coated with chitosan and nano-TiO_2. In the coated samples, polyphenol oxidase, peroxidase activity, and total flavonoid were observed to be high when compared to control samples. Treatment with a composite made of chitosan with nano-TiO_2 was found to provide desirable properties to maintain the firmness and decay index of mango fruit during storage, positively influence the phenol content in mango fruit, and thereby enhance mango quality. Coating mango fruits with chitosan/nano-TiO_2, hence, has the potential for commercial preservation (Xing et al., 2020). Deng et al. (2017) have claimed that cellulose nanocrystals have the ability to enhance the performance of chitosan coatings by improving the shelf life of postharvest pears under both cold and ambient storage conditions. The application of nanocomposite packaging material with nano-Ag, nano-TiO_2, and nano-SiO_2 on mushrooms was carried out by Donglu et al. (2016). Here, nanocomposite packaging was found to favor enhanced maintenance of nutrients and prevented the weight loss and respiration of mushrooms. In contrast, hydrogen peroxide and superoxide radicals were increased throughout storage time. Tian et al. (2019) have fabricated and investigated the chitosan film by incorporating 0.05% w/v nano-SiO_2 or 0.02% w/v nano-TiO_2 for their effect on maintaining the quality of *Ginkgo biloba* seeds. Here, postharvest qualities such as decay rate, firmness, and shrinkage rate could be efficiently preserved by preventing contaminant microbial growth. The nanocomposite coatings also controlled superoxide anion oxidation and the defense enzyme activities. Furthermore, higher antimicrobial activity and antioxidant enzyme activity were observed by using nanomaterials loaded chitosan matrix. The coated *G. biloba* seeds considerably showed higher firmness than control samples throughout the storage time. Consequently, they claimed that for prolonging the shelf life and preserving *G. biloba* seeds or other nuts, these types of nanocomposites could be utilized in industry. The use of Cu nanoparticles caused a greater firmness for tomato fruits. Moreover, bioactive compounds were found to get acutely accumulated in tomato fruits treated with Cu nanoparticles. The findings precisely demonstrate the ability of Cu nanoparticles to induce higher non-enzyme antioxidant compounds in the tomato fruit and thereby prolong its shelf life. Thus, the quality of tomato fruits could be retained for longer durations (López-Vargas et al., 2018). Hashim et al. (2019) have described the synthesis of chitosan nanoparticles, silica nanoparticles, Cu nanoparticles, and a combination of them. By studying the antifungal activity against *Botrytis cinerea*, nanomaterials such as silica and chitosan nanoparticles were found to be efficient antifungals with the potential to reduce gray mold on table grapes. Here, chitosan and silica nanoparticles were found to cause the loss of linearity and asymmetrical branching of *B. cinerea* hyphae in the apical part. In a recent study, Esfahani et al. (2020) have fabricated a sago starch–based film containing 3% titanium dioxide nanoparticles and 2% cinnamon essential oil and then applied it to fresh pistachio packaging (Table 11.1). The antimicrobial, physicochemical, and sensory properties of fresh pistachios over a 20-day storage time were further examined to evaluate the effect of bionanocomposite films under temperature and relative humidity of storage conditions. The findings

Table 11.1 Advantages of Nano-Packaging on Fruits and Vegetables

Matrix	*Advantages*	*Product*	*References*
Cu-chitosan NPs	Extended shelf life; reduced microbial decay, weight loss, respiration rate; preserved firmness, control microbial infection	Tomato	Meena et al., 2020
Chitosan/nano-TiO_2	Preserved quality and nutrient, higher total phenol content; improved firmness; decreased respiration rate	Mango	Xing et al., 2020
Nano-SiO_2	Prevented internal browning; delayed reduced titratable acidity, preserve quality; extended shelf life	Loquat	L. Wang et al., 2020
Cassava starch/zinc NPs	Preserved the quality of products; better positive effect β-carotene and ascorbic acid content association; prolonged shelf life	Cucumber and garden eggs	Fadeyibi et al., 2020
Sago starch/cinnamon EOs/TiO2 NPs	Enhanced physicochemical properties; the growth of A. *flavus* slowed; extended shelf life	Pistachio	Esfahani et al., 2020
10% LDPE-g-MA, 2% Cloisite® 20A (nanoclay), and 10% EVOH + MAP	Chemical, physical and mechanical properties enhanced; controlled respiration rate; extended shelf life	White button mushrooms	Gholami et al., 2020
Cinnamon essential oil, zinc oxide NPs+MAP	Prevented microbial growth; maintained sensorial properties; prolonged the shelf life	Pistachio	Kazemi et al., 2020

Matrix	Advantages	Product	References
Chitosan, silica and copper NPs	Decreased gray mold, potent antifungal agents	Grape	Hashim et al., 2019
Chitosan /Nano-SiOx	Reduced softness and weight loss; greater antimicrobial and antioxidant activity; prolonged shelf life	Green tomato	Zhu et al., 2019
Chitosan/nano-SiO_2 and chitosan/ nano-TiO_2	Better firmness; excellent antioxidant enzymes activity; improved quality; prevented mildew	Ginkgo biloba seeds	Tian et al., 2019
Nano-Ag, nano-TiO_2, nano-attapulgite, and SiO_2	Antimicrobial effect prevented mold growth; decreased fat and protein oxidation; enhanced maintaining quality	Nanjing 9108 rice	F. Wang et al., 2018
1-Methylcyclopropene/ nano-Ag PE bag	Decreased weight loss and respiration rate; preserved quality, extended shelf life	Pleurotus *eryngii.*	Xu et al., 2018
Chitosan/nano-sized titanium dioxide	Reduced quality change; improved barrier properties; delayed ripening process; prolonged shelf life	Cherry tomato	Kaewklin, Siripatrawan et al., 2018
Polylactic acid/ nano-Ag particles	Delayed decrease total phenol, reduced vitamin C loss, greater physical properties, maintained freshness	Strawberry	C. Zhang et al., 2018
Cu NPs	Better firmness, higher antioxidant capacity, prolonged shelf life	Tomato	López-Vargas et al., 2018
Nano-TiO_2/LDPE	Reduced firmness and titratable acid, higher antioxidant enzyme activity, extended shelf life	Strawberry	Li et al., 2017b
Chitosan/cellulose nanocrystal	Preserved firmness and postharvest quality, great antibacterial activity against gram-negative and gram-positive bacteria	Pear	Deng et al., 2017
Polyethylene/Ag/TiO_2	Positive effect of physicochemical properties and antimildew, reduced microbial population improve quality, prolonged shelf life	Rice	L. Li et al., 2017

(Continued)

Table 11.1 (Continued)

Matrix	*Advantages*	*Product*	*References*
Soy protein isolate/ nano-SiO_x	Higher physiological properties, decreased weight loss and titratable acid, preserved quality	Apple	R. Liu et al., 2017
LDPE/ nano-ZnO	Better firmness, preserved quality, decreased browning and decay rate, increased storage	Peach	Li et al., 2017a
Chitosan/nano-silica	Delayed weight loss and internal browning, improved antioxidant activity, extended shelf life	Loquat	Song et al., 2016
Nano-ZnO	Decreased microbial growth, higher antioxidant activity, delayed weight loss, preserved firmness	Strawberry	Sogvar et al., 2016
polyethylene/ nano-Ag, attapulgite, nano-SiO_2, nano-TiO_2	Prevented weight loss and respiration, antioxidant activity, maintained quality, prolonged shelf life	Mushrooms	Donglu, Wenjian, Kimatu, Xinxin, et al., 2016
Polyethylene/ nano-Ag, nano-TiO_2, nano-SiO_2, and attapulgite	Prevented microbial growth, reduced weight loss and cap opening, prolonged shelf life	Mushrooms	Donglu, Wenjian, Kimatu, Mariga, et al., 2016
LDPE/organically modified montmorillonite containing essential oils constituents (EOCs) mixture of carvacrol and thymol	Antimicrobial effect against *B. cinerea*	Strawberry	Campos-Requena et al., 2015
Ag/TiO_2/chitosan	Greater antimicrobial activity, preserved the quality, prolonged the shelf life	Cantaloupes	B. Lin et al., 2015
Chitosan/nano-silica	Decreased weight loss, prevented/reduced total soluble solids and titratable acidity, significantly increased shelf life	Longan	Shi et al., 2013
PVC/ nano-ZnO	Decreased decay rate, preserved the quality, extended shelf life	Fuji apple	X. Li et al., 2011

Matrix	Advantages	Product	References
Polyethylene/ nano-Ag/nano-TiO_2/ montmorillonite	Increased phenol contents, prevented weight loss, decreased fruit decay, preserved quality, prolonged shelf life	Kiwifruit	Hu et al., 2011
Hot air treatment, nano-TiO_2, and nano-Ag	Enhanced control of natural decay and green mold, prevented growth of *P. citrinum*, preserved firmness	Chinese bayberries	K. Wang et al., 2010
Polyethylene/nano-Ag, kaolin, anatase TiO_2, rutile TiO_2	Enhanced sensory and physicochemical quality, prevented weight loss and softening over 12 days.	Chinese jujube	H. Li et al., 2009

indicated that the fat, moisture, sensory properties, and color characteristics of fresh pistachios were protected, and the shrinkage of samples was reduced throughout the storage duration. In addition, bionanocomposite films decreased the aflatoxins and hydro-peroxide in pistachios. The growth of *A. flavus* was delayed because of the anti-oxidant and antifungal effects of the films. The results showed that by using these films, physicochemical properties can be improved, and the shelf life can be enhanced by decreasing the relative humidity (RH) and temperature of storage. Low-density polyethylene packaging incorporated with nano-TiO_2 was also studied for packaging strawberries and to investigate the effects of packaging on its quality and antioxidant capacity. This packaging formed a relatively higher CO_2 and lower O_2 air composition compared to low-density polyethylene packaging. Hence, the overall quality of the strawberry fruit was higher, with decreased decay and weight loss. Simultaneously, reactive oxygen species, such as hydrogen peroxide and superoxide anion, in packed strawberries decreased. Moreover, antioxidant enzyme activity in the packaged straw-berries was notably higher. The accumulation of anthocyanin was prevented, while a higher preservation of ascorbic and total phenolic material was observed. The results showed that these packaging types have a promising potential for preserving the post-harvest quality of strawberries (Li et al., 2017b). Zhang et al. (2018) fabricated an active nanocomposite packaging film containing nano-Ag particles and polylactic acid and then applied it to strawberry packaging for 10 days at $4 \pm 1°C$. Here, a greater preser-vation effect could be observed when compared with pure polylactic acid (PLA) film. Nano-Ag active films can minimize strawberries' weight loss rate throughout storage time efficiently and delay the reduced titratable acid content, soluble solids, and hard-ness. During the late storage period, strawberries had a greater antioxidant capacity; hence, the film could delay fruit aging and improve the preservation of total phenols and ascorbic acid. Thus, the freshness of strawberries can successfully be maintained by nanocomposite packaging film. Chitosan/nano-silica coating has increased chilling resistance and increased the shelf life of loquat fruit with quality preservation. After coating, a reduction in internal browning and weight loss was observed significantly in packed samples, along with the preservation of high levels of fructose, glucose, titratable acidity, and total soluble solids. Furthermore, increased antioxidant enzyme activities could also be observed (Song et al., 2016). Li et al. (2009) evaluated the preservation quality of Chinese jujube with nanocomposite packaging during the storage period. The findings revealed that the nanocomposite has a comparatively positive effect on this product's sensory and physicochemical properties during the storage period compared to normal packaging materials.

11.6 Conclusion

Packaging technology is continuously improving to meet the requirements of the ever-rising demand for high-quality, safe, and healthy products. Packaging technology is mainly designed to preserve sensory properties, protect from bacteria or fungi contamination, increase safety, maintain quality, and extend the shelf life of products. Nevertheless, nanotechnology developments offer tremendous potential for addressing existing issues associated with packaging materials. Horticultural crops undergo many changes from harvest stage to consumption due to various conditions. The challenges faced regarding the quality and safety of products generally lead to extensive economic losses. However, advances in packaging and nanotechnology have made it possible to minimize postharvest issues. The future scenario is challenging to anticipate, considering the vast possibilities available for packaging using bio-based nanomaterial. With proper nanoparticles, packaging with stronger mechanical, thermal, and barrier performance can be generated. Nanostructured materials can prevent microorganisms from entering or infiltrating the product's protection system. Nevertheless, for wider application of nanomaterial systems, regulatory approval should also be considered.

References

Abass, A. B., Ndunguru, G., Mamiro, P., Alenkhe, B., Mlingi, N., & Bekunda, M. (2014). Post-harvest food losses in a maize-based farming system of semi-arid savannah area of Tanzania. *Journal of Stored Products Research*, *57*, 49–57. https://doi.org/10.1016/j.jspr.2013.12.004

Babapour, H., Jalali, H., & Mohammadi Nafchi, A. (2021). The synergistic effects of zinc oxide nanoparticles and fennel essential oil on physicochemical, mechanical, and antibacterial properties of potato starch films. *Food Science & Nutrition*, *9*(7), 3893–3905. https://doi.org/10.1002/fsn3.2371

Barbosa-Pereira, L., Angulo, I., Lagarón, J. M., Paseiro-Losada, P., & Cruz, J. M. (2014). Development of new active packaging films containing bioactive nanocomposites. *Innovative Food Science & Emerging Technologies*, *26*, 310–318. https://doi.org/10.1016/j.ifset.2014.06.002

Campos-Requena, V. H., Rivas, B. L., Pérez, M. A., Figueroa, C. R., & Sanfuentes, E. A. (2015). The synergistic antimicrobial effect of carvacrol and thymol in clay/polymer nanocomposite films over strawberry gray mold. *LWT—Food Science and Technology*, *64*(1), 390–396. https://doi.org/10.1016/j.lwt.2015.06.006

Chegere, M. J. (2018). Post-harvest losses reduction by small-scale maize farmers: The role of handling practices. *Food Policy*, *77*, 103–115. https://doi.org/10.1016/j.foodpol.2018.05.001

Chen, S., Brahma, S., Mackay, J., Cao, C., & Aliakbarian, B. (2020). The role of smart packaging system in food supply chain. *Journal of Food Science*, *85*(3), 517–525. https://doi.org/10.1111/1750-3841.15046

Choudhury, A., Jeelani, P. G., Biswal, N., & Chidambaram, R. (2019). Application of bionanocomposites on horticultural products to increase the shelf life. In *Polymers for agri-food applications* (pp. 525–543): Springer.

Cruz-Romero, M. C., Murphy, T., Morris, M., Cummins, E., & Kerry, J. P. (2013). Antimicrobial activity of chitosan, organic acids and nano-sized solubilisates for potential use in smart antimicrobially-active packaging for potential food applications. *Food Control, 34*(2), 393–397. https://doi.org/10.1016/j.foodcont.2013.04.042

Dagdelen, C., & Aday, M. S. (2021). The effect of simulated vibration frequency on the physico-mechanical and physicochemical properties of peach during transportation. *LWT, 137*, 110497. https://doi.org/10.1016/j.lwt.2020.110497

Deng, Z., Jung, J., Simonsen, J., Wang, Y., & Zhao, Y. (2017). Cellulose nanocrystal reinforced chitosan coatings for improving the storability of postharvest pears under both ambient and cold storages. *Journal of Food Science, 82*(2), 453–462. https://doi.org/10.1111/1750-3841.13601

Dobrucka, R., & Cierpiszewski, R. (2014). Active and intelligent packaging food—research and development—a review. *Polish Journal of Food and Nutrition Sciences, 64*(1), 7–15.

Donglu, F., Wenjian, Y., Kimatu, B. M., Mariga, A. M., Liyan, Z., Xinxin, A., & Qiuhui, H. (2016). Effect of nanocomposite-based packaging on storage stability of mushrooms (*Flammulina velutipes*). *Innovative Food Science & Emerging Technologies, 33*, 489–497. https://doi.org/10.1016/j.ifset.2015.11.016

Donglu, F., Wenjian, Y., Kimatu, B. M., Xinxin, A., Qiuhui, H., & Liyan, Z. (2016). Effect of nanocomposite packaging on postharvest quality and reactive oxygen species metabolism of mushrooms (*Flammulina velutipes*). *Postharvest Biology and Technology, 119*, 49–57. https://doi.org/10.1016/j.postharvbio.2016.04.012

Duan, J., Wu, R., Strik, B. C., & Zhao, Y. (2011). Effect of edible coatings on the quality of fresh blueberries (Duke and Elliott) under commercial storage conditions. *Postharvest Biology and Technology, 59*(1), 71–79. https://doi.org/10.1016/j.postharvbio.2010.08.006

El-Ramady, H. R., Domokos-Szabolcsy, É., Abdalla, N. A., Taha, H. S., & Fári, M. (2015). Postharvest management of fruits and vegetables storage. In *Sustainable agriculture reviews* (pp. 65–152). Springer.

Emamhadi, M. A., Sarafraz, M., Akbari, M., Thai, V. N., Fakhri, Y., Linh, N. T. T., & Mousavi Khaneghah, A. (2020). Nanomaterials for food packaging applications: A systematic review. *Food and Chemical Toxicology, 146*, 111825. https://doi.org/10.1016/j.fct.2020.111825

Esfahani, A., Ehsani, M. R., Mizani, M., & Nafchi, A. M. (2020). Application of bionanocomposite filmsbased on nano-TiO_2 and cinnamon essential oil to improve the physiochemical, sensory, and microbial properties of fresh pistachio. *Journal of Nuts, 11*(3), 195–212. https://doi.org/10.22034/jon.2020.1903741.1091

Fadeyibi, A., Osunde, Z. D., & Yisa, M. G. (2020). Effects of period and temperature on quality and shelf-life of cucumber and garden-eggs packaged using cassava starch-zinc nanocomposite film. *Journal of Applied Packaging Research, 12*(1), 3.

Gholami, R., Ahmadi, E., & Ahmadi, S. (2020). Investigating the effect of chitosan, nanopackaging, and modified atmosphere packaging on physical, chemical, and mechanical properties of button mushroom during storage. *Food Science & Nutrition, 8*(1), 224–236. https://doi.org/10.1002/fsn3.1294

Gustavsson, J., Cederberg, C., Sonesson, U., Van Otterdijk, R., & Meybeck, A. (2011). *Global food losses and food waste*. FAO.

Han, J. H. (2014). Edible films and coatings: A review. In J. H. Han (Ed.), *Innovations in food packaging* (2nd ed., pp. 213–255). Academic Press.

Han, J.-W., Ruiz-Garcia, L., Qian, J.-P., & Yang, X.-T. (2018). Food packaging: A comprehensive review and future trends. *Comprehensive Reviews in Food Science and Food Safety, 17*(4), 860–877. https://doi.org/10.1111/1541-4337.12343

Hashim, A. F., Youssef, K., & Abd-Elsalam, K. A. (2019). Ecofriendly nanomaterials for controlling gray mold of table grapes and maintaining postharvest quality. *European Journal of Plant Pathology, 154*(2), 377–388. https://doi.org/10.1007/s10658-018-01662-2

Heller, M. C., & Keoleian, G. A. (2015). Greenhouse gas emission estimates of U.S. dietary choices and food loss. *Journal of Industrial Ecology, 19*(3), 391–401. https://doi.org/10.1111/jiec.12174

Hengsdijk, H., & de Boer, W. J. (2017). Post-harvest management and post-harvest losses of cereals in Ethiopia. *Food Security, 9*(5), 945–958. https://doi.org/10.1007/s12571-017-0714-y

Hosseini, S. F., Rezaei, M., Zandi, M., & Farahmandghavi, F. (2015). Fabrication of bio-nanocomposite films based on fish gelatin reinforced with chitosan nanoparticles. *Food Hydrocolloids, 44*, 172–182. https://doi.org/10.1016/j.foodhyd.2014.09.004

Hu, Q., Fang, Y., Yang, Y., Ma, N., & Zhao, L. (2011). Effect of nanocomposite-based packaging on postharvest quality of ethylene-treated kiwifruit (*Actinidia deliciosa*) during cold storage. *Food Research International, 44*(6), 1589–1596. https://doi.org/10.1016/j.foodres.2011.04.018

Hussein, Z., Fawole, O. A., & Opara, U. L. (2020). Harvest and postharvest factors affecting bruise damage of fresh fruits. *Horticultural Plant Journal, 6*(1), 1–13. https://doi.org/10.1016/j.hpj.2019.07.006

Jafarzadeh, S., Jafari, S. M., Salehabadi, A., Nafchi, A. M., Uthaya Kumar, U. S., & Khalil, H. P. S. A. (2020). Biodegradable green packaging with antimicrobial functions based on the bioactive compounds from tropical plants and their by-products. *Trends in Food Science & Technology, 100*, 262–277. https://doi.org/10.1016/j.tifs.2020.04.017

Jafarzadeh, S., Mohammadi Nafchi, A., Salehabadi, A., Oladzad-Abbasabadi, N., & Jafari, S. M. (2021). Application of bio-nanocomposite films and edible coatings for extending the shelf life of fresh fruits and vegetables. *Advances in Colloid and Interface Science, 291*, 102405. https://doi.org/10.1016/j.cis.2021.102405

Jafarzadeh, S., Salehabadi, A., Mohammadi Nafchi, A., Oladzadabbasabadi, N., & Jafari, S. M. (2021). Cheese packaging by edible coatings and biodegradable nanocomposites; improvement in shelf life, physicochemical and sensory properties. *Trends in Food Science & Technology, 116*, 218–231. https://doi.org/10.1016/j.tifs.2021.07.021

James, A., & Zikankuba, V. (2017). Postharvest management of fruits and vegetable: A potential for reducing poverty, hidden hunger and malnutrition in sub-Sahara Africa. *Cogent Food & Agriculture, 3*(1), 1312052.

Jung, H. M., Lee, S., Lee, W.-H., Cho, B.-K., & Lee, S. H. (2018). Effect of vibration stress on quality of packaged grapes during transportation. *Engineering in Agriculture, Environment and Food, 11*(2), 79–83. https://doi.org/10.1016/j.eaef.2018.02.007

Kaewklin, P., Siripatrawan, U., Suwanagul, A., & Lee, Y. S. (2018). Active packaging from chitosan-titanium dioxide nanocomposite film for prolonging storage life of tomato fruit. *International Journal of Biological Macromolecules, 112*, 523–529. https://doi.org/10.1016/j.ijbiomac.2018.01.124

Kasso, M., & Bekele, A. (2018). Post-harvest loss and quality deterioration of horticultural crops in Dire Dawa Region, Ethiopia. *Journal of the Saudi Society of Agricultural Sciences, 17*(1), 88–96. https://doi.org/10.1016/j.jssas.2016.01.005

Kazemi, M. M., Hashemi-Moghaddam, H., Mohammadi Nafchi, A., & Ajodnifar, H. (2020). Application of modified packaging and nano ZnO for extending the shelf life of fresh pistachio. *Journal of Food Process Engineering*, e13548. https://doi.org/10.1111/jfpe.13548

Kim, K. W., Min, B. J., Kim, Y.-T., Kimmel, R. M., Cooksey, K., & Park, S. I. (2011). Antimicrobial activity against foodborne pathogens of chitosan biopolymer films of different molecular weights. *LWT—Food Science and Technology*, *44*(2), 565–569. https://doi.org/10.1016/j.lwt.2010.08.001

Kumar, D. K., Basavaraja, H., & Mahajanshetti, S. (2006). An economic analysis of post-harvest losses in vegetables in Karnataka. *Indian Journal of Agricultural Economics*, *61*.

Ladaniya, M. S. (2008). Postharvest losses. In M. S. Ladaniya (Ed.), *Citrus fruit* (pp. 67–78). Academic Press.

Li, D., Li, L., Luo, Z., Lu, H., & Yue, Y. (2017a). Effect of nano-ZnO-packaging on chilling tolerance and pectin metabolism of peaches during cold storage. *Scientia Horticulturae*, *225*, 128–133. https://doi.org/10.1016/j.scienta.2017.07.003

Li, D., Ye, Q., Jiang, L., & Luo, Z. (2017b). Effects of nano-TiO_2-LDPE packaging on postharvest quality and antioxidant capacity of strawberry (*Fragaria ananassa* Duch.) stored at refrigeration temperature. *Journal of the Science of Food and Agriculture*, *97*(4), 1116–1123.

Li, H., Li, F., Wang, L., Sheng, J., Xin, Z., Zhao, L., . . . Hu, Q. (2009). Effect of nano-packing on preservation quality of Chinese jujube (*Ziziphus jujuba Mill.* var. *inermis (Bunge) Rehd*). *Food Chemistry*, *114*(2), 547–552. https://doi.org/10.1016/j.foodchem.2008.09.085

Li, X., Li, W., Jiang, Y., Ding, Y., Yun, J., Tang, Y., & Zhang, P. (2011). Effect of nano-ZnO-coated active packaging on quality of fresh-cut 'Fuji' apple. *International Journal of Food Science & Technology*, *46*(9), 1947–1955. https://doi.org/10.1111/j.1365-2621.2011.02706.x

Lin, B., Luo, Y., Teng, Z., Zhang, B., Zhou, B., & Wang, Q. (2015). Development of silver/titanium dioxide/chitosan adipate nanocomposite as an antibacterial coating for fruit storage. *LWT–Food Science and Technology*, *63*(2), 1206–1213. https://doi.org/10.1016/j.lwt.2015.04.049

Lin, W., Ni, Y., & Pang, J. (2019). Microfluidic spinning of poly (methyl methacrylate)/konjac glucomannan active food packaging films based on hydrophilic/hydrophobic strategy. *Carbohydrate Polymers*, *222*, 114986. https://doi.org/10.1016/j.carbpol.2019.114986

Liu, R., Liu, D., Liu, Y., Song, Y., Wu, T., & Zhang, M. (2017). Using soy protein SiOx nanocomposite film coating to extend the shelf life of apple fruit. *International Journal of Food Science & Technology*, *52*(9), 2018–2030. https://doi.org/10.1111/ijfs.13478

Liu, W., Zhang, M., & Bhandari, B. (2020). Nanotechnology—A shelf life extension strategy for fruits and vegetables. *Critical Reviews in Food Science and Nutrition*, *60*(10), 1706–1721. https://doi.org/10.1080/10408398.2019.1589415

López-Vargas, E. R., Ortega-Ortíz, H., Cadenas-Pliego, G., de Alba Romenus, K., Cabrera de la Fuente, M., Benavides-Mendoza, A., & Juárez-Maldonado, A. (2018). Foliar application of copper nanoparticles increases the fruit quality and the content of bioactive compounds in tomatoes. *Applied Sciences*, *8*(7), 1020. https://doi.org/10.3390/app8071020

Luís, Â., Domingues, F., & Ramos, A. (2019). Production of hydrophobic zein-based films bioinspired by the lotus leaf surface: Characterization and bioactive properties. *Microorganisms*, *7*(8), 267. https://doi.org/10.3390/microorganisms7080267

Madani, B., Mirshekari, A., & Imahori, Y. (2019). Physiological responses to stress. In E. M. Yahia (Ed.), *Postharvest physiology and biochemistry of fruits and vegetables* (pp. 405–423): Woodhead Publishing.

Mari, M., Bautista-Baños, S., & Sivakumar, D. (2016). Decay control in the postharvest system: Role of microbial and plant volatile organic compounds. *Postharvest Biology and Technology, 122*, 70–81. https://doi.org/10.1016/j.postharvbio.2016.04.014

Marsh, K., & Bugusu, B. (2007). Food packaging—roles, materials, and environmental issues. *Journal of Food Science, 72*(3), R39–R55. https://doi.org/10.1111/j.1750-3841.2007.00301.x

Meena, M., Pilania, S., Pal, A., Mandhania, S., Bhushan, B., Kumar, S., . . . Saharan, V. (2020). Cu-chitosan nano-net improves keeping quality of tomato by modulating physio-biochemical responses. *Scientific Reports, 10*(1), 21914. https://doi.org/10.1038/s41598-020-78924-9

Mishra, V. K., & Gamage, T. (2007). *Postharvest physiology of fruit and vegetables: MS Rahman*. CRC Press.

Moslehi, Z., Mohammadi Nafchi, A., Moslehi, M., & Jafarzadeh, S. (2021). Aflatoxin, microbial contamination, sensory attributes, and morphological analysis of pistachio nut coated with methylcellulose. *Food Science & Nutrition, 9*(5), 2576–2584. https://doi.org/10.1002/fsn3.2212

Mousavian, D., Mohammadi Nafchi, A., Nouri, L., & Abedinia, A. (2020). Physicomechanical properties, release kinetics, and antimicrobial activity of activated low-density polyethylene and orientated polypropylene films by Thyme essential oil active component. *Journal of Food Measurement and Characterization*. https://doi.org/10.1007/s11694-020-00690-z

Ncama, K., Magwaza, L. S., Mditshwa, A., & Tesfay, S. Z. (2018). Plant-based edible coatings for managing postharvest quality of fresh horticultural produce: A review. *Food Packaging and Shelf Life, 16*, 157–167. https://doi.org/10.1016/j.fpsl.2018.03.011

Nerín, C. (2010). Antioxidant active food packaging and antioxidant edible films. In E. A. Decker, R. J. Elias, & D. Julian McClements (Eds.), *Oxidation in foods and beverages and antioxidant applications* (pp. 496–515): Woodhead Publishing.

Nile, S. H., Baskar, V., Selvaraj, D., Nile, A., Xiao, J., & Kai, G. (2020). Nanotechnologies in food science: Applications, recent trends, and future perspectives. *Nano-Micro Letters, 12*(1), 45. https://doi.org/10.1007/s40820-020-0383-9

Nourbakhsh, S. M., Bai, Y., Maia, G. D. N., Ouyang, Y., & Rodriguez, L. (2016). Grain supply chain network design and logistics planning for reducing post-harvest loss. *Biosystems Engineering, 151*, 105–115. https://doi.org/10.1016/j.biosystemseng.2016.08.011

Oberdörster, G., Oberdörster, E., & Oberdörster, J. (2005). Nanotoxicology: An emerging discipline evolving from studies of ultrafine particles. *Environmental health perspectives, 113*(7), 823–839. https://doi.org/10.1289/ehp.7339

Pedreschi, R., Lurie, S., Hertog, M., Nicolaï, B., Mes, J., & Woltering, E. (2013). Postharvest proteomics and food security. *Proteomics, 13*(12–13), 1772–1783. https://doi.org/10.1002/pmic.201200387

Perrone, G., Ferrara, M., Medina, A., Pascale, M., & Magan, N. (2020). Toxigenic fungi and mycotoxins in a climate change scenario: Ecology, genomics, distribution, prediction and prevention of the risk. *Microorganisms, 8*(10), 1496. https://doi.org/10.3390/microorganisms8101496

Pessu, P., Agoda, S., Isong, I., & Ikotun, I. (2011). The concepts and problems of post-harvest food losses in perishable crops. *African Journal of Food Science, 5*(11), 603–613. https://doi.org/10.5897/AJFS.9000281

Santos, S. F. D., Cardoso, R. D. C. V., Borges, Í. M. P., Almeida, A. C. E., Andrade, E. S., Ferreira, I. O., & Ramos, L. D. C. (2020). Post-harvest losses of fruits and vegetables in supply centers in Salvador, Brazil: Analysis of determinants, volumes and reduction strategies. *Waste Management*, *101*, 161–170. https://doi.org/10.1016/j.wasman.2019.10.007

Shi, S., Wang, W., Liu, L., Wu, S., Wei, Y., & Li, W. (2013). Effect of chitosan/nano-silica coating on the physicochemical characteristics of longan fruit under ambient temperature. *Journal of Food Engineering*, *118*(1), 125–131. https://doi.org/10.1016/j.jfoodeng.2013.03.029

Siracusa, V., Rocculi, P., Romani, S., & Rosa, M. D. (2008). Biodegradable polymers for food packaging: A review. *Trends in Food Science & Technology*, *19*(12), 634–643. https://doi.org/10.1016/j.tifs.2008.07.003

Smith, W. L., Jr. (1962). Chemical treatments to reduce postharvest spoilage of fruits and vegetables. *Botanical Review*, *28*(3), 411–445. www.jstor.org/stable/4353655

Sogvar, O. B., Koushesh Saba, M., Emamifar, A., & Hallaj, R. (2016). Influence of nano-ZnO on microbial growth, bioactive content and postharvest quality of strawberries during storage. *Innovative Food Science & Emerging Technologies*, *35*, 168–176. https://doi.org/10.1016/j.ifset.2016.05.005

Song, H., Yuan, W., Jin, P., Wang, W., Wang, X., Yang, L., & Zhang, Y. (2016). Effects of chitosan/nano-silica on postharvest quality and antioxidant capacity of loquat fruit during cold storage. *Postharvest Biology and Technology*, *119*, 41–48. https://doi.org/10.1016/j.postharvbio.2016.04.015

Stathers, T., Holcroft, D., Kitinoja, L., Mvumi, B. M., English, A., Omotilewa, O., . . . Torero, M. (2020). A scoping review of interventions for crop postharvest loss reduction in sub-Saharan Africa and South Asia. *Nature Sustainability*, *3*(10), 821–835. https://doi.org/10.1038/s41893-020-00622-1

Tian, F., Chen, W., Wu, C. E., Kou, X., Fan, G., Li, T., & Wu, Z. (2019). Preservation of Ginkgo biloba seeds by coating with chitosan/nano-TiO_2 and chitosan/nano-SiO_2 films. *International Journal of Biological Macromolecules*, *126*, 917–925. https://doi.org/10.1016/j.ijbiomac.2018.12.177

Tripathi, P., & Dubey, N. K. (2004). Exploitation of natural products as an alternative strategy to control postharvest fungal rotting of fruit and vegetables. *Postharvest Biology and Technology*, *32*(3), 235–245. https://doi.org/10.1016/j.postharvbio.2003.11.005

Usman, M., Farooq, M., Wakeel, A., Nawaz, A., Cheema, S. A., Rehman, H. U., . . . Sanaullah, M. (2020). Nanotechnology in agriculture: Current status, challenges and future opportunities. *Science of The Total Environment*, *721*, 137778. https://doi.org/10.1016/j.scitotenv.2020.137778

Wang, F., Hu, Q., Mugambi Mariga, A., Cao, C., & Yang, W. (2018). Effect of nano packaging on preservation quality of Nanjing 9108 rice variety at high temperature and humidity. *Food Chemistry*, *239*, 23–31. https://doi.org/10.1016/j.foodchem.2017.06.082

Wang, K., Jin, P., Shang, H., Li, H., Xu, F., Hu, Q., & Zheng, Y. (2010). A combination of hot air treatment and nano-packing reduces fruit decay and maintains quality in postharvest Chinese bayberries. *Journal of the Science of Food and Agriculture*, *90*(14), 2427–2432. https://doi.org/10.1002/jsfa.4102

Wang, L., Shao, S., Madebo, M. P., Hou, Y., Zheng, Y., & Jin, P. (2020). Effect of nano-SiO_2 packing on postharvest quality and antioxidant capacity of loquat fruit under ambient temperature storage. *Food Chemistry*, *315*, 126295. https://doi.org/10.1016/j.foodchem.2020.126295

Wyrwa, J., & Barska, A. (2017). Packaging as a source of information about food products. *Procedia Engineering*, *182*, 770–779. http://dx.doi.org/10.1016/j.proeng.2017.03.199

Xing, Y., Yang, H., Guo, X., Bi, X., Liu, X., Xu, Q., . . . Zheng, Y. (2020). Effect of chitosan/Nano-TiO_2 composite coatings on the postharvest quality and physicochemical characteristics of mango fruits. *Scientia Horticulturae, 263*, 109135. https://doi.org/10.1016/j.scienta.2019.109135

Xu, F., Liu, Y., Shan, X., & Wang, S. (2018). Evaluation of 1-methylcyclopropene (1-MCP) treatment combined with nano-packaging on quality of pleurotus eryngii. *Journal of Food Science and Technology, 55*(11), 4424–4431. https://doi.org/10.1007/s13197-018-3354-0

Xue Mei, L., Mohammadi Nafchi, A., Ghasemipour, F., Mat Easa, A., Jafarzadeh, S., & Al-Hassan, A. A. (2020). Characterization of pH sensitive sago starch films enriched with anthocyanin-rich torch ginger extract. *International Journal of Biological Macromolecules, 164*, 4603–4612. https://doi.org/10.1016/j.ijbiomac.2020.09.082

Yildirim, S., Röcker, B., Pettersen, M. K., Nilsen-Nygaard, J., Ayhan, Z., Rutkaite, R., . . . Coma, V. (2018). Active packaging applications for food. *Comprehensive Reviews in Food Science and Food Safety, 17*(1), 165–199. https://doi.org/10.1111/1541-4337.12322

Yousefi, H., Su, H.-M., Imani, S. M., Alkhaldi, K., M. Filipe, C. D., & Didar, T. F. (2019). Intelligent food packaging: A review of smart sensing technologies for monitoring food quality. *ACS Sensors, 4*(4), 808–821. https://doi.org/10.1021/acssensors.9b00440

Youssef, A. M., & El-Sayed, S. M. (2018). Bionanocomposites materials for food packaging applications: Concepts and future outlook. *Carbohydrate Polymers, 193*, 19–27.

Yousuf, B., Qadri, O. S., & Srivastava, A. K. (2018). Recent developments in shelf-life extension of fresh-cut fruits and vegetables by application of different edible coatings: A review. *LWT, 89*, 198–209. https://doi.org/10.1016/j.lwt.2017.10.051

Yu, H., Park, J.-Y., Kwon, C. W., Hong, S.-C., Park, K.-M., & Chang, P.-S. (2018). An overview of nanotechnology in food science: Preparative methods, practical applications, and safety. *Journal of Chemistry, 2018*. https://doi.org/10.1155/2018/5427978

Zhang, C., Li, W., Zhu, B., Chen, H., Chi, H., Li, L., . . . Xue, J. (2018). The quality evaluation of postharvest strawberries stored in Nano-Ag packages at refrigeration temperature. *Polymers, 10*(8), 894. https://doi.org/10.3390/polym10080894

Zhang, S., Zheng, Q., Xu, B., & Liu, J. (2019). Identification of the fungal pathogens of postharvest disease on peach fruits and the control mechanisms of Bacillus subtilis JK-14. *Toxins, 11*(6), 322. https://doi.org/10.3390/toxins11060322

Zhu, Y., Li, D., Belwal, T., Li, L., Chen, H., Xu, T., & Luo, Z. (2019). Effect of nano-SiOx/chitosan complex coating on the physicochemical characteristics and preservation performance of green tomato. *Molecules, 24*(24), 4552. https://doi.org/10.3390/molecules24244552

Chapter 12

Nanotechnology in Shelf-Life Enhancement of Minimally Processed Horticultural Produce

Girish N. Mathad and Simple Kumar

List of Abbreviations

AgNPs	Silver nanoparticles
$CaCO_3$	Calcium carbonate
CA	Controlled atmosphere
CMC	Carboxymethyl cellulose
F and V	Fruits and vegetables
EVOH	Ethylene vinyl alcohol
iPP	Isotactic polypropylene
GG	Gaur gum
GRAS	Generally recognized as safe
HHP	High-pressure processing
LDPE	Low-density polyethylene
MAP	Modified atmospheric packaging
MPFV	Minimally processed fruit and vegetable
NCC	Nanocrystalline cellulose
NM	Nanomaterial
NP	Nanoparticle
PEF	Pulse electric field
PNC	Polymer nanocomposite
PVC	Polyvinyl chloride

DOI: 10.1201/9781003142287-12

RFID	Radiofrequency identification
RH	Relative humidity
US FDA	US Food and Drug Administration

12.1 Introduction

Nanotechnology is an emerging technology in which the study and alteration of materials from 1–100 nm (Curtis et al., 2006; Di Sia, 2017; Servin & White, 2016). The word *nanotechnology* was first used by Richard Feynman in 1960. In agriculture, nanotechnology has made a big impact, significantly improving agricultural and horticultural productivity and efficiency by reducing cost and decreasing agricultural waste (Kah, 2015; Scott & Chen, 2013). Significantly, this technology is used not only in the agriculture sector but also in other sectors like communication, food, defense, biotechnology, sports, and horticultural industries. This application in various sectors has raised safety issues about human and environment; sometimes, the use of nanomaterials (NMs) may affect biochemical mechanisms. Hence, efforts have been done by experts on the environmental, toxicological, and various physicochemical properties and parameters of nanoparticles (NPs) used in the horticulture field (Iavicoli et al., 2017; Kaphle et al., 2018).

Recently, the demand for MPFVs are increasing day by day, which has resulted in the availability of a variety of MPFVs in retail shops and supermarkets rapidly increasing (Gross et al., 2016). In India, 30–35% of vegetable production is lost due to poor postharvest practices (Singh et al., 2014). About 2% of vegetables are processed in India, whereas in Brazil and US, 70% and 65% of vegetables, respectively, are processed (Dutta et al., 2014).

Globally, F and V production has increased in recent years, reaching almost 1 billion tons. F and V play a major part in supplying nutrients to people in developing, as well as developed, countries (WHO, 2003). Thirty percent of the world's total production of F and V comes from India, China, and Brazil. With the aim of enhancing the production rate of crops and processing foods to improve their market value, preventing postharvest loss due to disease is necessary (WHO, 2012). In North America, the development of the horticulture industry has improved North American economic development, while in Asia, China and India are key producers in the horticulture sector, which is developing rapidly. In the European Union, the market expansion of F and V production has been profound, likewise in Africa and South America. This increased the use of MPFVs in retail and supermarkets (Nakai, 2018).

12.2 Minimally Processed Food

Minimally processed foods are natural foods altered by processes, such as drying, crushing, grinding, filtering, roasting, boiling, pasteurization, refrigeration, freezing, placing in containers, and others, to remove inedible or unwanted parts and provide easy and convenient food for human consumption. Horticulture produce, namely, F and V, a very good example of easy, healthy foods, may also require different processes (e.g., peeling, trimming, chopping, cleaning, disinfecting and rinsing, etc.) to obtain easy, good food, providing satisfaction to consumers and ensuring food safety too (Figure 12.1).

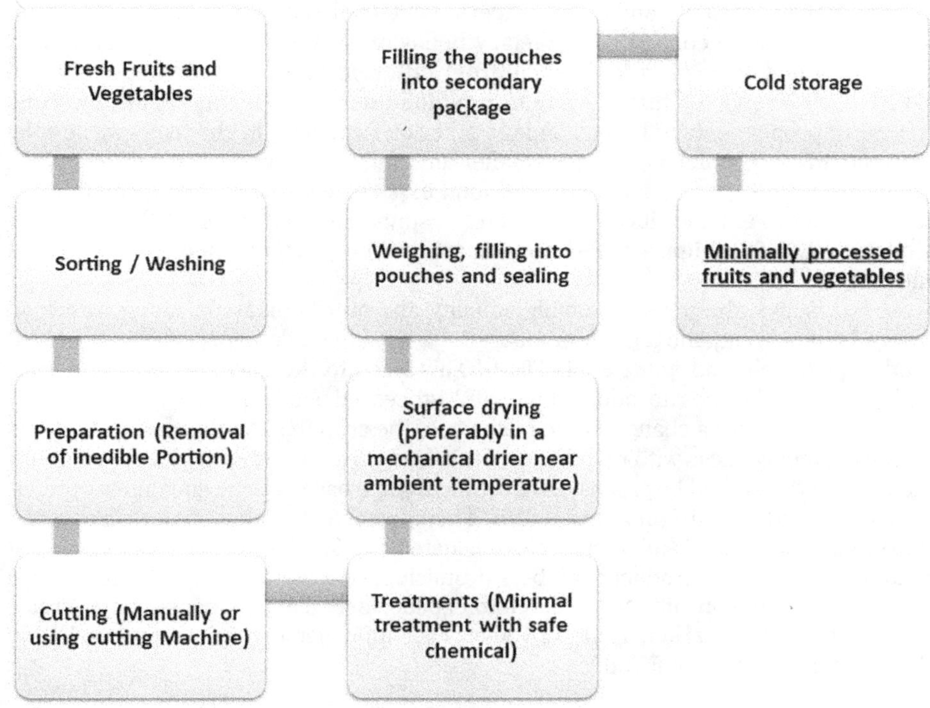

Figure 12.1 Flow diagram for the production of minimally processed F and V.

Fresh foods, which contain natural nutrients, along with good sensory attributes such as flavor, odor, taste, and texture, have been in consistent demand from consumers for decades (Huxley et al., 2004). The demand for minimally processed foods is increasing day by day, and the production of safe food is not so easy. Therefore, the necessary alteration of food quality is observed based on the standard requirements (Siddiqui et al., 2011).

The main aims of minimal processing are (1) to maintain the food safe from microbial and chemical contamination, (2) to maintain the sensory attributes of the commodity, and (3) to create easy-consumption food for customers.

12.3 Role of Minimally Processed F and V on Human Health

Horticulture is the best way to achieve nutritional security. F and V are rich diet sources that include macro- (e.g., carbohydrates, fats, protein) and micronutrients, as well as some phytochemical compounds (e.g., phenolic compounds, carotenoids, glucosinolates, etc.) that help in protecting people from certain specific diseases.

It has been observed that a person who consumes more green, healthy food products is likely to possess a lower risk for chronic diseases, strokes, various cardiovascular diseases, and some cancers (Lock et al., 2005; Sjöström et al., 2005). These diseases are connected to dietary behavior, so various vitamins and minerals from F and V help protect people from different diseases (Lock et al., 2005). (WHO, 2002) reported that more than 3 million premature deaths occur in Europe due to improper dietary habits. This is a serious concern in the average middle-class population, which consumes smaller amounts of healthy and nutritious foods or more unhealthy food. Nutrition and food experts suggest reducing the price of food in order to enhance its consumption (Gustavsen & Rickertsen, 2006; Kafatos & Codrington, 2000). Figure 12.2 shows the application of nanotechnology in the food industry.

Nowadays, the demand for high-quality and nutritious foods, along with food safety, is improving, and a good number of the urban population prefers ready-to-eat foods (Velderrain-Rodríguez et al., 2019). This leads to the need for food industries and expert authorities to adopt minimally processed food products. The nutrient intake changes with a change in the maturity of the crop (e.g., in the case of bananas, starch content reduces with ripening, and this leads to an increase in glucose content). Hence, F and V should be processed at the ideal stage based on the consumers' preferences (Velderrain-Rodríguez et al., 2019). The influence of minimal processing on the nutritional quality of F and V has been evaluated and studied. The results show that minimally processed products can be a promising, cheap alternative to improve the nutrient intake of consumers compared to expensive food supplements that are readily available in the market. Table 12.1 shows the influence of minimal processing on the nutritional quality of F and V.

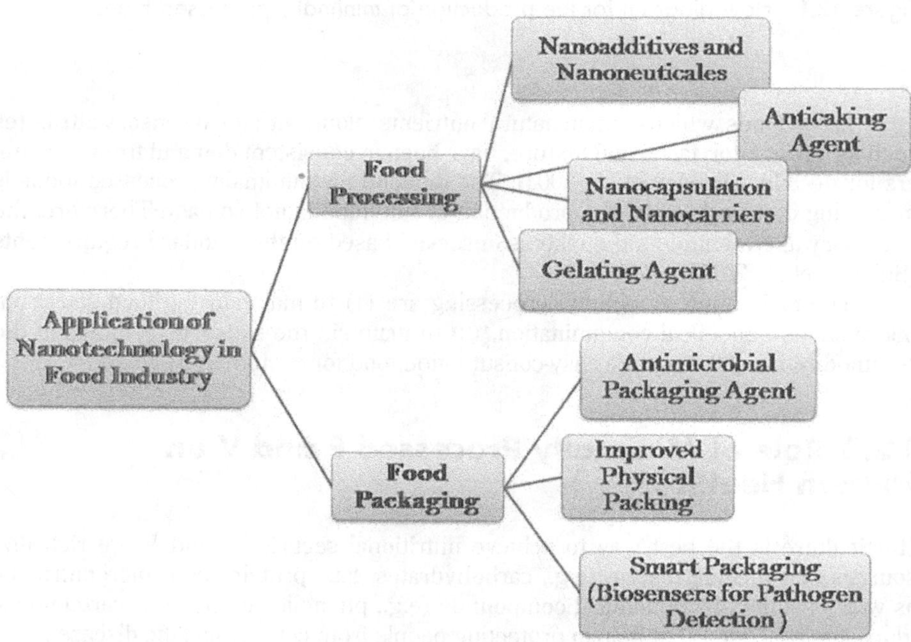

Figure 12.2 Nanotechnological application in food industries.

Table 12.1 Influence of Minimal Processing on Nutritional Quality of F and V

Commodity	*Nutritional component observed*	*Shelf life*	*Constraints*	*Reference*
Sliced apples	Phenolics	1 week	Refrigeration	Rocha & Morais, 2001
Peeled potatoes	Organic acids, phenolics, and sugar	1 week	Refrigeration	Rocha et al., 2003
Sliced apples	Phenolics	1 week	CA and additives	Rocha & De Morais, 2005
Shredded oranges	Carotenoids and phenolics	2 weeks	Refrigeration and modified atmosphere	Alasalvar et al., 2005
Peeled carrot	Sugars and carotenoids	1 week	vacuum and refrigeration	Rocha et al., 2007
Grated carrots	Phenolics and carotenoids	1–2 week	Vacuum and refrigeration	Rocha et al., 2007

12.4 Food Packaging: Technologies and Treatments

Food packaging technologies include innovative packaging technology that is said to improve the quality, safety, and sustainability of the food. Recently, emerging food packaging technologies, such as antimicrobial packaging, nanotechnology, controlled release packaging, biosensors, and RFID, have grabbed the attention of the food packaging industry. Food packaging involves various functions in protecting food from external and internal conditions like moisture, odors, dust, microorganisms, mechanical shock, and vibrations. Also, an advanced packaging system involves physical, chemical, and biological processes, whereas in intelligent packaging, it provides information and marketing functions on the packing system, and the active packaging targets providing protection and preservation of the food commodity (Figure 12.3; Mihindukulasuriya & Lim, 2014). Intelligent packing is a new method of packaging that boosts the communication part of the package that conveys information about the quality and condition of the package, along with the safety of packaging. These systems have novel communication methods such as nanosensors, O_2 sensors, and time–temperature and freshness indicators, among others (Bouwmeester et al., 2009; Majid et al., 2018). These nanosensors are used in food packaging to help determine spoilage, chemical contaminants, and other pathogens. Hence, it gives exact information about the food product. Generally, nanosensors are applied in the form of labels and coatings to ensure that the package is free from leakage, time–temperature indications, and microbial safety (Fuertes et al., 2016; Mahalik & Nambiar, 2010; Watson et al., 2011). Oxygen sensors in packaging work on luminescence, which observes changes caused by analyte and other optical sensors used in checking the quality of product influenced by hydrogen sulfide, and carbon dioxide (Biji et al., 2015). Hence, these kinds of novel technologies help customers and food regulators (Duncan, 2011).

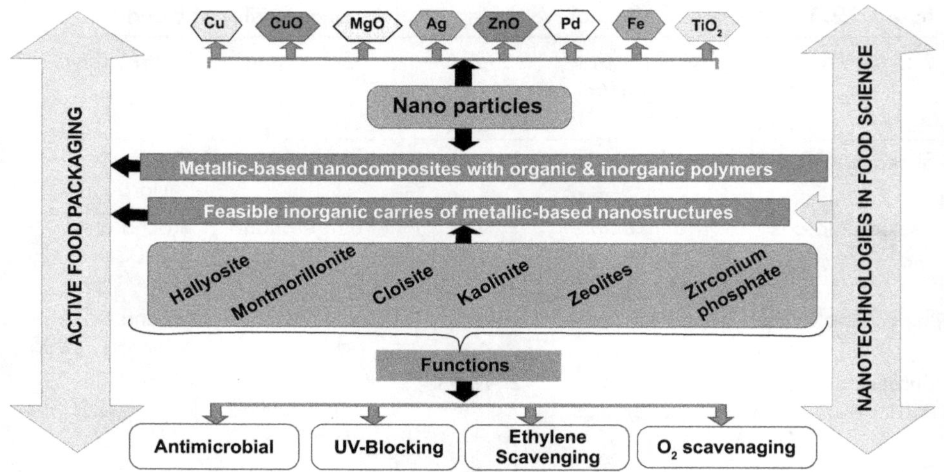

Figure 12.3 Food packaging and its characteristic functions.

The main intention of nano-packaging is to enhance the shelf life of packed food by reducing its moisture level and UV light exposure (Sorrentino et al., 2007). For food additives, NMs are used in the production of smart packaging to enhance its bioavailability in manufacturing biosensors in order to detect toxins, pathogens, and pesticides (Dasgupta & Ranjan, 2018; He et al., 2019). Among the NMs, nanoclay is most extensively used for food packaging because of its mechanical and thermal properties and lesser cost (Jose et al., 2014).

PNCs are a combination of organic and inorganic fillers composed with different kinds of particulates like fibers, flakes, spheres, and others; these novel packaging materials were recently introduced (Thakur & Gupta, 2016). These packaging fillers have a greater aspect ratio and retain the highest specific area in association with great reinforcing properties (Rafieian & Simonsen, 2014). NMs like silica (Bracho et al., 2012), clay (Schübbe et al., 2012), chitosan (Chang et al., 2010), cellulose (Sandquist, 2013), different types of graphene (Lee et al., 2013), polysaccharide nanocrystals (Lin et al., 2012), various carbon nanotubes (Swain et al., 2013), and other NMs like ZnO_2 (Esthappan et al., 2013), Ti (R. Li et al., 2011a), and colloidal Cu (Cárdenas et al., 2009) are extensively used as fillers.

Antimicrobial NPs like copper NPs have been used to restrict the development of *Saccharomyces cerevisiae*, *E. coli*, *L. monocytogenes*, and *S. aureus* and have shown good antibacterial effects (Cioffi et al., 2005; Sheikh et al., 2011). A similar study was conducted on Cu NPs by Chatterjee et al. (2014) that revealed that because of lipid peroxidation and DNA degradation, protein oxidation leads to antimicrobial activity. Various NP oxides like titanium dioxide (TiO_2), magnesium oxide (MgO), silicon oxide (SiO_2), and zinc oxide (ZnO) are used in food packaging because of their capacity to restrict the UV light and photocatalytic activity (Fujishima et al., 2000).

NMs exhibit various physiochemical properties that provide several opportunities that create an influence on food packaging and storage. Some NPs have various applications; for example, they act as antimicrobial components, O_2 scavengers, and different biosensors. Food packaging systems can be split into the following main

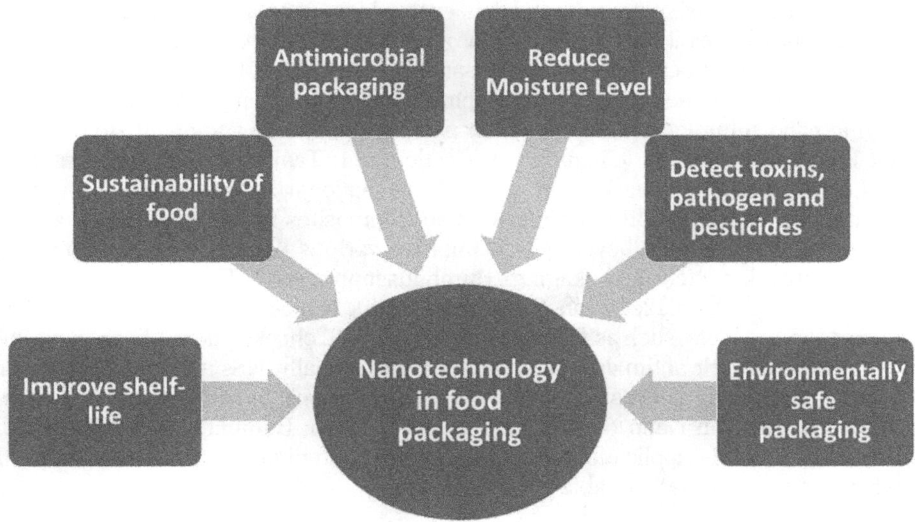

Figure 12.4 Advantages of nanotechnology in food packaging.

applications. nonreinforcement (i.e., NPs considerably increase the packaging properties, hence it enhances the shelf life of packaging materials), active food nanosystem packaging (i.e., due to the presence of NPs in packaging, they play a major role in the preservation of food), and smart food nanosystem packaging (i.e., it consists of nanodevices that sense the biochemical as well as microbial changes occurs in food and even helps in protecting against fake imitation. Another trend in packaging is nano-biodegradable packaging. Here, plant-based plastics, which are said to be bioplastics, are used as alternatives to plastic bags. Nanotitania particles and nanocrystalline titania films are used in the packaging of oxygen-sensitive food to protect food from UV light in beverage packaging, respectively (Abdullaeva, 2017).

Nanotechnology is an emerging tool associated with food packaging materials, which has various positive effects on improving the shelf life of food, along with its quality and safety, which benefits consumers and producers. However, more research is necessary on NMs that are used in food packaging systems regarding health, safety, and environmental concerns. Figure 12.4 shows the benefits of nanotechnology in food packaging systems.

12.5 Nanotechnology in Food Packaging

Globally, nanotechnological applications are increasing in various sectors like medicine, electronics, defense, food, and agriculture, among others, and even in food industries related to packaging, food safety, processing, fortification, pathogen detection, encapsulation, and many more (Ravichandran, 2010). Food packaging with nanotechnology has numerous advantages over traditional food packaging such as temperature control, barrier, improved durability, recycling, and so on at minimal cost and with fewer environmental issues. Nanotechnology application is even used in the meat and cheese industries and in confectionaries, and it is even widely used as an extrusion coating, which is applied in fruit juices and dairy products, and in the manufacturing of bottles for different kinds of beverages (Bumbudsanpharoke & Ko, 2015; Trujillo et al., 2016).

Recently, nanotechnology is developing globally, and many food companies are applying nanotechnology to improve their packaging material and quality to enhance shelf life and food safety (Chaudhry & Castle, 2011; Trujillo et al., 2016). International companies like Nanocor™, Honeywell International, Mitsubishi Gas Chemical Company Inc. USA, and others are using nanocomposites for food packaging and in beer and other alcoholic beverages to improve various properties (gas, moisture, strength, abrasion, etc.) of packaging (Bumbudsanpharoke & Ko, 2015; Costa et al., 2011; Cushen et al., 2012; Durán & Marcato, 2013). Nano-silvers (NSs) are used in various food products, such as fruits, vegetables, bread, cheese, meats, herbs, and the like, because of their antimicrobial properties, and globally, NSs are marketed under various names, such as Fresher Longer™ (USA), NS Food Container (South Korea), NS Food and Anson Nano (China), and Zeomic (Japan; Bumbudsanpharoke & Ko, 2015). Various NMs applications in food packaging studied by different authors in food systems are briefed in Table 12.2.

Table 12.2 Implementation of Nanomaterials in Food Packaging Systems

Nanomaterial	*Method of application*	*Food material*	*Impact*	*References*
TiO_2 coated with polypropylene	Antimicrobial	Lettuce	2 log decrease of *E. coli*	Chawengkijwanich & Hayata, 2008
Ag montmorillonite	Antimicrobial	Fresh fruit salad	Restricting the development of spoilage microorganisms and enhanced sensory attributes	Costa et al., 2011
		Minimally processed carrots	Restricting the development of spoilage microorganisms and improved shelf life of carrots by 60 days stored at ± 1°C	Costa et al., 2012
PVC with ZnO NPs	Antimicrobial	Chopped Apples	Delaying of fruit decay	Li et al., 2011b
LDPE films	Antimicrobial	Orange juice	Enhanced shelf life of orange juice and *Lactobacillus plantarum* was deactivated	Emamifar et al., 2010

Nanomaterial	*Method of application*	*Food material*	*Impact*	*References*
Cellulose silver NPs (AgNPs)	Antimicrobial	Kiwi and melon juices	Decrease in total variable count of yeast and bacteria by 99.9%	Lloret et al., 2012
LDPE films with AgNPs	Antimicrobial	Barberry	2.3 log decrease in molds and 2.84 log decrease in bacteria	Valipoor Motlagh et al., 2013
EVOH with AgNPs	Antimicrobial	Lettuce, apples, peels, eggshells	2 log decrease in *Salmonella* spp., *L. monocytogenes* count	Martínez-Abad et al., 2012
iPP with $CaCO_3$ nanofiller	Antimicrobial	Sliced apple	Enhanced shelf life up to 10 days	Volpe et al., 2015
Polyethylene with silver and TiO_2 NPs	Antimicrobial	Fresh apple, fresh carrot, orange juice, soft cheese	*Penicillium* and *Lactobacillus* growth was restricted	Metak, 2015

12.6 Nanotechnology in Minimally Processed Foods

The trend of nanotechnology application on F and V to improve shelf life has been increasing recently to overcome the shortcomings of traditional methods of processing. The use of nanotechnology in agriculture and food processing has been increasing recently (Bouwmeester et al., 2014; Iavicoli et al., 2017). MPFVs accelerate deterioration that leads to the production of off-flavors, browning, and loss of nutrients; many investigations have been done to inhibit these kinds of changes. Recently, the concept of nanomaterials has shown promising results in improving the previously mentioned properties (Rojas-Graü et al., 2009). The use of NPs is increasing rapidly in various food industries and has shown beneficial effects on packaging and enhanced shelf life of F and V, for example, Chinese jujube (Li et al., 2009), Chinese bayberries (Wang et al., 2010), and green asparagus (An et al., 2008). Nowadays, the use of nanomaterials like nano-ZnO, nano-SiO_2, nano-$CaCO_3$. Nanoclays are common in agricultural production and packaging (Adhikari et al., 2016; Peters et al., 2016). Nano-ZnO is known to restrict microbial growth and is extensively used in many applications in day-to-day life like the pharmaceutical, cosmetics, and medical fields and in various food sectors. Additionally, nano-ZnO is listed under GRAS by US FDA, because of its good mechanical and O_2 barriers (Sirelkhatim et al., 2015; Zhang & Xiong, 2015). X. Li et al. (2011b) and created a PVC film coated by nano-ZnO that enhanced the shelf life of sliced Fuji apples. Koushesh Saba and Amini (2017) studied nano-ZnO + CMC by coating it on pomegranate arils; the results showed that this combination of compounds extended the shelf life of pomegranate aril.

Nano-$CaCO_3$ is a polymer filler usually used in packaging because of its lower cost and nontoxicity. Nano-$CaCO_3$ polymer also reduced the permeability of oxygen and enhanced packaging performance (Avolio et al., 2013; Luo et al., 2014). The efficacy of nano-$CaCO_3$-based LDPE on Chinese yam was studied by Luo et al. (2015); the findings showed that nano-CaCO3 significantly reduced the browning index, malondialdehyde, and total phenolic content and enhanced the shelf life of fresh-cut Chinese yam. Henceforth, nano-$CaCO_3$–LDPE films act as packaging material for fresh cut F and V in food sector.

During the minimal processing of F and V, the regulation of three major factors is necessary to keep the quality of F and V: (1) controlling the respiration of F and V (i.e., controlling senescence), (2) controlling spoilage and microorganism growth on F and V, and (3) controlling the RH of surrounding environment where F and V have been stored. These are some of the disadvantages of traditional storage methods that need to be updated using nanotechnology in order to fulfill food freshness and safety (Song et al., 2009; Yan & Zhang, 2010).

Since NPs has a larger surface area-to-volume ratio (i.e., 5%. w/w), they have better barrier properties compare to pure polymer (Khan et al., 2012). Cheng et al. (2016) and co-authors studied GG and produced a biodegradable nanocomposite film NCC. The film was manufactured to reduce the oxygen barrier for food packaging as well as food preservation. Another study to improve the barrier properties was reported by (Sharma et al., 2014), nano-cellulose fibril was improved by thermal treatment, which leads to a decrease in the porosity and improved crystallinity when treated at 175°C for 3 h; then, permeability and water vapor were reduced significantly.

Nano-TiO_2 is one of the most frequently used photocatalysts to keep F and V fresh for a longer time (Brody et al., 2010). TiO_2 is used as a flavor enhancer and a food additive, which is used in a variety of foods like dried vegetables, seeds, and soups, including wine and beer. Recently, the use of TiO_2 in food products increased from 5–36% at a nano-size range (Weir et al., 2012). It leads to the deterioration of ethylene and other related gases produced at the time of storage of F and V and hinders the spoilage the postharvested F and V (Mihindukulasuriya & Lim, 2014). As NMs are mostly used in the preservation of F and V, many studies have reported on the preservation of F and V (Table 12.3).

In addition to this, in order to inhibit humidity, water, and gas, PNCs are used for minimally processed foods, including antimicrobial agents, which are nano-sized in nature. These PNCs are used in food packaging sector and in minimally processed foods to decrease cost and improve shelf life by inhibiting the oxygen activity and water vapor and enhancing the rate of biodegradation (Naffakh et al., 2013).

Chitosan extracted from crustaceans shell (GRAS) has received a lot of attention from researchers due to its nontoxicity, biodegradability, antimicrobial properties, and capacity to form gels and films (Feyzioglu & Tornuk, 2016; Wang et al., 2014). Mohammadi et al. (2016) studied cucumber shelf life, which showed nanochitosan in combination with *Zataria multiflora* essential oil (EO) on cucumbers, resulting in a decrease in respiration rate, color change, and weight loss of the chitosan-coated fruit. Along with this, it also improved the shelf life of cucumbers. A similar study was conducted by Hasheminejad and Khodaiyan (2020) on

Table 12.3 Different Studies on Nanomaterial Applications in F and V Preservation

Commodity	*Nanoparticles*	*Outcome*	*References*
Apple	Nano-ZnO	Storage period was enhanced by 1 week compared to PVC film	Li et al., 2011b
Kiwifruit	Nano-ZnO	Coating with nano-ZnO improved texture and reduced ethylene content compared to control treatment	Meng et al., 2014
Chinese yam	Nano-CaCO3	Ripening process was significantly delayed and browning; microbial growth was also hindered	Luo et al., 2015
Strawberries	Nano-ZnO	Treating with nano-ZnO showed delay in loss of weight and restricted microbial growth, and functional properties was maintained	Sogvar et al., 2016
Green soybean	Nano-ZnO	Nano-ZnO remarkably prevented growth of bacteria, yeast, molds, and coliforms.	Yu et al., 2015
Cucumber	Chitosan	Antioxidant activity of cucumber was improved; storage period extended	Mohammadi et al., 2016
Pomegranate	Nano-ZnO	Nano-ZnO reduced the total yeast, mold count, and weight loss	Saba & Amini, 2017

pomegranate arils using clove EO with chitosan NPs that showed promising results in controlling the microbial count, improved physicochemical and sensory attributes, and enhanced the shelf life of pomegranate arils. Chitosan has also been studied to see if it plays a significant role in preserving the color of various fruits like litchi (De Reuck et al., 2009), strawberries (Zhang & Quantick, 1998), and raspberries (Zhang & Quantick, 1998).

Cu, Ti, Ag, Zn, and Au are the leading metal NPs that have biocidal characteristics (Toker et al., 2013). Of these, AgNPs have been shown to have an effective bactericidal character against a range of microorganisms like bacteria, fungi, yeast, and some viruses (Martínez-Abad et al., 2012). Since AgNPs have antimicrobial properties, they are widely used as stabilizing agents and coating material in food packaging technology with lesser risks (Cushen et al., 2012; Toker et al., 2013). Minimally processed food products blended with nano packaging prove to be an encouraging method for enhancing the quality and shelf life of food products.

12.7 Sustainability and Challenges of Minimally Processed Foods

The food industry is consistently working to utilize novel technologies to produce healthy foods. Manufacturers are facing many constraints in the production of safe, nutritious, and convenient foods. The major challenge is to reduce the rapid deterioration of food products because of pathogens. Before the adoption of any method by industries or authorities for MPFs, they should be standardized and verified. Another major challenge is minimizing the cost and making them available to each level of the population in underdeveloped and developing countries like India. Emphasis should not only be placed on more production of fruit crops; rather, more work should be carried out for much more distribution and for educating poor and marginal populations about the health benefits of fresh produce and processed or minimally processed foods.

12.8 Conclusion

New technologies in food processing industries have allowed consumers to have easy, healthy food with an enhanced shelf life. Moreover, still further studies are needed to cover many areas of minimally processed products. This includes deteriorative reactions, physiochemical changes, and nutritional impacts on human health, along with a determination of toxicity levels to ensure consumer safety, which has been raised as a concern, to identify the feasibility of these new technologies. However, emerging technologies, namely, MAP, CA, HHP, PEF, and others, have shown encouraging and effective results. In this context, nanotechnology has great potential to improve the efficiency and impact of preexisting methods of food preservation and packaging. In recent years, the market of MPFs has grown significantly because of the large inclination of the population toward easy, healthy, and convenient food. The characteristic benefits of MPFV blended with nanotechnology may be an approach that has a high potential to successfully meet consumers' demands. There is a requirement to fill this knowledge gap and ensure nanotechnology's contribution to the availability of safe and nutritious foods with improved shelf life worldwide.

Acknowledgments

I would like to express my special thanks of gratitude to my teacher as well as second author of this book chapter, Dr. Simple Kumar, Lecturer, AICPHT & CCM, Amity University, Noida, who gave me the opportunity to write on this wonderful project, which also helped me in doing a lot of research during my PhD, and I came to know about so many new things.

My sincere thanks to Dr. Radhakrishnan E.K. Assistant Professor, School of Biosciences, Mahatma Gandhi University, Kottayam, Kerala, and entire review committee for giving their valuable comments and suggestions for improving the quality, coherence, and content presentation of this book chapter.

Last but not least, we express our immense thanks and appreciate CRC Press, Taylor & Francis Group publishing team.

References

Abdullaeva, Z. (2017). *Nanomaterials in daily life*. Springer.

Adhikari, T., Kundu, S., & Rao, A. S. (2016). Zinc delivery to plants through seed coating with nano-zinc oxide particles. *Journal of Plant Nutrition, 39*(1), 136–146. https://doi.org/10.1080/01904167.2015.1087562

Alasalvar, C., Al-Farsi, M., Quantick, P. C., Shahidi, F., & Wiktorowicz, R. (2005). Effect of chill storage and modified atmosphere packaging (MAP) on antioxidant activity, anthocyanins, carotenoids, phenolics and sensory quality of ready-to-eat shredded orange and purple carrots. *Food Chemistry, 89*(1), 69–76. https://doi.org/10.1016/j.foodchem.2004.02.013

An, J., Zhang, M., Wang, S., & Tang, J. (2008). Physical, chemical and microbiological changes in stored green asparagus spears as affected by coating of silver nanoparticles-PVP. *LWT-Food Science and Technology, 41*(6), 1100–1107.

Avolio, R., Gentile, G., Avella, M., Carfagna, C., & Errico, M. E. (2013). Polymer—filler interactions in PET/$CaCO_3$ nanocomposites: Chain ordering at the interface and physical properties. *European Polymer Journal, 49*(2), 419–427.

Biji, K., Ravishankar, C., Mohan, C., & Gopal, T. S. (2015). Smart packaging systems for food applications: A review. *Journal of Food Science and Technology, 52*(10), 6125–6135.

Bouwmeester, H., Brandhoff, P., Marvin, H. J. P., Weigel, S., & Peters, R. J. B. (2014). State of the safety assessment and current use of nanomaterials in food and food production. *Trends in Food Science & Technology, 40*(2), 200–210. https://doi.org/https://doi.org/10.1016/j.tifs.2014.08.009

Bouwmeester, H., Dekkers, S., Noordam, M. Y., Hagens, W. I., Bulder, A. S., De Heer, C., Ten Voorde, S. E., Wijnhoven, S. W., Marvin, H. J., & Sips, A. J. (2009). Review of health safety aspects of nanotechnologies in food production. *Regulatory Toxicology and Pharmacology, 53*(1), 52–62.

Bracho, D., Dougnac, V. N., Palza, H., & Quijada, R. (2012). Functionalization of silica nanoparticles for polypropylene nanocomposite applications. *Journal of Nanomaterials, 2012*.

Brody, A. L., Zhuang, H., & Han, J. H. (2010). *Modified atmosphere packaging for fresh-cut fruits and vegetables*. John Wiley & Sons.

Bumbudsanpharoke, N., & Ko, S. (2015). Nano-food packaging: An overview of market, migration research, and safety regulations. *Journal of food science, 80*(5), R910–R923.

Cárdenas, G., Meléndrez, M., & Cancino, A. G. (2009). Colloidal Cu nanoparticles/chitosan composite film obtained by microwave heating for food package applications. *Polymer bulletin, 62*(4), 511–524.

Chang, P. R., Jian, R., Yu, J., & Ma, X. (2010). Starch-based composites reinforced with novel chitin nanoparticles. *Carbohydrate Polymers, 80*(2), 420–425.

Chatterjee, A. K., Chakraborty, R., & Basu, T. (2014). Mechanism of antibacterial activity of copper nanoparticles. *Nanotechnology, 25*(13), 135101.

Chaudhry, Q., & Castle, L. (2011). Food applications of nanotechnologies: An overview of opportunities and challenges for developing countries. *Trends in Food Science & Technology, 22*(11), 595–603.

Chawengkijwanich, C., & Hayata, Y. (2008). Development of TiO2 powder-coated food packaging film and its ability to inactivate Escherichia coli in vitro and in actual tests. *International Journal of Food Microbiology, 123*(3), 288–292. https://doi.org/10.1016/j.ijfoodmicro.2007.12.017

Cheng, S., Zhang, Y., Cha, R., Yang, J., & Jiang, X. (2016). Water-soluble nanocrystalline cellulose films with highly transparent and oxygen barrier properties. *Nanoscale*, *8*(2), 973–978. https://doi.org/10.1039/C5NR07647A

Cioffi, N., Torsi, L., Ditaranto, N., Tantillo, G., Ghibelli, L., Sabbatini, L., Bleve-Zacheo, T., D'Alessio, M., Zambonin, P. G., & Traversa, E. (2005). Copper nanoparticle/polymer composites with antifungal and bacteriostatic properties. *Chemistry of Materials*, *17*(21), 5255–5262.

Costa, C., Conte, A., Buonocore, G. G., & Del Nobile, M. A. (2011). Antimicrobial silver-montmorillonite nanoparticles to prolong the shelf life of fresh fruit salad. *International Journal of Food Microbiology*, *148*(3), 164–167.

Costa, C., Conte, A., Buonocore, G. G., Lavorgna, M., & Del Nobile, M. A. (2012). Calcium-alginate coating loaded with silver-montmorillonite nanoparticles to prolong the shelf-life of fresh-cut carrots. *Food Research International*, *48*(1), 164–169. https://doi.org/10.1016/j.foodres.2012.03.001

Curtis, J., Greenberg, M., Kester, J., Phillips, S., & Krieger, G. (2006). Nanotechnology and Nanotoxicology. *Toxicological Reviews*, *25*(4), 245–260. https://doi.org/10.2165/00139709-200625040-00005

Cushen, M., Kerry, J., Morris, M., Cruz-Romero, M., & Cummins, E. (2012). Nanotechnologies in the food industry—recent developments, risks and regulation. *Trends in Food Science & Technology*, *24*(1), 30–46.

Dasgupta, N., & Ranjan, S. (2018). Food nanoemulsions: Stability, benefits and applications. In *An introduction to food grade nanoemulsions* (pp. 19–48). Springer.

De Reuck, K., Sivakumar, D., & Korsten, L. (2009). Effect of integrated application of chitosan coating and modified atmosphere packaging on overall quality retention in litchi cultivars. *Journal of the Science of Food and Agriculture*, *89*(5), 915–920.

Di Sia, P. (2017). Nanotechnology among innovation, health and risks. *Procedia—Social and Behavioral Sciences*, *237*, 1076–1080. https://doi.org/10.1016/j.sbspro.2017.02.158

Duncan, T. V. (2011). Applications of nanotechnology in food packaging and food safety: Barrier materials, antimicrobials and sensors. *Journal of Colloid and Interface Science*, *363*(1), 1–24.

Durán, N., & Marcato, P. D. (2013). Nanobiotechnology perspectives. Role of nanotechnology in the food industry: A review. *International Journal of Food Science & Technology*, *48*(6), 1127–1134.

Dutta, J. K., Boruah, R., & Das, S. (2014). Measuring the level of commercialization of farmers: A case in Kamrup District of Assam. *Indian Research Journal of Extension Education*, *15*(1), 35–39.

Emamifar, A., Kadivar, M., Shahedi, M., & Soleimanian-Zad, S. (2010). Evaluation of nanocomposite packaging containing Ag and ZnO on shelf life of fresh orange juice. *Innovative Food Science & Emerging Technologies*, *11*(4), 742–748. https://doi.org/10.1016/j.ifset.2010.06.003

Esthappan, S. K., Sinha, M. K., Katiyar, P., Srivastav, A., & Joseph, R. (2013). Polypropylene/zinc oxide nanocomposite fibers: Morphology and thermal analysis. *Journal of Polymer Materials*, *30*(1), 79.

Feynman, R. P. (1960). *There's plenty of room at the bottom*. California Institute of Technology, Engineering and Science Magazine.

Feyzioglu, G. C., & Tornuk, F. (2016). Development of chitosan nanoparticles loaded with summer savory (*Satureja hortensis* L.) essential oil for antimicrobial and antioxidant delivery applications. *LWT*, *70*, 104–110.

Fuertes, G., Soto, I., Carrasco, R., Vargas, M., Sabattin, J., & Lagos, C. (2016). Intelligent packaging systems: Sensors and nanosensors to monitor food quality and safety. *Journal of Sensors*, *2016*.

Fujishima, A., Rao, T. N., & Tryk, D. A. (2000). Titanium dioxide photocatalysis. *Journal of Photochemistry and Photobiology C: Photochemistry Reviews, 1*(1), 1–21.

Gross, K. C., Wang, C. Y., & Saltveit, M. E. (2016). *The commercial storage of fruits, vegetables, and florist and nursery stocks.* United States Department of Agriculture, Agricultural Research Service.

Gustavsen, G. W., & Rickertsen, K. (2006). A censored quantile regression analysis of vegetable demand: The effects of changes in prices and total expenditure. *Canadian Journal of Agricultural Economics/Revue canadienne d'agroeconomie, 54*(4), 631–645.

Hasheminejad, N., & Khodaiyan, F. (2020). The effect of clove essential oil loaded chitosan nanoparticles on the shelf life and quality of pomegranate arils. *Food Chemistry, 309*, 125520. https://doi.org/10.1016/j.foodchem.2019.125520

He, X., Deng, H., & Hwang, H.-M. (2019). The current application of nanotechnology in food and agriculture. *Journal of Food and Drug Analysis, 27*(1), 1–21. https://doi.org/10.1016/j.jfda.2018.12.002

Huxley, R., Lean, M., Crozier, A., John, J., & Neil, H. (2004). Effect of dietary advice to increase fruit and vegetable consumption on plasma flavonol concentrations: Results from a randomised controlled intervention trial. *Journal of Epidemiology & Community Health, 58*(4), 288–289.

Iavicoli, I., Leso, V., Beezhold, D. H., & Shvedova, A. A. (2017). Nanotechnology in agriculture: Opportunities, toxicological implications, and occupational risks. *Toxicology and Applied Pharmacology, 329*, 96–111. https://doi.org/https://doi.org/10.1016/j.taap.2017.05.025

Jose, T., George, S. C., Maria, H. J., Wilson, R., & Thomas, S. (2014). Effect of bentonite clay on the mechanical, thermal, and pervaporation performance of the poly (vinyl alcohol) nanocomposite membranes. *Industrial & Engineering Chemistry Research, 53*(43), 16820–16831.

Kafatos, A., & Codrington, C. (2000). *Eurodiet core report, nutrition & diet for healthy lifestyles in Europe, science & policy implications.* University of Crete, School of Medicine and European Commission DG Sanco.

Kah, M. (2015). Nanopesticides and nanofertilizers: Emerging contaminants or opportunities for risk mitigation? [Perspective]. *Frontiers in Chemistry, 3*(64). https://doi.org/10.3389/fchem.2015.00064

Kaphle, A., Navya, P. N., Umapathi, A., & Daima, H. K. (2018). Nanomaterials for agriculture, food and environment: Applications, toxicity and regulation. *Environmental Chemistry Letters, 16*(1), 43–58. https://doi.org/10.1007/s10311-017-0662-y

Khan, A., Khan, R. A., Salmieri, S., Le Tien, C., Riedl, B., Bouchard, J., Chauve, G., Tan, V., Kamal, M. R., & Lacroix, M. (2012). Mechanical and barrier properties of nanocrystalline cellulose reinforced chitosan based nanocomposite films. *Carbohydrate Polymers, 90*(4), 1601–1608. https://doi.org/10.1016/j.carbpol.2012.07.037

Koushesh Saba, M., & Amini, R. (2017). Nano-ZnO/carboxymethyl cellulose-based active coating impact on ready-to-use pomegranate during cold storage. *Food Chemistry, 232*, 721–726. https://doi.org/10.1016/j.foodchem.2017.04.076

Lee, Y., Kim, D., Seo, J., Han, H., & Khan, S. B. (2013). Preparation and characterization of poly (propylene carbonate)/exfoliated graphite nanocomposite films with improved thermal stability, mechanical properties and barrier properties. *Polymer International, 62*(9), 1386–1394.

Li, H., Li, F., Wang, L., Sheng, J., Xin, Z., Zhao, L., Xiao, H., Zheng, Y., & Hu, Q. (2009). Effect of nano-packing on preservation quality of Chinese jujube (*Ziziphus jujuba Mill.* var. *inermis (Bunge) Rehd*). *Food Chemistry, 114*(2), 547–552.

Li, R., Liu, C., Ma, J., Yang, Y., & Wu, H. (2011a). Effect of org-titanium phosphonate on the properties of chitosan films. *Polymer Bulletin*, *67*(1), 77–89.

Li, X., Li, W., Jiang, Y., Ding, Y., Yun, J., Tang, Y., & Zhang, P. (2011b). Effect of nano-ZnO-coated active packaging on quality of fresh-cut 'Fuji' apple. *International Journal of Food Science & Technology*, *46*(9), 1947–1955.

Lin, N., Huang, J., & Dufresne, A. (2012). Preparation, properties and applications of polysaccharide nanocrystals in advanced functional nanomaterials: A review. *Nanoscale*, *4*(11), 3274–3294.

Lloret, E., Picouet, P., & Fernández, A. (2012). Matrix effects on the antimicrobial capacity of silver based nanocomposite absorbing materials. *LWT – Food Science and Technology*, *49*(2), 333–338, https://doi.org/10.1016/j.lwt.2012.01.042

Lock, K., Pomerleau, J., Causer, L., Altmann, D. R., & McKee, M. (2005). The global burden of disease attributable to low consumption of fruit and vegetables: Implications for the global strategy on diet. *Bulletin of the World health Organization*, *83*, 100–108.

Luo, Z., Wang, Y., Jiang, L., & Xu, X. (2015). Effect of nano-CaCO3-LDPE packaging on quality and browning of fresh-cut yam. *LWT—Food Science and Technology*, *60*(2, Part 2), 1155–1161. https://doi.org/10.1016/j.lwt.2014.09.021

Luo, Z., Wang, Y., Wang, H., & Feng, S. (2014). Impact of nano-$CaCO_3$-LDPE packaging on quality of fresh-cut sugarcane. *Journal of the Science of Food and Agriculture*, *94*(15), 3273–3280.

Mahalik, N. P., & Nambiar, A. N. (2010). Trends in food packaging and manufacturing systems and technology. *Trends in Food Science & Technology*, *21*(3), 117–128.

Majid, I., Nayik, G. A., Dar, S. M., & Nanda, V. (2018). Novel food packaging technologies: Innovations and future prospective. *Journal of the Saudi Society of Agricultural Sciences*, *17*(4), 454–462.

Martinez-Abad, A., Lagaron, J. M., & Ocio, M. J. (2012). Development and characterization of silver-based antimicrobial ethylene—vinyl alcohol copolymer (EVOH) films for food-packaging applications. *Journal of Agricultural and Food Chemistry*, *60*(21), 5350–5359.

Meng, X., Zhang, M., & Adhikari, B. (2014). The effects of ultrasound treatment and nano-zinc oxide coating on the physiological activities of fresh-cut kiwifruit. *Food and Bioprocess Technology*, *7*, 126–132. https://doi.org/10.1007/s11947-013-1081-0

Metak, A. M. (2015). Effects of nanocomposite based nano-silver and nano-titanium dioxide on food packaging materials. *International Journal of Applied Science and Technology*, *5*(2), 26–40.

Mihindukulasuriya, S. D. F., & Lim, L. T. (2014). Nanotechnology development in food packaging: A review. *Trends in Food Science & Technology*, *40*(2), 149–167. https://doi.org/10.1016/j.tifs.2014.09.009

Mohammadi, A., Hashemi, M., & Hosseini, S. M. (2016). Postharvest treatment of nanochitosan-based coating loaded with Zataria multiflora essential oil improves antioxidant activity and extends shelf-life of cucumber. *Innovative Food Science & Emerging Technologies*, *33*, 580–588.

Naffakh, M., Diez-Pascual, A. M., Marco, C., Ellis, G. J., & Gomez-Fatou, M. A. (2013). Opportunities and challenges in the use of inorganic fullerene-like nanoparticles to produce advanced polymer nanocomposites. *Progress in Polymer Science*, *38*(8), 1163–1231.

Nakai, J. (2018). Food and Agriculture Organization of the United Nations and the sustainable development goals. *Sustainable Development*, *22*.

Peters, R. J. B., Bouwmeester, H., Gottardo, S., Amenta, V., Arena, M., Brandhoff, P., Marvin, H. J. P., Mech, A., Moniz, F. B., Pesudo, L. Q., Rauscher, H., Schoonjans,

R., Undas, A. K., Vettori, M. V., Weigel, S., & Aschberger, K. (2016). Nanomaterials for products and application in agriculture, feed and food. *Trends in Food Science & Technology, 54*, 155–164. https://doi.org/10.1016/j.tifs.2016.06.008

Rafieian, F., & Simonsen, J. (2014). Fabrication and characterization of carboxylated cellulose nanocrystals reinforced glutenin nanocomposite. *Cellulose, 21*(6), 4167–4180.

Ravichandran, R. (2010). Nanotechnology applications in food and food processing: Innovative green approaches, opportunities and uncertainties for global market. *International Journal of Green Nanotechnology: Physics and Chemistry, 1*(2), P72–P96.

Rocha, A., & De Morais, A. (2005). Polyphenoloxidase activity of minimally processed 'Jonagored' apples (*Malus domestica*). *Journal of Food Processing and Preservation, 29*(1), 8–19.

Rocha, A., & Morais, A. M. (2001). Influence of controlled atmosphere storage on polyphenoloxidase activity in relation to colour changes of minimally processed 'Jonagored' apple. *International Journal of Food Science & Technology, 36*(4), 425–432.

Rocha, A. M., Coulon, E. C., & Morais, A. M. (2003). Effects of vacuum packaging on the physical quality of minimally processed potatoes. *Food Service Technology, 3*(2), 81–88.

Rocha, A. M., Mota, C. C., & Morais, A. M. (2007). Physico-chemical qualities of minimally processed carrot stored under vacuum. *Journal of Foodservice, 18*(1), 23–30.

Rojas-Graü, M. A., Oms-Oliu, G., Soliva-Fortuny, R., & Martín-Belloso, O. (2009). The use of packaging techniques to maintain freshness in fresh-cut fruits and vegetables: A review. *International Journal of Food Science & Technology, 44*(5), 875–889.

Saba, M. K., & Amini, R. (2017). Nano-ZnO/carboxymethyl cellulose-based active coating impact on ready-to-use pomegranate during cold storage. *Food Chemistry, 232*, 721–726.

Sandquist, D. (2013). New horizons for microfibrillated cellulose. *Appita: Technology, Innovation, Manufacturing, Environment, 66*(2), 156.

Schübbe, S., Schumann, C., Cavelius, C., Koch, M., Müller, T., & Kraegeloh, A. (2012). Size-dependent localization and quantitative evaluation of the intracellular migration of silica nanoparticles in CACO-2 cells. *Chemistry of Materials, 24*(5), 914–923.

Scott, N., & Chen, H. (2013). Nanoscale science and engineering for agriculture and food systems. *Industrial Biotechnology, 9*(1), 17–18. https://doi.org/10.1089/ind.2013.1555

Servin, A. D., & White, J. C. (2016). Nanotechnology in agriculture: Next steps for understanding engineered nanoparticle exposure and risk. *NanoImpact, 1*, 9–12. https://doi.org/10.1016/j.impact.2015.12.002

Sharma, S., Zhang, X., Nair, S. S., Ragauskas, A., Zhu, J., & Deng, Y. (2014). Thermally enhanced high performance cellulose nano fibril barrier membrane. *RSC Advances, 4*(85), 45136–45142. https://doi.org/10.1039/C4RA07469F

Sheikh, F. A., Kanjwal, M. A., Saran, S., Chung, W.-J., & Kim, H. (2011). Polyurethane nanofibers containing copper nanoparticles as future materials. *Applied Surface Science, 257*(7), 3020–3026. https://doi.org/10.1016/j.apsusc.2010.10.110

Siddiqui, M. W., Chakraborty, I., Ayala-Zavala, J., & Dhua, R. (2011). Advances in minimal processing of fruits and vegetables: A review. *Journal of Scientific and Industrial Research, 70*(10), 823–834.

Singh, V., Hedayetullah, M., Zaman, P., & Meher, J. (2014). Postharvest technology of fruits and vegetables: An overview. *Journal of Postharvest Technology*, *2*(2), 124–135.

Sirelkhatim, A., Mahmud, S., Seeni, A., Kaus, N. H. M., Ann, L. C., Bakhori, S. K. M., Hasan, H., & Mohamad, D. (2015). Review on zinc oxide nanoparticles: Antibacterial activity and toxicity mechanism. *Nano-Micro letters*, *7*(3), 219–242.

Sjöström, M., Poortvliet, E., & Nelson, M. (2005). Monitoring public health nutrition in Europe: Nutritional indicators and determinants of health status. *Journal of Public Health*, *13*(2), 74–83.

Sogvar, O. B., Saba, M. K., Emamifar, A., & Hallaj, R. (2016). Influence of nano-ZnO on microbial growth, bioactive content and postharvest quality of strawberries during storage. *Innovative Food Science & Emerging Technologies*, *35*, 168–176.

Song, X.-J., Zhang, M., Mujumdar, A. S., & Fan, L. (2009). Drying characteristics and kinetics of vacuum microwave—dried potato slices. *Drying Technology*, *27*(9), 969–974. https://doi.org/10.1080/07373930902902099

Sorrentino, A., Gorrasi, G., & Vittoria, V. (2007). Potential perspectives of bio-nanocomposites for food packaging applications. *Trends in Food Science & Technology*, *18*(2), 84–95.

Swain, S. K., Pradhan, A. K., & Sahu, H. S. (2013). Synthesis of gas barrier starch by dispersion of functionalized multiwalled carbon nanotubes. *Carbohydrate Polymers*, *94*(1), 663–668.

Thakur, V. K., & Gupta, R. K. (2016). Recent progress on ferroelectric polymer-based nanocomposites for high energy density capacitors: Synthesis, dielectric properties, and future aspects. *Chemical Reviews*, *116*(7), 4260–4317.

Toker, R., Kayaman-Apohan, N., & Kahraman, M. (2013). UV-curable nano-silver containing polyurethane based organic—inorganic hybrid coatings. *Progress in Organic Coatings*, *76*(9), 1243–1250.

Trujillo, L. E., Ávalos, R., Granda, S., Guerra, L. S., & País-Chanfrau, J. M. (2016). Nanotechnology applications for food and bioprocessing industries. *Biology and Medicine*, *8*(3), 1.

Valipoor Motlagh, N., Hamed Mosavian, M. T., & Mortazavi, S. A. (2013). Effect of polyethylene packaging modified with silver particles on the microbial, sensory and appearance of dried barberry. *Packaging Technology and Science*, *26*(1), 39–49.

Velderrain-Rodríguez, G. R., López-Gámez, G. M., Domínguez-Avila, J. A., González-Aguilar, G. A., Soliva-Fortuny, R., & Ayala-Zavala, J. F. (2019). Minimal processing. In E. M. Yahia (Ed.), *Postharvest technology of perishable horticultural commodities* (pp. 353–374). Woodhead Publishing. https://doi.org/10.1016/B978-0-12-813276-0.00010-9

Volpe, M., Di Stasio, M., Paolucci, M., & Moccia, S. (2015). Polymers for food shelf-life extension. In *Functional polymers in food science* (pp. 9–66). Scrivener Publishing LLC.

Wang, K., Jin, P., Shang, H., Li, H., Xu, F., Hu, Q., & Zheng, Y. (2010). A combination of hot air treatment and nano-packing reduces fruit decay and maintains quality in postharvest Chinese bayberries. *Journal of the Science of Food and Agriculture*, *90*(14), 2427–2432.

Wang, L., Wu, H., Qin, G., & Meng, X. (2014). Chitosan disrupts *Penicillium expansum* and controls postharvest blue mold of jujube fruit. *Food Control*, *41*, 56–62.

Watson, S. B., Gergely, A., & Janus, E. R. (2011). Where is agronanotechnolgoy heading in the United States and European Union. *Natural Resources & Environment*, *26*, 8.

Weir, A., Westerhoff, P., Fabricius, L., Hristovski, K., & Von Goetz, N. (2012). Titanium dioxide nanoparticles in food and personal care products. *Environmental Science & Technology*, *46*(4), 2242–2250.

World Health Organization. (2002). *The world health report 2002: Reducing risks, promoting healthy life*. WHO.

World Health Organization. (2003). *Diet, nutrition, and the prevention of chronic diseases: Report of a joint WHO/FAO expert consultation* (Vol. 916). World Health Organization.
World Health Organization. (2012). *World health statistics: A snapshot of global health*. World Health Organization.
Yan, W., & Zhang, M. (2010). Studies on different combined microwave drying of carrot pieces. *International Journal of Food Science & Technology, 45*(10), 2141–2148.
Yu, N., Zhang, M., Islam, M. N., Lu, L., Liu, Q., & Cheng, X. (2015). Combined sterilizing effects of nano-ZnO and ultraviolet on convenient vegetable dishes. *LWT–Food Science and Technology, 61*(2), 638–643.
Zhang, D., & Quantick, P. C. (1998). Antifungal effects of chitosan coating on fresh strawberries and raspberries during storage. *The Journal of Horticultural Science and Biotechnology, 73*(6), 763–767.
Zhang, Z.-Y., & Xiong, H.-M. (2015). Photoluminescent ZnO nanoparticles and their biological applications. *Materials, 8*(6), 3101–3127.

Chapter 13

Modeling and Simulation of Gas Barrier Properties of Nanocomposites Used for Packaging Applications

Shasiya P.S., Neethumol Varghese, Minu Joys, and Ajalesh B. Nair

List of Abbreviations

2D & 3D	Two-dimensional and three-dimensional
BEM	Boundary element method
CNT	Carbon nanotube
FEM	Finite element method
FVM	Finite volume method
MC	Monte Carlo
PNC	Polymer nanocomposite
SBR	Styrene butadiene rubber
TEM	Transmission electron microscopy

13.1 Introduction

Nanotechnology is creating an inevitable field of development in our world. Its possibility paves the way to making superior products and benefits the quality of human life. On the other side, nanotechnology takes initiatives in reducing negative effects

DOI: 10.1201/9781003142287-13

on the environment. In the packaging industry, nanotechnology makes sense from protection to transport, along the stream from raw materials to consumers. Currently, nanomaterials used in the packaging industry have increased accordingly due to their groundbreaking and purposeful properties. It has been stated that about 500 nano-packaging materials are being used for various commercial purposes and that almost 25% of products will be used for food packaging in the near future [1].

The use of nanotechnology shows significant enhancement in packaging materials, and better knowledge about the opportunities and obstacles of utilizing nanomaterials for packaging needs to be resolved in a scientific manner. The progress of nanotechnology in packaging leads to good shelf life, transportability, enhanced processing, reduced cost, and eco-friendliness of the material [2]. The aids of employing nanomaterials for packaging involve their ability to reduce overall packaging can make interactive packaging materials and their antimicrobial properties, as well as the ability to achieve improved barrier properties.

Packaging materials have a significant role in physically protecting and maintaining a good physicochemical status of products to obtain a reasonable shelf life. The packaging system prevents product worsening due to physicochemical or biotic factors based on an appropriate choice of material for packaging endowed with water vapor and gas barriers, as well as mechanical properties. It thus keeps the overall quality of the product throughout storage and usage. It is essential to have a high degree of exfoliation for PNCs to enhance their gas barrier properties. After their use, regarding environmental concerns, the packaging materials should be biodegradable [3]. Biopolymers are the finest choice in packaging material in relation to environmental concerns. They have some advantages in bettering food quality and increasing shelf life of the product by reducing microbial growth. They assist as a barrier to water vapors, moisture, solutes, and gases and help act as transporters of some active elements. The research on biodegradable polymers is a hot topic. Food packaging materials based on polylactic acid, which is known to be the cheapest biomaterial, and poly (butylene adipate-co-terephthalate) (PBAT) are now used in many European countries, Japan, and the United States [4].

Introducing nanoparticles as fillers into a polymer can improve packaging performance. Clay and silicates have received significant attention due to their ease of processing, abundance, and so on [5]. They also show considerable enhancement in the material property of the polymer. Other nanoparticles put to use for packaging are zinc oxides, starch nanocrystals, cellulose nanofibers, copper oxides, silver, CNTs, and nanowhiskers, among others [6, 7]. Fillers such as zinc oxide, copper, copper oxide, and titanium dioxide are vital for their antimicrobial activities while CNTs and other carbon-based fillers are restricted in packaging applications owing to their poor dispersion and high cost [8, 9].

Barrier property is a measure of the ability of a material to allow the passage or exchange of low-molecular-weight particles through a mass transport phenomenon called permeation. The permeability process is decided by various elements, including solubility, morphology, and diffusivity. To define the permeability of a nanocomposite, one must consider the fillers in the matrix that control the diffusion and solubility of the penetrating molecules, particularly at the interface. The consumer demands good product quality and better shelf life, so enhancing the barrier properties in packaging for nutrition, cosmetics, pharmaceuticals, and beverages is becoming crucial. Hence, the requirement to study the gas barrier behavior of a material remains essential and must be considered in detail. Refining the barrier property of a material by introducing suitable nanofillers can decrease the thickness of the mercantile packaging material, and thus, we can make large savings in material costs. Furthermore, it also helps make polymeric materials lighter and transparent, which benefits commercial purposes.

Barrier properties are very important when designing materials for the application of packaging. The barrier properties of polymers can be enhanced by different methods, such as incorporating nanofillers, using high-barrier coating materials, and using polymer blends [10]. Nanocomposites improve the barrier properties of the packaging material along with an enhancement in their thermal and mechanical properties [11]. Here, the prime concern is the property change that occurs when applying nanomaterials in food packaging. The tiny size of the nanoparticles can impart chemical and physical properties dissimilar to that of their macroscale chemical counterparts [12]. However, various PNCs have already proven their applications in the packaging industry. They are already available as packaging material with gas diffusion control and provide prolonged shelf life to various products. PNCs could replace conventional materials like metals, ceramics, and paper in packaging applications owing to their functionality, lightweight, low cost, and ease of processing. The food packaging industry is constantly searching for packaging materials with a reduced gas permeability rate in order to stretch the shelf life of foodstuffs. Per a report on a global polymer market survey, more than about 40% of polymers are suitable for packaging materials, and 50% of them are good for food packaging available in various forms such as films, sheets, and bottles [13]. Even though nanotechnology has garnered much attraction in the packaging field, they are in the middle of the road. The causes behind this are regulatory concerns, high manufacturing costs, and their environmental impact.

The main concern about using nanomaterials is consumer safety. There are studies that have reported on the migration of nanoparticles from packaging into food materials or other material-sensitive products, and the migration power will be associated with the amount of nanomaterial present in the packaging system. Another is the environmental concern. Using biopolymers such as polylactic acid, chitosan, and starch impart biodegradability but, at the same time, due to the poor mechanical properties, retard their choice. However, incorporating nanomaterials increases the mechanical properties as far as it is concerned. But, the problem is that of the reduction in their biodegradability. Finally, care must be taken when evaluating the pros and cons of using nanomaterials for packaging. This is a very important criterion when dealing with human health and the environment, as well as the economy.

Herein, the chapter considers different models for predicting the barrier properties of PNCs used in packaging upon incorporating various nanofillers. Conventional methods used for predicting permeation reduction are found to be inappropriate for determining the exact microstructure of nanocomposites. Hence, along with the bulk-based mechanical characteristics of the PNC materials, the barrier performance also needs to be explored.

13.2 Barrier Properties of Material

The barrier property of a material refers to the transmission of penetrants such as water vapor, gases, liquid, or organic components from a high-density region to a low-density region. The process of transmission involves the initial sorption of permeable objects at the upstream layer of the membrane; then the diffusion, or spreading, of these particles more widely through the material cross-section under the effect of the pertained driving force (i.e., pressure gradient, which parallels to a chemical potential

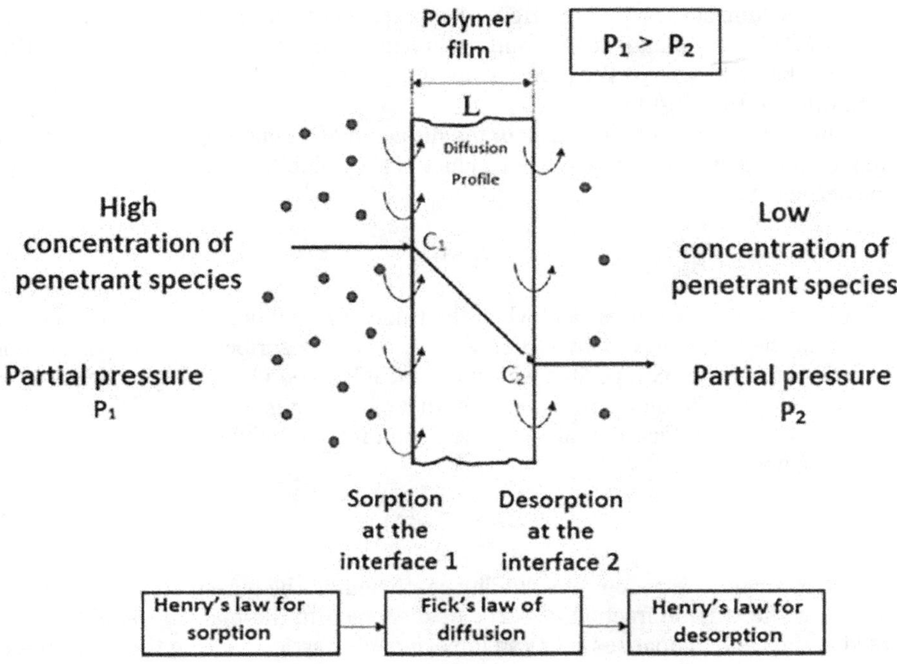

Figure 13.1 Permeation of penetrant species through polymer film [14].

gradient); and, finally, the release of substance at the downstream layer of the membrane (Figure 13.1). Using a Fickian transport process, that is, there are no boundaries between highly swollen regions from dry or glassy regions during diffusion, the time needed to attain interfacial equilibrium is much faster than the characteristic time of the diffusion process, which then governing the process of transport mechanism [14].

13.3 Theories of Permeation of Gas

Many theories have been proposed to explain the permeability behavior of PNCs. Three important parameters required to describe the barrier properties of a nanocomposite are diffusion coefficient, permeation coefficient, and sorption coefficient. In polymer films, steady-state diffusion can be studied by measuring the gas permeability with the application of a pressure gradient across the membrane of effective thickness $t_{m,}$ and the permeation coefficient is measured using the variable pressure–volume approach. Permeation of gas is generally considered a function of two variables: diffusion and sorption [15]. Based on Fick's law and Hendry's law, gas permeation (Q) is

$$Q = DSAt\left(\frac{Dp}{tm}\right). \tag{13.1}$$

In Equation 13.1, the diffusivity D reflects the kinetic part of permeation, and the solubility coefficient S describes the penetrant affinity as well as the thermodynamic part of gas permeation across the film. Here, A represents the effective area of gas permeation (nm^2), Q is the quantity of gas outflow or leakage through the membrane (nm^3), t is the time taken for gas permeation (s), and Dp gives the difference pressure on both side of the film (mmHg) [16].

The permeability coefficient P of the diffusion of a penetrant molecule across a homogeneous polymer matrix is the product of the solubility coefficient S and diffusion coefficient D.

$$P = DS \tag{13.2}$$

Equation 13.2 becomes valid when the value of D is independent of the concentration and the value of S follows Henry's law. It also describes the gas permeation properties of polymer composites with impermeable nanofillers. The barrier properties of nanocomposite films are better than that of homogeneous films. With the simplest model given by Picard et al. [17], the penetrant solubility in the nanocomposite film can be predicted as

$$S = S_0(1 - \Phi) \tag{13.3}$$

where S_0 indicates the penetrant solubility coefficient in the pure polymer matrix and ϕ indicates the volume fraction of particles dispersed in the matrix. Nanofillers dispersed in the nanocomposites act as an impermeable barrier, causing the penetrant to take a tortuous pathway across the film (Figure 13.2) [18]. This leads to an increase in the effective path of the gas that brings out the degradation of diffusion coefficient. By introducing a new factor, the tortuosity factor (τ), the effective diffusion coefficient (D) of nanocomposites can be expressed as

$$D = \left(\frac{D_o}{\tau}\right), \tag{13.4}$$

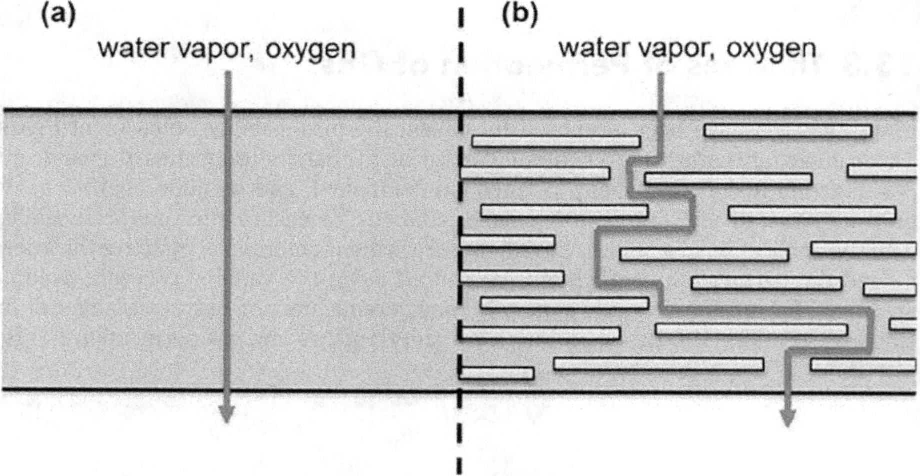

Figure 13.2 Tortuous pathway through a polymer nanocomposite [19].

where D_o gives the diffusion coefficient of the pure polymer matrix, and tortuosity factor (τ) can be expressed as

$$\tau = \frac{t_m}{t_{0m}}. \tag{13.5}$$

t_m is the distance between the tortuous pathways the penetrant follows across the film. The τ value depends on the shape and volume of the nanofillers. Also, tortuosity is expressed as

$$\tau = 1 + \frac{L}{2W}\Phi. \tag{13.6}$$

The term L/W is called aspect ratio α, that is the length-to-width ratio. Keeping the width constant, as the length of the nanoparticle increases, there will be a reduction in the gas permeability of the PNC.

Effective permeability coefficient P can be obtained from Equations 13.2–13.4 as

$$\left(\frac{P}{P_0}\right) = \frac{1-\Phi}{\tau}. \tag{13.7}$$

P_0 gives the permeability coefficient of the pure polymer matrix.

13.4 Mathematical Modeling for Barrier Properties

Many experiential models have been developed for predicting the enhancement of barrier properties of nanocomposite. All these approaches discussed are diverse mechanisms of transport that are normally related to the "tortuous path" resistance in the composite systems [20]. The barrier properties are expected to be improved by the tortuous path that was adopted by the penetrant molecule in the polymer matrix in the presence of a filler. The obstacle formed by the presence of nanofillers imparts an increased diffusion path length [21].

Using the rule of mixtures is the simplest way to represent any desired property of the composite. It is the general rule used to predict the properties of a nanocomposite material and for assigning a theoretical upper and lower limit to various properties like tensile strength, thermal conductivity, and elastic modulus [22]. However, PNCs generally do not obey this rule. Plasters with a high aspect ratio have more control of the permeability of gases across the membrane than the fillers with a lower aspect ratio. The alignment of fillers plays an important role (concerning the direction of gas permeation) in permeability in the bulk of the system. Certain mathematical models are briefly explained in the following subsections using various constraints, such as filler shape and content, filler location and distribution, filler size distribution, filler orientation, and filler stacking. Being a first model, Maxwell's theory [23] expects the effective gas permeability for a material having regularly arranged spherical impermeable fillers as discussed in the next section.

13.4.1 Maxwell's Model

Maxwell's theory [23] is used to find the effective permeability of nanocomposite with a periodic array of spherical impermeable fillers. The effective diffusivity of the particle decreases with an increase in the volume fraction of impermeable nanoparticles.

$$\left(\frac{D}{D_0}\right)=\frac{1}{\left(1+\frac{\Phi d}{2}\right)} \tag{13.8}$$

D is the effective diffusivity, and Φ is the volume fraction of fillers. This model further assumes that the connection between the filler and the surrounding matrix is flawless. This model works well for a diluted suspension of spherical fillers and may not always offer a good approximation for nanocomposites that contain graphene-like fillers.

13.4.2 Bruggeman's Model

The Bruggeman model suggested an equation to define the passage of molecules in biphasic media formed spherical dispersed fillers in a continuous matrix [24]:

$$\frac{P}{P_0}=\left(\frac{\frac{P_d}{P_0}-\frac{P}{P_0}}{(1-\Phi d)\left(\frac{P_d}{P_0}-1\right)}\right)^3. \tag{13.9}$$

P represents the effective permeability coefficient of the nanocomposite. P_0 and P_d are the permeability of the continuous phase and the permeability in the dispersed phase, respectively. Hence, the relative diffusivity is expressed as

$$\left(\frac{D}{D_0}\right)=\frac{1}{(1-\Phi d)}\cdot\left(\frac{\frac{P_d}{P_0}-\frac{P}{P_0}}{(1-\Phi d)\left(\frac{P_d}{P_0}-1\right)}\right)^3. \tag{13.10}$$

When an impermeable dispersion phase assumption is applied ($P_d=0$), the equation becomes

$$\left(\frac{D}{D_0}\right)=\frac{1}{(1-\Phi d)^{1/2}}. \tag{13.11}$$

Considering Maxwell's model, the tortuosity factor is expressed as

$$\tau=1+\frac{1+\frac{\Phi}{2}}{1-\Phi}. \tag{13.12}$$

The Maxwell model provides decent precision with experimental permeability information for a filler volume fraction up to 0.2, whereas the Bruggeman model could be used to study heavy filler loading. While using both the Maxwell and the Bruggeman model, the filler shape or filler distribution is not considered [25].

Initially, a large number of studies were focused on the dispersion of spherical fillers, and the attention was then waged on the impermeable fillers in the shape of disks, cylinders, ribbons, and other shapes.

The interest was focused on diluted or semi-diluted regimes for most of the theoretical model. Also, the studies were mainly focused on materials in which the fillers are adapted to a direction perpendicular to the gas flow.

13.4.3 Nielsen's Model

In this model, the extreme decrease in permeability was measured by the tortuosity arguments alone. The particles are supposed to be impermeable. Theoretically, the molecules have a longer and tortuous path, which causes a longer time to diffuse through the membrane [21]. The Nielsen model describes the relation between the increase of tortuosity along the gas diffusion path with aspect ratio and the volume fraction. It is further assumed that the fillers in the form of nanoplates are exfoliated and distributed completely along the direction perpendicular to the direction of diffusion.

The filler is assumed to be finally dispersed in the matrix and the matrix possessions are not affected by the filler. In complete dispersal, the presence of void and non-parallel alignment of particles will be more than that projected by this model. Thus, the relative permeability is given by

$$P/P_0 = (1-\Phi)/1+(\Phi\alpha/2). \tag{13.13}$$

Here, P and P_0 is the permeability of the composite and permeability of the neat polymer, respectively. Φ is the volume fraction of the particle, and α is the aspect ratio of the filler particles, which is specified by the equation

$$\alpha = w/t, \tag{13.14}$$

where w gives the transitional dimension of the particle and t is the lowest dimension of the particle [26].

The tortuosity factor is given by

$$\tau = 1 + \alpha D. \tag{13.15}$$

The difference from Maxwell's model is noted by the presence of the dimension of the particle in the tortuosity factor, which shows the significance of the aspect ratio of the particle on the diffusion direction.

13.4.4 Cussler's Model

The Cussler model adopts all the concepts from the Nielsen model except the concept of regularity of the array. Nielsen model holds good at dilute regime, but the volume fraction of the fillers is semi-dilute that is, $D \ll 1$ and can use the relation given by Cussler [27]. In the Cussler model, for nanocomposites containing nanoplates oriented perpendicular to the diffusion, the relative permeability of a nanocomposite with monodisperse elements arranged in a regular way is assumed by

$$P/P_0 = 1 - \Phi/(1-\Phi) + \alpha^2\Phi^2. \tag{13.16}$$

Here, α is the aspect ratio of the particle which is given by the equation

$$\alpha = R/t, \tag{13.17}$$

where R and t represent half of the intermediate dimension of the filler and the smallest dimension of the filler, respectively.

For a random array of monodisperse particles in a composite, the relative permeability according to the Cussler model is given by [28]

$$P/P_0 = 1 - \Phi/(1 + 2/3\,\Phi\alpha)^2. \tag{13.18}$$

For a film containing fillers oriented normally to the direction of diffusion, the expression for the tortuosity factor is

$$\tau = 1 + \frac{\alpha^2\Phi^2}{4(1-\Phi)}, \tag{13.19}$$

where $\alpha\ (= d/a)$ is the nanoplatelet aspect ratio, which is a representation of the nanoplatelet shape. Here, *a* gives the thickness of the nanoplatelet, and *d* is the distance between the nanoplatelets.

The uniformity of the films is not assured all the time.

It is not constantly probable to develop nanocomposites with fillers dispersed at regular intervals, and these sheets appear to be randomly arranged. Therefore, in the case of:

(1) two sequences of nanoplatelets with alignment and misalignment happening with equivalent probability, the tortuosity is given by

$$\tau = 1 + \frac{\alpha^2\Phi^2}{8(1-\Phi)}. \tag{13.20}$$

(2) random misalignment hexagonal nanoplatelets in successive layers τ:

$$\tau = 1 + \frac{\alpha^2\Phi^2}{54(1-\Phi)}. \tag{13.21}$$

13.4.5 Aris and Falla's Model

The tortuosity factor for a film containing a randomly arranged set of impermeable fillers is given by Aris and Falla model [29]:

$$\tau = 1 + \frac{\alpha^2 \Phi^2}{4(1-\Phi)} + \frac{\alpha \Phi}{2\sigma} + \frac{2\alpha \Phi}{\pi(1-\Phi)} + ln\left[\frac{\pi \alpha^2 \Phi}{4\sigma(1-\Phi)}\right] \tag{13.22}$$

The second term in the equation corresponds to the tortuous path for a diffusing molecule to take. The third term and the fourth term relate the constriction of slits formed between the nanoplates and the confrontation of a diffusion species to permit into and out of the slit, respectively. The square of alpha denotes the resistance for the diffusion, and such an effect contributes to the resistance through nanoplatelet-filled composites and is called wiggling [30].

13.4.6 Bharadwaj's Model

Bharadwaj modified the Nielsen model by correlating filler orientation, state of aggregation, length, and concentration of filler in the matrix. The key assumption in this model is that the sheet normally coincides with the direction of diffusion (Figure 13.3).

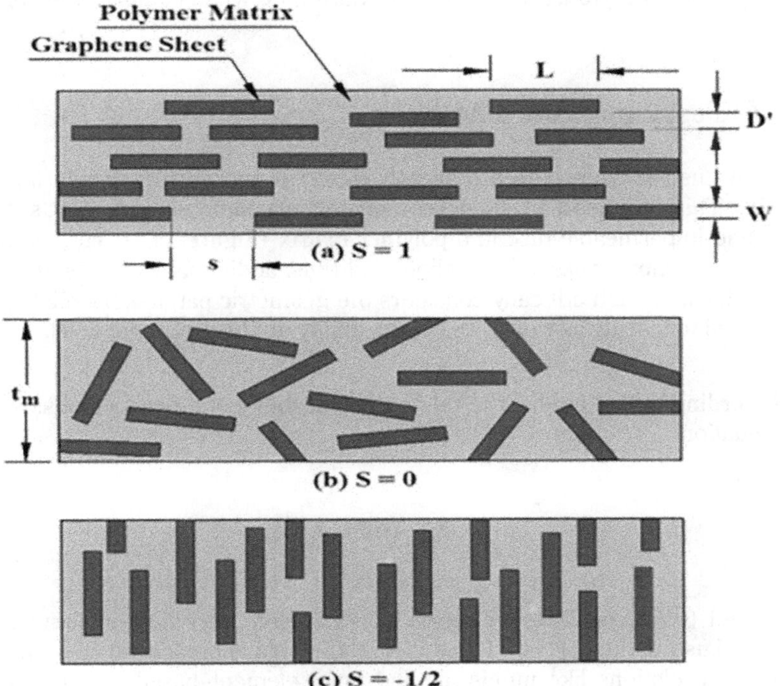

Figure 13.3 Different orientation of graphene layers in the nanocomposite film [32].

This causes the highest tortuosity. This model introduces a new orientation order factor S in the Nielsen model. Any deviation from this arrangement will cause a decline in barrier properties [31].

$$S = 1/2(3\mathrm{Cos}^2\theta - 1), \tag{13.23}$$

where S value ranges from 1 to -1/2.

$S = 1$ indicates the impeccable orientation of fillers along the path of diffusing gas, and $S = -1/2$ indicates a perpendicular orientation. A value of $S = 0$ specifies the random positioning of the filler in the matrix. This follow-on equation for diffusivity is given by

$$\frac{D}{D_0} = \frac{1}{1 + \left(\frac{L}{2}\right) W \Phi d \left(\frac{2}{3}\right)\left(S + \left(\frac{1}{2}\right)\right)}. \tag{13.24}$$

The barrier properties are lower for aggregates with a lower aspect ratio than those with a distinct platelet aspect ratio. Here the intra-platelet diffusions are neglected.

This equation holds well for explaining the permeability in polymer/layered silicate composites. The sheet-like morphology helps maximize the path length of the penetration owing to its high length-to-width ratio when compared to spheres or cubes.

13.4.7 Gusev and Lusti's Model

Gusev and Lusti [33] developed a slightly different approach for analyzing barrier properties. This assumes a 3-D model comprising of a randomly arranged, seamlessly parallel, and impermeable disc in a polymer matrix (Figure 13.4). The model originated from a computer-based simulation, and thus, additional variables are taken to account. This approach critically considers the geometric parameters related to with the improved tortuosity as well as the permeability on the molecular level.

According to this model, the relative permeability of a nanocomposite is given by the equation

$$\frac{P}{P_0} = exp(-(\alpha\Phi / x_0)^{\beta}). \tag{13.25}$$

Here x_0 and β are constants measured by regression procedure on simulated data. Gusev and Lusti defined $\beta = 0.71$ and $x_0 = 3.47$. This approach relies on numerical simulation calculations like multiscale and finite element–based permeability. This model assumes that the matrix is not affected by the gas permeation and that the system is made of a random array of the filler [35].

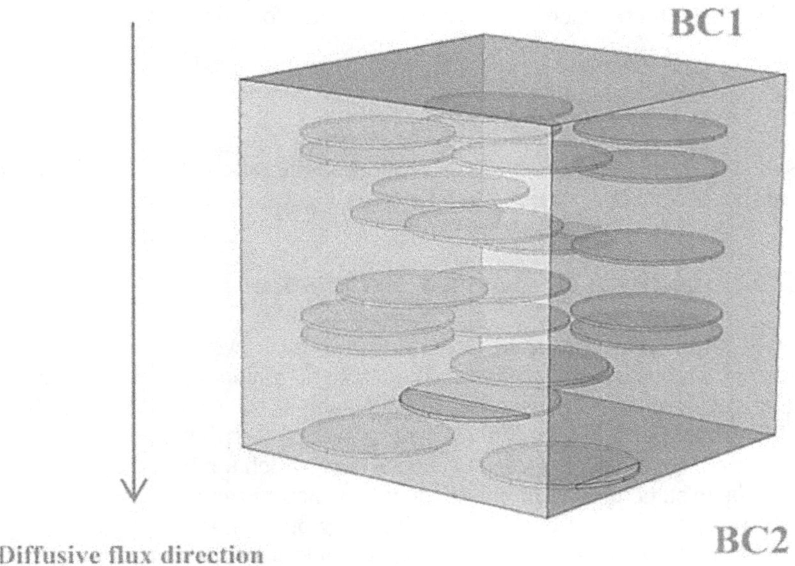

Figure 13.4 A three dimensional dispersion of homogeneous disks in the matrix [34].

13.4.8 Wakeham and Mason's Model

For diffusion through materials that have fillers in the form of nanoplates, a considerably diverse model was suggested by Wakeham and Mason [36].

$$\tau = 1 + \frac{\alpha^2\Phi^2}{4(1-\Phi)} + \frac{\alpha\Phi}{2\sigma} + 2(1-\Phi)ln\left[\frac{1-\Phi}{2\sigma\Phi}\right] \tag{13.26}$$

The modification between Equation 13.22 and lies in the fourth term of the right-hand side of the equation. The fourth term in Equation 13.22 depends on the factor α while it is not in Equation 13.26. The advancement in these equations is largely beneficial for comparing the enhancement in the barrier properties predicted theoretically with simulation or experimental results.

13.4.9 Fredrickson and Bicerano's Model

Fredrickson and Bicerano's model [37] considers particles that are disc-shaped. This particle forms nematic phases, there is no positional order for plates. This model can be applied equally to diluted and semi-diluted regimes. The equation for the Fredrickson and Bicerano model is given by

$$\frac{P}{P_0} = \frac{1}{4}\left(\frac{1}{\alpha 1\kappa\alpha\Phi} + \frac{1}{\alpha 2\kappa\alpha\Phi}\right)^2, \tag{13.27}$$

κ describes a geometric factor. The geometric factor for this model is given by

$$\kappa = \frac{\pi}{lnln\alpha}. \quad (13.28)$$

In this model, the aspect ratio is considered to be the ratio between the radial dimension and the thickness.

13.5 Simulation of Barrier Properties

A majority of the prototypes cited earlier include simple systems, which contain regularly or randomly distributed particles like platelets or ribbons in the polymer matrix. Furthermore, the influence of various factors, such as the volume fraction, the aspect ratio of the filler, particle orientation, dispersion, and distribution, on the barrier performance in these perspectives has been analyzed, which further upholds the simplest nanocomposite structures. A modified tortuosity factor is adequate to analyze the consequences on barrier properties. Moreover, using other modeling methods that allow the simulation of 2D and 3D complex structures, it is possible to examine the influence of such parameters on barrier properties. It is vital to solve these models using numerical methods to obtain approximate solutions and tools as otherwise it would be difficult to obtain a simple equation for the relative diffusivity. Methods that are commonly used include the MC method, the FVM, the BEM, and the FEM, among others [38].

In the MC method, numerical results are obtained by means of repeated random sampling. While the MC method offers numerical explanations for an inevitable problem through a microscopic and probabilistic method [38, 39], the other methods, such as BEM, FVM, and FEM, provide solutions in a macroscopic and deterministic means as they are grounded in the solutions of mathematical partial differential equations. In BEM, the numerical division of a continuous set into finite discrete elements is performed at abridged spatial dimensions. This method called as boundary method gives simpler linear equations and has fewer computer memory necessities [40]. FVM is a traditional method that is founded on a flux evaluation at the cell's margins. The advantages, such as accuracy and speed on regular engagements, seem to be less significant when the geometry turns more irregular and multifaceted, whereas the impact of the structural constraints, such as aspect ratio, volume fraction, intercalation level, and orientation angle, are analyzed and the diffusion process in nanocomposites well described in the FEM method. Also, the compatible, robust nature of representing various structures in this method is found to be superior to other methods [41].

Swannack et al. [42] presented the calculation of the diffusivity of gas molecules through a film comprising oriented platelets through 2D and 3D MC simulations of a polymer-clay nanocomposite. Furthermore, they noticed a significant difference between the simulation of 2D and 3D Brownian motion results. In most cases, it was found that the 2D simulation has a lower value of effective diffusivity than the 3D simulation. In addition, a greater value of the effective diffusion coefficient was noted in the 3D simulation due to the increased permeation as a result of the finite length of platelets in either direction normal to the direction of the solute carriage.

The FEM method used for evaluating the reduction of diffusivity In PNCs was reported by Statler and Gupta [43]. In this method, the impermeable fillers were measured as uniformly distributed platelets with a perpendicular alignment. While the results in the computational procedure, in which a relationship of concentration (C/C_0), lies between 0 and 1 at the left and right boundaries of the unit cell, appropriate agreement with the Cussler model for a tad filler volume element is exhibited. Nevertheless, the overprediction of relative diffusivity (D/D_0) in a similar region was noted with the Nielsen model.

Due to inaccuracy of the aspect ratio calculations of the diverse permeability representations in butyl rubber/vermiculate nanocomposites using TEM, Takahashi and coworkers [44] compared different permeability models. They realized the importance of each model in measuring the aspect ratio of nanofillers. Accordingly, it was reported that when using the Nielsen model, the average aspect ratio was found to be 480; with the regular Cussler model, 100; with the random Cussler model, 120; with the Fredrickson and Bicerano model, 475; and with the Gusev and Lusti model, 190. Through this analysis, Takashi et al. justified the insignificance of evaluating the models on the basis of their aspect ratios.

Zulhairun and Ismail [45] analyzed an increase in the particle aspect ratio with a decrease in filler loading in the case of PNCs such as polysulfone (PSF)/nanoclay, using the Nielsen's model and Cussler's model. In addition, their studies also revealed that Nielsen's model exhibited the highest aspect ratio in the PSF/nano clay.

Wang and coworkers [46] investigated the variation in permeability performance in styrene butadiene rubber (SBR)/rectorite and SBR/carbon black systems in which a comparison was carried out with morphological and free volume characteristics using these permeability results. The studies revealed that the values of impermeability, diffusion, and solubility coefficient reported in nano clay-loaded SBR were in good agreement. They noted that the aspect ratio values using Nielsen modeling in these composites were between 1 and 30, in which the aspect ratio result for Rectorite/SBR was found to be 30 and for SBR/CB the value was 1.

The applicability of these models was reflected in CNT-loaded systems and the materials having nanoclay and graphene as fillers. For instance, the part of the graphene oxide to the nitrogen permeability of carboxylated nitrile butadiene rubber (XNBR) systems was reported by Kang and coworkers [47] using the Nielson model, in which there was a 55% reduction in the gas permeability coefficient upon the addition of 1.9 volume percentage of graphene oxide.

Park et al. [48] carried out water vapor permeation studies of bio-nano composites composed of cellulose acetate, triethyl citrate (as plasticizer), and organically modified clay using the Cussler model. The results showed a decent agreement in the foreseen and tentative values of water vapour permeability (WVP) with $\alpha = 150$ at lower volume fractions ($\phi_f \leq 0.02$). However, at higher volume fractions ($\phi_f \geq 0.05$), the predicted and experimental values showed a good coincidence with a lower aspect ratio ($\alpha = 100$).

13.6 Conclusion

When developing a nanocomposite for barrier applications, there should be a complete understanding of the influence of the structural characteristics on gas permeation. This necessity has suggestively encouraged the progress of analytical and numerical modeling. The modeling data are either harmonizing or alternative approaches to investigational ones. Methodical approaches have been broadly used and have endorsed the first relationships between gas barrier properties and the size, shape, content, dispersion, and array of nanofillers. Although they agree well with experimental data, they grieve from limitations when considering complex systems. In this background, several numerical approaches by means of the most communal and proven computational approaches such as MC method, FVM, BEM, and FEM, among others, have been established. To improve the barrier properties of the nanocomposites, scientists are working toward spreading the approaches advanced to different works. Although studies on enhancing gas barrier properties are well known, the effect of orientation of filler, dispersion, exfoliation, or the state of aggregation has not been well studied yet. So the proper modeling methods and advancing the existing system also need to be analyzed and improved.

References

[1] Sozer N, Kokini JL, Nanotechnology and its applications in the food sector, Trends Biotechnol, 2009, 27(2), 82–89.

[2] Reynolds G, FDA recommends nanotechnology research, but not labelling, 2007. https://www.foodnavigator.com/Article/2007/07/26/FDA-recommends-nanotechnology-research-but-not-labeling

[3] Huang JY, Li X, Zhou W, Nanocomposites in food packaging applications and their risk assessment for health, Trends in Food Science & Technology, 2015, 45(2), 187–199.

[4] Dondero M, Tomba JP, Cisilino AP, The effect of flake orientational order on the permeability of barrier membranes: numerical simulations and predictive models, Journal of Membrane Science, 2016, 514, 95.

[5] Azeredo H, Mattoso LHC, Wood D, Williams TG, Avena-Bustillos RJ, McHugh TH, Recent advances on edible films based on fruits and vegetables—a review, Journal of Food Science, 2009, 74(5), 31–35.

[6] Mc Glashan SA, Halley P, Preparation and characterisation of biodegradable starch-based nanocomposite materials, Journal of Polymer International, 2003, 52(11), 1767–1773.

[7] Alexandre M, Dubois P, Polymer-layered silicate nanocomposites: preparation, properties and uses of a new class of materials, Materials Science and Engineering R Reports, 2000, 28(1), 1–63.

[8] Ramanathan T., Abdala A. A., Stankovich S., Dikin D. A., Herrera-Alonso M., Piner R. D., Adamson D. H., Schniepp, H. C., Chen X., Ruoff R. S., Nguyen S. T., Aksay I. A., Prud'Homme R. K., Brinson L. C., Functionalized graphene sheets for polymer nanocomposites. Nature Nanotechnology. 2008, 3(6),327–331.

[9] Duncan TV, Applications of nanotechnology in food packaging and food safety: barrier materials, antimicrobials and sensors, Journal of Colloid Science, 2011, 363(1), 1–24.

[10] Arora A, Padua GW, Review: nanocomposite in food packaging, Journal of Food Science, 2010, 75(1), 43–49.

[11] Avella M, De Vliger JJ, Errio ME, Fisher S, Vacca P, Volpe MG, Biodegradable starch/clay nanocomposites films for food pckaging application, Journal of Food Chemistry, 2005, 93(3), 467–474.

[12] Huang JY, Li X, Zhou W, Nanocomposites in food packaging applications and their risk assessment for health, Trends in Food Science & Technology, 2015, 45(2), 187–199.

[13] Dondero M, Tomba JP, Cisilino AP, The effect of flake orientational order on the permeability of barrier membranes: numerical simulations and predictive models, Journal of Membrane Science, 2016, 514, 9.

[14] Gokhale A, Lee I, Recent advances in the fabrication of nanostructured barrier films, Journal of Nanoscience and Technology, 2014, 14(3), 2157.
[15] Mokwena KK, Tang J, Dunne CP, Yang TCS, Chow E, Unlocking potentials of microwaves for food safety and quality, Journal of Food Science, 2009, 92(3), 291.
[16] Potts JR, Dreyer DR, Bielawski CW, Ruo RS, Graphene-based polymer nanocomposites, Polymer, 2011, 52, 5–25.
[17] Picard E, Vermogen A, Gerard J, Espuche E, Barrier properties of nylon 6-montmorillonite nanocomposite membranes prepared by melt blending: influence of the clay content and dispersion state. Consequences on modelling, Journal of Membrane Science, 2007, 292, 133–144.
[18] Zhang G, Lee PC, Jenkins S, Dooley J, Baer E, The effect of confined spherulite morphology of high-density polyethylene and polypropylene on their gas barrier properties in multilayered film systems, Polymer, 2014, 55, 4521–4530.
[19] Bugnicourt E, Kehoe T, Latorre M, Serrano C, Philippe S, Schmid M, Recent prospects in the inline monitoring of nanocomposites and nanocoatings by optical technologies, Nanomaterials, 2016, 6, 8, 150.
[20] Cui Y, Kumar S, Rao Kona B, van Houcke D, Gas barrier properties of polymer/clay nanocomposites, RSC Advances, 2015, 5, 63669–63690.
[21] Nielsen LE, Models for the permeability in filled polymer systems, Journal of Macromolecular Science, 1967, 1, 929.
[22] Alger, MSM. Polymer science dictionary (2nd edition), Chapman and Hall, London, 1996.
[23] Maxwell JC, A treatise on electricity and magnetism, Clarendon Press, Oxford, 1873, p. 489.
[24] Bruggeman DAG, Berechnung verschiedener physikalischer Konstanten von heterogenen Substanzen. I. Dielektrizitätskonstanten und Leitfähigkeiten der Mischkörper aus isotropen Substanzen, Annalen der Physik, 1935, 416, 636.
[25] Barrer RM, In Diffusion in Polymers, J Crank and GS Park (Eds.), Journal of applied polymer science, Academic Press, New York, 1968.
[26] Sinha Ray S, Okamoto M, Polymer/layered silicate nanocomposites: a review from preparation to processing, Progress in Polymer Science, 2003, 28, 1539.
[27] Yang C, Smyrl WH, Cussler EL, Flake alignment in composite coatings, Journal of Membrane Science, 2004, 231, 1.
[28] Lape NK, Nuxoll EE, Cussler EL, Polydisperse flakes in barrier film, Journal of Membrane Science, 2004, 236, 29–37.
[29] Cussler E, Hughes SE, Ward WJ, Aris R, Barrier membranes, Journal of Membrane Science, 1988, 38, 161–174.
[30] Kamal M, Jinnah I, Utracki L, Permeability of oxygen and water vapour through polyethylene/polyamide films, Polymer Engineering Science, 1984, 24, 1337–1347.
[31] Bharadwaj RK, Modeling the barrier properties of polymer-layered silicate nanocomposites, Macromolecules, 2001, 34(26), 9189–9192.
[32] Cui Y, Kundalwal SI, Kumar S, Gas barrier performance of graphene/polymer nanocomposites, Carbon, 2016, 98, 313–333.
[33] Gusev AA, Lusti HR, Rational design of nanocomposites for barrier applications, Advanced Materials, 2001, 13(21), 1641–1643.
[34] Zid S, Zenit M, Espuche E, Modeling diffusion mass transport in multiphase polymer systems for gas barrier applications: a review, Journal of Polymer Science, 2018, 1, 5253.
[35] Lape NK, Nuxoll EE, Cussler EL, Barrier properties of flake-filled membranes: review and numerical evaluation, Journal of Membrane Science, 2004, 236, 29–37.

[36] Wakeham WA, Mason EA, Diffusion through multiper-forate laminae, Industrial & Engineering Chemistry Fundamentals, 1979, 18(4), 301–305.
[37] Eitzman DM, Melkote RR, Cussler EL, Barrier membranes with tipped impermeable flakes, AICHE Journal, 1996, 42, 2–9.
[38] Bhunia K, Dhawan S, Sablani SS, Modeling the oxygen diffusion of nanocomposite based food packaging films, Journal of Food Science, 2012, 77, 29–39.
[39] Mittal V, Polymer nanocomposites: synthesis, microstructure, and properties, Journal of Composite Materials, 2008, 42, 2829.
[40] Chen X, Papathanasiou TD, Barrier properties of flake-filled membranes: review and numerical evaluation, Journal of Plastic Film and Sheeting, 2007, 23, 4, 319–346.
[41] Minelli M, Baschetti MG, Doghieri F, Analysis of modelling results for barrier properties in ordered nanocomposite systems, Journal of Membrane Science, 2009, 327, 208–215.
[42] Swannack C, Cox C, Liakos A, Hirt D, A 3-dimensional simulation of barrier properties of nanocomposite films, Journal of Membrane Science, 2005, 263, 47–56.
[43] Statler DL Jr, Gupta RK, A. Finite Element Analysis on the Modeling of Heat Release Rate, as Assessed by a Cone Calorimeter, of Char Forming Polycarbonate, Excerpt from the Proceedings of the COMSOL Conference. Boston, 2008.
[44] Takahashi S, Goldberg HA, Feeney CA, Karim DP, Farrell M, O'Leary K, Paul DR, Gas barrier properties of butyl rubber/vermiculite nanocomposite coatings, Polymer, 2006, 47, 3083–3093.
[45] Zulhairun AK, Ismail AF, Matsuura T, Abdullah MS, Mustafa A. Asymmetric mixed matrix membrane incorporating organically modified clay particle for gas separation. Chemical Engineering Journal. 2014;241:495–503.
[46] Wang ZF, Wang B, Qi N, Zhang HF, Zhang LQ, Influence of fillers on free volume and gas barrier properties in styrene-butadiene rubber studied by positrons, Polymer, 2005, 46, 719–724.
[47] Kang H, Zuo K, Wang Z, Zhang L, Liu L, Guo B, Using a green method to develop graphene oxide/elastomers nanocomposites with a combination of high barrier and mechanical performance, Composites Science and Technology, 2014, 92, 1–8.
[48] Park HM, Misra M, Drzal LT, Mohanty AK, "Green" nanocomposites from cellulose acetate bioplastic and clay; effect of ecofrendly triethyl citrate plasticizer, Biomacromolecule, 2004, 5(6), 2281–2288.

Chapter 14

Zinc-Oxide Nanoparticle-Based Methods for the Minimally Processed Horticultural Produce

Upekshya Welikala, Udari Wijesinghe, and Gobika Thiripuranathar

List of Abbreviations

Ag	Silver
AgNP	Silver nanoparticle
CMC	Carboxymethyl cellulose
FNB/IOM	Food and Nutrition Board/Institute of Medicine
GRAS	Generally recognized as safe
H_2O_2	Hydrogen peroxide
IZiNCG	International Zinc Nutrition Consultative Group
LDPE	Low-density polyethylene
MPVFs	Minimally processed vegetables and fruits
MWCNT	Multiwalled carbon nanotube
NOAEL	No observed adverse effects level
NP	Nanoparticle
PLA	Polylactic acid
PP	Polypropylene
PPO	Polyphenol oxidase
PVC	Polyvinyl chloride

DOI: 10.1201/9781003142287-14

PVP	Polyvinylpyrrolidone
ROS	Reactive oxygen species
SnO_2	Tin oxide
TiO_2	Titanium dioxide
USFDA	U.S. Food and Drug Administration
UV	Ultraviolet
WHO	World Health Organization
Zn^{2+}	Zinc ions
ZnO	Zinc oxide
ZnONP	Zinc-oxide nanoparticle

14.1 Introduction

In recent years, demand for minimally processed foods has increased rapidly due to the advantage of being convenient, easy to prepare, high quality, and producing little waste at a reasonable price. Postharvest vegetables and fruits tend to mature and deteriorate quickly due to biochemical and physiological changes. Per published literature, about 30–50% of fresh vegetables and fruits are unusable due to deterioration (Hasan et al. 2019). Therefore, there has been an increasing interest among the scientific community in developing and applying novel technologies to maintain fresh taste, nutritional value, and food appearance (De Corato 2020).

Minimal processing involves various steps, including removing inedible parts, drying, fractioning, ginning, washing, cutting, peeling, slicing, filtering, roasting, pasteurization, freezing, packaging, and storage steps (Guimarães et al. 2020; De Corato 2020; Jung and Zhao 2016; Espitia et al. 2012). These steps will reduce food wastage as all the inedible parts have already been removed (Sant'Anna et al. 2020). Every step must be taken care of from raw materials to the final product to ensure the better quality of MPVFs. To attain these necessities, the steps involved in the minimal process should be followed immediately after harvesting. Packaging acts as a barrier that shields food from external conditions such as chemical and physical hazards and pathogens, governing food safety and quality (Al-Naamani et al. 2018). The most effective way of achieving this is to incorporate antimicrobial and antifungal substances into packaging that aids in reducing the preservation steps (Al-Naamani et al. 2018; Sruthi et al. 2018). However, an increase in foodborne diseases by pathogens such as *Listeria monocytogenes*, *Salmonella*, and *Escherichia coli* from consumption of MPVFs has been reported worldwide (Sant'Anna et al. 2020; Agarwal et al. 2018). Different preservation methods practiced, such as drying, freezing, modified atmosphere storage, and others, do not yield most of the requirements of consumers, and therefore, the development of novel approaches has been discussed in past decades. Nanotechnology is a fast-escalating area with nanomaterials in various science fields and is immensely applied in the food industry as well. Nanomaterials can be used as antimicrobials and antioxidants and can be used to develop different sensors to detect, quantify, and extract contaminants, including heavy metals, dyes, pathogens, and toxins (Sant'Anna et al. 2020). The advent of nanotechnology in preserving minimally processed foods reduces food waste and upholds taste, freshness, appearance, and nutritional value. Most NPs such as magnesium oxide, TiO_2, Ag, and ZnO demonstrate excellent antimicrobial abilities (Al-Naamani et al. 2018; Zhu et al. 2014; Costa et al. 2011; Smijs and Pavel 2011; Wang et al. 2017). ZnONPs are semiconductors with unique physicochemical properties such as high photon absorption capacity, efficient transport of charge carriers, and antimicrobial and antifungal activities (Mooss et al. 2019; Xia et al. 2008; Sruthi et al. 2018). ZnONPs have the advantage of being less toxic to humans than other NPs, and the previously mentioned abilities depend on size, shape, and concentration (Kalpana and Devi Rajeswari 2018; Siddiqi et al. 2018; Berekaa 2015).

This chapter outlines the latest studies of ZnONP-mediated minimally processed horticulture products, packaging, coatings, different safety measurements, consumable concentrations, and toxicological and ecotoxicological risks.

14.2 ZnONPs

Versatile ZnONPs are biocompatible, inexpensive to produce, environmentally safe, physically and chemically stable, easily available, capable of catalyzing/oxidizing environmental pollutants when exposed to UV irradiation, and "GRAS" materials (Sabir et al. 2014). These particles received considerable interest in various fields, as these are n-type semiconductors with a wide bandgap of 3.37 eV and high excitation binding energy (60 meV) at room temperature (Sardella et al. 2019) and thus are promising candidates for developing piezo-electronics and optoelectronics devices (Wijesinghe et al. 2021; Sabir et al. 2014). ZnONPs have multifaceted applications due to their unique chemical and physical properties, such as stability at neutral pH, electronic band structures, mechanical properties, inclusive of piezoelectric constant and elastic constant, lattice dynamics, high thermal stability, electrical and thermal conductivity, optical absorption and transparency, bactericidal effect, and luminescent properties in the near-UV and the visible regions (Wijesinghe et al. 2021; Sabir et al. 2014). ZnONPs exhibit versatile applications in UV light emitters, catalysis, transparent and spin electronics, piezoelectric devices, antibacterial agents, gas sensors, biosensors, and storage (Sardella et al. 2019).

14.2.1 Mechanisms of Nanotoxicity in Bacteria

Due to the active target potential and photocatalytic properties, ZnONPs are one of the most effective antibacterial agents against Gram-negative and Gram-positive bacteria (Joe et al. 2017; Sruthi et al. 2018). The capacity of ZnONPs to penetrate cell membranes, which increases their antimicrobial activity, mainly depends on concentration, surface defects, shape, size, surface charge, pH, and temperature (Agarwal et al. 2018). Agarwal et al. (2018) proved that between spherical, rod, and triangular shape ZnONPs, triangular-shaped NPs showed better interaction with the cell membrane and higher toxicity against bacteria. Moreover, they demonstrated that the size of ZnONPs is inversely proportional, and the concertation of ZnONPs is directly proportional to the antibacterial activity as smaller particles easily penetrate through the membrane and leak intracellular components, leading to cell death (Berekaa 2015; Joe et al. 2017). Electrostatic attractions cause positively charged ZnONPs to better interact with the negatively charged cell membrane, improving nano-toxicity toward bacteria. Acidic pH and higher temperatures increase the dissolution and solubilization of ZnONPs, leading to higher antibacterial activity (Agarwal et al. 2018; Joe et al. 2017).

The antibacterial ability of ZnONPs is mainly accredited to the soluble zinc ions (Zn^{2+}) coming from ZnONPs. However, exact antibacterial mechanisms associated with ZnONPs have not been fully elucidated, although scientists have identified three probable mechanisms. Inhibition of the efflux pump is one of the primary mechanisms preventing the transportation of antibiotics inside the cell and removing toxic compounds and waste products from the cell, causing the accumulation and increasing toxicity inside the cell leading to cell expiry (Agarwal et al. 2018; Mooss et al. 2019; Joe et al. 2017). The generation of ROS like oxygen radicals and/or H_2O_2, when exposed to UV-Visible light, has been identified as another mechanism (Wijesinghe et al. 2021). ROS leads to alteration of DNA, protein structure, and lipid structure, and the mechanisms underlying the photocatalytic biocidal activity are outlined in Figure 14.1. When ZnONPs are irradiated under UV-Visible light, electrons (e^-) are excited from the valence band to the conduction band, subsequently leading to the formation of energized holes (h^+) in the conduction band. The photoexcited e^- and h^+ can diffuse to the photocatalyst (ZnONPs) surface reacting with the adsorbed molecules (O_2 and H_2O). The superoxide radical anions ($O_2^{\bullet -}$) and hydroxyl radicals ($^{\bullet}OH$) are derived from

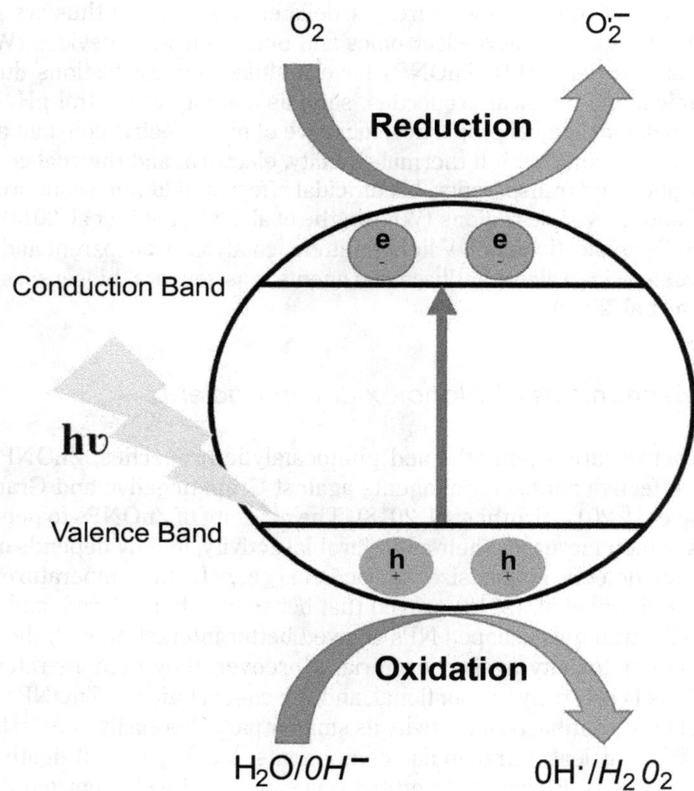

Figure 14.1 Schematic representation of mechanism of photocatalysis.

e^- and h^+, respectively, react with surface-adsorbed food adulterants/contaminants or microorganisms and undergo degradation.

The release and solubilization of Zn^{2+} from ZnONPs inside the cell are other mechanisms, and they prevent bacterial adhesion and colonization and cause variation in carbohydrate metabolism and biogenetic leakage of intracellular components. Moreover, Zn^{2+} retard mitochondria activities, causing respiratory failures and hinder metabolic enzymes in the gastrointestinal tract, disrupting the assimilation of food (Mooss et al. 2019; Joe et al. 2017). The release and dissolution of Zn^{2+} from ZnONPs are affected by the size and shape of the ZnONPs as they are critical factors of the surface area, determining the antibacterial activities as it is directly proportional to the antibacterial capacity. The size of ZnONPs is inversely proportional to the release and dissolution of Zn^{2+} (Agarwal et al. 2018; Rajput et al. 2018). Agarwal et al. (2018) showed that spherical ZnONPs release Zn^{2+} more effectively than flower and rod shapes. The previously mentioned three probable mechanisms proposed by the scientists are depicted in Figure 14.2.

Due to different membrane structures, gram-negative and gram-positive bacteria adopted different mechanisms. Gram-positive bacteria consist of a rigid cell membrane due to a thick layer of peptidoglycan with teichoic and lipoteichoic acids, and these acids act as chelating agents (Berekaa 2015; Sruthi et al. 2018; Rajput et al. 2018). Gram-negative bacteria have a triple layer of peptidoglycan, and the outer layer is composed of porins. Therefore, porins will facilitate the submissive circulation of ZnONPs inside the cell (Agarwal et al. 2018; Joe et al. 2017). Gram-negative bacteria are more vulnerable to ZnONP toxicity due to thin layers that govern a deficient resistance level. ZnONPs show complete inhibition of *Escherichia coli* (gram-negative) at ≥3.4 mM and *Staphylococcus aureus* (gram-positive) at ≥1 mM. ZnONPs also inhibit

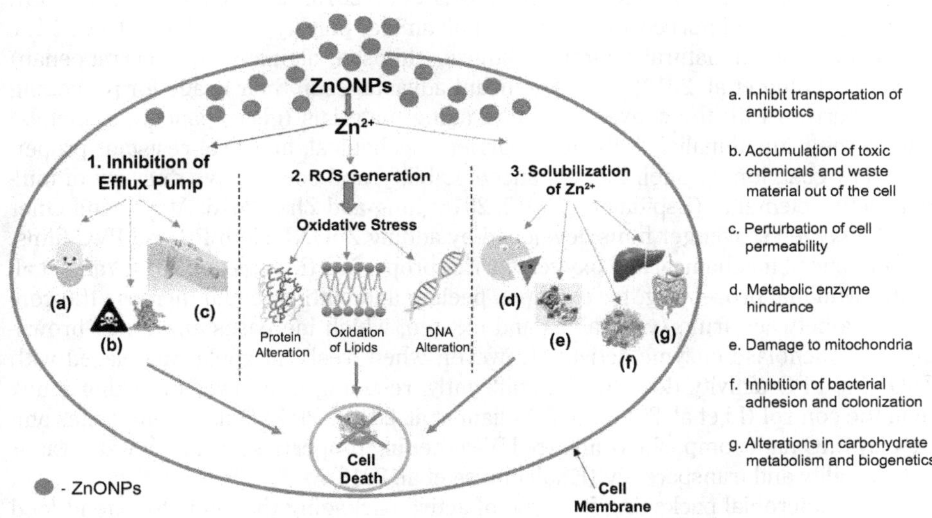

Figure 14.2 Mechanisms of antimicrobial activity of ZnONPs.

Chlorella vulgaris growth at 50 mg L^{-1} (Sruthi et al. 2018; Rajput et al. 2018; Suman et al. 2015).

14.3 Nano-Sized ZnONP-Based Active Packaging

Although MPVFs have many advantages, several factors diminish their storage time and shelf life. This is due to their high moisture content (>80%; Hasan et al. 2019) and fragile texture, which retard the quality and product perishability (high metabolic rate and oxidative reactions) and increases postharvest deterioration (Avella et al. 2007; Jung and Zhao 2016; Naknaen 2014). Therefore, the main aim of the MPVF industry is to extend shelf life while upholding the quality and nutritional factors (Hasan et al. 2019; Agarwal et al. 2018). The adoption of "good manufacturing practices" is necessary to maximize the preservation of prepared fruits and vegetables with minimal changes in their nutritional, sensorial, and flavor qualities (Al-Naamani et al. 2018; Espitia et al. 2012; Jung and Zhao 2016). Food packaging featuring innovative techniques/technology called "active packaging", has a desirable role in food preservation. Active packaging, defined by European Union's 1935/2004/E.C. and 450/2009/E.C. regulations, is an active material in contact with food, with the ability to change the composition of the food or its atmosphere (Restuccia et al. 2010).

Among all the NPs, ZnONPs are the most widely used components found in active food packaging, with higher biocidal effects on different species of food spoilage-related microorganisms. These particles abolish enzyme activities, disrupt membrane function, or damage the DNA of microbial cells and are capable of improving food stability by releasing desirable molecules like antioxidant agents or act as gas scavengers/absorbers (CO_2, ethylene, moisture and/or odor, and flavor taints) and/or maintain temperature control and thus have a broad range of applications in food science (Espitia et al. 2016; Jung and Zhao 2016; Restuccia et al. 2010; Realini and Marcos 2014). ZnO nanofillers reinforced in synthetic (polyamide, polystyrene, LDPE, PVC, PLA, polyolefin) and/or natural (starch, cellulose, chitosan, alginate, agar, carrageenan) polymers (Yang et al. 2010) have significant advantages in their usage for packaging in comparison with the conventional packaging materials (glass, paper, and metals) due to their functionality (enhanced barrier, mechanical, and heat-resistant properties), low cost, longer shelf life, enhanced stability, and consecutive delivery of multiple active elements (Espitia et al. 2012, 2016; Jung and Zhao 2016; Atayev and Oner 2014). Oxygen-scavenger films developed by adding ZnONPs into PP, and PVC films, had enhanced mechanical and oxygen barrier properties (Meng et al. 2014; Yang et al. 2010). Minimal processing, for example, peeling and cutting, could increase the contact area between fruits/vegetables and oxygen, which increases enzymatic browning and phenolase enzyme activity. However, when fresh-cut apple was placed with ZnONPs, PPO activity decreased significantly, resulting in a lower browning index than the control (Li et al. 2011, 2017; Yuliani et al. 2019). PP-ZnO nanocomposites and PP-clay-ZnO nanocomposites enhance UV-screening properties and affect water vapor permeability and transparency (Khoirunnisa et al. 2018).

Antimicrobial packaging is a type of active packaging that aims to extend food shelf life by reducing, inhibiting, or retarding the growth of microorganisms that may be present on minimally processed food surfaces (Espitia et al. 2012, 2016). ZnONPs display excellent antimicrobial activity, which allows the gradual diffusion of target bactericidal or bacteriostatic compounds into a food matrix in the form of sachets or films (Restuccia et al. 2010). From the previous study, it was demonstrated that polymeric nanocomposite film embedded with ZnONPs could be used in minimizing the growth of bacterial foodborne pathogens such as *Bacillus subtilis*, *Salmonella enteridis*, *Staphylococcus aureus*, *Escherichia coli*, *Pseudomonas aeruginosa*, *Campylobacter jejuni*, and *Lactobacillus plantarum* (Emamifar et al. 2011; Carlesso et al. 2018; Al-Naamani et al. 2018). Moreover, ZnONPs exhibit antifungal activity against *Botrytis cinerea*, *Penicillium expansum*, *Aspergillus nigar*, *Aspergillus flavus*, *Fusarium oxysporum*, *Penicillium notatum*, *Nigrospora oryzae*, and *Sclerotinia homoeocarpa* and extend the shelf life of food (Espitia et al. 2016; Hoffmanna et al. 2019; Ruffo Roberto et al. 2019). In a recent study, Ruffo Roberto et al. 2019 reported that PVC-ZnO nanocomposite showed biocidal activity against avocados, bananas, grapes, and soybeans. By affecting bacterial growth kinetics directly, ZnONP-loaded starch-coated polyethylene films showed promising *E. coli* inhibition (Tankhiwale and Bajpai 2012). Different bionanocomposites obtained by adding ZnONPs to agar, carrageenan, and CMC to develop antimicrobial nanocomposites inhibited the growth of various gram-negative and gram-positive foodborne pathogens (Kanmani and Rhim 2014). Besides biodegradable polymers, ZnONP-incorporated synthetic matrices such as polyethylene, PP, and LDPE have been developed (Ding et al. 2012; Atayev and Oner 2014; Tankhiwale and Bajpai 2012). Examples of different types of active packaging and their applications for minimally processed products are stated in Table 14.1.

Table 14.1 ZnONP-Incorporated Active Food Packaging Systems

Polymer matrix	*Tested fruit/ vegetable*	*Advantages*	*References*
ZnONPs	Caixin	Controlled total bacterial count and improved the product quality of a vacuum-packaged product	Liu et al. 2014
ZnONPs—Chitosan—Cellulose acetate phthalate	Black grape	Exhibited excellent UV-shielding ability, antimicrobial properties, biodegradability, optimum tensile strength, and stiffness; provided superior barrier to atmospheric agents; low surface wettability; and minimal invasion of food-spoiling bacteria and fungi on the fruit's extended shelf life up to 9 days	Indumathi et al. 2019
ZnONPs	Carrot	Controlled total bacterial count and improved the product quality of a vacuum-packaged product	Xu et al. 2017
ZnONPs–polybutylene succinate	Fresh-cut apple	Extended the shelf life for at least 18 days	Naknaen 2014

(Continued)

Table 14.1 (Continued)

Polymer matrix	*Tested fruit/ vegetable*	*Advantages*	*References*
ZnONPs–LDPE–AgNPs	Orange juice	Reduced the rate of microbial growth of *Lactobacillus plantarum*	Emamifar et al. 2011
ZnONPs–LDPE–AgNPs	Fresh-cut lotus root	Increased shelf life without any negative effects on sensory properties	Ding et al. 2012
ZnONPs–LDPE	Strawberry	Increased the shelf life and reduce microbial growth	
ZnONPs–PVC	Fresh-cut Fuji apple	Reduced fruit decay rate, slowed down the accumulation of malondialdehyde and ethylene content, maintained total soluble solids and titratable acid levels, and inhibited PPO and pyrogallol peroxidase activity	Li et al. 2011
ZnONPs–LDPE–chitosan	Okra	Reduced the total bacterial and fungal concentration and maintained quality attributes of the okra (pH, total soluble solids, moisture content, and weight loss)	Al-Naamani et al. 2018
ZnONPs–carrageenan	—	Increased water vapor barrier properties	Khoirunnisa et al. 2018
ZnONPs–AgNPs–chitosan	—	Decreased the growth of *Staphylococcus aureus* and *Escherichia coli*	Kaur et al. 2020
ZnONPs–PLA	—	Increased the elongation at break and elastic modulus; decrease tensile strength; exhibit good thermostability; higher water vapor permeability, opacity, and lower oxygen permeability	Li et al. 2017
ZnONPs–soybean polysaccharide	—	Showed good antibacterial *E. coli, Bacillus cereus*, and *S. aureus*), antimold (*Penicillium expansum*) activities, increased matrix tensile strength and elongation at break.	Salarbashi et al. 2016
ZnONPs–PP	—	Showed strong antibacterial activities against both gram-positive *S. aureus* and gram-negative *E. coli.*	Paisoonsin et al. 2013
ZnONPs–PP–clay	—	Enhanced O_2 barrier properties	Atayev and Oner 2014

14.4 Nanolaminated Coating for Preservation of Minimally Processed Horticulture

The coating can be formulated from a variety of substances in the form of liquid or viscous solutions. They are porous enough to permit respiration, be invisible, be capable of uniformly covering the surface, and improve the appearance (shine or gloss) of the coated products. Moreover, coatings sufficiently adhere to fruits/vegetables, preventing the removal of the applied coating during handling and transport. Coatings can be applied by different methods such as panning, spraying, dipping, or immersing the surface of whole or minimally processed foods to create thin layers of biopolymers (Guimarães et al. 2020; Kumar et al. 2020). Edible coatings are the most popular as they are biodegradable, renewable, available in nature, and harmless to the environment compared to conventional synthetic polymers (Anugrah et al. 2020; Kumar et al. 2020). In addition, these can be synthesized from hydrophilic and hydrophobic bases, such as polysaccharides (starch, xanthan gum, cellulose), proteins (gelatin, whey protein, zein), lipids (beeswax, carnauba wax), and resins to inhibit the deterioration of fresh-cut products (Anugrah et al. 2020; Dhall 2013; Guimarães et al. 2020; Kumar et al. 2020). Due to their edibility and excellent biocompatibility, hydrophilic bases are commonly used for MPVFs.

The incorporation of ZnONPs into coating material enhances good thermal, mechanical, structural, and photocatalytic properties, as well as controls the internal atmosphere of moisture and gases (O_2, CO_2, volatile substances), retard the respiration rate, and serve as a water vapor barrier, thereby minimizing dehydration. Moreover, ZnONPs act as a superior enzymatic anti-browning and antimicrobial substance (Anugrah et al. 2020; Guimarães et al. 2020; Jagadish et al. 2018), thereby lessening the deprivation of minimally processed fruits/vegetables and consequently increasing their postharvest shelf life (Anugrah et al. 2020; Yuliani et al. 2019). ZnO nanocomposite formulations prepared from chitosan, PLA, and alginate applied on guavas, apples, and strawberries indicated reduced weight loss, indicating reduced water loss, the ripening process, and the ripening index ratio (soluble solid/titratable acid) compared to the uncoated fruit. The firmness and color of the coated fruit were better maintained, which indicated enhanced storage life (Li et al. 2011, 2017). Thus, it was demonstrated that ZnO additives act as antioxidants and anti-ripening compounds and enhance organoleptic characteristics, functionality, and barrier capability of coatings. According to the researchers, the application of ZnONPs on fresh-cut MPVFs prolonged their shelf life without impairing sensory characteristics, enhanced overall quality, maintained structural integrity, improved the resistance of coated produce to mechanical injuries, and retarded of ripening and water loss without altering the color, taste, or original aromas. In addition, these are sufficiently cheap, nontoxic, and safe for consumption (Dhall 2013; Guimarães et al. 2020; Anugrah et al. 2020). Figure 14.3 depict the general flow diagram of the edible coating treatment of minimally processed fruit by using ZnONPs, and Table 14.2 illustrates the polymeric bases incorporated with ZnONPs used for coating formation and the function of such coating in minimally processed foods.

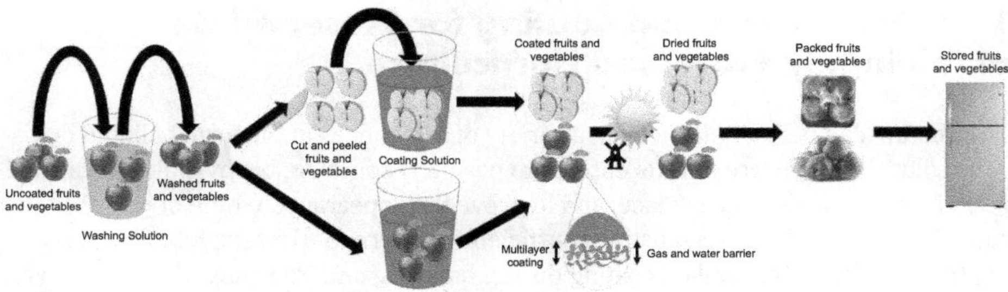

Figure 14.3 Schematic representation of minimally processed fruits coating treatment.

Table 14.2 Summary of Recent Studies on ZnONP Edible Food Coatings

Polymer matrix	*MP fruit/vegetable*	*Effect of the coating*	*References*
ZnONPs	Sweet potato (*Ipomea batatas*)	Protected from infections against *Rhizopus stolonifer*, maintained quality, and increased shelf life for up to 2 months in storage	Nafady et al. 2019
ZnONPs–Chitosan	Fresh-cut papaya (*Carica papaya. L*)	Reduced microbial growth of *Escherichia coli* and *Staphylococcus aureus*, increased shelf life up to 12 days	Lavinia et al. 2020
ZnONPs–PLA	Fresh-cut apple (*Malus domestica*)	Inhibited ethylene production, microbial growth, and browning; maintained tissue firmness, total phenolic, and sensory values; reduced PPO activity; and enhanced shelf life up to 6 days	Li et al. 2011, 2017
ZnONPs–chitosan–calcium gluconate	Strawberry (*Fragaria X ananassa*)	Inhibited microbial growth of *Botrytis cinerea* and *Rhizopus stolonifer*; delayed weight loss and maintained the functional nutrients of strawberries	Hernandez-Munoz et al. 2008
ZnONPs combined Ar gas treatment	Fresh cut orange *(Citrus X sinensis)*	Effectively reduced mass loss, titratable acid, and ascorbic acid contents; effectively inhibited total aerobic bacterial and fungal growth; and increased soluble solids	Wu et al. 2020
ZnONPs–LDPE	Strawberry (*Fragaria X ananassa*)	Maintained nutrients, anthocyanins, and soluble solids	Carlesso et al. 2018
ZnONPs–cassava starch–stearic acid	Fresh-cut mango (*Mangifera indica L*)	Maintained the quality of fresh-cut mango during the storage periods, suppressed weight loss and microbial counts (*Aspergillus niger*).	Yuliani et al. 2019

Polymer matrix	*MP fruit/vegetable*	*Effect of the coating*	*References*
ZnONPs–Alginate	Strawberry (*Fragaria X ananassa*)	Increased shelf life up to 20 days, reduced microbial growth and weight loss, better maintained firmness, lower increases in soluble solid and peroxidase activity, and lowered decreases in acidity, anthocyanin, phenolic, antioxidant activities, and superoxide dismutase activity	Emamifar et al. 2011
ZnONPs–alginate–chitosan	Guava (*Psidium guajava L*)	Reduced weight loss, color, ripening index ratio and *Phyllosticta psidicola* growth; no external injuries until end of storage and firmness were better maintained	Arroyo et al. 2020
ZnONPs–chitosan	Okra or ladyfinger (*Abelmoschus esculentus*)	Maintained the quality of the packed okra (pH, total soluble solids, moisture content, and weight loss), retarded microbial and fungal growth, and increased shelf life from 5 days to 12 days under room temperature storage	Al-Naamani et al. 2018
Ultrasound treatment with ZnONPs–chitosan–acetic acid solution	Fresh-cut kiwifruit (*Actinidia* sp.)	Maintained quality parameters and sensory quality (water loss, total soluble solids, titratable acidity, color, and firmness); reduced CO_2, ethylene production, and microbial population	Meng et al. 2014
ZnONPs combined with radio frequency heating	Carrot (*Daucus carota subsp. sativus*)	Controlled microorganism growth, improved product quality (hardness, color, carotenoids, microstructure), and extended the shelf life to nearly 60 days	Xu et al. 2017
ZnONPs–chitosan	Cut carrot (*Daucus carota subsp. sativus*)	Photocatalytically generated ROS, leading to high anti-diatom and antibacterial (*E. coli* and *S. aureus*) activities under sunlight and enhanced carrot pieces' shelf life by 9 days	Kumar et al. 2020
ZnONPs–alginate–chitosan	Guava (*Psidium guajava L*)	Firmness is better maintained and prevented external injuries	Arroyo et al. 2020

(Continued)

Table 14.2 (Continued)

Polymer matrix	*MP fruit/vegetable*	*Effect of the coating*	*References*
ZnONPs–carrageenan	Mango (*Mangifera indica L*)	Reduced weight loss and CO_2 production, and better maintained total acidity, color, and textural appearance	Meindrawan et al. 2018
ZnONPs–CMC	Pomegranate arils (*Punica granatum*)	Reduced weight loss, vitamin C loss, anthocyanin and phenolic content loss, total yeast and mold concentration, and higher antioxidant activities	Koushesh Saba and Amini 2017
ZnONPs–pectin	Starfruit (*Averrhoa carambola*)	Reduced weight loss, browning index, redness value, and physical damage	Anugrah et al. 2020; Widodo et al. 2020
ZnONPs–xanthan gum	Tomato (*Solanum lycopersicum*), Apple (*Malus domestica*)	Exhibited no weight loss and improved storage quality of fruits and vegetables	Joshy et al. 2020
ZnONPs–cellulose	Salak pondoh (*Salacca zalacca*)	Decreased the amount of total processed salicylic acid content; reduced the rate of evaporation of water vapor, fruit weight during storage, and the process of decomposition of complex compounds	Widodo et al. 2020

14.5 Intelligent Packaging for Food Freshness Detection

Ready-to-use or MPVFs are usually packed directly after harvesting and checked for quality prior to consumption. Therefore, advanced packaging systems are needed to address food safety—related problems and bioterrorism (Ding et al. 2012; Restuccia et al. 2010; Kumar et al. 2017). Smart/intelligent packaging uses various smart devices like labels, barcodes, time–temperature indicators, ripeness indicators, toxin indicators, gas indicators, and biosensors, which sense, detect, or monitor the physicochemical properties of food or its interior packaging environment parameters such as pH, temperature, and chemical contaminants (Kumar et al. 2017; Mihindukulasuriya and Lim 2014; Restuccia et al. 2010). Electro-polymerization and the development of ZnO nanosensors to detect contaminants present in food (bacteria, toxins, adulterants), volatile compounds (ethylene, ethanol, gaseous amine), and microorganisms are less time-consuming and more sensitive than other conventional methods that are able to notify consumers of the quality of minimally processed foods (Ansari et al. 2011; Restuccia et al. 2010; Sánchez-Acevedo et al. 2009).

ZnO nanocomposites in the form of printing inks (screen printing, flexography, inkjet, pad printing) are suitable for manufacturing new nanophotonic devices for packaging by printing techniques. ZnONPs have been used for barcoding and are in high demand for the simultaneous analysis of multiple targets. Due to their nanoscale dimensions, nano-barcodes have been implemented based on certain encoding elements including optical (fluorescent and nonfluorescent) with different shapes and sizes (Shikha et al. 2017). High-luminescent transparent films were obtained from the colloidal suspension of ZnONPs and PVP and stabilized by silicic acid anhydride and PVP in polymethylmethacrylate (Sarapulova et al. 2015). The compositions are nontoxic, safe, and suitable to be applied to the inner surface of active and intelligent packaging to obtain MPVFs' quality information for producers and consumers (Sarapulova et al. 2015). ZnONPs, being excellent catalysts, exhibit selective and sensitive gas-sensing properties toward different gases like ethanol, ethylene, acetone, and H_2O_2. Numerous reports indicated the use of catalysts or promoters by doping, impregnation, and fixation of ZnONPs to metal NPs or the polymer matrix. AgNP-embedded ZnO nanorods had an enhanced ethanol-sensing property compared with the traditional-doped ZnO or pure ZnO (Xiang et al. 2010). The development of pH sensing probes increases consumer acceptance as it ensures the safety of MPVFs. The composite polymeric film fabricated by incorporating ZnONPs, polyvinyl alcohol, starch, nutmeg oil, and Jamun extract showed effective pH sensing ability and is used as a pH sensor, utilizing colorimetric visual methods using anthocyanins pigments (Jayakumar et al. 2019). These biosensors also exhibited enhanced water barrier, UV barrier, mechanical properties, and antimicrobial properties (Jayakumar et al. 2019). Indicators formulated from ethylene-vinyl acetate nanocomposites containing ZnONPs, montmorillonite decorated with iron NPs, and rosemary extract with anthocyanin exhibited enhanced oxygen-absorbing capacity during storage of MPVFs. It also showed that enhanced antioxidant activity, thermal stability, and mechanical properties, and the color of anthocyanin films change with the alteration in pH of the surrounding environment, revealing the film's ability to visually display the deterioration of foodstuffs (Eskandarabadi et al. 2019). Due to the high mobility of conduction electrons, both ZnO and tin dioxide (SnO_2) are identified as good gas-sensing materials, used to measure both oxidative and reductive target gases. SnO_2-ZnO core–shell nanowires and SnO_2- $ZnSnO_3$ (zinc stannate) core–shell microsphere effectively sense ethanol gas with lower detection limits (Zhu et al. 2014). Table 14.3 depicts some of the additional intelligent food packaging systems modified with the use of ZnONPs.

Table 14.3 ZnONP-Based Nanosensors for Minimally Processed Produce

Type of NPs	*Method/Technique*	*Analyte*	*References*
ZnONPs	Electrochemical: cyclic voltammetry and impedance	Ochratoxin-A	Ansari et al. 2011
ZnO–SWCNTs–MWCNT nanocomposite	Electrochemical: cyclic voltammetry	Bisphenol A and Sudan I	Sánchez-Acevedo et al. 2009; Najafi et al. 2014
ZnONPs–MWCNT	Electrochemical: cyclic voltammetry and field-effect transistor	Bisphenol A, cadmium ions, sunset yellow, and tartrazine	Sánchez-Acevedo et al. 2009

(Continued)

Table 14.3 (Continued)

Type of NPs	*Method/Technique*	*Analyte*	*References*
ZnONPs	DNA probe sensor	*Escherichia coli* and *Staphylococcus aureus*	Kumar et al. 2017
ZnONPs–TiO_2NPs	Surface plasmon resonance	Microbes, pests, nutrient content, and pesticides	Baruah and Dutta 2009; Prasad and Kumar 2014
ZnONPs	Gas sensors	Ethanol	Ruchika 2015
ZnONPs–p-ABSA (p-aminobenzenesulfonic acid)	Voltametric sensor	Tartrazine	Karim-Nezhad et al. 2017
Ethylene-vinyl acetate–montmorillonite–ZnONPs	Time–temperature indicator	Film's ability to visually display the deterioration of foodstuffs	Eskandarabadi et al. 2019

14.6 Regulatory Aspects of ZnONPs

14.6.1 Legislation

Since ZnONPs have become a part of the food industry, such as in coatings and packaging, these are listed as GRAS by the USFDA. ZnONPs are nontoxic and safe, and therefore, USFDA recommends using ZnONPs as additives (Meng et al. 2014), and the European Food Safety Authority concluded that there are no adverse effects on eukaryotic cells below the upper limit of consumption (Meng et al. 2014; Siddiqi et al. 2018). A lack of knowledge about the safety regulations for both ZnONPs and Zn^{2+} leads to several problems. ZnO can release Zn^{2+}; therefore, regulations for Zn^{2+} should also be considered. Different criteria are available for safety measurements, giving different upper limits for Zn^{2+}, and these upper limits depend on the age, weight, and gender of the consumer. As age and weight increase, upper limits gradually increase, and men have more tolerance than women (China and Carvalho 2004). WHO, FNB/IOM, and the IZiNCG stated that different upper limits or NOAELs for zinc intake by life stage (China and Carvalho 2004).

14.6.2 Consumer's Health and Safety

According to the 2012 Scientific Committee on Consumer Safety report of the European Commission, ZnONPs show limited risk to consumers when they are present at a 25% concentration and its safe up to 500 µg per mL (Beegam et al. 2016; Sarkar et al. 2014). A study reported NOAEL for ZnO and Zn^{2+} as 50 and 0.83 mg per day, respectively (Subramaniam et al. 2019). The upper limit of Zn^{2+} by WHO, FNB/IOM, and IZiNCG is 13–48, 4–40, and 6–44 mg per day, respectively (China and Carvalho 2004).

The US Environmental Protection Agency expressed the bioaccessibility of Zn is 300 μg per kg body weight per day while the estimated daily intake is to be 8.5 μg per kg body weight per day. Overconsumption of regulatory threshold limits will cause significant harm to humans and wildlife (China and Carvalho 2004).

14.7 Toxicity of ZnONP-Incorporated Food Systems

Human exposure to ZnONPs has increased rapidly due to their vast application in various industries such as cosmetics, food, and daily care products (Sruthi et al. 2018). The enrichment of the shelf life of food by sustaining its quality, freshness, and safety can be attained with antibacterial additives with the ZnONPs incorporation. The use of ZnONPs must conform to consumer safety, environmental, ethical policy, and guideline matters, which is crucial (Šimon et al. 2008). Both ZnONPs and Zn^{2+} are toxic to humans and the environment, but the toxicity of ZnONPs is mainly based on the dissolution of ZnONPs into Zn^{2+} that is involved in the oxidant-injury paradigm (Xia et al. 2008). The toxicity of ZnONPs depends upon physicochemical properties like size, shape, concentration, route of exposure, chemical composition, aggregation, pH, the salinity of aqueous exposure medium, and coating material (Bai et al. 2010; Rajput et al. 2018; Sruthi et al. 2018). As the size of ZnONPs decreases, toxicity increases because they can penetrate through the cell rapidly, which increases the ZnONP quantity inside the cell. Studies reported nanorods and nano-sticks shapes show higher toxicity than other shapes, such as flowers and spheres. Toxicity on secondary organs mainly depends on the route of exposure (Sruthi et al. 2018). For example, ZnONPs entered via injection do not lead to neurotoxicity, but ingestion and inhalation cause neurotoxicity (Sruthi et al. 2018). The toxicity of ZnONPs is time- and dose-dependent, and generally, toxicity, dosage, and time have a direct relationship. However, few studies show an inverse relationship with different organs (Pasupuleti et al. 2012; Sruthi et al. 2018). The coating reduces the toxicity of ZnONPs as it acts as a barrier for the dissolution of ZnONPs into Zn^{2+} and surface contact with the cell and restricts the formation of radicals (Sruthi et al. 2018). However, this can vary depending on the substance in the coating material.

NPs enter the human body through different methods, and the bioavailability depends on the route of exposure (Agarwal et al. 2018). Administration via the gastrointestinal tract generates partial absorption of ZnONPs, and intracutaneous inoculation causes the direct entrance into the systematic transmission. The solubility, circulation, absorption, and release of Zn^{2+} from ZnONPs inside the body depend on the particle size, charge, and concentration. However, negatively charged ZnONPs have high absorption efficiency than positively charged ZnONPs, and therefore, positively charged ZnONPs are comparatively low in toxicity (Agarwal et al. 2018).

The migration of NPs from packaging to food, which allows oral exposure, should be a critical factor in packaging production. Migration is a function of the size and shape of ZnONPs, polymer material, food stimulants, temperature, and storage time. Food stimulants such as oily, acidic, and aqueous conditions show significant migration, which increases with storage time and temperature. However, the tendency of ZnONPs to migrate to food is low compared to other metal-oxide NPs (Liu et al. 2016).

Leaching Zn^{2+} from ZnONP-incorporated packaging materials into the environment cause the accumulation of Zn^{2+} in water streams, aquifers, soil, and sediments (Rajput et al. 2018). In the environment, ZnONPs remain as individual particles in the suspension, dissolve, aggregate, form larger particles and eventually sediment, transform chemically or biologically, and absorb onto water constituents (Beegam et al. 2016). Due to poor colloidal stability, ZnONPs tend to precipitate and increase the accumulation in soil and sediments. Accumulation of ZnONPs and Zn^{2+} could alter the pH, hardness, and dissolved organic carbon content in water bodies, which increases the toxicity levels more than desired levels causing undesirable environmental pollution (Al-Naamani et al. 2018). Upper levels of ZnONPs for soil and water are 3.1–31 µg per kg and 76–760 µg per L, respectively (Rajput et al. 2018). Wildlife and aquatic life can consume free Zn^{2+} and ZnONPs, causing translocation and accumulation and eventually entering the food chain. The buildup of Zn^{2+} and ZnONPs in plants causes the prevention of seed germination and root growth, differences in plant morphology, the reduction of biomass and leaf, damage to plant hormones, and many more, which is a function of the shape and size of ZnONPs and the nature of the plant (Imran et al. 2010; Al-Naamani et al. 2018). These lead to reduced water uptake, plant growth, and inhibition of chlorophyll synthesis, leading to wildlife expiry (Imran et al. 2010; Al-Naamani et al. 2018).

14.8 Conclusion

The change in the ever-growing human population's dietary habits requires efficient, more environmentally friendly, and sustainable methods to lessen perishable food spoilage to meet daily requirements. Apart from traditional methods, the advent of nanomaterials showed remarkable potential against microbial spillages. As a recent advancement of food and agriculture, ZnONP- and nanocomposite- (ZnONP-mediated polymers or natural matrices) incorporated coatings and food packings showed excellent antimicrobial and antifungal properties via different mechanisms. Depending on the process and environmental conditions, different matrices show various properties. However, toxicity, consumable concentrations, and health effects have to be the determining factors for when these are implemented.

References

Agarwal, Happy, Soumya Menon, S Venkat Kumar, and S Rajeshkumar. 2018. Mechanistic study on antibacterial action of zinc oxide nanoparticles synthesized using green route. *Chemico-Biological Interactions* 286:60–70.

Al-Naamani, Laila, Joydeep Dutta, and Sergey Dobretsov. 2018. Nanocomposite zinc oxide-chitosan coatings on polyethylene films for extending storage life of okra (Abelmoschus esculentus). *Nanomaterials* 8 (7):479.

Ansari, S, E Hadavi, M Salehi, and P Moradi. 2011. Application of microorganisms compared with nanoparticles of silver, humic acid and gibberellic acid on vase life of cut gerbera goodtimming. *Journal of Ornamental and Horticultural Plants* 1 (1):27–33.

Anugrah, Daru Seto Bagus, Hugo Alexander, Rianita Pramitasari, Dwi Hudiyanti, and Christyowati Primi Sagita. 2020. A review of polysaccharide-zinc oxide nanocomposites as safe coating for fruits preservation. *Coatings* 10 (10):988.

Arroyo, Betty Jarma, Anderson Campos Bezerra, Lara Lins Oliveira, Sara Jarma Arroyo, Enayde Almeida de Melo, and Andrelina Maria Pinheiro Santos. 2020.

Antimicrobial active edible coating of alginate and chitosan add ZnO nanoparticles applied in guavas (Psidium guajava L.). *Food Chemistry* 309:125566.

Atayev, Parahat, and Mualla Oner. 2014. Effect of incorporation of clay and zinc oxide nanoparticles on oxygen barrier properties of polypropylene sheet. *Journal of Plastic Film & Sheeting* 30 (3):248–265.

Avella, M, G Bruno, ME Errico, G Gentile, N Piciocchi, A Sorrentino, and MG Volpe. 2007. Innovative packaging for minimally processed fruits. *Packaging Technology and Science: An International Journal* 20 (5):325–335.

Bai, Wei, Zhiyong Zhang, Wenjing Tian, Xiao He, Yuhui Ma, Yuliang Zhao, and Zhifang Chai. 2010. Toxicity of zinc oxide nanoparticles to zebrafish embryo: a physicochemical study of toxicity mechanism. *Journal of Nanoparticle Research* 12 (5):1645–1654.

Baruah, Sunandan, and Joydeep Dutta. 2009. Nanotechnology applications in pollution sensing and degradation in agriculture: a review. *Environmental Chemistry Letters* 7 (3):191–204.

Beegam, Asfina, Parvathy Prasad, Jiya Jose, Miguel Oliveira, Fernando G. Costa, Amadeu M.V.M. Soares, Paula P. Gonçalves, Tito Trindade, Nandakumar Kalarikkal, Sabu Thomas, and Maria de Lourdes Pereira. 2016. Environmental fate of zinc oxide nanoparticles: risks and benefits. *Toxicology-New Aspects to This Scientific Conundrum* 5:81–112.

Berekaa, Mahmoud M. 2015. Nanotechnology in food industry; advances in food processing, packaging and food safety. *Int J Curr Microbiol App Sci* 4 (5):345–357.

Carlesso, LC, GLD Luz, CR Lajus, LL Silva, M Fiori, C Rossoni, SC Fernandes, and H Riella. 2018. Physical-chemical properties of strawberry pseudofruits submitted to applications of zinc oxide nanoparticles. *International Journal of Advanced Engineering Research and Science* 5 (7).

China, WE, and JCM Carvalho. 2004. Chapter 1 overview of zinc nutrition. *Food & Nutrition Bulletin* 25 (Suppl. 2):99S–129S.

Costa, Conte, Amalia Conte, Giovanna Giuliana Buonocore, and Matteo Alessandro Del Nobile. 2011. Antimicrobial silver-montmorillonite nanoparticles to prolong the shelf life of fresh fruit salad. *International Journal of Food Microbiology* 148 (3):164–167.

De Corato, Ugo. 2020. Improving the shelf-life and quality of fresh and minimally-processed fruits and vegetables for a modern food industry: as comprehensive critical review from the traditional technologies into the most promising advancements. *Critical Reviews in Food Science and Nutrition* 60 (6):940–975.

Dhall, RK. 2013. Advances in edible coatings for fresh fruits and vegetables: a review. *Critical Reviews in Food Science and Nutrition* 53 (5):435–450.

Ding, Yulong, Juan Yun, Xihong Li, Yao Tang, and Yuqian Jiang. 2012. Evaluation of nano-packing on the shelf life of fresh-cut lotus root (NelumbonuciferaGaerth). *Advances in Technology and Management*:775–780.

Emamifar, Aryou, Mahdi Kadivar, Mohammad Shahedi, and Sabihe Soleimanian-Zad. 2011. Effect of nanocomposite packaging containing Ag and ZnO on inactivation of Lactobacillus plantarum in orange juice. *Food Control* 22 (3–4):408–413.

Eskandarabadi, Somayeh Mahmoudi, Mehdi Mahmoudian, Kave Rahmani Farah, Arezoo Abdali, Ehsan Nozad, and Mojtaba Enayati. 2019. Active intelligent packaging film based on ethylene vinyl acetate nanocomposite containing extracted anthocyanin, rosemary extract and ZnO/Fe-MMT nanoparticles. *Food Packaging and Shelf Life* 22:100389.

Espitia, Paula Judith Perez, Nilda de Fátima Ferreira Soares, Jane Sélia dos Reis Coimbra, Nélio José de Andrade, Renato Souza Cruz, and Eber Antonio Alves

Medeiros. 2012. Zinc oxide nanoparticles: synthesis, antimicrobial activity and food packaging applications. *Food and Bioprocess Technology* 5 (5):1447–1464.

Espitia, PJP, CG Otoni, and NFF Soares. 2016. Zinc oxide nanoparticles for food packaging applications. *Antimicrobial Food Packaging*:425–431.

Guimarães, Maria Clara, Joyce Fagundes Gomes Motta, Dayana Ketrin Silva Francisco Madella, Lívia de Aquino Garcia Moura, Carlos Eduardo de Souza Teodoro, and Nathália Ramos de Melo. 2020. Edible coatings used for conservation of minimally processed vegetables: a review. *Research, Society and Development* 9 (8):e756986018.

Hasan, Mahmood Ul, Aman Ullah Malik, Sajid Ali, Amna Imtiaz, Anjum Munir, Waseem Amjad, and Raheel Anwar. 2019. Modern drying techniques in fruits and vegetables to overcome postharvest losses: a review. *Journal of Food Processing and Preservation* 43 (12):e14280.

Hernandez-Munoz, Pilar, Eva Almenar, Valeria Del Valle, Dinoraz Velez, and Rafael Gavara. 2008. Effect of chitosan coating combined with postharvest calcium treatment on strawberry (Fragaria× ananassa) quality during refrigerated storage. *Food Chemistry* 110 (2):428–435.

Hoffmanna, Tuany Gabriela, Daniel Amaral Petersa, Betina Louise Angiolettia, L Péres, M Reiter, and C De Souza. 2019. Potentials nanocomposites in food packaging. *Chemical Engineering* 75.

Imran, Muhammad, Anne-Marie Revol-Junelles, Agnieszka Martyn, Sávio Leandro Bertolia, Leonardo Vieira Péresb, Mercedes Gabriela Ratto Reiterc, and Carolina Krebs de Souzaa. 2010. Active food packaging evolution: transformation from micro-to nanotechnology. *Critical Reviews in Food Science and Nutrition* 50 (9):799–821.

Indumathi, MP, K Saral Sarojini, and GR Rajarajeswari. 2019. Antimicrobial and biodegradable chitosan/cellulose acetate phthalate/ZnO nano composite films with optimal oxygen permeability and hydrophobicity for extending the shelf life of black grape fruits. *International Journal of Biological Macromolecules* 132:1112–1120.

Jagadish, Krishnegowda, Yallappa Shiralgi, Bananakere N Chandrashekar, Bhadrapura L Dhananjaya, and Shivanna Srikantaswamy. 2018. Ecofriendly synthesis of metal/metal oxide nanoparticles and their application in food packaging and food preservation. *Impact of Nanoscience in the Food Industry*:197–216.

Jayakumar, Aswathy, KV Heera, TS Sumi, Meritta Joseph, Shiji Mathew, G Praveen, Indu C Nair, and EK Radhakrishnan. 2019. Starch-PVA composite films with zinc-oxide nanoparticles and phytochemicals as intelligent pH sensing wraps for food packaging application. *International Journal of Biological Macromolecules* 136:395–403.

Joe, Ara, Se-Ho Park, Kyu-Dong Shim, Da-Jung Kim, Kwang-Hwan Jhee, Hyo-Won Lee, Cheol-Ho Heo, Hwan-Myung Kim, and Eue-Soon Jang. 2017. Antibacterial mechanism of ZnO nanoparticles under dark conditions. *Journal of Industrial and Engineering Chemistry* 45:430–439.

Joshy, KS, Jiya Jose, Tianduo Li, Merlin Thomas, Aruna M. Shankregowda, Sreejith Sreekumaran, Nandakumar Kalarikkal, and Sabu Thomas. 2020. Application of novel zinc oxide reinforced xanthan gum hybrid system for edible coatings. *International Journal of Biological Macromolecules* 151:806–813.

Jung, J, and Y Zhao. 2016. Antimicrobial packaging for fresh and minimally processed fruits and vegetables. *Antimicrobial Food Packaging*:243–256.

Kalpana, VN, and V Devi Rajeswari. 2018. A review on green synthesis, biomedical applications, and toxicity studies of ZnO NPs. *Bioinorganic Chemistry and Applications* 2018.

Kanmani, Paulraj, and Jong-Whan Rhim. 2014. Properties and characterization of bionanocomposite films prepared with various biopolymers and ZnO nanoparticles. *Carbohydrate Polymers* 106:190–199.

Karim-Nezhad, G, Z Khorablou, M Zamani, P Seyed Dorraji, and M Alamgholiloo. 2017. Voltammetric sensor for tartrazine determination in soft drinks using poly (p-aminobenzenesulfonic acid)/zinc oxide nanoparticles in carbon paste electrode. *J Food Drug Anal* 25 (2):293–301.

Kaur, Jasreen, Kanika Sood, Neha Bhardwaj, Shailendra Kumar Arya, and Madhu Khatri. 2020. Nanomaterial loaded chitosan nanocomposite films for antimicrobial food packaging. *Materials Today: Proceedings* 28 (3):1904–1909.

Khoirunnisa, Assifa Rahma, I Made Joni, Camellia Panatarani, Emma Rochima, and Danar Praseptiangga. 2018. UV-screening, transparency and water barrier properties of semi refined iota carrageenan packaging film incorporated with ZnO nanoparticles. *AIP Conference Proceedings* 1927 (1):030041.

Koushesh Saba, Mahmoud, and Rasoul Amini. 2017. Nano-ZnO/carboxymethyl cellulose-based active coating impact on ready-to-use pomegranate during cold storage. *Food Chemistry* 232:721–726.

Kumar, Santosh, Avik Mukherjee, and Joydeep Dutta. 2020. Chitosan based nanocomposite films and coatings: emerging antimicrobial food packaging alternatives. *Trends in Food Science & Technology* 97:196–209.

Kumar, Vineet, Praveen Guleria, and Surinder Kumar Mehta. 2017. Nanosensors for food quality and safety assessment. *Environmental Chemistry Letters* 15 (2):165–177.

Lavinia, M, SN Hibaturrahman, H Harinata, and AA Wardana. 2020. Antimicrobial activity and application of nanocomposite coating from chitosan and ZnO nanoparticle to inhibit microbial growth on fresh-cut papaya. *Food Research* 4 (2):307–311.

Li, Wenhui, Lin Li, Yun Cao, Tianqing Lan, Haiyan Chen, and Yuyue Qin. 2017. Effects of PLA film incorporated with ZnO nanoparticle on the quality attributes of fresh-cut apple. *Nanomaterials* 7 (8):207.

Li, Xihong, Weili Li, Yunhong Jiang, Yulong Ding, Juan Yun, Yao Tang, and Peipei Zhang. 2011. Effect of nano-ZnO-coated active packaging on quality of fresh-cut 'Fuji'apple. *International Journal of Food Science & Technology* 46 (9):1947–1955.

Liu, J, J Hu, M Liu, G Cao, J Gao, and Y Luo. 2016. Migration and characterization of nano-zinc oxide from polypropylene food containers. *Am. J. Food Technol* 11:159–164.

Liu, Qian, Min Zhang, Zhong-xiang Fang, and Xiao-hong Rong. 2014. Effects of ZnO nanoparticles and microwave heating on the sterilization and product quality of vacuum-packaged Caixin. *Journal of the Science of Food and Agriculture* 94 (12):2547–2554.

Meindrawan, Bayu, Nugraha Edhi Suyatma, Ata Aditya Wardana, and Vega Yoesepa Pamela. 2018. Nanocomposite coating based on carrageenan and ZnO nanoparticles to maintain the storage quality of mango. *Food Packaging and Shelf Life* 18:140–146.

Meng, Xiangyong, Min Zhang, and Benu Adhikari. 2014. The effects of ultrasound treatment and nano-zinc oxide coating on the physiological activities of fresh-cut kiwifruit. *Food and Bioprocess Technology* 7 (1):126–132.

Mihindukulasuriya, SDF, and L-T Lim. 2014. Nanotechnology development in food packaging: a review. *Trends in Food Science & Technology* 40 (2):149–167.

Mooss, Vandana A, Faseela Hamza, Smita S Zinjarde, and Anjali A Athawale. 2019. Polyurethane films modified with polyaniline-zinc oxide nanocomposites for biofouling mitigation. *Chemical Engineering Journal* 359:1400–1410.

Nafady, Nivien A, Saad AM Alamri, Elhagag Ahmed Hassan, Mohamed Hashem, Yasser S Mostafa, and Kamal AM Abo-Elyousr. 2019. Application of ZnO-nanoparticles to manage Rhizopus soft rot of sweet potato and prolong shelf-life. *Folia Horticulturae* 31 (2):319–329.

Najafi, Maryam, Mohammad A. Khalilzadeh, and Hassan Karimi-Maleh. 2014. A new strategy for determination of bisphenol A in the presence of Sudan I using a ZnO/CNTs/ionic liquid paste electrode in food samples. *Food Chemistry* 158:125–131.

Naknaen, Phisut. 2014. Utilization possibilities of antimicrobial biodegradable packaging produced by poly (butylene succinate) modified with zinc oxide nanoparticles in fresh-cut apple slices. *International Food Research Journal* 21 (6).

Paisoonsin, Sutida, Orathai Pornsunthorntawee, and Ratana Rujiravanit. 2013. Preparation and characterization of ZnO-deposited DBD plasma-treated PP packaging film with antibacterial activities. *Applied Surface Science* 273:824–835.

Pasupuleti, Surekha, Srinivas Alapati, Selvam Ganapathy, Goparaju Anumolu, Neelakanta Reddy Pully, and Balakrishna Murthy Prakhya. 2012. Toxicity of zinc oxide nanoparticles through oral route. *Toxicology and Industrial Health* 28 (8):675–686.

Prasad, Kumar Suranjit, and Vivek Kumar. 2014. Nanotechnology in sustainable agriculture: present concerns and future aspects. *African Journal of Biotechnology* 13:705–713.

Rajput, Vishnu D, Tatiana M Minkina, Arvind Behal, Svetlana N. Sushkova, Saglara Mandzhieva, Ritu Singh, Andrey Gorovtsov, Viktoriia S. Tsitsuashvili, William O. Purvis, Karen A. Ghazaryan, and Hasmik S. Movsesyan. 2018. Effects of zinc-oxide nanoparticles on soil, plants, animals and soil organisms: a review. *Environmental Nanotechnology, Monitoring & Management* 9:76–84.

Realini, Carolina E, and Begonya Marcos. 2014. Active and intelligent packaging systems for a modern society. *Meat Science* 98 (3):404–419.

Restuccia, Donatella, U Gianfranco Spizzirri, Ortensia I Parisi, Giuseppe Cirillo, Manuela Curcio, Francesca Iemma, Francesco Puoci, Giuliana Vinci, and Nevio Picci. 2010. New EU regulation aspects and global market of active and intelligent packaging for food industry applications. *Food Control* 21 (11):1425–1435.

Ruchika, Ashok Kumar. 2015. Performance analysis of zinc oxide based alcohol sensors. *International Journal of Applied Sciences and Engineering Research* 4 (4):427–436.

Ruffo Roberto, Sergio, Khamis Youssef, Ayat Farghily Hashim, and Antonio Ippolito. 2019. Nanomaterials as alternative control means against postharvest diseases in fruit crops. *Nanomaterials* 9 (12):1752.

Sabir, Sidra, Muhammad Arshad, and Sunbal Khalil Chaudhari. 2014. Zinc oxide nanoparticles for revolutionizing agriculture: synthesis and applications. *The Scientific World Journal* 2014.

Salarbashi, Davoud, Seyed Ali Mortazavi, Mostafa Shahidi Noghabi, Bibi Sedigheh Fazly Bazzaz, Naser Sedaghat, Mohammad Ramezani, and Iman Shahabi-Ghahfarrokhi. 2016. Development of new active packaging film made from a soluble soybean polysaccharide incorporating ZnO nanoparticles. *Carbohydrate Polymers* 140:220–227.

Sánchez-Acevedo, Zayda C, Jordi Riu, and F Xavier Rius. 2009. Fast picomolar selective detection of bisphenol A in water using a carbon nanotube field effect transistor functionalized with estrogen receptor-α. *Biosensors and Bioelectronics* 24 (9):2842–2846.

Sant'Anna, Pedro B, Bernadette DG de Melo Franco, and Daniele F Maffei. 2020. Microbiological safety of ready-to-eat minimally processed vegetables in Brazil: an overview. *Journal of the Science of Food and Agriculture* 100:4664–4670.

Sarapulova, Olha, Valentyn Sherstiuk, Vitaliy Shvalagin, and Aleksander Kukhta. 2015. Photonics and nanophotonics and information and communication technologies in modern food packaging. *Nanoscale Research Letters* 10 (1):229.

Sardella, Davide, Ruben Gatt, and Vasilis P Valdramidis. 2019. Metal nanoparticles for controlling fungal proliferation: quantitative analysis and applications. *Current Opinion in Food Science* 30:49–59.

Sarkar, Joy, Manosij Ghosh, Anita Mukherjee, Dipankar Chattopadhyay, and Krishnendu Acharya. 2014. Biosynthesis and safety evaluation of ZnO nanoparticles. *Bioprocess and Biosystems Engineering* 37 (2):165–171.

Shikha, Swati, Thoriq Salafi, Jinting Cheng, and Yong Zhang. 2017. Versatile design and synthesis of nano-barcodes. *Chemical Society Reviews* 46 (22):7054–7093.

Siddiqi, Khwaja, Azizur Rahman, Tajuddin Tajuddin, and Azamal Husen. 2018. Properties of zinc oxide nanoparticles and their activity against microbes. *Nanoscale Research Letters* 13.

Siddiqi, Khwaja Salahuddin, Aziz Rahman, and Azamal Husen. 2018. Properties of zinc oxide nanoparticles and their activity against microbes. *Nanoscale Research Letters* 13 (1):1–13.

Šimon, Peter, Qasim Chaudhry, and Dušan Bakoš. 2008. Migration of engineered nanoparticles from polymer packaging to food–a physicochemical view. *Journal of Food & Nutrition Research* 47 (3).

Smijs, Threes G, and Stanislav Pavel. 2011. Titanium dioxide and zinc oxide nanoparticles in sunscreens: focus on their safety and effectiveness. *Nanotechnology, Science and Applications* 4:95.

Sruthi, S, J Ashtami, and PV Mohanan. 2018. Biomedical application and hidden toxicity of zinc oxide nanoparticles. *Materials Today Chemistry* 10:175–186.

Subramaniam, Vimala Devi, Suhanya Veronica Prasad, Antara Banerjee, Madhumala Gopinath, Ramachandran Murugesan, Francesco Marotta, XiaoFeng Sun, and Surajit Pathak. 2019. Health hazards of nanoparticles: understanding the toxicity mechanism of nanosized ZnO in cosmetic products. *Drug and Chemical Toxicology* 42 (1):84–93.

Suman, TY, SR Radhika Rajasree, and R Kirubagaran. 2015. Evaluation of zinc oxide nanoparticles toxicity on marine algae Chlorella vulgaris through flow cytometric, cytotoxicity and oxidative stress analysis. *Ecotoxicology and Environmental Safety* 113:23–30.

Tankhiwale, R, and SK Bajpai. 2012. Preparation, characterization and antibacterial applications of ZnO-nanoparticles coated polyethylene films for food packaging. *Colloids Surf B Biointerfaces* 90:16–20.

Wang, Linlin, Chen Hu, and Longquan Shao. 2017. The antimicrobial activity of nanoparticles: present situation and prospects for the future. *International Journal of Nanomedicine* 12:1227.

Widodo, Abel Kristanto, Dita Baeti Pridiana, Nafi Arrizqi, Muthia Nur Fajrina, Deka Setyawan, and Dimaz Aji Laksono. 2020. Combining cellulose and ZnO for extend the shelf life of fruit. *Materials Science Forum* 990:267–271.

Wijesinghe, Udari, Gobika Thiripuranathar, Haroon Iqbal, and Farid Menaa. 2021. Biomimetic synthesis, characterization, and evaluation of fluorescence resonance energy transfer, photoluminescence, and photocatalytic activity of zinc oxide nanoparticles. *Sustainability* 13 (4):2004.

Wu, Dan, Min Zhang, Baoguo Xu, and Zhimei Guo. 2020. Fresh-cut orange preservation based on nano-zinc oxide combined with pressurized argon treatment. *LWT* 135:110036.

Xia, Tian, Michael Kovochich, Monty Liong, Lutz Mädler, Benjamin Gilbert, Haibin Shi, Joanne I Yeh, Jeffrey I Zink, and Andre E. Nel. 2008. Comparison of the mechanism of toxicity of zinc oxide and cerium oxide nanoparticles based on dissolution and oxidative stress properties. *ACS Nano* 2 (10):2121–2134.

Xiang, Qun, Guifang Meng, Yuan Zhang, Jiaqiang Xu, Pengcheng Xu, Qingyi Pan, and Weijun Yu. 2010. Ag nanoparticle embedded-ZnO nanorods synthesized via a photochemical method and its gas-sensing properties. *Sensors and Actuators B: Chemical* 143 (2):635–640.

Xu, Jicheng, Min Zhang, Bhesh Bhandari, and Robert Kachele. 2017. ZnO nanoparticles combined radio frequency heating: a novel method to control microorganism and improve product quality of prepared carrots. *Innovative Food Science & Emerging Technologies* 44:46–53.

Yang, Zhe, Hongdan Peng, Weizhi Wang, and Tianxi Liu. 2010. Crystallization behavior of poly (ε-caprolactone)/layered double hydroxide nanocomposites. *Journal of Applied Polymer Science* 116 (5):2658–2667.

Yuliani, Sri, Aditya Ata, Ata Wardana, Bayu Meindrawan, Nugraha Suyatma, and Tien Muchtadi. 2019. Nanocomposite edible coating from cassava starch, stearic acid and ZnO nanoparticles to maintain quality of fresh-cut mango cv. Arumanis. *Annals of the University Dunarea de Jos of Galati* 42:49–58.

Zhu, Zhen, Cheng-Tse Kao, and Ren-Jang Wu. 2014. A highly sensitive ethanol sensor based on Ag@ TiO2 nanoparticles at room temperature. *Applied Surface Science* 320:348–355.

Chapter 15

Phyto-Nanocomposites for Minimally Processed Horticultural Produce

Maya Mathew, Ashitha Jose, Sandhya C., and Radhakrishnan E.K.

List of Abbreviations

Ag	Silver
Au	Gold
CMC	Carboxymethyl cellulose
CuO	Copper (II) oxide
FM	Film solubility
MC	Moisture content
MMT	Montmorillonite
NP	Nanoparticle
PE	Polyethylene
PHA	Polyhydroxyalkanoates
PLA	Polylactic acid
PVA	Polyvinyl alcohol
RFID	Radiofrequency identification
ROS	Reactive oxygen species
TTI	Time–temperature integrator
WHC	Water-holding capacity
ZnO	Zinc oxide
ZnONP	Zinc nanoparticle

DOI: 10.1201/9781003142287-15

15.1 Introduction

Horticultural produce plays a key role in human health and diet. The postharvest deterioration of the fresh produces owing to both natural and manmade causes in addition to poor management is critical to consider. The usual practices for the management of harvested produce include wrapping them in papers or various plastic wraps available (Choudhury *et al.*, 2019). Packaging elements that are derived from various plastic precursors have been widely used over the past decades as they are easily available and cost-effective and offer safe handling. Plastics, however, are permeable to the exchange of gases and vapors, in contrast to other traditional packaging aids like glass and metals. The luxury of molding them into different forms and shapes to meet consumer preferences spikes up their popularity (Armentano *et al.*, 2015). Even though petroleum-based products act as affordable packaging materials with sufficient barrier properties to meet the market requirements, they lack innate antimicrobial properties. This often leads to spoilage of the minimally processed items packed within these materials owing to various microbial infestations in addition to the over-ripening of fruits and vegetables and wilting of fresh flowers. Thus, suitable containment methodologies and antimicrobial strategies are to be put forward to prevent the enumeration of microbial flora and to ensure the safeguarding of the freshness, nutritional quality, and economical value of the freshly harvested and packed produces. However, being a petroleum-based product, traditional plastic packaging, in its innate form or recycled, has the potential to generate various toxicity issues. To date, petroleum-based polymers have taken up the role as the major polymer in the packaging industry, resulting in the generation of massive amounts of nonbiodegradable waste and environmental contamination (Kuswandi and Moradi, 2019; Wang *et al.*, 2018). Replacing plastics with environmentally friendly and biodegradable polymers that can, in turn, reduce pollution, preserve fossil fuel resources, and have a plethora of advantages when compared with traditional materials has gained attention of the scientific communities across the globe (Han *et al.*, 2018).

Recently, to overcome the downhill properties of the native materials, numerous bionanocomposites and polymer blends are reportedly being used as effective packaging/coating materials for horticultural produce to ensure the retainment of freshness, nutritive quality, and preserve economical value (Basumatary *et al.*, 2020). These bionanocomposites are combinations of NPs and biopolymer blends, wherein the NPs act as agents that enrich the barrier, mechanical, and antimicrobial properties of the final product. The modulation of various physical parameters such as increasing elasticity, tensile strength, and elongation to meet the tailor-made properties is an added advantage with the incorporated NPs (Choudhury *et al.*, 2019). The most extensive and appealing fact here is the prospect of adding specific functionality to the packaging material, making them an effective passive barrier as well as an active component (Armentano *et al.*, 2015). The development of various antimicrobial packaging materials thus serves as a successful approach strategy for guaranteeing food safety and extending shelf life (Suo *et al.*, 2017).

Various renewable polymers such as CMC, PLA, PHAs, PVA, and others are often used in the food packaging sector to meet consumer demands. Carbohydrate polymers like starch, cellulose, gums, chitosan, and pectin are commonly used in the food packaging, pharmaceutical, and textile sectors. Their amorphous state has good mechanical and chemical properties, as well as gas barrier features (Kjellgren *et al.*

2006). One of the most extensively used composite coatings in the food, pharmaceutical, and other industries is CMC, which has unique water-retention, film-forming, and dispersion characteristics and can reduce oil pickup (Tharanathan, 2003). Alternative packaging materials derived from renewable resources such as PLA, PHAs, starch, and chitosan have been suggested as possible alternatives to conventional polymers. However, PVA, which is derived from the hydrolysis of polyvinyl acetate, has properties such as biodegradability, biocompatibility, water solubility, nontoxicity, and lower cost and is produced as semicrystalline polymer (Abdullah *et al.*, 2019). This polymer has excellent thermomechanical properties, thermal strength, and flexibility, as well as optical and physical characteristics that are essential in packaging applications. There are lipid–polymer hybrid NPs (LPHNPs), which are next-generation core–shell nanostructures that are conceptually derived from both liposomes and polymeric NPs, with a lipid coating encasing the polymer core (Mukherjee *et al.*, 2019).

PE is the most widely used polymer in packaging due to its lower cost, high-impact toughness, and barrier resistance toward chemicals (Carrion *et al.*, 2007). Although it seems to possess high gas permeability and lower sensitivity to oils, it lacks odor resistance and cannot prevent microbial spoilage (Munteanu *et al.*, 2018). Antimicrobial PE films are developed by the incorporation of various antimicrobial agents into particular polymer-forming solutions. However, this application is limited due to the instability and incompatibility of the selected antimicrobial compounds with the PE–polymer matrix (Theapsak *et al.*, 2012). Coating plastic films with a polymer-based solution is a low-cost approach that often results in enhanced stability and adhesiveness of the antimicrobial molecule used in the study (Lin *et al.*, 2020).

15.2 Phyto-Nanocomposites

The food sector is always looking for new, cost-effective, and environmentally friendly packaging materials that can retain and monitor food quality (Han *et al.*, 2018). Starch is a widely available biocompatible polymer that is employed in the development of novel active packaging materials due to its high biodegradability and low cost, as well as ease of access and preparation. However, starch-based biodegradable films have poor mechanical and barrier properties when exposed to high relative humidity, which, in turn, limits their use in food packaging (Espitia *et al.*, 2013). There are numerous strategies to improve the thermal stability, mechanical strength, and barrier function of the biodegradable material developed; incorporating various types of reinforcing nanofillers into the starch matrix is one of them (Oun and Rhim, 2017). Furthermore, nanocomposite films can be used as transporters of active components such as bacteriostatic/bactericidal agents, antioxidants, plant extracts, and enzymes that, in turn, can extend the shelf life of the material packed within and ensure food safety under demanding situations.

Metallic NPs with antimicrobial properties have been extensively explored in this context. Many studies have already been done to describe the property enhancement of various nanocomposite and bionanocomposite materials through the incorporation of various types of metallic nanofillers (Dehghani *et al.*, 2022; Ebrahimi *et al.*, 2019). The use of inorganic and metal-oxide NPs such as nanoclay, MMT, Ag, ZnO, and CuO NPs in various nanocomposite materials developed have showcased to improve the mechanical and barrier properties, such as elasticity, gas barrier characteristics, and stability under various temperature and moisture conditions (Youssef *et al.*, 2018)

Researchers are intrigued by the wide range of properties exhibited by various phytochemicals, including their antibacterial and antioxidant properties. This has spurred the possibility of developing various phyto-nanocomposite films in the food industry. Phytochemicals are exploited in the food industry for their antibacterial and antioxidant properties, along with the feasibility of aroma protection, imparting freshness, and the capability of removing foul odors. These properties of the plant extracts vary with the presence of different secondary metabolites such as phenolic compounds, tannins, flavonoids, and others. Furthermore, phytochemicals' environmentally friendly approach, safety aspects, ease of availability, and so on raise the likelihood of their exploitation as a primary component in the preservation of fresh fruits and vegetables. These phytochemicals along with various custom-built nanocomposites play a significant role in the development and validation of new food packaging options such as films and edible coatings. The efficacy of food packaging films and coatings can be improved by combining a wide range of phytochemicals with various NPs, thereby bringing out the synergy (Cui *et al.*, 2020). The color-producing compounds present in the plant extract such as polyphenols can be used as an indicator of spoilage (Liu *et al.*, 2019a). Another promising phyto-nano combination yet to be explored to the fullest is the essential oil–based nanocomposites. The use of essential oil can ensure the antimicrobial and antioxidant properties of the developed material in addition to the pleasant aroma it has to offer. The study of Hai Chi *et al.*, (2019) showed that the TiO_2 and Ag NPs, along with bergamot essential oil, could be used for the shelf-life enhancement of fruits like mangoes. Essential oils, along with NPs, are also able to inhibit mycotoxins like aflatoxins (Kumar *et al.*, 2020). Many other essential oils are used to improve the quality of fresh horticultural produce and are widely used as a component of both intelligent and active food packaging films (Alexandre *et al.*, 2016; Alinaqi *et al.*, 2021).

15.3 Essential Characteristics of Phyto-Nanocomposite-Based Active Packaging Materials

15.3.1 Barrier Properties

Edible coatings are recommended as feasible alternatives to preserve fresh fruits and vegetables due to their capacity to prevent moisture loss, loss of odor, water uptake by the food material, and oxygen penetration. Moreover, they produce minimal to no waste, thereby reducing the environmental burden. By functioning as a moisture barrier, the phyto-nanocomposite films/coatings additionally help reduce weight loss. Furthermore, edible coatings improve food texture, appearance, and shelf life by forming semi-permeable barriers to moisture and gases such as carbon dioxide and oxygen (Duran *et al.*, 2016; Jafarzadeh *et al.*, 2018). In a study carried out by Emamifar and Bavaisi (2020), a novel formulation of sodium alginate and nano-ZnO was developed and applied to strawberries. The results showed that nano-ZnO considerably boosted the moisture barrier of films and hence reduced the weight loss of strawberries studied. Whereas the uncoated fruits showed higher weight loss by the end of the 20-day storage period. The possible combination of phyto-nanocomposite includes the use of plant-based extracts rich in alkaloids, flavonoids, and terpenes, along with the metal nanocomposites, which could effectively act as a potential candidate in the food packaging sector (Figure 15.1).

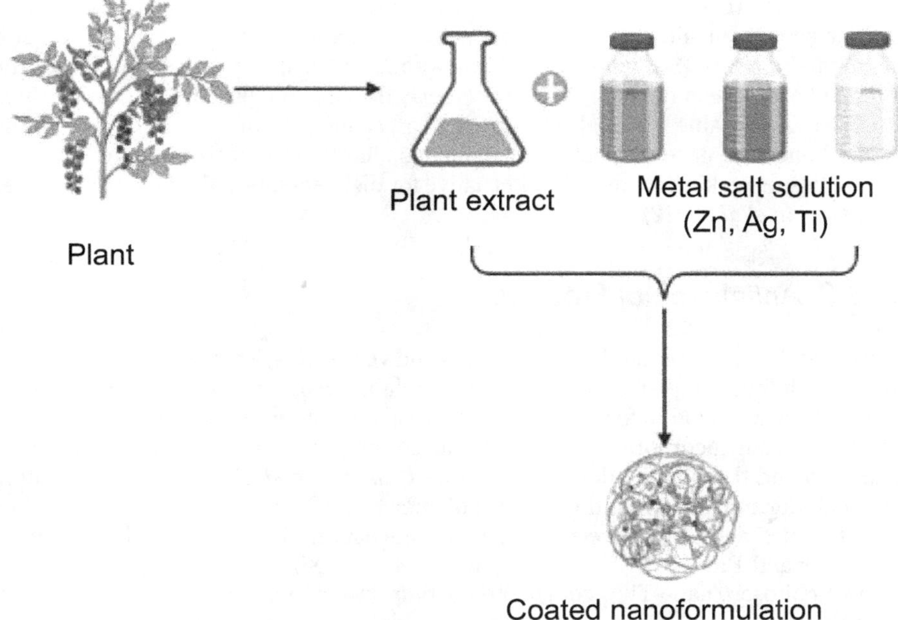

Figure 15.1 Formulation of phyto-nanocomposites.
Source: Adapted from Dikshit *et al.* (2021).

15.3.2 Scavenging Properties

Packed fruits and vegetables can be significantly harmed by the presence of oxygen trapped within the packaging. It can induce a variety of negative effects, including oxidative damage, nutritional loss, color changes, and the advancement of off-flavors (Lee *et al.*, 2015). It also has a significant impact on the rate of respiration and the formation of ethylene in packed fruits and vegetables. The oxygen scavengers thus ensure the proper protection of the fruits and vegetables. The exclusion or removal of oxygen has been a query worth attention to since it can ensure the delayed onset of degradation and thus extend the shelf life of the material packed within (Pereira and Malfeito-Ferreira, 2015). Nanoformulated packaging, edible films, and coatings containing oxygen scavengers and humidity control systems play a pivotal role in minimizing both the production and release of gases and other compounds that may accelerate fruit and vegetable decay (Restuccia *et al.*, 2010). In recent times, it has been proposed that the development of polymer films with the inclusion of NPs can strengthen the gas barrier properties and limit the rate of respiration, transpiration, and other metabolic activities occurring within the packed fruits (Kaur *et al.*, 2017). These scavengers prevent the growth of aerobic bacteria and molds in fruits and vegetables through the generation of a near-anaerobic condition. Along with these, controlling the quantity of ethylene produced or retained within the packaging during the storage period also plays a vital role in extending the shelf life of the packed food. Ethylene accelerates ripening, softens fruits, and degrades chlorophyll, thus eventually causing the decay of fruits and vegetables (De *et al.*, 2015). Hence, ethylene scavengers are used to safeguard ethylene-sensitive fruits and vegetables, such as bananas, apples, mangos, tomatoes, onions, and carrots (Siripatrawan and Kaewklin, 2018). It is possible to control the amount of ethylene in the fruits and hence delay the ripening process by coating them with a semipermeable material. The coating can reduce the amount of oxygen in the air, and hence the pace of respiration, while also limiting the distribution of CO_2 outside the tissue, making it a feasible alternative for disinfectants and chemical therapies (Moalemiyan *et al.*, 2012).

15.3.3 Antimicrobial Properties

With the spike in the demand for fresh fruits and vegetables, it is critical to ensure prolonged shelf life and quality maintenance, which, in turn, can result in minimal wastage of the packed food before it reaches the final consumer. Antimicrobial packaging systems contain incorporated antimicrobial compounds that can reduce microbial infiltration and thereby the deterioration rate (Jafarzadeh *et al.*, 2021). Incorporating phyto-nanoformulations into the classic polymer blends strengthens its antimicrobial characteristics and hence preserves packed horticultural produce for longer times (Emamifar and Bavaisi, 2020). A study conducted by Xing *et al.* (2020) showed the effect of chitosan/nano-TiO_2 composite coating on mangoes disclosed the potential of the developed material in preventing microbial growth and preserving the innate nutritional composition and fruit quality at a storage temperature of 13°C.

15.4 Types of NPs Used in the Development of Phyto-Nanocomposites

15.4.1 ZnO NPs

ZnO NPs, considered to be GRAS (generally recognized as safe), are efficient antimicrobial NPs commonly exploited in the food sector (Espitia *et al.*, 2012). Moreover, due to the nonspecific action of inorganic antimicrobial agents, using ZnO NPs to combat microbial resistance has gained a lot of attention. The small particle size and large surface area of ZnO NPs have resulted in their enhanced antimicrobial activity and surface reactivity (Da Silva *et al.*, 2019). Different types of polymer nanocomposites are developed with ZnO NPs as an antimicrobial additive, and the packaging materials developed through the incorporation of the same are found to have superior antimicrobial properties than the conventional films, revealing their potential in being exploited in food preservation. ZnO NPs also provide mechanical stability to the packaging film (Shi *et al.*, 2014). ZnO offers a wider range of advantages over other NPs owing to its superior antimicrobial potential and biocompatibility. Moreover, it has the best demonstrated photocatalytic effectiveness among all inorganic photocatalytic materials studied so far. Furthermore, the superior selectivity, durability, and heat resistance of ZnO NPs have led to their acceptance in fighting a wide range of microorganisms, including *S. aureus*, *E. coli*, and *Candida albicans* (Raghupathi *et al.*, 2011; Manoharan *et al.*, 2015; Janaki *et al.*, 2015; Sirelkhatim *et al.*, 2015). The antibacterial activity of ZnO NPs is usually dependent on their particle size; that is, a particle with a smaller size will have higher antibacterial activity (Raghupathi *et al.*, 2011).

Thus, novel bionanocomposites containing ZnO NPs and other active components like phytochemicals and nanoclays could be the future of the food sector (Jayakumar *et al.*, 2019; Avella *et al.*, 2005; Abbas *et al.*, 2019) Traditional methods of synthesis of ZnONPs are arduous and exorbitant. The huge amount of energy processed during the synthesis causes a significant rise in the environmental temperature and the generation of large quantities of secondary waste (Kruis *et al.*, 2000).

Nanocomposite development through the exploitation of natural components such as plants is a less expensive, environmentally friendly strategy, which ensures negligible adversities. The different phytochemical components present in plants, such as aldehydes, polysaccharides, flavonoids, phenolics, terpenoids, amines, and proteins, can act as both reducing and stabilizing agents in the formulation of phyto-mediated nanocomposites (Ramesh *et al.*, 2015). Fruits packed with various phyto-nanocomposite films were revealed to have the least microbial burden and deterioration rate. The produced films were transparent, with a moderate antioxidant content but a significant amount of antimicrobial activity. Biologically synthesized NPs can be employed as effective nanofillers to extend the shelf life and improve the quality of the packed fruits more successfully than the native polymeric films that do not contain any nanomaterials (Kalia *et al.*, 2021). In a study conducted by Kalia *et al.* (2021), the evaluation of the effect of packing guava fruits in chitosan–ZnO nanocomposite films were carried out. The study pointed out that fruit weight loss is a frequent fruit quality characteristic. The mechanism of action of ZnO NPs against the bacterial cells (Figure 15.2) are as follows: (1) the release Zn^{2+} ions from ZnO NPs, (2) ROS production, and (3) the internalization and direct interaction within the bacterial cell (Da Silva *et al.*, 2019).

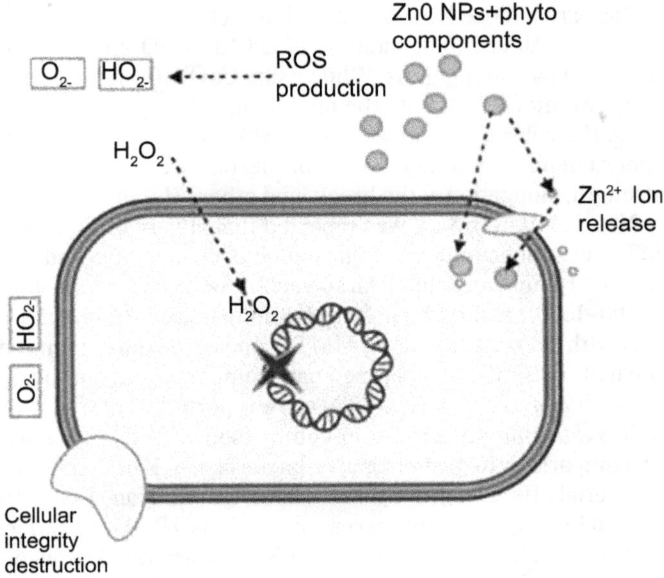

Figure 15.2 Mechanisms of action of ZnO NPs against bacterial cells.
Source: Modified from Da Silva *et al.* (2019).

The weight loss that transpired during the storage period of the fruits is mostly caused by evapotranspiration and respiration (Vishwasrao *et al.*, 2016). However, in fruits that exhibit climacteric behavior, the postharvest ripening process continues, resulting in ethylene-hormone-induced breakdown of the hemicellulose content in the fruits and thereby softening of the fruit, whereas cherries, grapes, oranges, raspberries, cucumbers, pomegranate, and watermelon are non-climacteric fruits that should be left on the tree to reach full physiological maturity before being harvested, and thus, the rise in ethylene and respiration rate is not observable (Chen *et al.*, 2018). It has also been discovered that the MC and the FS have a direct relationship. The films with high MC exhibit high WHC and FS and low water barrier properties. As a result of the increased water and gas barrier capabilities of the ZnO NP-chitosan films, the MC, WHC, and FS properties of the produced chitosan nanocomposite films decreased (Kalia *et al.*, 2021). In another study carried out by Steffy *et al.* (2018), the reduced particle size of a ZnO phyto-nanocomposite was evaluated using X-ray diffraction (XRD), which resulted in a significant increase in surface area, favoring the production of free excitons. These excitons are favored in the production of ROS that are fatal to bacterial cells. ZnO phyto-nanocomposite may also penetrate bacterial cells on a nanoscale, resulting in the generation of harmful oxygen radicals that damage DNA, cell membranes, and cell proteins, eventually inhibiting bacterial development and eventually resulting in bacterial cell death (Lipovsky *et al.*, 2009).

15.4.2 CuO NPs

CuO NPs are one of the most widely studied NPs in the food industry. The antimicrobial property is the most efficient characteristic of CuO NPs. Due to their large membrane surface area, these NPs can easily interact with microorganisms and show antimicrobial effects (Almasi, Jafarzadeh *et al.*, 2018). CuO NPs are widely used for preparing active food packaging films. While using CuO NPs in the food system, there is a chance of migrating Cu NPs onto the food items. Mostly, the antibacterial activity of CuO is due to the release of copper ions (Cu^{2+}). CuO-added nanocomposite films also show higher chemical and mechanical properties, such as water barrier properties, tensile strength, elongation at the break, and others (Li, Jin *et al.*, 2017). In a study conducted by Nouri *et al.* (2018), it was reported that at varied alkaline–ion exchange times, an MMT–CuO nanocomposite was created using a simple and straightforward process that did not require any chemical solvents. Due to its low cost, availability, high aspect ratios (100–1500), and high surface-to-volume ratio (700–800m2/g), as well as good miscibility with cationic polymers, MMT is one of the most commonly employed layered silicate materials in chitosan-based nanocomposites. In another study, carried out by Esmailzadeh *et al.* (2021), CuO NPs were reported to have a firmer antibacterial effect on *B. subtilis* and *S. aureus* in comparison with *E. coli* and *Pseudomonas aeruginosa*. In comparison to gram-negative bacteria, Cu-doped ZnO nanorods had a stronger antibacterial effect on gram-positive bacteria (Bhuyan *et al.*, 2015). The addition of Cu ONPs to biopolymer films increased the films' UV-light barrier, mechanical, water vapor barrier, and thermal stability. CuO NP–incorporated nanocomposite films with superior mechanical properties, water vapor barriers, UV screening, and thermal stability, as well as antibacterial properties, have a promising future as active food packaging films (Shankar *et al.*, 2015). In another study conducted by Ouzounidou and Gaitis (2011) focusing on its shelf life extension, the effects of colloidal copper (nano-copper) as a postharvest spray on raw parsley that was stored at a temperature of 20°C for 10 days followed by a storage temperature of 5°C for 23 days.

15.4.3 Ag NPs

The broad-spectrum antibacterial action of Ag NPs is well recognized. Costa *et al.* (2011) developed an efficient antimicrobial active packaging film with enhanced mechanical properties by incorporating Ag-MMT through an ion-exchange reaction. The Ag-MMT NPs were found to be capable of extending the shelf life of carrots to more than 2 months, as confirmed by sensory and microbiological quality evaluations. Many nanocomposite materials include Ag NPs as an addition to improve the mechanical and barrier properties of the polymer composite film. The antimicrobial activity of Ag NPs can be explained by three different mechanisms: (1) Ag NPs with sizes ranging from 1 to 10 nm cause cellular damage by binding to cell membrane surfaces, (2) cell damage caused by interacting with bacterial cell DNA, and (3) binding of Ag ions from Ag NPs to the cellular membrane surface (Simbine *et al.*, 2019). Ag NPs are widely used for the storage of postharvest fruits and vegetables to maintain the quality of minimally processed horticulture produce. In a study conducted by Sarwar *et al.* (2018), the cytotoxicity of the composite films was analyzed in HepG2 cells using the MTT assay. The obtained result showed that the cell viability was more than 70%, which indicated that nanocomposite films have strong non-cytotoxicity effects. The studied films were also found to have a high moisture retention capacity, which is essential for fresh fruits and vegetables having a high respiration rate. There is a chance for the migration of these Ag NPs into the packaged products, so many researchers are working on the safety aspects of using Ag NPs in the packaging material, and they are also concerned with the toxic effects of Ag NPs (Panea *et al.*, 2014; Istiqola and Syafiuddin, 2020). Another study revealed the use of laponite as a thickening agent for preparing nanocomposite films, along with Ag NPs in order to keep litchi fruit fresh (Wu *et al.*, 2018). The intention of packaging film production using Ag NPs became effective after the successful implementation of the same packaging material in different varieties of fruits and vegetables. One such example is the study conducted by Jiang *et al.* (2013), where they developed a unique alginate/nano-Ag composite, which improves the quality in preservation, better extends the storage period, and suppresses microbiological growth in the case of cold-storage shiitake mushrooms. The experimental data analysis also showed that the cold-storage shiitake mushroom with alginate/nano-Ag coating material formed a more satisfactory texture and freshness effect on the mushroom than the control group.

15.5 Phytocompound-Based Films in Packaging

Phenolic compounds are considered the most diverse groups of secondary metabolites found in edible plants (Marrez *et al.*, 2019). They aid in plant growth and reproduction, give resistance to pathogens and predators, and protect crops from disease and seed germination prior to harvest (Ross and Kasum, 2002). They are produced in response to environmental and physiological challenges such as disease and insect attack, UV radiation, and injury (Chung *et al.*, 2003). Flavonoids, tannins, and lignin are a few of the most widely distributed polyphenol classes that have been found in vegetables commonly consumed by humans (Marrez *et al.*, 2019). Dietary flavonoids have an important role in human health, and their wide variety of bioactivity has piqued attention (Pandey and Rizvi, 2009). Flavonoids are potent antioxidants, free radical scavengers, and metal chelators that reduce lipid peroxidation and have a variety of physiological effects, including anti-inflammatory, anti-allergic, anticarcinogenic, anti-hypertensive, anti-arthritic, and antibacterial properties (Aly *et al.*, 2016). Anthocyanins derived from roselle, purple sweet potato, black soybean seed coat, red cabbage, and mulberry have been used to enhance the antioxidants, thereby making intelligent pH-sensitive films (Liu *et al.*, 2019b; Chen *et al.*, 2020). These films have excellent pH-sensitive properties, mechanical properties, and water vapor permeability along with antioxidant properties. When exposed to extreme conditions such as light, temperature, oxygen, and enzymes, anthocyanins are easily destroyed (Tarone *et al.*, 2020). Hence, anthocyanin stability must be improved in order to broaden the use of anthocyanin-based films in smart packaging. Several micro/nanoencapsulation strategies, such as complexes, emulsions, freeze-drying, coacervation, liposomes, molecular inclusion, electrospraying, and spray-drying, have been shown to increase the stability of anthocyanins in recent years (Sharif *et al.*, 2020; Tarone *et al.*, 2020). From the mentioned studies it was reported that not only does encapsulation protect anthocyanins from extreme conditions, but it can also offer a prolonged release of anthocyanins under certain conditions. Nanocomplexes, in particular, are useful for stabilizing anthocyanins by entrapping them in nano-sized complexes among the several encapsulation strategies (Ye *et al.*, 2020). Another important flavanol found in buckwheat, tea, apple, and passionflower is rutin, also known as vitamin P, which has a large range of pharmacological properties along with nutritional characteristics along with antimicrobial properties (Narasagoudr *et al.*, 2020). In the mentioned study, rutin was exploited as a bioactive constituent to make bioactive films along with PVA and chitosan with better physicochemical and functional qualities. The bioactive films were made using a simple, cost-effective, and environmentally friendly solvent-casting approach. The addition of rutin has improved the tensile properties of the film as well as UV-light barrier qualities and water resistance. Quercetin is also a significant dietary flavanol that has excellent antioxidant and antibacterial properties.

15.6 Phytosynthesis of Metal NPs

Concerns about the environmental impact of plastic polymers in food packaging have prompted a greater focus on creating biodegradable alternatives. The synthesis of metal NPs using a green protocol has created the packaging of postharvest produce to become nontoxic, economical, environmentally benign, and compatible with food (Basumatary *et al.*, 2018). Earlier studies showed that green approaches are more effective for the generation of NPs because they have fewer risks of failure, are less expensive, and are easier to characterize (Abdelghany *et al.*, 2018). Because of the toxic metabolites produced by physical and chemical ways of synthesizing NPs, the environment has been subjected to a number of stresses. Plant-based NP synthesis is a simple technique in which a metal salt is synthesized with plant extract, and the reaction is finished in minutes to a few hours at ambient temperature. Various studies have been carried out in the past about the same with different matrixes and NPs (Table 15.1). In the last decade, this technique has gotten a lot of attention, especially for Ag and Au NPs, which are more secure than other metallic

Table 15.1 Combination of Film Matrix along with NPs to the Packaging of Fresh Harvest Produce (modified from Jafarzadeh *et al.*, 2021)

Matrix	*Applications*	*Suggested NPs*	*Fruits/ Vegetables*	*Reference*
Soy protein isolate (SPI)	Delay ripening rate and weight loss of the fruit	ZnO	Banana	Li *et al.*, 2019
PLA	Minimal loss of fruit firmness	Ag	Mango	Chi *et al.*, 2019
Chitosan-rice	Decrease in the surface microbial flora count & increase in the shelf life of fruits	ZnO and Ag NPs	Peach	Kaur *et al.*, 2017
PLA	Preservation of texture	ZnO	Cut apple	Naknaen, 2014
Carboxymethyl-cellulose	Prevention of weight loss & increase in vitamin C and antioxidant content.	ZnO	Pomegranate	Saba and Amini, 2017
Chitosan/ cellulose acetate phthalate	Increase in the shelf life up to 9 days	ZnO	Black grapefruits	Indumathi *et al.*, 2019
Chitosan	Reduction of respiration rate & blocking of fruit softening,	Ag	Fresh melon	Ortiz-Duarte *et al.*, 2019
Sodium alginate-gum	Inhibition of the growth of foodborne pathogens and extension of the shelf life.	Ag	Black grapes	Kanikireddy *et al.*, 2019
CMC	Delay in the disease severity in fruits and enhancement of firmness, and reduction of weight loss and respiration rate.	ZnO	Tomato fruits	Saekow *et al.*, 2019
Cassava starch	Improvement of the quality of product,	ZnO	Fresh-sliced okra	Fadeyibi, 2019

(Continued)

Table 15.1 (Continued)

Matrix	*Applications*	*Suggested NPs*	*Fruits/ Vegetables*	*Reference*
Chitosan	Effective control in reducing weight loss, firmness and delayed changes in the respiration rate during 3 weeks.	CuO	Strawberry	Eshghi *et al.*, 2014
CMC Guar gum	Antimicrobial activity on gram-positive, gram-negative bacteria, and fungi.	Ag	Strawberry	Kanikireddy *et al.*, 2020
Chitosan	Improve the OP barrier properties of films.	Montmorillonite (MMT)	Tangerine	Xu *et al.*, 2018

NPs (Aman and Narendra, 2019). Green synthesis methods are appealing because they have the potential to lessen the toxicity of NPs. As a result, the use of vitamins, amino acids, and plant extracts is becoming increasingly popular (Baruwati *et al.*, 2009). In a study conducted by Marrez *et al.* (2019), a cellulose acetate film with Ag NPs was prepared as an active food packaging system. In the study, Ag NPs were synthesized by the reduction of Ag ions by phenolic compounds, such as gallic and pyrogallic acid, rutin, and quercetin (Marrez *et al.*, 2019).

At the distribution and storage stages, the appropriate selection of packaging techniques and technologies has a significant impact on the product's freshness and quality. Fruit and vegetable packaging is typically used to protect them against mechanical stress during transportation, as well as oxygen, moisture, water vapor, light, dust, and germs (Jafarzadeh *et al.*, 2019). As a result, appropriate mechanical performance and protection against oxygen, moisture, water vapor, UV light, ethylene, and microbiological development are the most important requirements for materials used in fruit and vegetable packaging. Various systems have been created for this purpose, as detailed earlier.

15.7 Sensors

Smart food packaging technology has enabled better monitoring and exchange of information regarding food quality by combining indicators, sensors, and RFID into packaging. Manufacturers and consumers may now trace a product's history via critical stages in the food supply chain thanks to this technology. Many types of indicators and sensors have been created for smart packaging to monitor the integrity of food products, such as time–temperature integrators (TTIs) and freshness indicators/sensors that use factors that indicate quality as analytes.

An analytical device that detects or quantifies problem or visual signals in order to quantify chemical or physical qualities is known as a sensor. Sensors that are incorporated in food-quality detection methods include gas sensors, fluorescence-based oxygen sensors, and biosensors. Gas sensors respond to the presence of a gas by changing the physical constraint of the sensor, which can then be verified by an external instrument (Grist *et al.*, 2010). For example, the change in gas composition caused by fresh produce respiration might result in the growth of spoilage bacteria or gas transfer packing material. Carbon dioxide sensors work in a different way, they have optical filters mounted on the detector surface which works by enumerating the difference in the amount of light between the two filters. Fluorescence-based oxygen sensors can be used to detect the presence of microbial spoilage, as well as pesticide traces by fluorescent derivatization method (Jiménez-López *et al.*, 2016). Biosensors are receptors such as enzymes, antigens, microbes, hormones, and nucleic acids, which are used individually or in combination.

15.8 Applications of Bionanocomposites as Packaging Material for Fresh Harvest Produce

Organic acids and their salts, enzymes, fatty acids, metal NPs, antibiotics, essential oils, antioxidants, plant volatiles, fungicides, and bacteriocin are examples of active ingredients used in food packaging. In order to maintain and promote the quality and safety of fruits and vegetables, active packaging techniques use releasing mechanisms, absorbers, and other systems Because ethylene control during storage is critical for increasing the postharvest life of fresh fruits and vegetables, suitable packaging material for fruits and vegetables should absorb ethylene (Duran *et al.*, 2016). The study carried out by Chiabrando *et al.* (2019) explained that the coating materials maintained the quality of fresh strawberries and enhanced the shelf life compared to untreated fruits (Figure 15.3). The prepared packaging prevented postharvest senescence; reduced

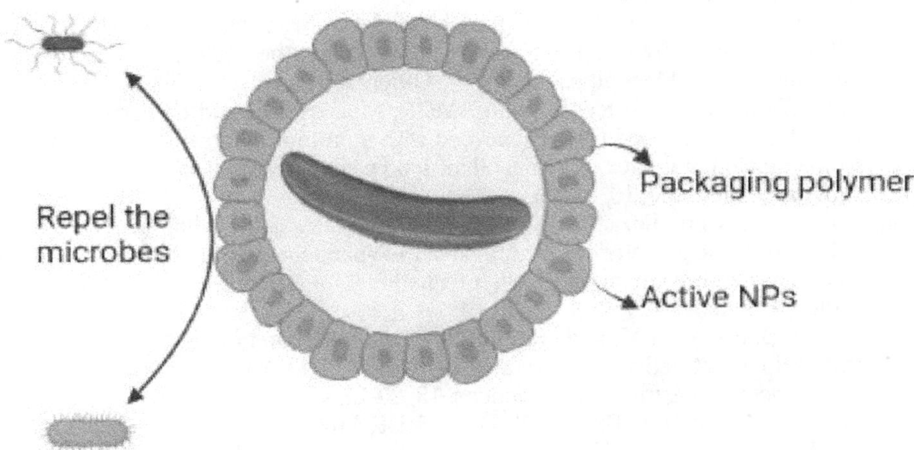

Figure 15.3 The repel mechanism of phyto-nanocomposites against microbes on fresh courgettes.

Source: Modified from Jafarzadeh *et al.* (2021).

weight loss; preserved color, acidity, and vitamin C content; and improved sensory attributes, according to the findings. Costa *et al.* (2011) stated that the combination of Ag-MMT used in the active packaging to limit microbiological development has improved the sensory characteristics of the film and extended the shelf life of fresh fruit salad.

15.9 Conclusion

The metal NPs together with various biopolymers being exploited as a food packaging material effectively reduces the postharvest loss of fruits and vegetables. The phyto-nanocomposites containing a variety of metal NPs along with the phyto-formulation are considered the best packaging options currently available for the plethora of postharvest produce like fruits and vegetables to improve their shelf life. The innate antimicrobial and scavenging properties of the plant extract and nanomaterials act as a barrier to the packed material, thereby enhancing the shelf life of the material. The vision of the future lies where packaging operates as an advanced system having both smart and conventional materials, thereby creating value and benefits to the postharvest produce. For an effective and proper usage of nanomaterials in postharvest produce packaging, standard guidelines should be issued by the respective authorities.

References

Abbas M, Buntinx M, Deferme W, and Peeters R. 2019. (Bio) polymer/ZnO nanocomposites for packaging applications: A review of gas barrier and mechanical properties. Nanomaterials, 9(10), 1494.

Abdelghany TM, Al-Rajhi AM, Al Abboud MA, Alawlaqi MM, Ganash Magdah A, Helmy EA, and Mabrouk AS. 2018. Recent advances in green synthesis of silver nanoparticles and their applications: about future directions. A review. BioNanoScience, 8, 5–16.

Abdullah ZW and Dong Y. 2019. Biodegradable and water resistant poly (vinyl) alcohol (PVA)/starch (ST)/glycerol (GL)/halloysite nanotube (HNT) nanocomposite films for sustainable food packaging. Frontiers in Materials, 6, 58.

Alexandre EMC, Lourenço RV, Bittante AMQB, Moraes ICF, and do Amaral Sobral PJ. 2016. Gelatin-based films reinforced with montmorillonite and activated with nanoemulsion of ginger essential oil for food packaging applications. Food Packaging and Shelf Life, 10, 87–96.

Alinaqi Z, Khezri A, and Rezaeinia H. 2021. Sustained release modeling of clove essential oil from the structure of starch-based bio-nanocomposite film reinforced by electrosprayed zein nanoparticles. International Journal of Biological Macromolecules, 173, 193–202.

Aly S, Sabry B, Shaheen M, and Hathout A. 2016. Assessment of antimycotoxigenic and antioxidant activity of star anise (*Illicium verum*) in vitro. Journal of the Saudi Society of Agricultural Sciences, 15, 20–27.

Armentano I, Fortunati E, Burgos N, Dominici F, Luzi F, Fiori S, Jiménez A, Yoon K, Ahn J, Kang S, and Kenny JM. 2015. Bio-based PLA_PHB plasticized blend films: Processing and structural characterization. LWT-Food Science and Technology, 64(2), 980–988.

Avella M, De Vlieger JJ, Errico ME, Fischer S, Vacca P, and Volpe MG. 2005. Biodegradable starch/clay nanocomposite films for food packaging applications. Food Chemistry, 93(3), 467–474.

Baruwati B, Polshettiwar V and Varma RS. 2009. Glutathione promoted expeditious green synthesis of silver nanoparticles in water using microwaves. Green Chemistry, 11, 926–930.

Basumatary IB, Mukherjee A, Katiyar V, and Kumar S. 2020. Biopolymer-based nanocomposite films and coatings: Recent advances in shelf-life improvement of fruits and vegetables. Critical Reviews in Food Science and Nutrition, 1–24.

Basumatary K, Daimary P, Das SK, Thapa M, Singh M, Mukherjee A, and Kumar S. 2018. Lagerstroemia speciosa fruit-mediated synthesis of silver nanoparticles and its application as filler in agar based nanocomposite films for antimicrobial food packaging. Food Packaging and Shelf Life, 17, 99–106.

Bhuyan T, Khanuja M, Sharma R, Patel S, Reddy M, Anand S, and Varma A. 2015. A comparative study of pure and copper (Cu)-doped ZnO nanorods for antibacterial and photocatalytic applications with their mechanism of action. Journal of Nanoparticle Research, 17(7), 288.

Carrion FG, Sanes J, and Bermudez MD. 2007. Influence of ZnO nanoparticle filler on the properties and wear resistance of polycarbonate. Wear, 262, 1504.

Chen S, Wu M, Lu P, Gao L, Yan S, and Wang S. 2020. Development of pH indicator and antimicrobial cellulose nanofibre packaging film based on purple sweet potato anthocyanin and oregano essential oil. International Journal of Biological Macromolecules, 149, 271–280.

Chen Y, Grimplet J, David K, Castellarin SD, Terol J, Wong DC, Luo Z, Schaffer R, and Chervin C. 2018. Ethylene receptors and related proteins in climacteric and non-climacteric fruits. Plant Science, 276, 63–72.

Chi H, Song S, Luo M, Zhang C, Li W, Li L, and Qin Y. 2019. Effect of PLA nanocomposite films containing bergamot essential oil, TiO2 nanoparticles, and Ag nanoparticles on shelf life of mangoes. Scientia Horticulturae, 249, 192–198.

Chiabrando V, Garavaglia L, and Giacalone G. 2019. The postharvest quality of fresh sweet cherries and strawberries with an active packaging system. Foods, 8(8), 335.

Choudhury A, Jeelani PG, Biswal N, and Chidambaram R. 2019. Application of bionanocomposites on horticultural products to increase the shelf life. In: Polymers for Agri-Food Applications. Edition 1, Springer, Cham, pp. 525–543.

Chung IM, Park MR, Chun JC, and Yun SJ. 2003. Resveratrol accumulation and resveratrol synthase gene expression in response to abiotic stresses and hormones in peanut plants. Plant Science, 164, 103–109.

Costa C, Conte A, Buonocore GG, and Del Nobile MA. 2011. Antimicrobial silver-montmorillonite nanoparticles to prolong the shelf life of fresh fruit salad. International Journal of Food Microbiology, 148(3), 164–167.

Cui H, Surendhiran D, Li C, and Lin L. 2020. Biodegradable zein active film containing chitosan nanoparticle encapsulated with pomegranate peel extract for food packaging. Food Packaging and Shelf Life, 24, 100511.

da Silva BL, Abuçafy MP, Manaia EB, Junior JAO, Chiari-Andréo BG, Pietro RCR, and Chiavacci LA. 2019. Relationship between structure and antimicrobial activity of zinc oxide nanoparticles: An overview. International Journal of Nanomedicine, 14, 9395.

De Chiara MLV, Pal S, Licciulli A, Amodio ML, and Colelli G. 2015. Photocatalytic degradation ofethylene on mesoporous TiO_2/SiO_2 nanocomposites: effects on the ripening of mature green tomatoes. Biosystems Engineering, 132, 61–70.

Dehghani S, Rezaei K, Hamishehkar H, and Oromiehie A. 2022. The effect of electrospun polylactic acid/chitosan nanofibers on the low density polyethylene/ploy lactic acid film as bilayer antibacterial active packaging films. Journal of Food Processing and Preservation, 46(9), e15889.

Dikshit PK, Kumar J, Das AK, Sadhu S, Sharma S, Singh S, Gupta PK, and Kim BS. 2021. Green synthesis of metallic nanoparticles: Applications and limitations. Catalysts, 11(8), 902.

Duran M, Aday MS, Zorba NND, Temizkan R, Büyükcan MB, and Caner C. 2016. Potential of antimicrobial active packaging 'containing natamycin, nisin, pomegranate and grape seed extract in chitosan coating' to extend shelf life of fresh strawberry. Food and Bioproducts Processing, 98, 354–363.

Ebrahimi F, Sadeghizadeh A, Neysan F, and Heydari M. 2019. Fabrication of nanofibers using sodium alginate and Poly (Vinyl alcohol) for the removal of Cd2+ ions from aqueous solutions: Adsorption mechanism, kinetics and thermodynamics. Heliyon, 5(11).

Emamifar A and Bavaisi S. 2020. Nanocomposite coating based on sodium alginate and nano-ZnO for extending the storage life of fresh strawberries (*Fragaria×ananassa* Duch.). Journal of Food Measurement and Characterization, 14, 1012–1024.

Eshghi S, Hashemi M, Mohammadi A, Badii F, Mohammadhoseini Z, and Ahmadi K. 2014. Effect of nano chitosan-based coating with and without copper loaded on physicochemical and bioactive components of fresh strawberry fruit (*Fragaria x ananassa* Duchesne) during storage. Food and Bioprocess Technology, 7, 2397–409.

Esmailzadeh H, Sangpour, P, Shahraz, F, Eskandari, A, Hejazi, J, and Khaksar, R. 2021. CuO/LDPE nanocomposite for active food packaging application: A comparative study of its antibacterial activities with ZnO/LDPE nanocomposite. Polymer Bulletin, 78(3), 1671–1682.

Espitia PJP, Soares NDFF, Coimbra JSDR, de Andrade NJ, Cruz RS, and Medeiros EAA. 2012. Zinc oxide nanoparticles: synthesis, antimicrobial activity and food packaging applications. Food and Bioprocess Technology, 5, 1447–1464.

Espitia PJP, Soares NDFF, Teófilo RF, dos Reis Coimbra JS, Vitor DM, Batista RA, Ferreira SO, de Andrade NJ, and Medeiros EAA. 2013. Physical–mechanical and antimicrobial properties of nanocomposite films with pediocin and ZnO nanoparticles. Carbohydrate Polymers, 94(1), 199–208.

Fadeyibi A. 2019. Optimization of processing parameters of nanocomposite film for fresh sliced okra packaging. The Journal of Applied Packaging Research, 11, 1.

Gour A and Jain NK. 2019. Advances in green synthesis of nanoparticles. Artificial Cells, Nanomedicine, and Biotechnology, 47(1), 844–851. DOI:10.1080/21691401.2019.1577878

Grist SM, Chrostowski L, and Cheung KC. 2010. Optical oxygen sensors for applications in microfluidic cell culture. Sensors, 10, 9286–9316.

Han JW, Ruiz-Garcia L, Qian JP, and Yang XT. 2018. Food packaging: A comprehensive review and future trends. Comprehensive Reviews in Food Science and Food Safety, 17(4), 860–877.

Indumathi MP, Sarojini KS, and Rajarajeswari GR. 2019. Antimicrobial and biodegradable chi-tosan/cellulose acetate phthalate/ZnO nano composite films with optimal oxygen permeability and hydrophobicity for extending the shelf life of black grape fruits. International Journal of Biological Macromolecules, 132, 1112–1120.

Istiqola A. and Syafiuddin A. 2020. A review of silver nanoparticles in food packaging technologies: Regulation, methods, properties, migration, and future challenges. Journal of the Chinese Chemical Society, 67(11), 1942–1956.

Jafarzadeh S, Alias AK, Ariffin F, and Mahmud S. 2018. Physico-mechanical and microstructural properties of semolina flour films as influenced by different sorbitol/glycerol concentrations. International Journal of Food Properties, 21(1), 983–995.

Jafarzadeh S, Rhim JW, Alias AK, Ariffin F, and Mahmud S. 2019. Application of antimicrobial active packaging film made of semolina flour, nano zinc oxide and nano-kaolin to maintain the quality of low-moisture mozzarella cheese during low-temperature storage. Journal of the Science of Food and Agriculture, 99(6), 2716–2725.

Jafarzadeh S, Nafchi AM, Salehabadi A, Oladzad-Abbasabadi N, and Jafari SM. 2021. Application of bio-nanocomposite films and edible coatings for extending the shelf life of fresh fruits and vegetables. Advances in Colloid and Interface Science, 102405.

Janaki AC, Sailatha E, and Gunasekaran S. 2015. Synthesis, characteristics and antimicrobial activity of ZnO nanoparticles. Spectrochim Acta Part A, 144, 17–22. doi:10.1016/j.saa.2015.02.041

Jayakumar A, Heera KV, Sumi TS, Joseph M, Mathew S, Praveen G, Indu C. Nair, and Radhakrishnan, EK. 2019. Starch-PVA composite films with zinc-oxide nanoparticles and phytochemicals as intelligent pH sensing wraps for food packaging application. International Journal of Biological Macromolecules, 136, 395–403.

Jiang T, Feng L, and Wang Y. 2013. Effect of alginate/nano-Ag coat-ing on microbial and physicochemical characteristics of shiitake mushroom (*Lentinus edodes*) during cold storage. Food Chemistry, 141(2), 954–960. doi:10.1016/j.foodchem.2013.03.093.

Jiménez-López J, Ortega-Barrales P, and Ruiz-Medina A. 2016. Development of an semi-automatic and sensitive photochemically induced fluorescence sensor for the determination of thiamethoxam in vegetables. Talanta, 149, 149–155.

Kalia A, Kaur M, Shami A, Jawandha SK, Alghuthaymi MA, Thakur A, and Abd-Elsalam KA. 2021. Nettle-leaf extract derived ZnO/CuO nanoparticle-biopolymer-based antioxidant and antimicrobial nanocomposite packaging films and their impact on extending the post-harvest shelf life of guava fruit. Biomolecules, 11(2), 224.

Kanikireddy V, Kanny K, Padma Y, Velchuri R, Ravi G, Jagan Mohan Reddy B, and Vithal M. 2019. Development of alginate-gum acacia-Ag0 nanocomposites via green process for inactivation of foodborne bacteria and impact on shelf life of black grapes (Vitis vinifera). Journal of Applied Polymer Science, 136(15), 47331.

Kanikireddy V, Varaprasad K, Rani MS, Venkataswamy P, Reddy BJM, and Vithal M. 2020. Bio-synthesis of CMC-Guar gum-Ag0 nanocomposites for inactivation of food patho-genic microbes and its effect on the shelf life of strawberries. Carbohydrate Polymers, 236, 116053.

Kaur M, Kalia A, and Thakur A. 2017. Effect of biodegradable chitosan—rice-starch nanocom-positefilms on post-harvest quality of stored peach fruit. Starch/Staerke, 69.

Kjellgren H, Gallstedt M, Engstrom G, and Jarnstron L. 2006. Barrier and surface properties of chitosan-coated greaseproof paper. Carbohydrate Polymers, 64, 453–460.

Kruis FE, Fissan H, and Rellinghaus B. 2000. Sintering and evaporation characteristics of gas-phase synthesis of size-selected PbS nanoparticles. Materials Science and Engineering: B, 69, 329–334.

Kumar A, Singh PP, and Prakash B. 2020. Unravelling the antifungal and anti-aflatoxin B1 mechanism of chitosan nanocomposite incorporated with Foeniculum vulgare essential oil. Carbohydrate Polymers, 236, 116050.

Kuswandi B and Moradi M. 2019. Improvement of food packaging based on functional nanomaterial. Nanotechnology: Applications in Energy, Drug and Food, 309–344.

Lee SY, Lee SJ, Choi DS, and Hur SJ. 2015. Current topics in active and intelligent food packaging for preservation of fresh foods. Journal of the Science of Food and Agriculture, 95, 2799–2810.

Li J, Sun Q, Sun Y, Chen B, Wu X, and Le T. 2019. Improvement of banana postharvest quality using a novel soybean protein isolate/cinnamaldehyde/zinc oxide bionanocomposite coating strategy. Scientia Horticulturae (Amsterdam), 258, 108786.

Li W, Li L, Cao Y, Lan T, Chen H, and Qin Y. 2017. Effects of PLA film incorporated with ZnOnanoparticle on the quality attributes of fresh-cut apple. Nanomaterials, 7, 207.

Lin F, Duan QY, and Wu FG. 2020. Conjugated polymer-based photothermal therapy for killing microorganisms. ACS Applied Polymer Materials, 2(10), 4331–4344.

Lipovsky A, Tzitrinovich Z, Friedmann H, Applerot G, Gedanken A, and Lubart R. 2009. EPR study of visible light-induced ROS generation by nanoparticles of ZnO. The Journal of Physical Chemistry C, 113(36), 15997–16001.

Liu J, Wang H, Guo M, Li L, Chen M, Jiang S, Li X, and Jiang S. 2019a. Extract from Lycium ruthenicum Murr. Incorporating κ-carrageenan colorimetric film with a wide pH–sensing range for food freshness monitoring. Food Hydrocolloids, 94, 1–10.

Liu Y, Qin Y, Bai R, Zhang X, Yuan L, and Liu J. 2019b. Preparation of pH-sensitive and antioxidant packaging films based on κ-carrageenan and mulberry polyphenolic extract. International Journal of Biological Macromolecules, 134, 993–1001.

Manoharan C, Pavithra G, Dhanapandian S, Dhamodaran P, and Shanthi B. 2015. Properties of spray pyrolised ZnO: Sn thin films and their antibacterial activity. Spectrochim Acta Part A, 141, 292–299.

Marrez DA, Abdelhamid AE, and Darwesh OM. 2019. Eco-friendly cellulose acetate green synthesized silver nano-composite as antibacterial packaging system for food safety. Food Packaging and Shelf Life, 20, 100302.

Moalemiyan M, Ramaswamy HS, and Maftoonazad N. 2012. Pectin-based edible coating for shelf-life extension of ataulfo mango. Journal of Food Process Engineering, 35(4), 572–600.

Mukherjee A, Waters AK, Kalyan P, Achrol AS, Kesari S, and Yenugonda VM. 2019. Lipid—polymer hybrid nanoparticles as a next-generation drug delivery platform: State of the art, emerging technologies, and perspectives. International Journal of Nanomedicine, 14, 1937.

Munteanu BS, Sacarescu L, Vasiliu AL, Hitruc GE, Pricope GM, Sivertsvik M, Rosnes JT, and Vasile C. 2018. Antioxidant/antibacterial electrospun nanocoatings applied onto PLA films. Materials, 11(10), 1973.2

Naknaen P. 2014. Utilization possibilities of antimicrobial biodegradable packaging produced by poly (*butylene succinate*) modified with zinc oxide nanoparticles in fresh-cut apple slices. International Food Research Journal, 21.

Narasagoudr SS, Hegde VG, Chougale RB, Masti SP, Vootla S, and Malabadi RB. 2020. Physico-chemical and functional properties of rutin induced chitosan/poly (vinyl alcohol) bioactive films for food packaging applications. Food Hydrocolloids, 109, 106096.

Nouri A, Yaraki MT, Ghorbanpour M, Agarwal S, and Gupta VK. 2018. Enhanced antibacterial effect of chitosan film using montmorillonite/CuO nanocomposite. International Journal of Biological Macromolecules, 109, 1219–1231.

Ortiz-Duarte G, Pérez-Cabrera LE, Artés-Hernández F, and Martínez-Hernández GB. 2019. Ag-chitosan nanocomposites in edible coatings affect the quality of fresh-cut melon. Postharvest Biology and Technology, 147, 174–184.

Oun AA and Rhim JW. 2017. Effect of oxidized chitin nanocrystals isolated by ammonium persulfate method on the properties of carboxymethyl cellulose-based films. Carbohydrate Polymers, 175, 712–720.

Ouzounidou G and Gaitis F. 2011. The use of nano-technology in shelf life extension of green vegetables. Journal of Innovation Economics Management, 2, 163–171.

Pandey K and Rizvi S. 2009. Plant polyphenols as dietary antioxidants in human health and disease. Oxidative Medicine and Cellular Longevity, 2(5), 270–278.

Panea B, Ripoll G, González J, Fernández-Cuello Á, and Albertí P. 2014. Effect of nanocomposite packaging containing different proportions of ZnO and Ag on chicken breast meat quality. Journal of Food Engineering, 123, 104–112.

Pereira M and Malfeito-Ferreira M. 2015. A simple method to evaluate the shelf life of refrigerated rabbit meat. Food Control, 49, 70–74.

Raghupathi KR, Koodali RT, and Manna AC. 2011. Size-dependent bacterial growth inhibition and mechanism of antibacterial activity of zinc oxide nanoparticles. Langmuir, 27(7), 4020–4028.

Ramesh M, Anbuvannan M, and Viruthagiri G. 2015. Green synthesis of ZnO nanoparticles using Solanum nigrum leaf extract and their antibacterial activity. Spectrochimica Acta Part A: Molecular and Biomolecular Spectroscopy, 136, 864–870.

Restuccia D, Spizzirri UG, Parisi OI, Cirillo G, Curcio M, Iemma F, Puoci F, Vinci G, and Picci N. 2010. New EU regulation aspects and global market of active and intelligent packaging for food industry applications. Food Control, 21, 1425–1435.

Ross J and Kasum C. 2002. Dietary flavonoids: Bioavailability, metabolic effects, and safety. Annual Review of Nutrition, 22(1), 19–34.

Saba Mka and Amini R. 2017. Nano-ZnO/carboxymethyl cellulose-based active coating impacton ready-to-use pomegranate during cold storage. Food Chemistry, 232, 721–726.

Saekow M, Naradisorn M, Tongdeesoontorn W, and Hamauzu Y. 2019. Effect of carboxymethyl cellulose coating containing ZnO-nanoparticles for prolonging shelf life of persimmon and tomato fruit. Journal of Agricultural Science and Food Technology, 5, 41–48.

Sarwar MS, Niazi MBK, Jahan Z, Ahmad T, and Hussain A. 2018. Preparation and characterization of PVA/nanocellulose/Ag nanocomposite films for antimicrobial food packaging. Carbohydrate Polymers, 184, 453–464.

Shankar S, Teng X, Li G, and Rhim JW. 2015. Preparation, characterization, and antimicrobial activity of gelatin/ZnO nanocomposite films. Food Hydrocolloids, 45, 264–271.

Sharif N, Khoshnoudi-Nia S, and Jafari SM. 2020. Nano/microencapsulation of anthocyanins; a systematic review and meta-analysis. Food Research International, 132, 109077.

Shi LE, Li ZH, Zheng W, Zhao YF, Jin YF, and Tang, ZX. 2014. Synthesis, antibacterial activity, antibacterial mechanism and food applications of ZnO nanoparticles: A review. Food Additives & Contaminants: Part A, 31(2), 173–186.

Simbine EO, Rodrigues LDC, Lapa-Guimaraes J, Kamimura ES, Corassin CH, and Oliveira, CAFD. 2019. Application of silver nanoparticles in food packages: A review. Food Science and Technology, 39, 793–802.

Sirelkhatim A, Mahmud S, Seeni A, Kaus NH, Ann LC, and Bakhori SK. 2015. Review on zinc oxide nanoparticles: antibacterial activity and toxicity mechanism. Nano-Micro Letters, 7(3), 219–242. Doi:10.1007/s40820-015-0040-x

Siripatrawan U and Kaewklin P. 2018. Fabrication and characterization of chitosan-titaniumdioxide nanocompositefilm as ethylene scavenging and antimicrobial active foodpackaging. Food Hydrocolloids, 84, 125–134.

Steffy K, Shanthi G, Maroky AS, and Selvakumar S. 2018. Synthesis and characterization of ZnO phytonanocomposite using Strychnos nux-vomica L.(Loganiaceae) and antimicrobial activity against multidrug-resistant bacterial strains from diabetic foot ulcer. Journal of Advanced Research, 9, 69–77.

Suo B, Li H, Wang Y, Li Z, Pan Z, and Ai Z. 2017. Effects of ZnO nanoparticle-coated packaging film on pork meat quality during cold storage. Journal of the Science of Food and Agriculture, 97(7), 2023–2029.

Tarone AG, Cazarin CBB, and Junior MRM. 2020. Anthocyanins: New techniques and challenges in microencapsulation. Food Research International, 133, 109092.

Tharanathan R. 2003. Biodegradable films and composite coatings: Past, present and future. Trends in Food Science and Technology, 14, 71–78.

Theapsak S, Watthanaphanit A, and Rujiravanit R. 2012. Preparation of chitosan-coated polyethylene packaging films by DBD plasma treatment. ACS Applied Materials & Interfaces, 4(5), 2474–2482.

Vishwasrao C and Ananthanarayan L. 2016. Postharvest shelf-life extension of pink guavas (*Psidium guajava* L.) using HPMC-based edible surface coatings. Journal of Food Science and Technology, 53, 1966–1974.

Wang H, Qian J, and Ding F. 2018. Emerging chitosan-based films for food packaging applications. Journal of Agricultural and Food Chemistry, 66(2), 395–413.

Wu Z, Huang X, Li YC, Xiao H, and Wang X. 2018. Novel chitosan films with laponite immobilized Ag nanoparticles for active food packaging. Carbohydrate Polymers, 199, 210–218.

Xing Y, Yang H, Guo X, Bi X, Liu X, Xu Q, Wang Q, Li W, Li X, Shui Y, and Zheng Y. 2020. Effect of chitosan/Nano-TiO_2 composite coatings on the postharvest quality and physicochemical characteristics of mango fruits. Scientia Horticulturae, 263, 109135.

Xu D, Qin H, and Ren D. 2018. Prolonged preservation of tangerine fruits using chitosan/montmorillonite composite coating. Postharvest Biology and Technology, 143, 50–57. https://doi.org/10.1016/j.postharvbio.2018.04.013.

Ye W, Pei F, Lan X, Cheng Y, Fang X, Zhang Q, Zheng Y, Chen C, Peng DL and Wang, MS. 2020. Stable nano-encapsulation of lithium through seed-free selective deposition for high-performance Li battery anodes. Advanced Energy Materials, 10(7), 1902956.

Youssef AM, El-Sayed SM, El-Sayed HS, Salama HH, Assem FM, and Abd El-Salam MH. 2018. Novel bionanocomposite materials used for packaging skimmed milk acid coagulated cheese (Karish). International Journal of Biological Macromolecules, 115, 1002–1011.

Chapter 16

Toxicity Aspects of Nanomaterials and Their Composites Being Used in the Postharvest Nanotechnology

Yolanda González García, Mari Carmen López Pérez, Fabián Pérez Labrada, Gregorio Cadenas Pliego, and Antonio Juárez Maldonado

List of Abbreviations

Ag NPs	Silver nanoparticles
APX	Ascorbate peroxidase
CAT	Catalase
LOX	Lipoxygenase
MDA	Malondialdehyde
NM	Nanomaterial
NP	Nanoparticle
PAL	Phenylalanine ammonia-lyase
PD	Parkinson's disease
POD	Peroxidase
PPO	Polyphenol oxidase
PS	Plastic nanoparticle
rGO	Graphene-oxide nanocompound
RH	Relative humidity

DOI: 10.1201/9781003142287-16

ROS	Reactive oxygen species
SOD	Superoxide dismutase
TSS	Total soluble solid

16.1 Introduction

One of the main sources of food for humans is of plant origin, and hence, the production of agricultural crops is one of the most important activities of humans. However, once the fruits and vegetables are harvested, they follow a natural degradation process due to their active metabolic activity. This results in the loss of water, loss of firmness, discoloration, browning, and susceptibility to microorganisms, among other effects (Gouda et al., 2020; Gundewadi et al., 2018). These changes, which decrease the postharvest quality and the shelf life of fruits and vegetables, result in considerable economic losses (10–40%; Bose et al., 2021; Mossa et al., 2021). Also, a large part of these foods when they reach the consumers are not in their optimal quality (Chen et al., 2020).

To counteract the negative impacts derived from the physiological activity of fruits and vegetables, various alternatives have been sought, such as the use of organic and inorganic coatings, packaging, modified atmospheres, or other treatments that reduce losses, and increase postharvest quality and shelf life (Munhuweyi et al., 2020; Riva et al., 2020; Wang and Zhang, 2020). For this, the unique characteristics of NMs (size, shape, surface/volume ratio, etc.) have been used (Juárez-Maldonado et al., 2019, 2021) and have been applied in two main ways, through NM-based coatings and NM-based packaging. The use of NMs has made it possible to reduce the respiration rate, loss of water, softening of the fruits and increase the shelf life and quality, as well as the control of microbial activity in fruits and vegetables (Saqib et al., 2020; Saravanakumar et al., 2020; Vieira et al., 2020; Zambrano-Zaragoza et al., 2020; Zhao et al., 2021).

However, these NMs in some cases can get consumed by humans together with the food in which they were applied as edible coatings. Also, they can move through trophic transfer from the waste that is generated by the packaging and ultimately reach living organisms (Morales-Díaz et al., 2017). This can represent certain risks not only to humans when consuming NM-coated fruits and vegetables but also to other organisms at different trophic levels due to the NM accumulation in the soil or water.

This chapter describes the NMs that have been applied in order to improve the postharvest characteristics of fruits and vegetables, through different technologies such as coatings, packaging or others. Based on this, the possible risks with the use of these NMs in food and packaging are reviewed. This is intended to put the application of nanotechnology in food postharvest into perspective in order to improve the scientific development of this area without putting human health and the environment at risk.

16.2 Nanotechnology

Nanotechnology is the application of various materials that have at least one dimension on a scale of 1–100 nm. Materials at this scale have particular characteristics such as shape, size, dimensions, origin, and composition. Due to the zeta potential and high density of surface charge, the interaction of NMs with the cell walls and membranes causes the generation of a series of responses at the biochemical and physiological levels (Juárez-Maldonado et al., 2019; Manzoor et al., 2020; Wang et al., 2014).

The interaction between NMs and different organelles can improve the physiological processes and also impact the nutritional quality and postharvest life of fruits and vegetables (Mali et al., 2020). In addition, NMs have a great advantage to be used in agriculture and postharvest, since the materials such as TiO_2, Ag, CuO, and graphene represent a low environmental risk, which together with the multiple benefits that they can induce can be very useful when applied in different ways (Zhao et al., 2020).

16.3 Postharvest in Foods

Fruits and vegetables are the main foods for human consumption, so their production depends directly on the demand of the population, requiring more and more food of better quality (Maringgal et al., 2020). However, fruits and vegetables, once they are harvested, present losses as well as deterioration that leads to a decrease in quality, representing up to a third of those that are not of adequate quality for consumption (Chen et al., 2020).

Fruits and vegetables show an active metabolism even after harvest, especially the climacteric fruits that continue to ripen after being harvested due to the production of CO_2 and C_2H_4 (Gouda et al., 2020; Gundewadi et al., 2018). In particular, the respiration rate is one of the most important factors in postharvest, since it induces the decomposition of organic molecules causing the "softening" of the tissues due to the loss of water (Gundewadi et al., 2018).

The ripening and senescence of the fruit are regulated mainly by ethylene and abscisic acid, and these processes influence several important quality parameters in postharvest, such as firmness, color, aroma, sugar content, and organic acids (Leng et al., 2014; Tang et al., 2020). During the ripening process, the loss of water in the fruits that originates the softening occurs which increases the susceptibility to pathogens and decreases the postharvest quality (Posé et al., 2019), thereby generating economic losses ranging from 10 to 40% (Bose et al., 2021; Mossa et al., 2021).

Most fruits are not marketed with a water loss of 5 to 10% of their initial weight. In addition to causing marketable weight loss, this promotes "browning", loss of firmness, texture, flavor, accelerated senescence, susceptibility to cold damage, and membrane disintegration (Lufu et al., 2020). Browning is an important physiological disorder that causes economic losses and is due to the enzyme PPO. This enzyme is characterized by containing copper, which, under the presence of oxygen, hydroxylates monophenols to o-diphenols, and subsequently, these are oxidized to o-quinones, finally causing brown, red, and black pigments (Karimirad et al., 2018; Lo'ay and Ameer, 2019; Queiroz et al., 2008).

Pathogens and pests can also induce a reduction in the postharvest quality of fruits and vegetables (Sharma et al., 2020). Postharvest diseases cause the loss of 10% to 30% of total crop yield, and they occur during the handling, transport, storage, and marketing (Mossa et al., 2021). In addition, infected fruits can be risky for human health due to the fact that some pathogens such as *Penicillium expansum* produce toxic secondary metabolites, such as patulin and citrinin as in the case of apple fruits (Yu et al., 2020).

Poor handling of fruits and vegetables, storage in inappropriate packaging, and physiological changes themselves are the main problems that must be controlled postharvest (Chen et al., 2020; Maringgal et al., 2020). Mechanical damage such as cuticular cracking, pitting, and bruising can lead to serious damage to the epidermis and excessive loss of water in the fruit. This can also increase the production of ethylene and favor contamination with pathogenic microorganisms (Hussein et al., 2020; Jansasithorn et al., 2014). Therefore, strategies to improve the postharvest quality of fruits and vegetables are aimed at reducing the impact of these factors and increasing the shelf life.

16.4 Nanotechnology in Postharvest

The application of nanotechnology in postharvest is oriented in two main areas, its use as coatings either of a single NM or in combination with some other compounds and the use of NMs for the production and/or improvement of packaging such as bags, boxes, and others (Tables 16.1 and 16.2). The main problem observed in fruits or other foods is the loss of weight derived from the loss of water during the respiration process, and this can also result in the softening of the foods. The softening is caused by the activity of enzymes that degrade the cell membrane (phospholipase-D and pectin methyl esterase), in addition to other compounds such as MDA and the activity of PPO (Ashitha et al., 2020; Nguyen and Nguyen, 2020). The respiration process and other metabolic processes, such as ripening, remain active after harvest and especially in high-temperature conditions (Malek et al., 2020; Melo et al., 2020). Hence, these factors must be controlled in order to maintain the postharvest quality. In addition to these problems, the browning of food also decreases the postharvest quality of food. The browning results from the oxidation of phenols to o-quinone through the activity of PPO, resulting in the production of brown, red, and black pigments (Karimirad et al., 2018; Lo'ay and Ameer, 2019; Queiroz et al., 2008; Ranjbar et al., 2018).

Derived from this problem that affects the postharvest quality of foods, alternatives have been sought to control the main factors, namely, weight loss, fruit softening, ripening, and enzymatic browning. For this, various NMs have been used for the production of coatings, packaging, or other applications. For example, the application of NM-based coatings to fruits can decrease the degradation of the cell wall and enzymatic browning, in addition to increasing the content of antioxidant compounds (Karimirad et al., 2018). In addition, the production of ethylene by climacteric fruits can be reduced, as well as the production of MDA during ripening (Nguyen et al., 2020), resulting in a longer shelf life.

16.4.1 NM-Based Coatings

Among the possible applications of nanotechnology in postharvest, the use of NMs as coatings is the main one, and they can be applied individually or in combination with other materials (Table 16.1). This is possible because NMs have particular and novel characteristics (size, mono-dispersion, zeta potential, architecture) that improve the application of the coatings, making them more uniform and firm, even in fruits with waxy surfaces (Meena et al., 2020).

NM-based coatings have the ability to form semipermeable barriers on the surface of the fruits, modifying the gas exchange between the fruits and the environment (Amiri et al., 2020; Melo et al., 2020). This results in a variety of responses such as decreased weight and water loss, decreased respiration rate, and reduced electrolyte leakage, among others (Shah et al., 2015). Also, pigments oxidation, the production

Table 16.1 NM-Based Coatings and Their Effects on Postharvest of Foods

NM	*Foods*	*Storage condition*	*Effects*	*References*
Ag NPs + Sodium alginate	Carrots (*Daucus carota* L.) Pears (*Pyrus communis* L)	Room temperature at 27°C for 10 days	Weight loss reduction	Fayaz et al., 2009
Agar–Ag NPs	Lime (*Citrus aurantifolium* L.) Apple (*Pyrus malus* L.)	9 days	Weight loss reduction. Increase in antimicrobial capacity.	Gudadhe et al., 2014
Nanochitosan	Asiatic wild rice (*Zizania latifolia* (Griseb.) Turcz. ex Stapf.)	Stems was stored at 1°C for 12 days	Browning, lignification and the increase in total color difference were retarded. Lower POD and MDA, maintain SOD, improved CAT, and APX.	Luo et al., 2013
Solid lipid NP + xanthan gum	Guava (*Psidium guajava* L.)	10°C and 85% RH for 30 days	Weight loss reduction and preserved the best quality	Zambrano-Zaragoza et al., 2013
Chitosan–surfactant nanostructure	Tomato (*Solanum lycopersicum* L.)	Stored at 15±2°C and 70–80 % RH for 20 days	Decrease titratable acidity and chlorophyll content. Increased soluble solids and phenol content. Lower level of respiration and weight loss.	Mustafa et al., 2014
Carboxymethyl cellulose–Ag NPs and guar gum–Ag NPs	Kinnow (*Citrus reticulata* L. cv. Blanco)	4–10°C, 85–90% RH for 120 days	Increased ascorbic acid, total phenols, and antioxidant activity. Lower incidence of fungal diseases.	Shah et al., 2015

(Continued)

Table 16.1 (Continued)

NM	*Foods*	*Storage condition*	*Effects*	*References*
Nano-Zn	Strawberries (*Fragaria* x *anannasa* Duch. cv. Parous.)	Stored at 1°C with 95% RH for 18 days	Decrease in microbial load. Weight loss delay, firmness maintenance. Increased anthocyanin, ascorbic acid, phenol, and antioxidant activity.	Sogvar et al., 2016
Chitosan/ nano-silica	Loquat (*Eriobotrya japonica* Lindl. cv. Baiyu)	5°C for 40 days.	Delayed internal browning and weight loss. Maintain TSSs and titratable acidity. Increased glucose, fructose, SOD, CAT, APX, PAL, LOX. Decreased MDA and membrane permeability.	Song et al., 2016
Ag NPs	Cherry tomatoes (*Solanum lycopersicum* L. cv. cerasiforme)	Room temperature for 18 days	Reduced weight loss. Increased TSSs and ascorbic acid.	Gao et al., 2017
Solid lipid NPs from candeuba wax and xanthan gum	Guava (*Psidium guajava* L. cv. Media china)	10°C, 85% RH for 5 weeks	Lower O_2 and CO_2 respiration rate. Less weight loss. Better retention of ascorbic acid and total phenols.	García-Betanzos et al., 2017
Curcumin-loaded zein NPs	Grapes (*Vitis vinifera* L. cv. Benitaka)	25°C and 50% RH for 7 days	There was no alteration in titratable acidity. The presence of the coating did not reduce the texture properties and in turn increased the intensity of the fruit's color.	Lemes et al., 2017

NM	Foods	Storage condition	Effects	References
Nano-ZnO + carboxymethyl cellulose	Pomegranate arils (*Punica granatum* L. cv. Malas)	4°C for 12 days	Decreased total yeast + mold and total mesophilic bacteria. Decreased weight loss and soluble solids. Higher total anthocyanin, ascorbic acid, and antioxidant capacity.	Saba and Amini, 2017
Nanochitosan	Peach (*Prunus persica.* L. cv. Early Swelling)	4 ± 2°C for 7 weeks	Weight loss reduction. Improved fruit firmness.	Ali and Toliba, 2018
1. Chitosan-thyamol/tripolyphosphate NPs 2. Gelatin–coconut fiber powder/titanium dioxide (TiO_2) NPs 3. Chitosan-methyl cellulose/silica (SiO_2) NPs 4. Gelatin-Chitosan/(Ag/ZnO) NPs 5. Gelatin-Anthocyanin/kafirin NPs	Fruits: apples (*Malus Domestica* L. cv. Anna) and red grapes (*Vitis vinifera* L. cv. Flame Seedless) Vegetables: tomatoes (*Lycopersicon lycopersicum* L. cv. Suber Strain B') and sweet green pepper (*Capsicum annuum* L. cv. California Wonde)	Fruit: 0–2°C, 90–95% RH (63 and 56 days, respectively, for apples and grapes) Vegetables: 8°C, 90–95% RH for 42 and 35 days, respectively, for tomatoes and peppers.	Weight loss reduction. Increased TSSs, firmness, total acidity, and ascorbic acid.	Bakhy et al., 2018
Chitosan-TiO_2 NPs	Cherry tomatoes (*Solanum lycopersicum* L.)	Stored at 20°C and 85 % RH	Exhibited ethylene photodegradation activity and delayed the ripening process, and changes in the quality of fruit.	Kaewklin et al., 2018

(Continued)

Table 16.1 (Continued)

NM	*Foods*	*Storage condition*	*Effects*	*References*
ZnO NPs	Fig (*Ficus carica* L.)	Room temperature for 6 days	Delayed maturation. Weight loss reduction. Reduced color change. Increased firmness and inhibited the growth of microorganisms.	Lakshmi et al., 2018
Chitosan NPs	Cavendish banana (*Musa acuminata* AAA group)	25 ± 1°C for 15 days.	Shelf life was extended due to modifications in starch content, weight loss, pulp-peel ratio, and TSSs.	Lustriane et al., 2018
Carrageenan—ZnO NPs	Mango (*Mangifera indica* L. cv. Gedong Gincu)	20°C, 61% RH for 33 days	Decreased transmission of water vapor. Increased antimicrobial activity. Delayed discoloration.	Meindrawan et al., 2018
Chitosan NPs	Grapes (*Vitis labrusca* L.)	25°C for 12 days. 12°C for 24 days.	Delayed grape ripening. Decreased weight loss soluble solids and reduced sugar content. Greater retention of moisture and preservation of valuable acidity values. Inhibition of microorganism growth.	Melo et al., 2018
Thymol nanoemulsions incorporated in quinoa protein/ chitosan	Cherry tomatoes (*Solanum lycopersicum* L.)	5°C for16 days.	Decreased growth of *Botrytis cinerea*	Robledo et al., 2018

NM	*Foods*	*Storage condition*	*Effects*	*References*
Chitosan NPs	Cavendish banana (*Musa acuminata* AAA group)	22 ± 1°C	Slower skin discoloration.	Esyanti et al., 2019
Soybean protein isolate + cinnamaldehyde + ZnO NPs	Banana (Musaceae)	25°C, 40% RH for 7 days	Less weight loss, more firmness. Reduction of the content of total soluble sugars and increase in titratable acidity.	Li et al., 2019
Ca NPs	Mango (*Mangifera indica* L. cv Hindi Be-Sennara)	6 ± 1°C, 95 ± 1% RH for 35 days	Reduction of the internal browning index. Higher content of total phenolic compounds and flavonoids. Lower activity of the PPO, PAL, and LOX and lower content of MDA. Reduction of pectinase and cellulase activity and less electrolyte leakage from the cell membrane. Higher antioxidant activity.	Lo'ay and Ameer, 2019
Polymeric nanocapsules containing *Thymus vulgaris* L. essential oil	Grapes (*Vitis vinifera* L.)	25°C for 13 days	Increased antioxidant capacity and maintained characteristics of color, firmness, titratable acidity, and TSS for longer time. Acting as a barrier that reduced the metabolism of fruits.	Pina-Barrera et al., 2019

(Continued)

Table 16.1 (Continued)

NM	*Foods*	*Storage condition*	*Effects*	*References*
Ag NPs	Loquat (*Eriobotrya japonica* Lindl.)	Storage at 4 and 8°C for 1 month.	Reduced weight loss, TSSs and sugars. Increased content of ascorbic acid, phenols, antioxidants and titratable acidity.	Ali et al., 2020
Chitosan NPs	Pomegranates arils (*Punica granatum* L. cv. Rabab)	5°C for 18 days	Reduction of microbial counts. Higher content of total phenol and ascorbic acid.	Amiri et al., 2020
Chitosan + nano ZnO	Guava (*Psidium guajava* L.)	21 ± 1°C, 80 ± 2% RH for 15 days	Inhibition of rot, delayed ripening process, less weight loss.	Arroyo et al., 2020
Chitosan + nano SiO_2	*Blueberry (Vaccinium myrtillus* L.)	Room temperature for 8 days	Inhibition of microbial populations. Increased anthocyanin content.	Eldib et al., 2020
Chitosan NPs loaded with clove essential oil	Pomegranate arils (*Punica granatum* L. cv. Malase Saveh)	5°C	Extended the shelf life of the aril for 54 days. Weight, TSSs, titratable acidity, pH, total phenol, and total anthocyanin content, as well as antioxidant activity and sensory quality, were maintained.	Hasheminejad and Khodaiyan, 2020
Chitosan NPs with α-pineno	Bell peppers (*Capsicum annuum* L. cv. California)	12 ± 2°C for 21 days followed by 20 ± 2°C for 5 days	Alterations in weight loss and preservation of physicochemical quality. Inhibition of *Alternaria alternata*.	Hernández-López et al., 2020

NM	*Foods*	*Storage condition*	*Effects*	*References*
Chitosan NPs	Bell pepper (*Capsicum annuum* L. cv. grossum)	5°C for12 days	Less weight loss (protection against water loss). Inhibition of pathogens.	Hu et al., 2020
Chitosan combined with ZnO NPs	Papaya (*Carica papaya*. L cv. California)	10°C for 12 days.	Suppression of microbial growth.	Lavinia et al., 2020
Natamycin-loaded zein NPs stabilized by carboxymethyl chitosan	Hongjia strawberries (*Fragaria × ananassa* Duch. cv. Benihoppe)	28°C 90% RH during 8 days	Reduction of the occurrence of rot, mildew, and gray mold.	Lin et al., 2020
Chitosan NPs	Strawberries (*Fragaria × ananassa* Duch. cv. Camarosa)	4 ± 1°C for 8 days	Less weight loss, more firmness. Maintenance of antioxidant compounds and antioxidant capacity.	Martínez-González et al., 2020
ZnO NPs	Mango (*Mangifera indica* L. cv. Golden Lily)	27 and 32°C for 7 days.	Ripening delayed, weight loss slowdown, and increased firmness.	Malek et al., 2020
Chitosan NPs	Strawberry (*Fragaria* x *ananassa* Duch.)	25 ± 2°C for 6 days 10 ± 2°C for 12 days.	Antifungal action. Reduction in weight loss, TSSs, maturity index, and moisture loss.	Melo et al., 2020
Cu-chitosan NPs	Tomato (*Solanum lycopersicum* L.)	27 ± 2°C, 55 ± 2% RH for 21 days	Microbial decomposition reduction. Reduction in weight loss and respiration rate. Maintained fruit firmness. Delay of loss of titratable acidity. Retention of TSSs, total and reducing sugars, lycopene, ascorbic acid. Inhibition of PPO.	Meena et al., 2020

(Continued)

Table 16.1 (Continued)

NM	*Foods*	*Storage condition*	*Effects*	*References*
Nano-chitosan	Strawberry (*Fragaria* x *ananassa* Duch.)	2°C for 21 days	Preservation of the quality index. Reduction of weight loss. Retention of firmness, valuable acidity, and ascorbic acid. Increased anthocyanin content. MDA production delay and inhibition of PPO activity.	Nguyen and Nguyen, 2020
$CaCl_2$+nano-chitosan	Strawberry (*Fragaria* x *ananassa* Duch.)	4°C for 15 days	Weight loss reduction. Preservation of ascorbic acid and total anthocyanins. Malonaldehyde production delay.	Nguyen et al., 2020
Ag NP-polyvinylpyrrolidone-based glycerosomes	Red and yellow bell pepper (*Capsicum annuum* L. var. grossum (L.) Sendt)	4°C and 15°C, 90–95 % RH for 15 days.	Extension of the shelf life of bell peppers (red and yellow).	Saravanakumar et al., 2020
Ag NPs	Apricot (*Prunus armeniaca* L. cv. Canino)	25 ± 3°C and 6 ± 2°C for 4 days.	Less weight loss, TSSs and MDA concentration. High content of total acidity and total carotenoids.	Shahat et al., 2020a
Chitosan NPs	Strawberry (*Fragaria* x *ananassa* Duch. cv. Camarosa)	25 ± 3°C for 6 days and 6 ± 2°C for 16 days	Lower values for weight loss, decay TSS and MDA concentration. Higher levels of total acidity and anthocyanin. Less of total aerobic, mold and yeast.	Shahat et al., 2020b

NM	Foods	Storage condition	Effects	References
Chitosan + iron oxide NPs	Peach (*Prunus persica* L.)	4°C for 25 days	Reduction of weight loss. Inhibition of microbial growth on the fruit surface.	Saqib et al., 2020
Ag NPs + hydroxypropyl-methylcellulose	Papaya (*Carica papaya* L. cv. Golden)	10°C for 7 days, 88.5% RH, followed by 20°C for 7 days, 88.5% RH	Reduced the incidence and severity of *Colletotrichum gloeosporioides* and weight loss.	Vieira et al., 2020
Chitosan/ nano-TiO_2	Mango (*Mangifera indica* L.)	13°C for 20 days.	Reduced decomposition index, respiration, TSSs, and MDA. Increased firmness, POD, PPO, phenol, and flavonoids.	Xing et al., 2020
Beeswax Solid Lipid NPs + xanthan gum and propylene glycol	Strawberry (*Fragaria* x *ananassa* Duch. cv. camarosa)	4°C for 21 days.	Less weight loss, low decay rate, and less loss of firmness.	Zambrano-Zaragoza et al., 2020
S-nitrosoglutathione-chitosan NPs	Apple (*Malus domestica* L. cv. Fuji)	4°C for 4 days	It preserved the quality of the freshly cut slices. Higher polyphenol content, APX, POD, CAT, and SOD activity. Maintained the surface color and weight. Reduced the respiration rate and ROS.	Zhao et al., 2021

of ethylene, and the ripening process, in addition to inhibiting the enzymes related to hydrolysis of starch are decreased (Melo et al., 2020; Sorrentino et al., 2007). All of this results in increased shelf life and postharvest quality. However, the final effect depends on the NM used, concentration, type, size, and exposure time (Liu et al., 2020).

Possibly, the decrease in respiration rate is one of the main mechanisms that NM-based coatings have, but they also have the ability to reduce the degradation of key antioxidants related to postharvest quality, such as the ascorbic acid and phenols, which can decrease the activity of the PPO responsible for the enzymatic browning (Ali et al., 2020; Chi et al., 2019; Meena et al., 2020).

The joint application of organic compounds, such as chitosan, and NMs, such as Cu NPs, has shown positive results in postharvest quality. Cu NP–chitosan coating has been reported to decrease the respiration rate and weight loss in strawberry (*Fragaria x ananassa* Duch.) and tomato fruits (*Solanum lycopersicum* L.; Meena et al., 2020; Nguyen et al., 2020).

The NM-based coatings have the ability to damage the cell membrane of microorganisms and even affect their DNA, thereby limiting microbial growth of these in foods (Morones et al., 2005). This effect is also due to the fact that NMs can induce the production of ROS and oxidize the cellular components of microorganisms, causing cell death (Dakal et al., 2016). For example, ZnO NPs can limit the activity of microorganisms in fruits due to the induction of ROS and the release of Zn^{2+} ions that can penetrate the cell wall and react with the cytoplasm (Sun et al., 2018).

NMs such as chitosan can inhibit the growth of microorganisms (Ali and Toliba, 2018). Nanochitosan has been observed to penetrate the cell wall of bacteria, thereby providing a protective effect on grapes (*Vitis labrusca* L.; Melo et al., 2018). In addition, nanochitosan decreased the respiration and metabolism of the fruits, reduced water loss, and reduced oxidation reactions by blocking the pores of the fruits (Amiri et al., 2020; Melo et al., 2018, 2020; Shah et al., 2015).

In addition to its antimicrobial effect, nanochitosan-based coatings can increase the content of antioxidants such as phenols and anthocyanins. This may be due to the fact that nanochitosan can react with free radicals through the residual free amino groups that can originate ammonium groups (NH^{+3}) by taking hydrogen from the cell (Amiri et al., 2020; Melo et al., 2020). Other NMs of organic origin with antimicrobial effects, due to their particle size and concentration, are the beeswax solid lipid NPs with xanthan gum and propylene glycol. These coatings were evaluated in strawberries (*Fragaria x ananassa* Duch. Cv. Camarosa) with positive effects on the reduction of fungal growth (Zambrano-Zaragoza et al., 2020).

16.4.2 NM-Based Packaging

The use of NM-based packaging to increase the postharvest life, as well as maintain or even improve some characteristics of food during postharvest, is one of the main applications of nanotechnology (Table 16.2). For this, various NMs such as Ag, Ag_2O, TiO_2, ZnO, and SiO_2 have been used as well as some NMs of organic origin, such as chitosan, and, in some cases, a combination of both types of materials, such as chitosan-TiO_2. NM-based packaging works very similarly to coatings since they are a physical barrier with physical-chemical characteristics that improve postharvest quality and shelf life (Basumatary et al., 2020). Derived from this, positive effects have been observed with the use of NMs in the packaging of some foods, mainly because they can maintain quality for longer and therefore increase the shelf life. This is mainly due to the reduction or inhibition of weight loss, decreased loss of firmness, delayed maturation, lower browning index, and less microbial activity (Table 16.2).

The application of NMs together with polyethylene has been one of the main uses of nanotechnology in order to improve the postharvest of foods. Polyethylene bags with NPs of TiO_2, Ag_2O, Ag, and SiO_2 have been developed for the packaging of

Table 16.2 NM-Based Packaging and Their Effects on Postharvest of Foods

NM	*Packaging and storage*	*Foods*	*Effects*	*References*
Polyethylene with nano-Ag, nano-TiO_2, montmorillonite	Bags. Stored at 4 ± 1°C by 42 days.	Kiwifruit (*Actinidia deliciosa* L. cv. Qinmei)	Inhibition of weight loss, softening, content of TSSs. Lower concentration of MDA. Increase in the content of ascorbic acid and phenols.	Hu et al., 2011
Polyethylene with Ag_2O NPs	Bags. Stored at 5 and 15 ± 1°C in darkness up to 24 days	Apple (*Malus domestica* L. cv. Fuji)	Decreased weight loss. Microbial spoilage prevention.	Zhou et al., 2011
Polyethylene with Ag NPs	Bags. Stored at 10°C (±2°C) and RH 88 % for 10 days.	Carrot (*Daucus carota* L. cv. Brasília)	Higher content of ascorbic acid, lower weight loss and lower content of soluble solids.	Becaro et al., 2016
Polyethylene with nano-TiO_2	Bags. Stored at 4°C for 10 days	Strawberry (*Fragaria* x *ananassa* Duch. cv. Akihime)	Internal reduction of O_2 and increase of CO_2. Lower decomposition rate and weight loss, delay in the decrease of firmness and titratable acidity. ROS ($O_2^{\bullet-}$ and H_2O_2) decrease. Increased antioxidant activity. Retention in content of ascorbic acid and total phenols.	Li et al., 2017

(Continued)

Table 16.2 (Continued)

NM	*Packaging and storage*	*Foods*	*Effects*	*References*
Polyethylene with TiO_2 NPs	Bags. Stored at 4°C for 15 days.	Button mushrooms (*Agaricus bisporus*)	Reduction of weight loss, phenolic compounds, ascorbic acid. Lower browning index. Lower microbial count of fungi.	Almasi et al., 2019
Polylactic acid film + bergamot essential oil impregnated with nano-TiO_2 and nano-Ag.	Bags. Store at room temperature for 15 days.	Mango (*Magnifera indica* L.)	Less weight loss, delayed loss of firmness. Higher content of ascorbic acid.	Chi et al., 2019
Carboxymethylcellulose and cinnamaldehdye with ZnO NPs	Individual wrap. Store at ~25°C and 45% RH for 10 days.	Cherry tomatoes (*Solanum lycopersicum* L.)	Greater firmness of the fruit, reduced weight loss.	Guo et al., 2020
Polyethylene with nano-SiO_2	Bags. Stored at 20 ± 1°C, 90% RH for 12 days.	Loquat fruit (*Eriobotrya japonica* L. cv. Dawuxing and Qingzhong)	Increased content of phenolic compounds, SOD, CAT, ascorbic acid. Inhibited internal browning, delayed decrease in TSSs.	Wang et al., 2020

various foods such as kiwifruit, apples, carrots, strawberries, button mushrooms, and loquat fruit. Positive effects were observed in all the cases, the main one being the reduction in weight loss, in addition to increasing antioxidant activity and the content of some antioxidants such as ascorbic acid and phenols (Almasi et al., 2019; Becaro et al., 2016; Hu et al., 2011; Li et al., 2017; Wang et al., 2020; Zhou et al., 2011).

Another material used to make bags that will serve as food packaging is polylactic acid in combination with nano-TiO_2 and nano-Ag. This package was evaluated in mangoes (*Magnifera indica* L.) stored for 15 days at room temperature. It was observed in the fruits less weight loss and therefore delayed loss of firmness; in addition, the content of ascorbic acid was increased (Chi et al., 2019).

Another material that has been used as individual fruit packaging is carboxymethylcellulose and cinnamaldehdye with ZnO NPs. This was evaluated in cherry tomatoes (*Solanum lycopersicum* L.) stored at 25°C and 45% RH for 10 days. The results observed were greater firmness of the fruits and reduced weight loss (Guo et al., 2020).

16.4.3 Other Applications of NMs

In addition to their use as coatings and packaging, other ways of using NMs in postharvest that do not belong to these two types of applications have also been reported (Table 16.3). The most common has been to impregnate some carrier material such as filter paper with essential oils loaded with some type of NM, the carrier material is placed inside the package along with the food to be preserved (Karimirad et al., 2018, 2019; Taheri et al., 2020).

Table 16.3 Other Applications of NMs and Their Effects on Postharvest of Food

NM	*Foods*	*Storage conditions*	*Effects*	*References*
Oil of *Citrus aurantium* loaded with chitosan NPs	Button mushroom (*Agaricus bisporus*)	Filter paper was impregnated with NMs and placed inside the container. Stored at 4°C for 20 days.	Color change rate slowdown, weight loss, and firmness. It promoted accumulation of phenolic compounds and ascorbic acid. Increased CAT and SOD and decreased PPO activity.	Karimirad et al., 2018
Oil of *Cuminum cyminum* loaded with chitosan NPs	Button mushroom (*Agaricus bisporus* cv. Sylvan A15)	Filter paper was impregnated with NMs and placed inside the container. Stored 4°C for 20 days.	Maintained color, firmness, and acceptability. Inhibition of bacteria and fungi. Higher SOD and APX activity, antioxidant capacity, and total phenolic content. PPO activity reduction.	Karimirad et al., 2019

(Continued)

Table 16.3 (Continued)

NM	*Foods*	*Storage conditions*	*Effects*	*References*
Heracleum persicum oil loaded with chitosan NPs	Red sweet bell pepper (*Capsicum annuum* L.)	Filter paper was impregnated with NMs and placed inside the container. Stored at 9 ± 1°C, 95% RH for 30 days.	Reduction in the change of color, firmness, and weight. Higher content of phenolic compounds, flavonoids, and ascorbic acid. Increased SOD and CAT and reduced POD activity.	Taheri et al., 2020
Pullulan packaging films with cinnamaldehyde and solid lipid NPs	Strawberry (*Fragaria* x *ananassa* Duch.)	Molded fiber basket with pullulan film at the bottom. Stored at 3°C for 10 days and 12°C for 6 days.	Reduction of yeast and mold population, reduction of metabolic rate and maintenance of desirable color, delay in loss of firmness.	Trinetta et al., 2020

Red sweet bell peppers (*Capsicum annuum* L.) packaged together with filter paper impregnated with *Heracleum persicum* oil loaded with chitosan NPs showed positive results. Here, the color change rate, weight loss, and firmness were reduced, and the accumulation of phenolic compounds and ascorbic acid, CAT, and SOD were increased; PPO activity was decreased (Taheri et al., 2020).

It has been shown that with this method of applying nanotechnology in postharvest, it was possible to reduce the loss of firmness and weight in button mushrooms (*Agaricus bisporus*). This was done by impregnating filter paper with oil of *Citrus aurantium* loaded with chitosan NPs (Karimirad et al., 2018). In addition, the same authors observed an increase in phenolic compounds and ascorbic acid, as well as the activity of CAT and SOD; on the contrary, the PPO activity was decreased. With this same method but using oil of *Cuminum cyminum* loaded with chitosan NPs, it was observed in button mushrooms (*Agaricus bisporus*) that the color, firmness, and acceptability were maintained; also, enzymatic and non-enzymatic antioxidants increased (SOD and APX activity, antioxidant capacity, total phenolic content); and PPO activity was reduced, and bacteria and fungi were inhibited (Karimirad et al., 2019).

Another way to use NMs postharvest has been to place some composite material with NMs in the lower part of the container. Using this method, it was possible to reduce the loss of firmness of strawberry fruits. In addition to maintaining the desirable color of the fruits, a reduction in yeast and mold population could also be observed (Trinetta et al., 2020).

16.5 Toxicity and Risks of NMs

With the rapid development of nanotechnology and its widespread application in multiple areas of knowledge such as physical, chemical, biological, engineering, and electronic sciences, the commercial demand for NMs has increased dramatically, and it is inevitable that NMs will be released into the environment or directly absorbed by the human body (Handford et al., 2014; Wu et al., 2020). There are several types of NMs with different dimensions (0D, 1D, 2D, 3D), morphologies, states, and chemical compositions, which play a key role in their properties and toxicity, and the latter can depend on several parameters such as mass, number, surface area, size, volume, surface chemistry, aggregation, surface coating, and functionalization (Saleh, 2020).

NMs do have a positive impact on postharvest quality and shelf life. However, it is necessary to determine their impact not only regarding the use but also with respect to human health and environmental toxicity. As documented here, the majority of NM application is through food coatings, resulting in its intake through food. Furthermore, in the case of NM-based packaging, the waste from this packaging can move through different trophic levels since it can be consumed by organisms such as protozoa, arthropods, annelids, mollusks, fish, insects, and possibly birds and mammals (Morales-Díaz et al., 2017). The risk of NMs will be determined by the concentration that can be reached in a certain environment; for example, if the NM concentration level is exceeded in an organism, it can induce negative responses or even inhibition of growth or death in extreme cases (Juárez-Maldonado et al., 2021). Therefore, it is of the utmost importance to properly review the use of NMs in postharvest, since in some cases they will be ingested together with food, or else, move through the food chain by residues originating from NM-based packaging, which can have ultimate negative impact on human health and the environment.

16.5.1 Risks of NMs Used in Postharvest Technology

In order to increase world food production, with nutritional value, quality, and safety, nanotechnology has been applied in various areas of the agri-food sector. Nanotechnology applications can be classified into nano-interiors (use of NMs in primary production or food processing using mainly nutrient supply and bioavailability systems) and nano-exterior (use of NMs in packaging materials, packaging, and edible films; Durán and Marcato, 2013; Handford et al., 2014). This extensive use of NMs in food products, as well as pharmaceuticals, cosmetics, and textiles, has led to direct contact with human skin and entry into the body through the gastrointestinal tract and intravenous injection (Wu et al., 2020).

In the food packaging industry, nanotechnology has reached broad prospects, as it can effectively solve the insufficiency of the general packaging properties of polymers due to its superior bacteriostatic properties, high mechanical strength, and good gas barrier properties to prolong the shelf life of the product by limiting oxygen penetration or preventing CO_2 leakage that allows increased postharvest quality (Deng et al., 2021; Eleftheriadou et al., 2017; Wang et al., 2020). However, these benefits are accompanied by safety concerns due to their potential toxicology to consumer health (Enescu et al., 2019). The potential risks are related to the fact that NMs, which can be organic and inorganic in nature, can migrate from food packaging into food products (Zhou et al., 2021).

16.5.1.1 Risks to Human Health

The nature of NMs plays an important role within the gastrointestinal tract, since some of the organic NMs such as lipid, protein, and starch can be digested by lipases, proteases, and amylases in the mouth, stomach, small intestine, or colon; however, some substances may not be digested in the upper digestive tract like inorganic NMs (McClements and Xiao, 2017). Any NM that is not digested or absorbed in the gastrointestinal tract can interact with substances in the digestive tract and directly affect the microbiota and may lead to the development of clinical disorders such as colitis, obesity, and immune dysfunctions (Pietroiusti et al., 2016; Zhang et al., 2019). Therefore, the main concern is that some of the NMs used in postharvest technology cannot be metabolized by the body. Also, NPs can enter cells, tissues, and organs more easily than particles of a larger size and thereby increasing risk (Enescu et al., 2019).

Human exposure to food-related NMs will largely depend on their specific use in the food and food industry. For example, the incorporation of NMs into multi-composite packaging materials in which the NM layer is coated with other materials is likely to result in little or no transfer to food (Magnuson et al., 2011). On the contrary, the use of NMs as carriers of nutrients or bioactive compounds that will be added directly into food products can result in higher levels of exposure that will depend on the concentration in the food and the amount of food that will be consumed (Handford et al., 2014; Magnuson et al., 2011). Examples of these are some NMs such as silicon and titanium that are used as food additives in the form of edible white pigments in confectionery products, dairy products, beverages, baked products, and flavorings, among others (Deng et al., 2021). However, there is a controversy, and more research is needed to corroborate whether they can be digested by the human digestive tract and if they meet food safety standards (Wang et al., 2020).

When the human body absorbs some type of NPs, it has been found that these are distributed in various organs, such as the stomach, kidney, liver, heart, and bone marrow (Shotop and Al-Suwiti, 2021). This generates oxidative stress in the first instance by increasing the production of free radicals such as hydrogen peroxide (H_2O_2), the superoxide anion radical ($O_2^{\bullet -}$), the hydroxyl radical ($\bullet OH$), hypochlorous acid (HOCl), and organoid hydroperoxides (ROOH; Kennel and Greten, 2021). These free radicals cause the modification of enzymes and proteins involved in antioxidant defense and cause DNA damage and cell apoptosis (Božinović et al., 2020).

Studies carried out in human breast cancer cells exposed to doses of 40, 80, 120, or 160 $\mu g\ mL^{-1}$ of iron NPs have been reported to induce the formation of ROS, with its distribution mainly in mitochondria and plasma, causing partial cell death. In addition to activating the ferroptosis pathways in the cytomembrane and plasma, and it activated the autophagy pathway in plasma (Zhang et al., 2021). Božinović et al. (2020) have conducted ***in vitro*** studies on the biological effects of MoO_3 NPs in human keratinocytes, at a concentration of 1 $mg\ mL^{-1}$ for period up to 6 h. Here it did not find to have a significant effect on cell death, ROS production, or the expression of proteins involved in antioxidant defense, but the cell signaling cascades that are involved in cell proliferation and inflammation were activated. In general, the toxicity of ingested NPs could be depend on their ability to damage cells or organs within humans (Zhang et al., 2019).

16.5.1.2 Impact of NMs on Animal Species

To better understand the possible impact of NMs on human health, studies have been conducted in some animal species. Ag NPs at concentrations of 25 and 50 μM have been reported to have negative effects on the blastocyst stage of mouse embryos, reducing the number of cells and promoting apoptosis in the inner cell mass and the trophectoderm of blastocysts of mice and consequently caused a reduction in implantation rates, mortality, and delayed development in mouse embryos both *in vitro* and *in vivo* (Li et al., 2010). Ag NPs can also stop the growth of spermatozoa cells, inhibit the acrosomal reaction of spermatozoa, induce oxidative damage to DNA, and reduce the number of spermatozoa, their normal morphology and motility percentage (Moradi-Sardareh et al., 2018). Hadrup et al. (2020) have evaluated the toxicity of Fe_2O_3, $ZnFe_2O_4$, $NiFe_2O_4$, and $NiZnFe_4O_8$ NMs in mice, and found all the NMs studied to cause inflammation in the lungs after 24-h exposure.

The effect of different concentrations of MoS_2 NPs (5×10^{-1}, 5×10^{-2}, and 5×10^{-3} mg mL^{-1}) on the embryonated eggs of *Gallus domesticus* has also been studied. The results showed all the MoS_2 NPs concentrations to cause death or a drastic embryonic reduction. Likewise, they presented morphological alterations, such as reduced size and visceral ectopy with exposure of the abdominal intestine due to a lack of closure of the abdominal walls. They also caused extensive hemorrhagic areas, especially in the liver and lung tissue of embryos treated with these NPs (Scalisi et al., 2020).

16.5.1.3 Risks of NMs in Soil

In addition to the risks derived from the direct contact of NMs in the food consumed, these NMs can also be released into various environmental compartments. In the soil, it has been shown that NMs can affect fertility due to the impact on the microbiota that includes plant growth–promoting bacteria, mycorrhizae, microorganisms in general, and nematodes, which affect the biogeochemical cycle of nutrients and carbon, detoxification, and the minimization and degradation of soil pollutants, and therefore have a direct or indirect impact on plant growth (Ameen et al., 2021). In addition, several parameters such as the content of organic matter, humic or fulvic acids, pH, and ionic strength have been identified to influence the NMs and can enhance their toxicity (Tortella et al., 2020).

There are several studies on the toxicity of NMs performed on soil microorganisms, since these materials have the potential to severely impact key microbial processes such as nitrification in the environment. Beddow et al. (2014) have demonstrated significant inhibition of the nitrification potential rate of *Nitrosomonas Europaea*, *Nitrospira multiformis* and *Nitrosococcus oceani* by NMs. In addition, they inhibited the growth of *Escherichia coli* and *Bacillus subtilis*, when applied at a dose range from 0.5 to 50 mg L^{-1} of Ag NPs.

CuO NPs at concentrations from 5 to 158 mg kg^{-1} of soil can induce toxic effects in *Folsomia candida* in clay soils, since the nanometric form of Cu tends to form aggregates, which changes its bioavailability and can generate toxicity in cells of this species (Fischer et al., 2021). Hou et al. (2017) have reported the Cu NPs to be toxic to different microorganisms such as *Saccharomyces cerevisiae, Vibrio fischeri, E. coli, Staphylococcus aureus, B. subtilis, Staphylococcus epidermidis, Proteus* spp., *Pseudomonas aeruginosa, Pseudomonas fluorescens, Janthinobacterium* spp., and *Microbacteria testaceum.*

16.5.1.4 Risks of NMs in Water

The direct and indirect release of NMs and their ecotoxicological impact in the aquatic environment are of the utmost importance since it is estimated that more than 10,000 tons per year of NMs are released into the environment, and by 2050, it is estimated that the concentration of NMs in fresh and marine waters, sediments, and soils will be 1000 times higher (Pereira et al., 2020; Salieri et al., 2018).

In the aquatic environment, NMs can be modified by physical, chemical, and biological transformations, such as reduction, oxidation, dissolution, sulfurization, degradation of the surface coating, interaction with macromolecules present in the environment, homoaggregation and heteroaggregation, photooxidation, biooxidation, and biodegradation that can harm aquatic organisms (Caixeta et al., 2021). Therefore, the contamination of the aquatic environments caused by these materials can be harmful to aquatic organisms in addition to the fact that drinking water can also be eventually affected by these materials and thereby seriously impact human health (Zhu et al., 2019).

ZnO NPs in different doses (0.1, 1, 10, 50, and 100 μg mL^{-1}) generate toxicity in embryos of zebrafish (*Danio rerio*), since they induce neurotoxicity and as a consequence the embryos developed symptoms similar to those of PD. They also generated excessive ROS production; triggered abundant mitochondrial damage, as well as apoptosis through the mitochondrial-mediated pathway; and thereby contributed to the appearance of symptoms similar to PD (Jin et al., 2021). In addition, Ag and Au NPs at concentrations of 3, 10, 50, and 100 nM caused the induction of various morphological malformations in the development of zebrafish (*Danio rerio*) embryos (Bar-Ilan et al., 2009).

Cu NPs and TiO_2 NPs in different concentrations (0.1, 1, 10, 100, and 1000 mM) can decrease the fronds per colony of *Lemna minor*. They also generate a gradual reduction in the growth rate of the roots depending on the concentration of NPs and caused morphological changes, such as decreased length of the roots and the fresh weight, while at the biochemical level, they decreased the chlorophyll content (Dolenc-Koce, 2017).

In the microalgae *Chlamydomonas reinhardtii*, the toxicity of six graphene oxide nanocompounds (rGO) was evaluated (rGO-Au, rGO-Ag, rGO-Pd, rGO-Fe_3O_4, rGO-Co_3O_4, rGO-SnO_2). The results showed that microalgae cells with abundant functional groups and hydrophobic surfaces exhibited more metal adsorption and interactions with NMs. Furthermore, heterointerfaces and ions released from rGO nanocomposites caused various toxicities by interfering with interactions between algae cells (Yin et al., 2020).

In the green microalgae *Dunaliella tertiolecta* and the brine shrimp *Artemia franciscana* that represent the two levels at the base of the marine trophic web, as prey and predator, respectively, the effects of the negative (–COOH) and positive (–NH_2) surface-charged plastic NPs (PSs) were evaluated. PS–COOH formed microscale aggregates in the medium and did not affect the growth of microalgae at a concentration of up to 50 μg mL^{-1} and neither of brine shrimp at a concentration of 10 μg mL^{-1}. However, PS–COOH were found adsorbed on microalgae, as well as accumulated and excreted in brine shrimp, suggesting a possible trophic transfer along marine trophic webs. On the contrary, PS–NH_2 were found as nanometric aggregates in both media, which caused the inhibition of algae growth and mortality in brine shrimp in addition to significantly inducing physiological alterations, such as molting, organogenesis, and tissue remodeling in early larvae (Bergami et al., 2017).

16.6 Conclusion

The application of nanotechnology in the postharvest of fruits and vegetables undoubtedly presents a series of advantages compared to traditional technologies, especially the use of NMs in coatings and packaging. It has been clearly observed that with the use of NMs, it has been possible to reduce the loss of postharvest quality and increase the shelf life; this mainly by reducing the respiration rate, the loss of water, and the inhibition of microbial activity on the surface of fruits and vegetables.

However, it is also true that the use of nanotechnology presents some risks that must be considered. Although the toxicity of NMs depends on their concentration in the body, it must be taken into account that continuous exposure to these NMs can be risky in the long term. As already described, not only does it involve risks to human health, but it can also affect different organisms through trophic transfer.

Therefore, it is extremely important to carry out more scientific studies to adequately understand the possible implications of the use of nanotechnology in the agri-food sector. Studies at different omic levels are required to assess the different impacts of NMs on organisms and thereby ensure the application of nanotechnology to be safe for human health and the environment.

References

Ali, A.A. and Toliba, A.O. 2018. Effect of organic calcium spraying and nano chitosan fruits coating on yield, fruit quality and storability of peach cv 'early swelling.' *Curr. Sci. Int.* 7, 737–749.

Ali, M., Ahmed, A., Shah, S.W.A., Mehmood, T. and Abbasi, K.S. 2020. Effect of silver nanoparticle coatings on physicochemical and nutraceutical properties of loquat during postharvest storage. *J. Food Process. Preserv.* 44, e14808. https://doi.org/10.1111/jfpp.14808

Almasi, H., Pourfathi, B. and Zonouzi, R.M. 2019. Studying of the effect of low density poly ethylene (LDPE) based antimicrobial nanocomposite packaging containing TiO_2 nanoparticles on the shelf-life extension of button mushroom (*Agaricus bisporus*) HADI. *Iran. J. Biosyst. Eng.* 51, 195–209. https://doi.org/10.22059/IJBSE.2019.287876.665216

Ameen, F., Alsamhary, K., Alabdullatif, J.A. and ALNadhari, S. 2021. A review on metal-based nanoparticles and their toxicity to beneficial soil bacteria and fungi. *Ecotoxicol. Environ. Saf.* 213, 112027. https://doi.org/10.1016/j.ecoenv.2021.112027

Amiri, A., Ramezanian, A., Mortazavi, S.M.H., Hosseini, S.M.H. and Yahia, E. 2020. Shelf-life extension of pomegranate arils using chitosan nanoparticles loaded with *Satureja hortensis* essential oil. *J. Sci. Food Agric.* https://doi.org/10.1002/jsfa.11010

Arroyo, B.J., Bezerra, A.C., Oliveira, L.L., Arroyo, S.J., Melo, E.A. and Santos, A.M.P. 2020. Antimicrobial active edible coating of alginate and chitosan add ZnO nanoparticles applied in guavas (*Psidium guajava* L.). *Food Chem.* 309, 125566. https://doi.org/10.1016/j.foodchem.2019.125566

Ashitha, G.N., Sunny, A.C. and Nisha, R. 2020. Effect of pre-harvest and post-harvest hexanal treatments on fruits and vegetables: A review. *Agric. Rev.* 41, 124–131. https://doi.org/10.18805/ag.r-1928

Bakhy, E.A., Zidan, N.S. and Aboul-Anean, H.E.D. 2018. The effect of nano materials on edible coating and films' improvement. *Int. J. Pharm. Res. Allied Sci.* 7, 20–41.

Bar-Ilan, O., Albrecht, R.M., Fako, V.E. and Furgeson, D.Y. 2009. Toxicity assessments of multisized gold and silver nanoparticles in zebrafish embryos. *Small* 5, 1897–1910. https://doi.org/10.1002/smll.200801716

Basumatary, I.B., Mukherjee, A., Katiyar, V. and Kumar, S. 2020. Biopolymer-based nanocomposite films and coatings: Recent advances in shelf-life improvement of fruits and vegetables. *Crit. Rev. Food Sci. Nutr.* https://doi.org/10.1080/10408398.2020.1848789

Becaro, A.A., Puti, F.C., Panosso, A.R., Gern, J.C., Brandão, H.M., Correa, D.S. and Ferreira, M.D. 2016. Postharvest quality of fresh-cut carrots packaged in plastic films containing silver nanoparticles. *Food Bioprocess Technol.* 9, 637–649. https://doi.org/10.1007/s11947-015-1656-z

Beddow, J., Stolpe, B., Cole, P., Lead, J.R., Sapp, M., Lyons, B.P., Colbeck, I. and Whitby, C. 2014. Effects of engineered silver nanoparticles on the growth and activity of ecologically important microbes. *Environ. Microbiol. Rep.* 6, 448–458. https://doi.org/10.1111/1758-2229.12147

Bergami, E., Pugnalini, S., Vannuccini, M.L., Manfra, L., Faleri, C., Savorelli, F., Dawson, K.A. and Corsi, I. 2017. Long-term toxicity of surface-charged polystyrene nanoplastics to marine planktonic species *Dunaliella tertiolecta* and *Artemia franciscana*. *Aquat. Toxicol.* 189, 159–169. https://doi.org/10.1016/j.aquatox.2017.06.008

Bose, S.K., Howlader, P., Wang, W. and Yin, H. 2021. Oligosaccharide is a promising natural preservative for improving postharvest preservation of fruit: A review. *Food Chem.* 341, 128178. https://doi.org/10.1016/j.foodchem.2020.128178

Božinović, K., Nestić, D., Centa, U.G., Ambriović-Ristov, A., Dekanić, A., de Bisschop, L., Remškar, M. and Majhen, D. 2020. In-vitro toxicity of molybdenum trioxide nanoparticles on human keratinocytes. *Toxicology* 444, 152564. https://doi.org/10.1016/j.tox.2020.152564

Caixeta, M.B., Araújo, P.S., Rodrigues, C.C., Gonçalves, B.B., Araújo, O.A., Bevilaqua, G.B., Malafaia, G., Silva, L.D. and Rocha, T.L. 2021. Risk assessment of iron oxide nanoparticles in an aquatic ecosystem: A case study on *Biomphalaria glabrata*. *J. Hazard. Mater.* 401, 123398. https://doi.org/10.1016/j.jhazmat.2020.123398

Chen, T., Ji, D., Zhang, Z., Li, B., Qin, G. and Tian, S. 2020. Advances and strategies for controlling the quality and safety of postharvest fruit. *Engineering* 7, 1177–1184 https://doi.org/10.1016/j.eng.2020.07.029

Chi, H., Song, S., Luo, M., Zhang, C., Li, W., Li, L. and Qin, Y. 2019. Effect of PLA nanocomposite films containing bergamot essential oil, TiO_2 nanoparticles, and Ag nanoparticles on shelf life of mangoes. *Sci. Hortic.* 249, 192–198. https://doi.org/10.1016/j.scienta.2019.01.059

Dakal, T.C., Kumar, A., Majumdar, R.S. and Yadav, V. 2016. Mechanistic basis of antimicrobial actions of silver nanoparticles. *Front. Microbiol.* 7, 1831. https://doi.org/10.3389/fmicb.2016.01831

Deng, J., Ding, Q.M., Jia, M.X., Li, W., Zuberi, Z., Wang, J.H., Ren, J.L., Fu, D., Zeng, X.X. and Luo, J.F. 2021. Biosafety risk assessment of nanoparticles: Evidence from food case studies. *Environ. Pollut.* 116662. https://doi.org/10.1016/j.envpol.2021.116662

Dolenc-Koce, J. 2017. Effects of exposure to nano and bulk sized TiO_2 and CuO in *Lemna minor*. *Plant Physiol. Biochem.* 119, 43–49. https://doi.org/10.1016/j.plaphy.2017.08.014

Durán, N. and Marcato, P.D., 2013. Nanobiotechnology perspectives. Role of nanotechnology in the food industry: A review. *Int. J. Food Sci. Technol.* 48, 1127–1134. https://doi.org/10.1111/ijfs.12027

Eldib, R., Khojah, E., Elhakem, A., Benajiba, N. and Helal, M. 2020. Chitosan, nisin, silicon dioxide nanoparticles coating films effects on blueberry (*Vaccinium myrtillus*) quality. *Coatings* 10, 962. https://doi.org/10.3390/coatings10100962

Eleftheriadou, M., Pyrgiotakis, G. and Demokritou, P. 2017. Nanotechnology to the rescue: Using nano-enabled approaches in microbiological food safety and quality. *Curr. Opin. Biotechnol.* 44, 87–93. https://doi.org/10.1016/j.copbio.2016.11.012

Enescu, D., Cerqueira, M.A., Fucinos, P. and Pastrana, L.M. 2019. Recent advances and challenges on applications of nanotechnology in food packaging. A literature review. *Food Chem. Toxicol.* 134, 110814. https://doi.org/10.1016/j.fct.2019.110814

Esyanti, R.R., Zaskia, H., Amalia, A. and Nugrahapraja, D.H. 2019. Chitosan nanoparticle-based coating as post-harvest technology in banana. *J. Phys. Conf. Ser.* 1204, 012109. https://doi.org/10.1088/1742-6596/1204/1/012109

Fayaz, A.M., Girilal, M., Kalaichelvan, P.T. and Venkatesan, R. 2009. Mycobased synthesis of silver nanoparticles and their incorporation into sodium alginate films for vegetable and fruit preservation. *J. Agric. Food Chem.* 57, 6246–6252. https://doi.org/10.1021/jf900337h

Fischer, J., Evlanova, A., Philippe, A. and Filser, J. 2021. Soil properties can evoke toxicity of copper oxide nanoparticles towards springtails at low concentrations. *Environ. Pollut.* 270, 116084. https://doi.org/10.1016/j.envpol.2020.116084

Gao, L., Li, Q., Zhao, Y., Wang, H., Liu, Y., Sun, Y., Wang, F., Jia, W. and Hou, X. 2017. Silver nanoparticles biologically synthesised using tea leaf extracts and their use for extension of fruit shelf life. *IET Nanobiotechnol* 11, 637–643. https://doi.org/10.1049/iet-nbt.2016.0207

García-Betanzos, C.I., Hernández-Sánchez, H., Bernal-Couoh, T.F., Quintanar-Guerrero, D. and Zambrano-Zaragoza, M.L. 2017. Physicochemical, total phenols and pectin methylesterase changes on quality maintenance on guava fruit (*Psidium guajava* L.) coated with candeuba wax solid lipid nanoparticles-xanthan gum. *Food Res. Int.* 101, 218–227. https://doi.org/10.1016/j.foodres.2017.08.065

Gouda, M.H.B., Zhang, C., Wang, J., Peng, S., Chen, Y., Luo, H. and Yu, L. 2020. ROS and MAPK cascades in the post-harvest senescence of horticultural products. *J. Proteomics Bioinform.* 13, 1–7. https://doi.org/10.35248/0974-276x.1000508

Gudadhe, J.A., Yadav, A., Gade, A., Marcato, P.D., Durán, N. and Rai, M. 2014. Preparation of an agar-silver nanoparticles (A-AgNp) film for increasing the shelf-life of fruits. *IET Nanobiotechnol* 8, 190–195. https://doi.org/10.1049/iet-nbt.2013.0010

Gundewadi, G., Reddy, V.R. and Bhimappa, B. 2018. Physiological and biochemical basis of fruit development and ripening-a review. *J. Hill Agric.* 9, 7–21. https://doi.org/10.5958/2230-7338.2018.00003.4

Guo, X., Chen, B., Wu, X., Li, J. and Sun, Q. 2020. Utilization of cinnamaldehyde and zinc oxide nanoparticles in a carboxymethylcellulose-based composite coating to improve the postharvest quality of cherry tomatoes. *Int. J. Biol. Macromol.* 160, 175–182. https://doi.org/10.1016/j.ijbiomac.2020.05.201

Hadrup, N., Saber, A.T., Kyjovska, Z.O., Jacobsen, N.R., Vippola, M., Sarlin, E., Ding, Y., Schmid, O., Wallin, H., Jensen, K.A. and Vogel, U. 2020. Pulmonary toxicity of Fe_2O_3, $ZnFe_2O_4$, $NiFe_2O_4$ and $NiZnFe_4O_8$ nanomaterials: Inflammation and DNA strand breaks. *Environ. Toxicol. Pharmacol.* 74, 103303. https://doi.org/10.1016/j.etap.2019.103303

Handford, C.E., Dean, M., Henchion, M., Spence, M., Elliott, C.T. and Campbell, K. 2014. Implications of nanotechnology for the agri-food industry: Opportunities, benefits and risks. *Trends Food Sci. Technol.* 40, 226–241. https://doi.org/10.1016/j.tifs.2014.09.007

Hasheminejad, N. and Khodaiyan, F. 2020. The effect of clove essential oil loaded chitosan nanoparticles on the shelf life and quality of pomegranate arils. *Food Chem.* 309, 125520. https://doi.org/10.1016/j.foodchem.2019.125520

Hernández-López, G., Ventura-Aguilar, R.I., Correa-Pacheco, Z.N., Bautista-Baños, S. and Barrera-Necha, L.L. 2020. Nanostructured chitosan edible coating loaded with α-pinene for the preservation of the postharvest quality of *Capsicum annuum* L.

and *Alternaria alternata* control. *Int. J. Biol. Macromol.* 165, 1881–1888. https://doi.org/10.1016/j.ijbiomac.2020.10.094

Hou, J., Wang, X., Hayat, T. and Wang, X. 2017. Ecotoxicological effects and mechanism of CuO nanoparticles to individual organisms. *Environ. Pollut.* 221, 209–217. https://doi.org/10.1016/j.envpol.2016.11.066

Hu, Q., Fang, Y., Yang, Y., Ma, N. and Zhao, L., 2011. Effect of nanocomposite-based packaging on postharvest quality of ethylene-treated kiwifruit (*Actinidia deliciosa*) during cold storage. *Food Res. Int.* 44, 1589–1596. https://doi.org/10.1016/j.foodres.2011.04.018

Hu, X., Saravanakumar, K., Sathiyaseelan, A. and Wang, M.H., 2020. Chitosan nanoparticles as edible surface coating agent to preserve the fresh-cut bell pepper (*Capsicum annuum* L. var. grossum (L.) Sendt). *Int. J. Biol. Macromol.* 165, 948–957. https://doi.org/10.1016/j.ijbiomac.2020.09.176

Hussein, Z., Fawole, O.A. and Opara, U.L. 2020. Harvest and postharvest factors affecting bruise damage of fresh fruits. *Hortic. Plant J.* 6, 1–13. https://doi.org/10.1016/j.hpj.2019.07.006

Jansasithorn, R., East, A.R., Hewett, E.W. and Heyes, J.A. 2014. Skin cracking and postharvest water loss of Jalapeño chilli. *Sci. Hortic.* 175, 201–207. https://doi.org/10.1016/j.scienta.2014.05.037

Jin, M., Li, N., Sheng, W., Ji, X., Liang, X., Kong, B., Yin, P., Li, Y., Zhang, X. and Liu, K. 2021. Toxicity of different zinc oxide nanomaterials and dose-dependent onset and development of Parkinson's disease-like symptoms induced by zinc oxide nanorods. *Environ. Int.* 146, 106179. https://doi.org/10.1016/j.envint.2020.106179

Juárez-Maldonado, A., Ortega-Ortiz, H., González-Morales, S., Morelos-Moreno, Á., Cabrera-de la Fuente, M., Sandoval-Rangel, A., Cadenas-Pliego, G. and Benavides-Mendoza, A. 2019. Nanoparticles and nanomaterials as plant biostimulants. *Int. J. Mol. Sci.* 20, 1–19. https://doi.org/10.3390/ijms20010162

Juárez-Maldonado, A., Tortella, G., Rubilar, O., Fincheira, P. and Benavides-Mendoza, A. 2021. Biostimulation and toxicity: The magnitude of the impact of nanomaterials in microorganisms and plants. *J. Adv. Res.* 31, 113–126. https://doi.org/10.1016/j.jare.2020.12.011

Kaewklin, P., Siripatrawan, U., Suwanagul, A. and Lee, Y.S. 2018. Active packaging from chitosan-titanium dioxide nanocomposite film for prolonging storage life of tomato fruit. *Int. J. Biol. Macromol.* 112, 523–529. https://doi.org/10.1016/j.ijbiomac.2018.01.124

Karimirad, R., Behnamian, M. and Dezhsetan, S. 2019. Application of chitosan nanoparticles containing *Cuminum cyminum* oil as a delivery system for shelf life extension of *Agaricus bisporus*. *Lwt—Food Sci. Technol.* 106, 218–228. https://doi.org/10.1016/j.lwt.2019.02.062

Karimirad, R., Behnamian, M., Dezhsetan, S. and Sonnenberg, A. 2018. Chitosan nanoparticles loaded *Citrus aurantium* essential oil: A novel delivery system for preserving the postharvest quality of *Agaricus bisporus*. *J. Sci. Food Agric.* 98, 5112–5119. https://doi.org/https://doi.org/10.1002/jsfa.9050

Kennel, K.B. and Greten, F.R. 2021. Immune cell—produced ROS and their impact on tumor growth and metastasis. *Redox Biol.* 101891. https://doi.org/10.1016/j.redox.2021.101891

Lakshmi, S.J. and Bai, R.R.S., Sharanagouda, H., Ramachandra, C.T., Nadagouda, S., Nidoni, U. 2018. Effect of biosynthesized zinc oxide nanoparticles coating on quality parameters of fig (*Ficus carica* L.) fruit. *J. Pharmacogn. Phytochem.* 7, 10–14.

Lavinia, M., Hibaturrahman, S.N., Harinata, H. and Wardana, A.A. 2020. Antimicrobial activity and application of nanocomposite coating from chitosan and ZnO nanoparticle to inhibit microbial growth on fresh-cut papaya. *Food Res.* 4, 307–311. https://doi.org/10.26656/fr.2017.4(2).255

Lemes, G.F., Marchiore, N.G., Moreira, T.F.M., Da Silva, T.B.V., Sayer, C., Shirai, M.A., Gonçalves, O.H., Gozzo, A.M. and Leimann, F.V. 2017. Enzymatically crosslinked gelatin coating added of bioactive nanoparticles and antifungal agent: Effect on the quality of Benitaka grapes. *Lwt—Food Sci. Technol.* 84, 175–182. https://doi.org/10.1016/j.lwt.2017.05.050

Leng, P., Yuan, B. and Guo, Y. 2014. The role of abscisic acid in fruit ripening and responses to abiotic stress. *J. Exp. Bot.* 65, 4577–4588. https://doi.org/10.1093/jxb/eru204

Li, D., Ye, Q., Jiang, L. and Luo, Z. 2017. Effects of nano-TiO_2-LDPE packaging on postharvest quality and antioxidant capacity of strawberry (*Fragaria ananassa* Duch.) stored at refrigeration temperature. *J. Sci. Food Agric.* 97, 1116–1123. https://doi.org/10.1002/jsfa.7837

Li, J., Sun, Q., Sun, Y., Chen, B., Wu, X. and Le, T. 2019. Improvement of banana postharvest quality using a novel soybean protein isolate/cinnamaldehyde/zinc oxide bionanocomposite coating strategy. *Sci. Hortic.* 258, 108786. https://doi.org/10.1016/j.scienta.2019.108786

Li, P.-W., Kuo, T.-H., Chang, J.-H., Yeh, J.-M. and Chan, W.-H. 2010. Induction of cytotoxicity and apoptosis in mouse blastocysts by silver nanoparticles. *Toxicol. Lett.* 197, 82–87. https://doi.org/10.1016/j.toxlet.2010.05.003

Lin, M., Fang, S., Zhao, X., Liang, X. and Wu, D. 2020. Natamycin-loaded zein nanoparticles stabilized by carboxymethyl chitosan: Evaluation of colloidal/chemical performance and application in postharvest treatments. *Food Hydrocoll.* 106, 105871. https://doi.org/10.1016/j.foodhyd.2020.105871

Liu, W., Zeb, A., Lian, J., Wu, J., Xiong, H., Tang, J. and Zheng, S. 2020. Interactions of metal-based nanoparticles (Mbnps) and metal-oxide nanoparticles (monps) with crop plants: A critical review of research progress and prospects. *Environ. Rev.* 28, 294–310. https://doi.org/10.1139/er-2019-0085

Lo'ay, A.A. and Ameer, N.M. 2019. Performance of calcium nanoparticles blending with ascorbic acid and alleviation internal browning of 'Hindi Be-Sennara' mango fruit at a low temperature. *Sci. Hortic.* 254, 199–207. https://doi.org/10.1016/j.scienta.2019.05.006

Lufu, R., Ambaw, A. and Opara, U.L. 2020. Water loss of fresh fruit: Influencing preharvest, harvest and postharvest factors. *Sci. Hortic.* 272, 109519. https://doi.org/10.1016/j.scienta.2020.109519

Luo, H., Jiang, L., Bao, Y., Wang, L. and Yu, Z., 2013. Effect of chitosan/nano-chitosan composite coating on browning and lignification of fresh-cut *Zizania latifolia*. *J. Food Qual.* 36, 426–431. https://doi.org/10.1111/jfq.12056

Lustriane, C., Dwivany, F.M., Suendo, V. and Reza, M. 2018. Effect of chitosan and chitosan-nanoparticles on post harvest quality of banana fruits. *J. Plant Biotechnol.* 45, 36–44. https://doi.org/10.5010/JPB.2018.45.1.036

Magnuson, B.A., Jonaitis, T.S. and Card, J.W. 2011. A brief review of the occurrence, use, and safety of food-related nanomaterials. *J. Food Sci.* 76, R126–R133. https://doi.org/10.1111/j.1750-3841.2011.02170.x

Malek, N.S.A., Rosman, N., Mahmood, M.R., Khusaimi, Z. and Asli, N.A. 2020. Effects of storage temperature on shelf-life of mango coated with zinc oxide nanoparticles. *Sci. Lett.* 14, 47–57. https://doi.org/10.24191/sl.v14i2.9542

Mali, S.C., Raj, S. and Trivedi, R. 2020. Nanotechnology a novel approach to enhance crop productivity. *Biochem. Biophys. Reports* 24, 100821. https://doi.org/10.1016/j.bbrep.2020.100821

Manzoor, A., Bashir, M.A. and Hashmi, M.M. 2020. Nanoparticles as a preservative solution can enhance postharvest attributes of cut flowers. *Italus Hortus* 27, 1–14. https://doi.org/10.26353/j.itahort/2020.2.0114

Maringgal, B., Hashim, N., Mohamed Amin Tawakkal, I.S. and Muda Mohamed, M.T. 2020. Recent advance in edible coating and its effect on fresh/fresh-cut fruits quality. *Trends Food Sci. Technol.* 96, 253–267. https://doi.org/10.1016/j.tifs.2019.12.024

Martínez-González, M.C., Bautista-Baños, S., Correa-Pacheco, Z.N., Corona-Rangel, M.L., Ventura-Aguilar, R.I., Del Río-García, J.C. and Ramos-García, M.L. 2020. Effect of nanostructured chitosan/propolis coatings on the quality and antioxidant capacity of strawberries during storage. *Coatings* 10, 90. https://doi.org/doi:10.3390/coatings10020090

McClements, D.J. and Xiao, H. 2017. Is nano safe in foods? Establishing the factors impacting the gastrointestinal fate and toxicity of organic and inorganic food-grade nanoparticles. *NPJ Sci. Food* 1, 6. https://doi.org/10.1038/s41538-017-0005-1

Meena, M., Pilania, S., Pal, A., Mandhania, S., Bhushan, B., Kumar, S., Gohari, G. and Saharan, V. 2020. Cu-chitosan nano-net improves keeping quality of tomato by modulating physio-biochemical responses. *Sci. Rep.* 10, 21914. https://doi.org/10.1038/s41598-020-78924-9

Meindrawan, B., Suyatma, N.E., Wardana, A.A. and Pamela, V.Y. 2018. Nanocomposite coating based on carrageenan and ZnO nanoparticles to maintain the storage quality of mango. *Food Packag. Shelf Life* 18, 140–146. https://doi.org/10.1016/j.fpsl.2018.10.006

Melo, C.B.N.F., de MendonçaSoares, B.L., Diniz, M.K., Leal, F.C., Canto, D., Flores, M.A.P., Tavares-Filho, H. da C.J., Galembeck, A., Stamford, M.T.L., Stamford-Arnaud, M.T. and Stamford, M.T.C. 2018. Effects of fungal chitosan nanoparticles as eco-friendly edible coatings on the quality of postharvest table grapes. *Postharvest Biol. Technol.* 139, 56–66. https://doi.org/10.1016/j.postharvbio.2018.01.014

Melo, N.F.C.B., Pintado, M.M.E., Medeiros, J.A. da C., Galembeck, A., Vasconcelos, M.A. da S., Xavier, V.L., de Lima, M.A.B., Stamford, T.L.M., Stamford-Arnaud, T.M., Flores, M.A.P. and Stamford, T.C.M. 2020. Quality of postharvest strawberries: Comparative effect of fungal chitosan gel, nanoparticles and gel enriched with edible nanoparticles coatings. *Int. J. Food Stud.* 9, 373–393. https://doi.org/10.7455/ijfs/9.2.2020.a9

Moradi-Sardareh, H., Basir, H.R.G., Hassan, Z.M., Davoudi, M., Amidi, F. and Paknejad, M. 2018. Toxicity of silver nanoparticles on different tissues of Balb/C mice. *Life Sci.* 211, 81–90. https://doi.org/10.1016/j.lfs.2018.09.001

Morales-Díaz, A.B., Ortega-Ortíz, H., Juárez-Maldonado, A., Cadenas-Pliego, G., González-Morales, S. and Benavides-Mendoza, A. 2017. Application of nanoelements in plant nutrition and its impact in ecosystems. *Adv. Nat. Sci. Nanosci. Nanotechnol.* 8, 1–13. https://doi.org/10.1088/2043-6254/8/1/013001

Morones, J.R., Elechiguerra, J.L., Camacho, A., Holt, K., Kouri, J.B., Ramírez, J.T. and Yacaman, M.J. 2005. The bactericidal effect of silver nanoparticles. *Nanotechnology* 16, 2346–2353. https://doi.org/10.1088/0957-4484/16/10/059

Mossa, A.-T.H., Mohafrash, S.M.M., Ziedan, E.-S.H.E., Abdelsalam, I.S. and Sahab, A.F. 2021. Development of eco-friendly nanoemulsions of some natural oils and evaluating of its efficiency against postharvest fruit rot fungi of cucumber. *Ind. Crops Prod.* 159, 113049. https://doi.org/10.1016/j.indcrop.2020.113049

Munhuweyi, K., Mpai, S. and Sivakumar, D., 2020. Extension of avocado fruit postharvest quality using non-chemical treatments. *Agronomy* 10. https://doi.org/10.3390/agronomy10020212

Mustafa, M.A., Ali, A., Manickam, S. and Siddiqui, Y., 2014. Ultrasound-assisted chitosan-surfactant nanostructure assemblies: Towards maintaining postharvest quality of tomatoes. *Food Bioprocess Technol.* 7, 2102–2111. https://doi.org/10.1007/s11947-013-1173-x

Nguyen, H.V.H. and Nguyen, D.H.H. 2020. Effects of nano-chitosan and chitosan coating on the postharvest quality, polyphenol oxidase activity and malondialdehyde content of strawberry (*Fragaria x ananassa* Duch.). *J. Hortic. Postharvest Res.* 3, 11–24. https://doi.org/10.22077/jhpr.2019.2698.1082

Nguyen, V.T.B., Nguyen, D.H.H. and Nguyen, H.V.H., 2020. Combination effects of calcium chloride and nano-chitosan on the postharvest quality of strawberry (*Fragaria x ananassa* Duch.). *Postharvest Biol. Technol.* 162, 111103. https://doi.org/10.1016/j.postharvbio.2019.111103

Pereira, A.C., Gonçalves, B.B., Brito, R. da S., Vieira, L.G., Lima, E.C. de O. and Rocha, T.L. 2020. Comparative developmental toxicity of iron oxide nanoparticles and ferric chloride to zebrafish (*Danio rerio*) after static and semi-static exposure. *Chemosphere* 254, 126792. https://doi.org/10.1016/j.chemosphere.2020.126792

Pietroiusti, A., Magrini, A. and Campagnolo, L. 2016. New frontiers in nanotoxicology: Gut microbiota/microbiome-mediated effects of engineered nanomaterials. *Toxicol. Appl. Pharmacol.* 299, 90–95. https://doi.org/10.1016/j.taap.2015.12.017

Pina-Barrera, A.M., Alvarez-Roman, R., Baez-Gonzalez, J.G., Amaya-Guerra, C.A., Rivas-Morales, C., Gallardo-Rivera, C.T. and Galindo-Rodriguez, S.A. 2019. Application of a multisystem coating based on polymeric nanocapsules containing essential oil of *Thymus vulgaris* L. to increase the shelf life of table grapes (*Vitis vinifera* L.). *IEEE Trans. Nanobioscience* 18, 549–557. https://doi.org/10.1109/TNB.2019.2941931

Posé, S., Paniagua, C., Matas, A.J., Gunning, A.P., Morris, V.J., Quesada, M.A. and Mercado, J.A. 2019. A nanostructural view of the cell wall disassembly process during fruit ripening and postharvest storage by atomic force microscopy. *Trends Food Sci. Technol.* 87, 47–58. https://doi.org/10.1016/j.tifs.2018.02.011

Queiroz, C., Lopes, M.L.M., Fialho, E. and Valente-Mesquita, V.L. 2008. Polyphenol oxidase: Characteristics and mechanisms of browning control. *Food Rev. Int.* 24, 361–375. https://doi.org/10.1080/87559120802089332

Ranjbar, S., Rahemi, M. and Ramezanian, A. 2018. Comparison of nano-calcium and calcium chloride spray on postharvest quality and cell wall enzymes activity in apple cv. red delicious. *Sci. Hortic.* 240, 57–64. https://doi.org/10.1016/j.scienta.2018.05.035

Riva, S.C., Opara, U.O. and Fawole, O.A. 2020. Recent developments on postharvest application of edible coatings on stone fruit: A review. *Sci. Hortic.* 262, 109074. https://doi.org/10.1016/j.scienta.2019.109074

Robledo, N., Vera, P., López, L., Yazdani-Pedram, M., Tapia, C. and Abugoch, L. 2018. Thymol nanoemulsions incorporated in quinoa protein/chitosan edible films; antifungal effect in cherry tomatoes. *Food Chem.* 246, 211–219. https://doi.org/10.1016/j.foodchem.2017.11.032

Saba, M.K. and Amini, R. 2017. Nano-ZnO/carboxymethyl cellulose-based active coating impact on ready-to-use pomegranate during cold storage. *Food Chem.* 232, 721–726. https://doi.org/10.1016/j.foodchem.2017.04.076

Saleh, T.A. 2020. Nanomaterials: Classification, properties, and environmental toxicities. *Environ. Technol. Innov.* 20, 101067. https://doi.org/10.1016/j.eti.2020.101067

Salieri, B., Turner, D.A., Nowack, B. and Hischier, R. 2018. Life cycle assessment of manufactured nanomaterials: Where are we? *NanoImpact* 10, 108–120. https://doi.org/10.1016/j.impact.2017.12.003

Saqib, S., Zaman, W., Ayaz, A., Habib, S., Bahadur, S., Hussain, S., Muhammad, S. and Ullah, F. 2020. Postharvest disease inhibition in fruit by synthesis and characterization of chitosan iron oxide nanoparticles. *Biocatal. Agric. Biotechnol.* 28, 101729. https://doi.org/10.1016/j.bcab.2020.101729

Saravanakumar, K., Hu, X., Chelliah, R., Oh, D.H., Kathiresan, K. and Wang, M.H. 2020. Biogenic silver nanoparticles-polyvinylpyrrolidone based glycerosomes coating to expand the shelf life of fresh-cut bell pepper (*Capsicum annuum* L. var. grossum (L.) Sendt). *Postharvest Biol. Technol.* 160, 111039. https://doi.org/10.1016/j.postharvbio.2019.111039

Scalisi, E.M., Salvaggio, A., Antoci, F., Messina, A., Pecoraro, R., Cantarella, M., Gorrasi, G., Impellizzeri, G. and Brundo, M.V. 2020. Toxicity assessment of two-dimensional nanomaterials molybdenum disulfide in *Gallus gallus* domesticus. *Ecotoxicol. Environ. Saf.* 200, 110772. https://doi.org/10.1016/j.ecoenv.2020.110772

Shah, S.W.A., Jahangir, M., Qaisar, M., Khan, S.A., Mahmood, T., Saeed, M., Farid, A. and Liaquat, M. 2015. Storage stability of kinnow fruit (*Citrus reticulata*) as affected by CMC and guar gum-based silver nanoparticle coatings. *Molecules* 20, 22645–22661. https://doi.org/10.3390/molecules201219870

Shahat, M., Ibrahim, M., Osheba, A. and Taha, I. 2020a. Preparation and characterization of silver nanoparticles and their use for improving the quality of apricot fruits. *Al-Azhar J. Agric. Res.* 45, 33–43. https://doi.org/10.21608/ajar.2020.126625

Shahat, M., Ibrahim, M., Osheba, A. and Taha, I. 2020b. Improving the quality and shelf-life of strawberries as coated with nano-edible films during storage. *Al-Azhar J. Agric. Res.* 45, 1–13.

Sharma, R.R., Nagaraja, A., Goswami, A.K., Thakre, M., Kumar, R. and Varghese, E. 2020. Influence of on-the-tree fruit bagging on biotic stresses and postharvest quality of rainy-season crop of 'Allahabad Safeda' guava (*Psidium guajava* L.). *Crop Prot.* 135, 105216. https://doi.org/10.1016/j.cropro.2020.105216

Shotop, Y.M. and Al-Suwiti, I.N. 2021. The possible role of vitamins E and C in reducing the toxicity of copper nanoparticles in the kidney and liver of the rats (*Rattus norvegicus*). *J. King Saud Univ.—Sci.* 101357. https://doi.org/10.1016/j.jksus.2021.101357

Sogvar, O.B., Koushesh Saba, M., Emamifar, A. and Hallaj, R. 2016. Influence of nano-ZnO on microbial growth, bioactive content and postharvest quality of strawberries during storage. *Innov. Food Sci. Emerg. Technol.* 35, 168–176. https://doi.org/10.1016/j.ifset.2016.05.005

Song, H., Yuan, W., Jin, P., Wang, W., Wang, X., Yang, L. and Zhang, Y. 2016. Effects of chitosan/nano-silica on postharvest quality and antioxidant capacity of loquat fruit during cold storage. *Postharvest Biol. Technol.* 119, 41–48. https://doi.org/10.1016/j.postharvbio.2016.04.015

Sorrentino, A., Gorrasi, G. and Vittoria, V. 2007. Potential perspectives of bionanocomposites for food packaging applications. *Trends Food Sci. Technol.* 18, 84–95. https://doi.org/10.1016/j.tifs.2006.09.004

Sun, Q., Li, J. and Le, T. 2018. Zinc oxide nanoparticle as a novel class of antifungal agents: Current advances and future perspectives. *J. Agric. Food Chem.* 66, 11209–11220. https://doi.org/10.1021/acs.jafc.8b03210

Taheri, A., Behnamian, M., Dezhsetan, S. and Karimirad, R. 2020. Shelf life extension of bell pepper by application of chitosan nanoparticles containing *Heracleum persicum* fruit essential oil. *Postharvest Biol. Technol.* 170, 111313. https://doi.org/10.1016/j.postharvbio.2020.111313

Tang, N., An, J., Deng, W., Gao, Y., Chen, Z. and Li, Z. 2020. Metabolic and transcriptional regulatory mechanism associated with postharvest fruit ripening and senescence in cherry tomatoes. *Postharvest Biol. Technol.* 168, 111274. https://doi.org/10.1016/j.postharvbio.2020.111274

Tortella, G.R., Rubilar, O., Durán, N., Diez, M.C., Martínez, M., Parada, J. and Seabra, A.B. 2020. Silver nanoparticles: Toxicity in model organisms as an overview of its hazard for human health and the environment. *J. Hazard. Mater.* 390, 121974. https://doi.org/10.1016/j.jhazmat.2019.121974

Trinetta, V., McDaniel, A., Batziakas, K.G., Yucel, U., Nwadike, L. and Pliakoni, E. 2020. Antifungal packaging film to maintain quality and control postharvest diseases in strawberries. *Antibiotics* 9, 1–12. https://doi.org/10.3390/antibiotics9090618

Vieira, A.C.F., de Matos Fonseca, J., Menezes, N.M.C., Monteiro, A.R. and Valencia, G.A. 2020. Active coatings based on hydroxypropyl methylcellulose and silver nanoparticles to extend the papaya (*Carica papaya* L.) shelf life. *Int. J. Biol. Macromol.* 164, 489–498. https://doi.org/10.1016/j.ijbiomac.2020.07.130

Wang, A., Ng, H.P., Xu, Y., Li, Y., Zheng, Y., Yu, J., Han, F., Peng, F. and Fu, L. 2014. Gold nanoparticles: Synthesis, stability test, and application for the rice growth. *J. Nanomater.* 451232. https://doi.org/10.1155/2014/451232

Wang, G. and Zhang, X. 2020. Evaluation and optimization of air-based precooling for higher postharvest quality: Literature review and interdisciplinary perspective. *Food Qual. Saf.* 4, 59–68. https://doi.org/10.1093/fqsafe/fyaa012

Wang, L., Shao, S., Madebo, M.P., Hou, Y., Zheng, Y. and Jin, P. 2020. Effect of nano-SiO_2 packing on postharvest quality and antioxidant capacity of loquat fruit under ambient temperature storage. *Food Chem.* 315, 126295. https://doi.org/10.1016/j.foodchem.2020.126295

Wu, D., Ma, Y., Cao, Y. and Zhang, T. 2020. Mitochondrial toxicity of nanomaterials. *Sci. Total Environ.* 702, 134994. https://doi.org/10.1016/j.scitotenv.2019.134994

Xing, Y., Yang, H., Guo, X., Bi, X., Liu, X., Xu, Q., Wang, Q., Li, W., Li, X., Shui, Y., Chen, C. and Zheng, Y. 2020. Effect of chitosan/Nano-TiO_2 composite coatings on the postharvest quality and physicochemical characteristics of mango fruits. *Sci. Hortic.* 263, 109135. https://doi.org/10.1016/j.scienta.2019.109135

Yin, J., Dong, Z., Liu, Y., Wang, H., Li, A., Zhuo, Z., Feng, W. and Fan, W. 2020. Toxicity of reduced graphene oxide modified by metals in microalgae: Effect of the surface properties of algal cells and nanomaterials. *Carbon* 169, 182–192. https://doi.org/10.1016/j.carbon.2020.07.057

Yu, L., Qiao, N., Zhao, J., Zhang, H., Tian, F., Zhai, Q. and Chen, W. 2020. Postharvest control of *Penicillium expansum* in fruits: A review. *Food Biosci.* 36, 100633. https://doi.org/10.1016/j.fbio.2020.100633

Zambrano-Zaragoza, M.L., Mercado-Silva, E., Ramirez-Zamorano, P., Cornejo-Villegas, M.A., Gutiérrez-Cortez, E. and Quintanar-Guerrero, D. 2013. Use of solid lipid nanoparticles (SLNs) in edible coatings to increase guava (*Psidium guajava* L.) shelf-life. *Food Res. Int.* 51, 946–953. https://doi.org/10.1016/j.foodres.2013.02.012

Zambrano-Zaragoza, M.L., Quintanar-Guerrero, D., Del Real, A., González-Reza, R.M., Cornejo-Villegas, M.A. and Gutiérrez-Corte, E. 2020. Effect of nano-edible coating based on beeswax solid lipid nanoparticles on strawberry's preservation. *Coatings* 10, 253. https://doi.org/10.3390/coatings10030253

Zhang, Y., Hai, Y., Miao, Y., Qi, X., Xue, W., Luo, Y., Fan, H. and Yue, T. 2021. The toxicity mechanism of different sized iron nanoparticles on human breast cancer (MCF7) cells. *Food Chem.* 341, 128263. https://doi.org/10.1016/j.foodchem.2020.128263

Zhang, Z., Zhang, R., Xiao, H., Bhattacharya, K., Bitounis, D., Demokritou, P. and McClements, D.J. 2019. Development of a standardized food model for studying the impact of food matrix effects on the gastrointestinal fate and toxicity of ingested nanomaterials. *NanoImpact* 13, 13–25. https://doi.org/10.1016/j.impact.2018.11.002

Zhao, H., Fan, Z., Wu, J. and Zhu, S. 2021. Effects of pre-treatment with S-nitrosoglutathione-chitosan nanoparticles on quality and antioxidant systems of fresh-cut apple slices. *LWT* 139, 110565. https://doi.org/10.1016/j.lwt.2020.110565

Zhao, J., Lin, M., Wang, Z., Cao, X. and Xing, B. 2020. Engineered nanomaterials in the environment: Are they safe? *Crit. Rev. Environ. Sci. Technol.* 1–36. https://doi.org/10.1080/10643389.2020.1764279

Zhou, H., Liu, J., Dai, T., Muriel Mundo, J.L., Tan, Y., Bai, L. and McClements, D.J. 2021. The gastrointestinal fate of inorganic and organic nanoparticles in vitamin D-fortified plant-based milks. *Food Hydrocoll.* 112, 106310. https://doi.org/10.1016/j.foodhyd.2020.106310

Zhou, L., Lv, S., He, G., He, Q. and Shi, B. 2011. Effect of PE/Ag_2O nano-packaging on the quality of apple slices. *J. Food Qual.* 34, 171–176. https://doi.org/10.1111/j.1745-4557.2011.00385.x

Zhu, Y., Liu, X., Hu, Y., Wang, R., Chen, M., Wu, J., Wang, Y., Kang, S., Sun, Y. and Zhu, M. 2019. Behavior, remediation effect and toxicity of nanomaterials in water environments. *Environ. Res.* 174, 54–60. https://doi.org/10.1016/j.envres.2019.04.014

Chapter 17

Life-Cycle Assessment

A Tool to Evaluate the Environmental Impact of Nanomaterials Used in Packaging

Alexandre R. Lima, Nathana L. Cristofoli, and Margarida C. Vieira

List of Abbreviations

AgNP	Silver nanoparticle
Al	Aluminum
CAGR	Compound annual growth rate
CNM	Carbon-based nanomaterial
CNT	Carbon nanotube
CO_2	Carbon dioxide
Cu	Copper
EC	European Commission
EU-OSHA	European Agency for Safety and Health at Work
EOL	End of life
EU	European Union
FU	Functional unit
GHG	Greenhouse gas
ILCD	Life Cycle Data System
ISO	International Organization for Standardization
Kg/ton	Kilogram per tonne
LCA	Life-cycle assessment
LCI	Life-cycle inventory
LCIA	Life-cycle impact assessment
Mg/L	Milligrams per liter

DOI: 10.1201/9781003142287-17

MgO	Magnesium oxide
MJ/ton	Megajoule per tonne
MMT	Montmorillonite
MWCNT	Multiwalled carbon nanotubes
nm	Nanometer
NM	Nanomaterial
NP	Nanoparticles
PA6	Polyamide 6
PAS	Publicly available specification
PE	Polyethylene
PET	Polyethylene terephthalate
PLA	Polylactic acid
PP	Polypropylene
PVC	Polyvinyl chloride
SCENIHR	Scientific Committee on Emerging and Recently Identified Health Risks
SiO_2	Silicon dioxide
TiN	Titanium nitride
TiO_2	Titanium dioxide
UN	United Nations
UV	Ultraviolet
ZnO	Zinc oxide

17.1 Introduction

Nanotechnology has great potential for application in the packaging industry. Nanomaterials have properties and applicability that stand out in the development of new processes for the generation of innovative and intelligent packaging. In this sense, the new perspectives for the packaging industry from the use of nanomaterials are mainly associated with the production of nanocomposites—polymers reinforced with particles that can have one or more dimensions, in the order of 100 nm or even less, allowing the development of new materials with the most advanced optical, thermal, mechanical properties; tolerance to high temperatures; oxygen barrier; changes in chemical reactivity; and dimensional stability (ISO, 2015; Rawtani et al., 2020).

In the food packaging industry, nanoparticles, nanolaminates, and nanotubes can be developed and associated with the most diverse functions, for example, by incorporating nanosensors capable of detecting pathogens, chemicals, and toxins into packaging; nanoparticles with bioactive potential (e.g. with an antimicrobial capacity to maintain ideal food conditions) or nanocomposites that improve the gas and moisture barrier capacity, flexibility properties, and UV irradiation absorption capacity (Duncan, 2011; Emamhadi et al., 2020).

Along with the exponential growth of nanotechnology and its benefits, the debate inherent to the potential risks of nanomaterials during their production, use, and disposal remains and, especially concerning human health and environmental impact, keeps growing. One of the major challenges in monitoring the evolution of nanomaterials is their risk management due to the diversity of shapes and types based on chemical composition and physical structure. Regarding the good management practices of these matters, the EU-OSHA has published different approaches to identify and manage risks. The industrial sectors of the EU and the United States have produced directives that mainly cover the issues related to occupational safety and consumer exposure (EU-OSHA, 2020). Thus, sustainable strategies to reduce the environmental impact of nanomaterials in packaging have been promoted for sustainable development, as proposed by the UN in the 2030 Agenda (UN, 2015).

The information we have about the advantages of nanomaterials in packaging production is far superior to the one we have about their effects on health and the environment. According to SCENIHR, nanomaterials, even those supposedly nontoxic, can bring concrete health risks (Auvinen et al., 2010). Currently, a new approach to the recycling concept is to reduce energy consumption by reusing raw materials that would otherwise be disposed of as waste is being implemented. Recycling also helps preserve valuable natural resources and reduces the amount of waste sent to landfills and incineration facilities (Pauer et al., 2019). Thus, the cost is reduced, and environmental protection can be ensured by reducing the leaching of toxic chemical products from landfills and decreasing GHG emissions from the incineration of plastics.

The idea of reusing traditional residues in nanomaterials has gained interest in the last decade. For example, in a recent project, plastic bags were used to create carbon dots, small carbon nanoparticles (less than 10 nm in size) with interesting optical properties and potential applications as imaging agents (Hu et al., 2014). In this sense, evaluating the life cycle of packaging nanomaterials becomes essential to add value to the nanoparticles and recover and reuse these materials.

LCA is one of the best methodologies for evaluating environmental impacts that currently exist and following standardized processes under the ISO standards (e.g., ISO 14044: 2006 — Environmental management—Life cycle assessment—Requirements and guidelines; ISO 14044, 2006; Rey-Álvarez et al., 2022). The methodology involves four steps: goal and scope, inventory analysis, impact assessment, and interpretation. In this manner, it is possible to obtain a better prediction of the impacts and a better understanding of how to interpret the data and handle the methodology (Westh et al., 2015). For nanomaterials used in packaging, the LCA is useful for quantifying potential environmental impacts—not only as a single approach but also as a framework in combination with other methods to develop safe nanomaterials (Som et al., 2010).

17.2 LCA Methodology

LCA is a methodology to evaluate the environmental impacts caused by products throughout their life cycle. The potential effects of raw material acquisition, production, distribution, use, EOL treatment, final disposal, and recycling on the environment are expected of LCA and are indicators for the real effects on the local, regional, and global levels (Ligthart and Ansems, 2012).

The end use of waste should be considered in LCA due to the various alternatives, including prevention, preparing for reuse, recycling, other recovery ways, and disposal. Furthermore, the materials and energy used and the emissions released in the whole process also have a relevant environmental impact. LCA is structured with four main phases: goal and scope, inventory analysis, impact assessment and interpretation (Figure 17.1), and guidelines for conducting and reviewing studies.

The main standard framework is ISO 14040 and ISO 14044 with more specified requirements. Still, other guidelines may assist in the processes that involve the environment, carbon footprint, circular economy, and recycling (Table 17.1).

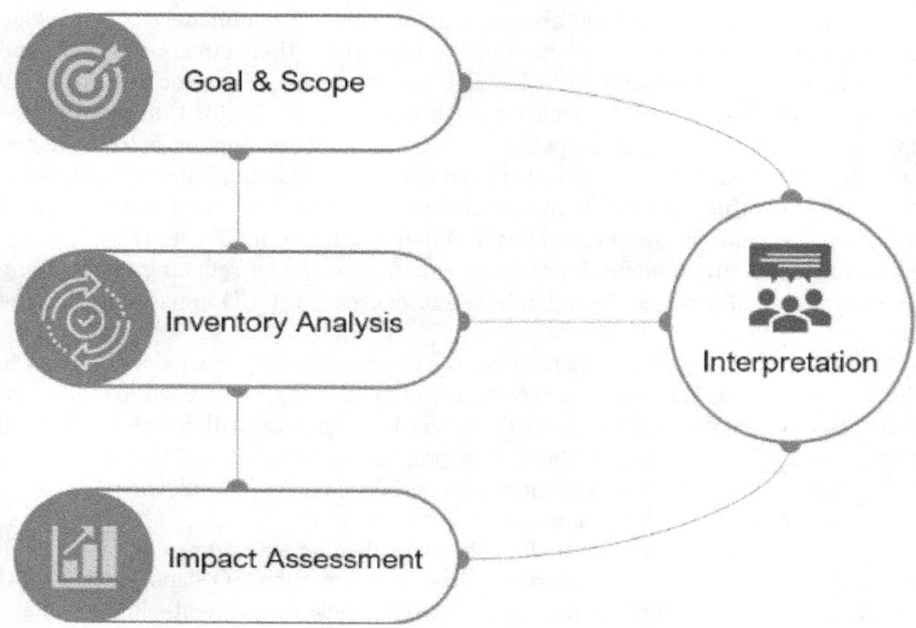

Figure 17.1 Stages of an LCA according to EN ISO 14040.

Table 17.1 Guidelines Related to LCA

Guidelines	*Details*	*Reference*
ISO 14040:2006	Principles and framework for life cycle assessment (LCA): goal and scope definition; life cycle inventory analysis (LCI); life cycle impact assessment (LCIA); life cycle interpretation.	ISO 14040, 2006
ISO 14044:2006	Specifies requirements and guidelines for life cycle assessment (LCA): goal and scope definition; life cycle inventory analysis (LCI); life cycle impact assessment (LCIA); life cycle interpretation.	ISO 14044, 2006
ISO 14021:2016	Environmental labels and declarations: materials used to produce packaging; the fate of the materials after disposal of packaging.	ISO 14021, 2016

Guidelines	Details	Reference
ISO 14025:2006	Environmental labels and declarations: principles and specifies the procedures for developing Type III environmental declaration; an addition to the ISO 14020:2000.	ISO 14025, 2006
ISO 14064:2006	Measure and control GHG emissions: determination of boundaries, quantification, mitigation, and removal.	ISO 14064, 2006
ISO 14067:2018	Carbon footprint of a product: principles; requirements; guidelines for the quantification and reporting.	ISO 14067, 2018
EU 2018/852	Circular economy of plastics: life cycle thinking; preventing the leakage of plastics into the environment.	EU Directive, 2018
EC No 282/2008	Recycling process: quality of the recycled materials guaranteed.	EU Regulation, 2008
Life Cycle Data System (ILCD)	Life cycle assessment: guidance for the development of Life Cycle Inventory (LCI) data.	European Commission, 2007
PAS 2050	Carbon footprints: quantifying product carbon footprints.	British Standards Institution, 2008

17.2.1 Goals and Scope Definitions

Goal and scope definitions are among the most important phases of an LCA study, focused on formulating questions and analyzing the context to answer. The intent of the application must be explored, analyzing the reason for conducting the study (product improvement, strategic planning, policy formulation for sustainability). The entire life cycle of a product or process must be considered in all the necessary inputs for production, energy consumption, transportation, EOL, and its environmental impacts. The results are expressed as an FU, a reference unit for comparison between the systems (ISO 14040, 2006; ISO 14044, 2006).

Considering the complexity of nanomaterials, the LCA is recognized as an important tool to identify potential life-cycle impacts of nanoproducts, and its application has increased in the last decade (Salieri et al., 2018).

17.2.2 Inventory Analysis

The main objective of the inventory analysis of an LCA is the data collection, quantifying the inputs (energy, water, raw material) and outputs (products, co-products, emissions, waste generation) for the system under assessment and using the information of the entire life cycle to establish the product system. The data can be provided by manufacturers, suppliers, LCA analysts, literature, databases, and estimations (Ligthart and Ansems, 2012).

The inventory analysis also records the economic and elementary flows, gathering data on the natural resources extracted from the environment as well as on the chemicals emitted into the natural environment. The LCI results in details of all the impact categories usually reported in an inventory table with the flow data of the inputs and outputs between the product and the environment. For nanomaterials and nanoproducts, this table must be improved with detailed information (composition, size, activity), relevant to the evaluation of the product, environmental impact, and disposal (Cucurachi et al., 2019).

17.2.3 Impact Assessment

In this phase, the inventory analysis results are processed in terms of potential impacts on the environment (global warming, acidification, and ecotoxicity). The LCIA consists of classification, characterization, normalization, and weighting. Classification is the inventory data aggregation into impact categories. Characterization is the relevance of potential impacts, where the LCI results are related to numbers and expressed in the same unit. The characterization results could be normalized to facilitate comparisons of the potential impacts and relate to a reference value. Weighting is performed to evaluate the relative importance of the environmental impact identified in previous stages; therefore, weights are assigned to obtain the final score to be compared. For NMs, special attention is needed in the LCI for toxic impacts (Cucurachi et al., 2019; Escamilla and Paul, 2014).

17.2.4 Interpretation

The last phase of LCA consists of a discussion and interpretation of the inventory and impact assessment results, in which significant environmental issues and potential areas for improvement are identified. The final assessment of the LCA involves a review to check the consistency and quality of the data according to the goal and scope and suggests quantitative and qualitative improvement measures. For NMs, it is crucial to consider sensitivity and uncertainty analysis, such as user behavior and interactions with other materials in interpretation (Escamilla and Paul, 2014).

17.3 Recycling and LCA in Packaging Systems

17.3.1 Types and Applications of NMs in the Packaging Industry

NMs are at the forefront of food and packaging development research. From the basic function of storage to tracking, NMs have been used at various stages of the food chain for different purposes. The nanotechnological advancements in the packaging sector can be attributed to the versatility and unique properties of the nanomaterials being exploited (Bumbudsanpharoke et al., 2015). According to a report by the Persistence Market Research in 2014, the global market for nano-enabled food and beverage packaging was estimated at US$6.5 billion in 2013 and growing at a CAGR of up to 12.7% to reach about US$15.0 billion in the next 10 years (Persistence Market Research, 2014).

Nanocomposites have improved technological properties, such as the extension of products' shelf life, mechanical resistance, UV protection, and barriers to moisture, gases, and volatile components (Ghanbarzadeh et al., 2015; Emamhadi et al., 2020). Among the latest innovations in the food packaging industry, biodegradable polymers reinforced with nanofillers are highlighted for meeting the real demand from consumers looking for ecologically sustainable products (Abdollahi et al., 2012; Souza and Fernando, 2016). Besides these, several researchers have studied specific applications for NMs in food packaging, namely, as functional additives, including nanoclay, starch nanocrystals, cellulose nanowhiskers, CNTs, AgNPs, nano-ZnO, and nano-TiO_2 (Bumbudsanpharoke et al., 2015; Rossi et al., 2017). Some of the main applications of NMs in food packaging are shown in Table 17.2.

Table 17.2 Nanomaterials Used in Food Packaging

Nanomaterials	*Function*	*Polymer type*	*Application*	*Reference*
AgNP	Antimicrobial and sterilization, antioxidant	PP, silicon, co-polyester, PE, PS, ABS	Apple juice, food containers, fruits, vegetables, and food	Echegoyen and Nerín, 2013; Marambio-Jones and Hoek, 2010; Mennini, 2013
Nanoclay (MMT)	Barrier properties	Nylon 6, Starch	Lettuce, spinach, beer bottles, trays for chocolate	Bumbudsanpharoke and Ko, 2019; Mennini, 2013
Nano-ZnO	Photocatalysis agent, antimicrobial and fungistatic agent	Plastic glasses, plastic films, PE film	Orange juice, meat Plastic glasses, plastic films	Alfei et al., 2020; Mustafa and Andreescu, 2020
Nano-TiO_2	Photocatalysis agent, thermal properties	PP	Lettuce	Alfei et al., 2020
TiN nanoparticle	Mechanical and thermal properties, water resistance, antimicrobial	PET	Beverage bottle, soft white cheese	Bumbudsanpharoke and Ko 2015; Chaudhry and Castle 2011
Cellulosic NMs	Intelligent packaging			Tsagkaris et al., 2018
Chitosan NPs	Mechanical properties		Edible films	Rubilar et al., 2013
CNTs	Mechanical properties			Tsagkaris et al., 2018
Graphene-Cu NPs	Oxidation of carbohydrate		Banana and bovine milk	De Silva et al., 2017

(Continued)

Table 17.2 (Continued)

Nanomaterials	*Function*	*Polymer type*	*Application*	*Reference*
Cu NPs	Oxidation of carbohydrate		Soft drink	De Silva et al., 2017
MgO	Mechanical properties, antimicrobial activity, moisture barrier properties			De Silva et al., 2017
Zein NPs	Mechanical properties, water vapor barrier		Meat	Alfei et al., 2020; Oymaci and Altinkaya, 2016
Zein NPs	Mechanical properties and thermal stability			Q. Li et al., 2020

The different characteristics and chemical structures of NMs provide additional properties to the receiving material, which leads to further functional applications in packaging. For example, AgNP and nano-ZnO, which are commonly used in nano-enabled food packaging, have greater antimicrobial activity than the macro- or microparticles (Espitia et al., 2012; Sarsar et al., 2014). Another example is starch nanocrystals that show great use as binders or adhesives in the paper-making industry due to their excellent dispersibility in water, with a surface area 400-fold larger than regular starch (Bel Haaj et al., 2016). The larger specific surface area of the NM makes it highly active compared to its macro and micro-scale particles (Bumbudsanpharoke et al., 2015). Nanoclay or CNTs, when composed in conventional thermoplastics, are uniformly dispersed in the polymeric matrix. NPs improve thermal and mechanical properties through the ability to change molecular mobility and the relaxation of the plastic polymer.

However, despite NMs promoting and driving major revolutions in the packaging sector, there is little proven information on toxicity to human health through the possible migratory potential of nanoparticles incorporated in food packaging. In recent years, the migration of NPs has been the focus of some studies with simulations in several types of foods and experimental conditions that classified the migratory limit as low and acceptable (Coles and Frewer, 2013). Besides, in the field of ecotoxicology, there are few assessments on the environmental impact possibly caused by NMs, with urgency in studies that clarify the complete life cycle of food packaging containing NPs (Souza and Fernando, 2016). By comparison, there are no appropriate standardized methods and adequate legislation. Hence, these issues need to be addressed by further research (Ndukwu et al., 2020).

17.3.2 NMs Through the Recycling Process and LCA

The knowledge of LCA of NMs used in food packaging is limited, and it is important to know the transformations that occur in these products and the related environmental impact. In the context of sustainable production, the concept of circularity and recyclability describes the environmental footprint and the innovative and preservative character of a product. Unlike linear packaging, circular packaging has reusable, recyclable, and often compostable content and is produced using renewable energy (Ligthart and Ansems, 2012).

LCAs can be used to assess the potential benefits of packaging NMs recycling programs by assessing energy and materials consumption, emissions in the environment, and the final destination of waste, as well as accompanying each activity from the extraction of raw materials to the return of waste to the soil (Bataineh, 2020; Figure 17.2).

In 2018, 33.45 kg per capita of packaging waste was generated in the EU-28; Eurostat, 2020). As a solution to reduce plastic waste, recycling has been applied extensively. The data collected by NCES (2020) between 2006 and 2018 estimated that the volume of plastic packaging waste sent to landfills was reduced by more than 50 percent while recycling increased, approaching 3.3 million metric tons (Figure 17.3). In this sense, combined with recycling, LCA can be used as an alternative tool to reuse packaging waste, either to produce new plastic packaging or to obtain NMs with high application potential.

Among the different alternatives to packaging disposal, recycling is the most conventional; however, only recycling the discarded material is not enough. Regarding this issue, the LCA of plastic packaging assumes that the recycled material can replace 100 percent of the virgin material, reducing its production. Moreover, the CO_2 emissions

Figure 17.2 Example of LCA approach for food packaging with NMs.

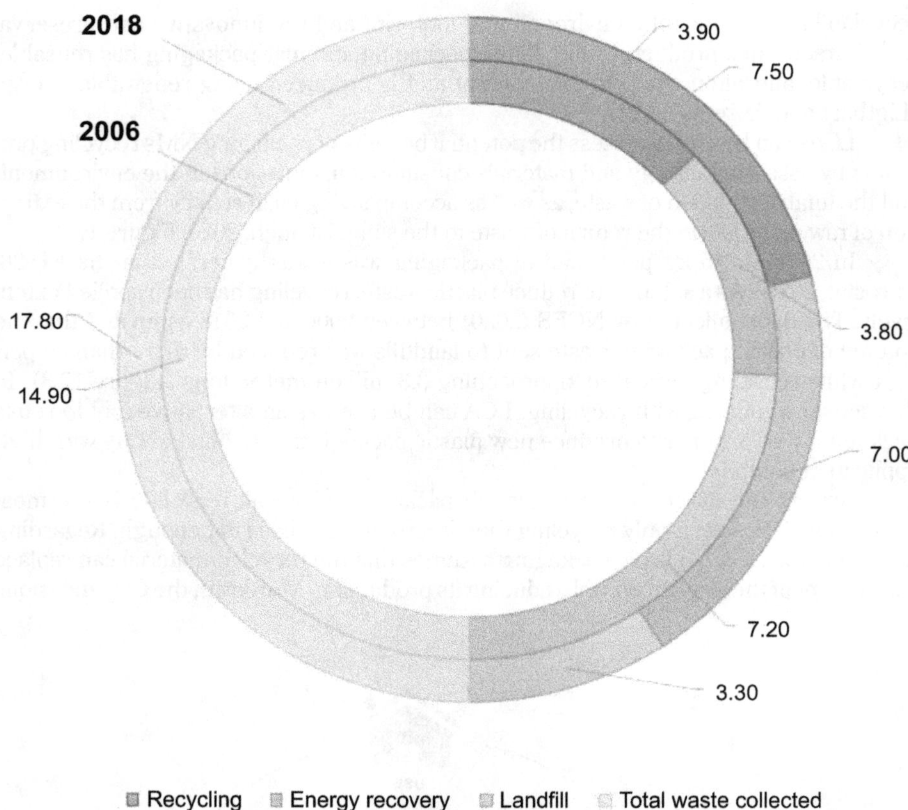

Figure 17.3 Plastic packaging waste treatment in the EU-28 by treatment type (in million metric tons). Values outside the figure represent 2018, and values inside the figure represent 2006.

for PE, PP, PET, and PVC can reduce approximately 620 kg per ton of mechanically recycled polymer (Ligthart and Ansems, 2012; Barlow and Morgan, 2013).

Plastic/PET packaging is used in several products, mainly as a bottle. Being a widely marketed product worldwide, it has become a significant problem for the environment. The recycling of PET bottles can present a significant net benefit in GHG emissions of up to 15 tons of CO_2 equivalent per ton of recycled PET (Patel et al., 2000). LCA studies of PET packaging demonstrate that recycled PET fibers offer considerable environmental benefits over virgin PET for bottle production. Table 17.3 shows the benefits of using recycled PET.

In addition, the introduction of nanotechnology in the packaging industries with the development of active, functional, and reinforcing materials reduces the impacts related to the production, use, and disposal of packaging (Pauer et al., 2019). As an example, it is possible to produce CNMs from recycled PET fibers. Recently, carbon nanostructure materials were obtained through thermal dissociation in a closed system under autogenic pressure (El Essawy et al., 2017). As a result, it was possible to produce and control the microstructures of various forms of carbon NPs from PET

Table 17.3 Environmental Impact of Using Recycled PET Fibers Compared to Virgin PET

	Virgin PET	*Recycled PET*	*References*
Energy consumption for production	70830 MJ/ton	7970 MJ/ton	Perugini et al., 2005 Chilton et al., 2010
Reduction of GHG emissions		-27%	Shen et al., 2010
Open-loop recycling			
Fuel energy savings		40–85%	Chilton et al., 2010
Global warming savings		25–75%	
Closed-loop recycling and incineration		Reduce CO_2, carbon monoxide, acid gases, particulate matter, heavy metals, and dioxins	Rajendran et al., 2012

packaging waste. Another study obtained similar results using the rotating cathode arc discharge technique (Joseph Berkmans et al., 2014).

Although the conversion of low-value waste into high-value CNMs contributes to sustainable recycling and processing technology, it is ideal to look for green synthesis methods with energetically economical reactions from cheap materials that can be easily extended and with potential for application on an industrial scale (Deng et al., 2016).

Nano-enabled packaging allows a longer shelf life for food than traditional plastic. The application of PLA film revealed that the incorporation of MMT-Na enhanced the water barrier property, extending the shelf life of meat products (Vilarinho et al., 2018). A study using nanoclay and AgNp as potential reinforcements to improve the properties of PLA packaging designed for fruit salad retards the diffusion of oxygen into the packaging (Lorite et al., 2017).

An LCA study proved that ZnO-reinforced active packaging reduces the environmental impact by extending the shelf life of the products (Vigil et al., 2020). These authors also compared the EOL of PLA and PP packaging with ZnO-reinforced. PP packaging has advantages due to recycling, unlike PLA, which is meant to be composted, impacting the EOL result due to the recommended industrial composting. By comparison, recycling presents some factors that are not addressed in the study of LCA, such as marine waste and microplastics present in the environment.

Lorite et al. (2017) conducted an LCA study of PLA formulations enhanced with nanoclays, reporting the improvement of the physicochemical and antimicrobial properties, bringing the characteristics and behavior close to PET packages, and extending the product's shelf life. The impact assessment shows that packaging manufacturing is the major contributor to the environmental impacts, where high energy consumption and emissions are required for PLA/nanoclay. However, with the shelf life extended by approximately 30 percent, the new polymers can be considered more environmentally friendly than PET packaging.

Although the current waste scenario (60 percent of PET is incinerated, 40 percent goes to landfill) provides an advantage to PET due to the energy produced by incineration, this packaging presents a lower environmental aspect of the desired scenario (PET 100 percent incinerated) because the database considers incineration less eco-friendly than landfilling.

NANOPOLYTOX, an EU-funded research project, has studied the life-cycle impacts of different nanomaterials, such as SiO_2, TiO_2, ZnO, MWCNTs, and nanoclays, modeling all life stages of the materials and predicting the environmental impacts according to additional characteristics (size, composition, morphology, etc.). At the endpoint level, all NMs showed a high impact on climate change, due to high energy consumption, human health damage (84–80 percent), and ecosystem diversity damage (97–95 percent). In waste treatment, for the nanocomposite Polyamide 6 + TiO_2 (PA6+TiO_2) the higher impact was in chemical recycling and for MWCNT in mechanical recycling, whereas the NM synthesis had lower impacts.

Comparison of NPs is intricated, due to the different approaches in the applied pretreatments to the materials before incorporation in the packaging, such as TiO_2 which needs to be functionalized before incorporation presenting very high impacts on the LCA. In the worst scenario, it contributed 24 percent to human damage and 0.03 percent to damage to the ecosystem. For ZnO, this damage was 7 and 0.04 percent, respectively, and MWCNTs contributed only to 1 and 0.003 percent, respectively. Since nanoclays are a natural material, release into the environment had no significance (Grieger et al., 2012; Hischier and Walser, 2012).

17.3.3 Environmental Impact Versus NMs

Toxicology analysis and risk assessment are required to prove the security of nanomaterials on the environment and human health. Also, the comprehension of the migration from the packaging nanomaterial to the food is important to the safety assessment. The chemistry data and the acting of NMs must be investigated due to the different nature of the particles, size, behavior, and exposure control which may spoil the mechanisms of action (Cushen et al., 2012).

Failure to apply the LCA or the incorrect recycling of NMs promotes the irresponsible disposal of nano-packaging. As a result of the decomposition, nanoparticles available in the environment contaminate the ecosystem and consequently the human body. Many studies describe the NMs used in food packaging and the environmental impact of their disposal into the environment. Still, few works evaluate the risk of the absorption of these materials by the human body (Rhim et al., 2013). Figure 17.4 describes the possible pathways of NPs discarded in the ecosystems.

A simulation with AgNP added as an antimicrobial compound to the packaging showed this nanoparticle can slowly migrate to the surface of the food and attack the microorganisms. However, the final results quantified 3 percent of silver in the food matrix (Song et al., 2011). Another study with AgNP-PVC detected that the highest rate of silver migration was 8.85 mg/Kg (Cushen et al., 2013). Emamifar et al. (2011) studied the application of NPs to extend the shelf life of orange juice and showed that Zn and Ag might accelerate the degradation of ascorbic acid.

Nanoclay is the most widely used nanomaterial as a reinforced barrier because of the poor mechanical properties of biopolymers. Moreover, it is a natural compound with low toxicity and is easily found in nature. Schmidt et al. (2009) did not detect the migration of nanoclay from the composite film to food, confirming the safety of this nanomaterial for use in food packaging.

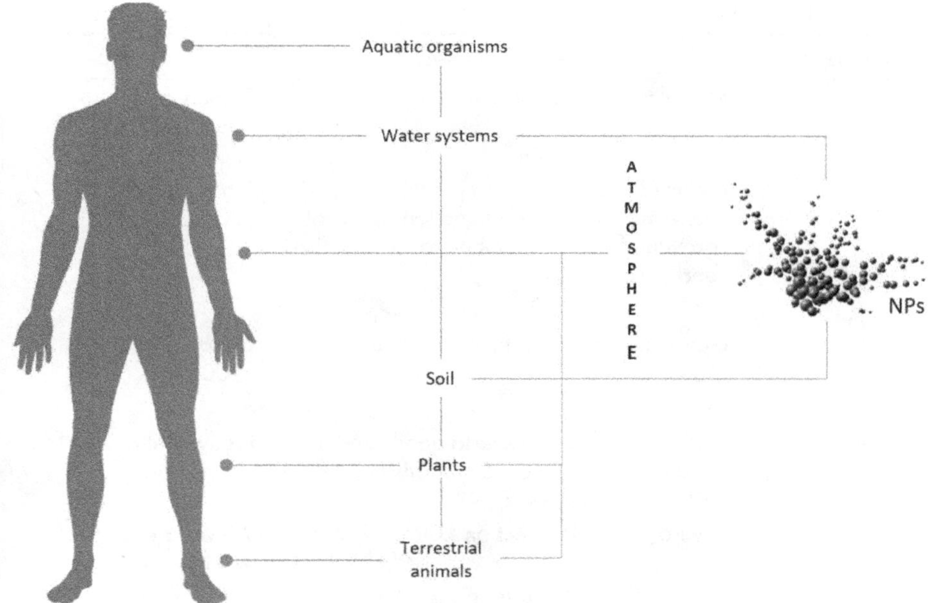

Figure 17.4 Description of possible contamination pathways of NPs by the human body.

Table 17.4 Negative Impacts of NMs on Environmental and Human Health

Nanomaterial	*Negative impact*	*Reference*
Fullerene C60	Antibacterial to a broad range of bacteria; Inhibits growth of *Escherichia coli* and *Mycobacteria*, antimutagenic in *Salmonella typhimurium*	Klaine et al., 2008
Fullerene C60 and derivative	Potential to damage biological membranes.	Kamat et al., 2000
Fullerene C60	Delayed zebrafish (*Danio rerio*) embryo and larval development, decreased survival and hatching rates, pericardial edema at 1.5 mg/L over 96 h.	Gimbert et al., 2006
Fullerene C60	Reduction in number of bacteria. No effect on respiration, microbial biomass C, and protozoan abundance.	Zhang and Selim, 2007
CNTs, carbon nanofibers, graphene	Formation of dust clouds	Turkevich et al., 2015
CNTs	Antibacterial to *E. coli*, Cytotoxic to microbes	Klaine et al., 2008
CNTs	Mortality of *D. magna* exposed to concentrations up to 20 mg/L over 96 h.	Roberts et al., 2007

(Continued)

Table 17.4 (Continued)

Nanomaterial	*Negative impact*	*Reference*
CNTs	Mortality of some freshwater crab species	Francis and Devasena, 2018
CNTs	Reduced root growth of ryegrass. No effect on seed germination and root growth of corn, cucumber, lettuce, radish, and mustard rape.	Goodman et al., 2004
CNTs	Genotoxic to human bronchial cells	Lindberg et al., 2009
Graphene	Selected intestinal bacteria unaffected against different concentrations (10 to 500 ug/ml) for 24 h,	Nguyen et al., 2015
Graphene	In vitro cytotoxic effect on ROMI 8226 cells	Y. Wang et al., 2014
TiO_2 NPs	Induced DNA damage in *Nicotina tabacum*	Ghosh et al., 2010
MgO	Antibacterial activity against *Bacillus subtilis* and *Staphylococcus aureus*	Zhang and Selim, 2007
Nano-ZnO	Damage to bacterial cell membrane and decrease bacterial ATP level	Tong et al., 2015
ZnO NPs	Reduced seed germination of corn. Reduced root growth of corn, cucumber, lettuce, radish, mustard rape, and ryegrass.	Goodman et al., 2004
ZnO NPs	Toxic to green alga *Tetraselmis suecica* and diatom *Phaeodactylum tricornutum*	J. Li et al., 2017
Graphene	In vitro cytotoxic effect on ROMI 8226 cells	Y. Wang et al., 2014
Ag NPs	Bactericidal, viricidal	Morones et al., 2005
Ag NPs	Reduced plant growth, upregulation of plant defense system	P. Wang et al., 2017
Nano Cd	Oxidative damage to *D. magna*	Gao et al., 2018
Gold NPs	Low toxicity to *E. coli* and *S. aureus*	Goodman et al., 2004
Cu NPs	Reduced growth of seedlings of mung bean (*Phaseolus radiatus*) and wheat (*Triticum aestivum*)	Lee et al., 2008
Nano-Al	Increased nano-Al concentration in leaves of Lolium perenne	Doshi et al., 2008
Au NPs	Toxic to human endothelial cells	Fede et al., 2015

Table 17.5 Positive Impacts of NMs on the Environment

Impact category	*Positive impacts*
Boost chemical reaction	Reduce pollution and expenses (e.g., automotive exhaust pollution, petroleum refining cost).
CO_2 isolation	Reduce GHG
Life cycle of packaging	Increase the biodegradability of food packaging; Saving energy in packaging production
Low resistance conductor	Increase power transmission efficiency.
More efficient conductor	Increase performance and reduce cost of solar energy generation.
Precision manufacturing	Generation of less waste; Reduce the requirement of large industrial plants.
Purification, filtration and oxidation	Reduce and control pollution; Convert toxic elements into nontoxic.
Self-cleaning	Save water, energy, and cleaning agents.

Source: Adapted from Kabir et al. (2018).

NMs play a vital role in reducing manufacturing costs and increasing packaging efficiency and are directly related to food quality and shelf life. In addition, with NMs, the packaging can significantly improve biodegradability, reducing the environmental impact of these products. In general, recent studies have shown beneficial results in several sectors that use nanotechnology, and some of these positive impacts are shown in Table 17.5.

17.4 Trends and Future Prospects for LCA in Packaging

Overall, a detailed assessment of nanomaterials and their potential use in the packaging industry shows that the flow of NMs in this sector will solve several problems associated with food and offer economic benefits in the medium to long term. The possibility of producing innovative packaging with intelligent systems and adjustable storage properties to improve food protection can revolutionize the global chain of distribution and storage of perishable products. Thus, through a perspective focused on packaging, nanotechnology will have a huge impact on sustainability and will bring benefits to the environment and the quality of product processing if adequately managed and regulated. It is hoped that in the future, labeling and regulations of nano-packaging may influence the marketing and acceptance by the consumer. Moreover, with the novel developments with nanomaterials, it is essential to develop reliable tools for risk assessment as well as a careful system for assessing the life cycle of nano-packaging. Thus, in a standardized way, research is expected regarding issues such as lack of resources and environmental degradation and not only the benefits of the packaging sector. In conclusion, the pace of the introduction of nanotechnology in the food packaging sector should be well balanced with the evolution of guidelines that guarantee the safety of this innovative technology without leaving behind issues such as sustainability and environmental footprint.

References

Abdollahi, M., Rezaei, M., and Farzi, G. 2012. A novel active bionanocomposite film incorporating rosemary essential oil and nanoclay into chitosan. *Journal of Food Engineering*, 111, 343–350.

Alfei, S., Marengo, B., and Zuccari, G. 2020. Nanotechnology application in food packaging: A plethora of opportunities versus pending risks assessment and public concerns. *Food Research International*, 137, 109664.

Auvinen, A., Bridges, J. W., Dawson, K., Jong, W. D., Hartemann, P., Hoet, P., Jung, T., Mattsson, M. O., Norppa, H., Pagès, J. M., Proykova, A., Rodríguez-Farré, E., Schulze-Osthoff, K., Schüz, J., Thomsen, M., and Vermeire, T. 2010. *Scientific Committee on Emerging and Newly Identified Health Risks SCENIHR: Scientific Basis for the Definition of the Term "Nanomaterial."* European Commission DG Health & Consumers, Brussels.

Barlow, C. Y., and Morgan, D. C. 2013. Polymer film packaging for food: An environmental assessment. *Resources, Conservation and Recycling*, 78, 74–80.

Bataineh, K. M. 2020. Life-cycle assessment of recycling postconsumer high-density polyethylene and polyethylene terephthalate. *Advances in Civil Engineering*, 2020, 1–15.

Bel Haaj, S., Thielemans, W., Magnin, A., and Boufi, S. 2016. Starch nanocrystals and starch nanoparticles from waxy maize as nanoreinforcement: A comparative study. *Carbohydrate Polymers*, 143, 310–317.

British Standards Institution, B. 2008. PAS 2050:2008. *Specification for the Assessment of the Life Cycle Greenhouse Gas Emissions of Goods and Services*. BSI, London, United Kingdom.

Bumbudsanpharoke, N., Choi, J., and Ko, S. 2015. Applications of nanomaterials in food packaging. *Journal of Nanoscience and Nanotechnology*, 15, 6357–6372.

Bumbudsanpharoke, N., and Ko, S. 2015. Nano-food packaging: An overview of market, migration research, and safety regulations. *Journal of Food Science*, 80, 910–923.

Bumbudsanpharoke, N., and Ko, S. 2019. Nanoclays in food and beverage packaging. *Journal of Nanomaterials*, 2019, 1–13.

Chaudhry, Q., and Castle, L. 2011. Food applications of nanotechnologies: An overview of opportunities and challenges for developing countries. *Trends in Food Science and Technology*, 22, 595–603.

Chilton, T., Burnley, S., and Nesaratnam, S. 2010. A life cycle assessment of the closed-loop recycling and thermal recovery of post-consumer PET. *Resources, Conservation and Recycling*, 54, 1241–1249.

Coles, D., and Frewer, L. J. 2013. Nanotechnology applied to European food production—A review of ethical and regulatory issues. *Trends in Food Science and Technology*, 34, 32–43.

Cucurachi, S., Scherer, L., Guinée, J., and Tukker, A. 2019. Life cycle assessment of food systems. *One Earth Primer*, 1, 292–297.

Cushen, M., Kerry, J., Morris, M., Cruz-Romero, M., and Cummins, E. 2012. Nanotechnologies in the food industry—Recent developments, risks and regulation. *Trends in Food Science and Technology*, 24, 30–46.

Cushen, M., Kerry, J., Morris, M., Cruz-Romero, M., and Cummins, E. 2013. Migration and exposure assessment of silver from a PVC nanocomposite. *Food Chemistry*, 139, 389–397.

Deng, J., You, Y., Sahajwalla, V., and Joshi, R. K. 2016. Transforming waste into carbon-based nanomaterials. *Carbon*, 96, 105–115.

De Silva, R. T., Mantilaka, M. M. M. G. P. G., Ratnayake, S. P., Amaratunga, G. A. J., and Nalin de Silva, K. M. N. 2017. Nano-MgO reinforced chitosan nanocomposites for high performance packaging applications with improved mechanical, thermal and barrier properties. *Carbohydrate Polymers*, 157, 739–747.

Doshi, R., Braida, W., Christodoulatos, C., Wazne, M., and O'Connor, G. 2008. Nanoaluminum: Transport through sand columns and environmental effects on plants and soil communities. *Environmental Research*, 106, 296–303.

Duncan, T. V. 2011. Applications of nanotechnology in food packaging and food safety: Barrier materials, antimicrobials and sensors. *Journal of Colloid and Interface Science*, 363, 1–24.

Echegoyen, Y., and Nerín, C. 2013. Nanoparticle release from nano-silver antimicrobial food containers. *Food and Chemical Toxicology*, 62, 16–22.

El Essawy, N. A., Konsowa, A. H., Elnouby, M., and Farag, H. A. 2017. A novel one-step synthesis for carbon-based nanomaterials from polyethylene terephthalate (PET) bottles waste. *Journal of the Air & Waste Management Association*, 67, 358–370.

Emamhadi, M. A., Sarafraz, M., Akbari, M., Thai, V. N., Fakhri, Y., Linh, N. T. T., and Mousavi Khaneghah, A. 2020. Nanomaterials for food packaging applications: A systematic review. *Food and Chemical Toxicology*, 146, 111825.

Emamifar, A., Kadivar, M., Shahedi, M., and Soleimanian-Zad, S. 2011. Effect of nanocomposite packaging containing Ag and ZnO on inactivation of Lactobacillus plantarum in orange juice. *Food Control*, 22, 408–413.

Escamilla, M., and Paul, R. 2014. Methodology for Testing the Life Cycle Sustainability of Flame Retardant Chemicals and Nanomaterials. pp. 891–905. In: C. D. Papaspyrides and P. Kiliaris (Eds.), *Polymer Green Flame Retardants*. Elsevier, Amsterdam.

Espitia, P. J. P., Soares, N. D. F. F., Coimbra, J. S. D. R., de Andrade, N. J., Cruz, R. S., and Medeiros, E. A. A. 2012. Zinc oxide nanoparticles: Synthesis, antimicrobial activity and food packaging applications. *Food and Bioprocess Technology*, 5, 1447–1464.

EU Directive, 2018/852. 2018. *Directives Related to Circular Economy (EU) 2018/852*. European Commission, Brussels.

EU-OSHA. 2020. *Nanomaterials in the Workplace : Risks, Health Effects and Prevention*. European Commission, Brussels.

EU Regulation, 282/2008. 2008. *Regulation on Recycled Plastic (EU) 282/2008*. European Commission, Brussels.

European Commission, J. 2007. *International Reference Life Cycle Data System (ILCD) Handbook*. European Commission, Ispra, VA, Italy.

Eurostat. 2020. *Packaging Waste by Waste Management Operations*. Eurostat, Luxembourg.

Fede, C., Fortunati, I., Weber, V., Rossetto, N., Bertasi, F., Petrelli, L., Guidolin, D., Signorini, R., De Caro, R., Albertin, G., and Ferrante, C. 2015. Evaluation of gold nanoparticles toxicity towards human endothelial cells under static and flow conditions. *Microvascular Research*, 97, 147–155.

Francis, A. P., and Devasena, T. 2018. Toxicity of carbon nanotubes: A review. *Toxicology and Industrial Health*, 34, 200–210.

Gao, M., Zhang, Z., Lv, M., Song, W., and Lv, Y. 2018. Toxic effects of nanomaterial-adsorbed cadmium on Daphnia magna. *Ecotoxicology and Environmental Safety*, 148, 261–268.

Ghanbarzadeh, B., Oleyaei, S. A., and Almasi, H. 2015. Nanostructured materials utilized in biopolymer-based plastics for food packaging applications. *Critical Reviews in Food Science and Nutrition*, 55, 1699–1723.

Ghosh, M., Bandyopadhyay, M., and Mukherjee, A. 2010. Genotoxicity of titanium dioxide (TiO2) nanoparticles at two trophic levels: Plant and human lymphocytes. *Chemosphere*, 81, 1253–1262.

Gimbert, L. J., Haygarth, P. M., Beckett, R., and Worsfold, P. J. 2006. The influence of sample preparation on observed particle size distributions for contrasting soil suspensions using flow field-flow fractionation. *Environmental Chemistry*, 3, 184–191.

Goodman, C. M., McCusker, C. D., Yilmaz, T., and Rotello, V. M. 2004. Toxicity of gold nanoparticles functionalized with cationic and anionic side chains. *Bioconjugate Chemistry*, 15, 897–900.

Grieger, K. D., Laurent, A., Miseljic, M., Christensen, F., Baun, A., and Olsen, S. I. 2012. Analysis of current research addressing complementary use of life-cycle assessment and risk assessment for engineered nanomaterials: Have lessons been learned from previous experience with chemicals? *Journal of Nanoparticle Research*, 14.

Hischier, R., and Walser, T. 2012. Life cycle assessment of engineered nanomaterials: State of the art and strategies to overcome existing gaps. *Science of the Total Environment*, 425, 271–282.

Hu, Y., Yang, J., Tian, J., Jia, L., and Yu, J. S. 2014. Green and size-controllable synthesis of photoluminescent carbon nanoparticles from waste plastic bags. *RSC Advances*, 4, 47169–47176.

ISO. 2015. Technical Specifications ISO/TS 800004–2:2015 (E): Nanotechnologies —Vocabulary—Part 2: Nano-Objects. ISO, The International Organization for Standardization.

ISO 14021, 2006. 2016. Environmental Labels and Declarations—Self-Declared Environmental Claims (Type II Environmental Labelling).

ISO 14025, 2006. 2006. Environmental Labels and Declarations—Type III Environmental Declarations—Principles and Procedures.

ISO 14040, 2006. 2006. Environmental Management—Life Cycle Assessment—Requirements and Guidelines.

ISO 14044, 2006. 2006. Environmental Management—Life Cycle Assessment—Principles and Framework.

ISO 14064, 2006. 2006. Greenhouse Gases—Part 1: Specification with Guidance at the Organization Level for Quantification and Reporting of Greenhouse Gas Emissions and Removals.

ISO 14067, 2018. 2018. Greenhouse Gases—Carbon Footprint of Products—Requirements and Guidelines for Quantification.

Joseph Berkmans, A., Jagannatham, M., Priyanka, S., and Haridoss, P. 2014. Synthesis of branched, nano channeled, ultrafine and nanocarbon tubes from PET wastes using the arc discharge method. *Waste Management*, 34, 2139–2145.

Kabir, E., Kumar, V., Kim, K. H., Yip, A. C. K., and Sohn, J. R. 2018. Environmental impacts of nanomaterials. *Journal of Environmental Management*, 225, 261–271.

Kamat, J. P., Devasagayam, T. P. A., Priyadarsini, K. I., and Mohan, H. 2000. Reactive oxygen species mediated membrane damage induced by fullerene derivatives and its possible biological implications. *Toxicology*, 155, 55–61.

Klaine, S. J., Alvarez, P. J. J., Batley, G. E., Fernandes, T. F., Handy, R. D., Lyon, D. Y., Mahendra, S., McLaughlin, M. J., and Lead, J. R. 2008. Nanomaterials in the environment: Behavior, fate, bioavailability, and effects. *Environmental Toxicology and Chemistry*, 27, 1825–1851.

Lee, W. M., An, Y. J., Yoon, H., and Kweon, H. S. 2008. Toxicity and bioavailability of copper nanoparticles to the terrestrial plants mung bean (Phaseolus radiatus) and wheat (Triticum aestivum): Plant agar test for water-insoluble nanoparticles. *Environmental Toxicology and Chemistry*, 27, 1915–1921.

Li, J., Schiavo, S., Rametta, G., Miglietta, M. L., La Ferrara, V., Wu, C., and Manzo, S. 2017. Comparative toxicity of nano ZnO and bulk ZnO towards marine algae Tetraselmis suecica and Phaeodactylum tricornutum. *Environmental Science and Pollution Research*, 24, 6543–6553.

Li, Q., Gao, R., Wang, L., Xu, M., Yuan, Y., Ma, L., Wan, Z., and Yang, X. 2020. Nanocomposites of bacterial cellulose nanofibrils and zein nanoparticles for food packaging. *ACS Applied Nano Materials*, 3, 2899–2910.

Ligthart, T. N., and Ansems, T. A. M. M. 2012. Modelling of Recycling in LCA. pp. 185–210. In: E. Damanhuri (Ed.), *Post-Consumer Waste Recycling and Optimal Production*. IntechOpen, London.

Lindberg, H. K., Falck, G. C. M., Suhonen, S., Vippola, M., Vanhala, E., Catalán, J., Savolainen, K., and Norppa, H. 2009. Genotoxicity of nanomaterials: DNA damage and micronuclei induced by carbon nanotubes and graphite nanofibres in human bronchial epithelial cells in vitro. *Toxicology Letters*, 186, 166–173.

Lorite, G. S., Rocha, J. M., Miilumäki, N., Saavalainen, P., Selkälä, T., Morales-Cid, G., Gonçalves, M. P., Pongrácz, E., Rocha, C. M. R., and Toth, G. 2017. Evaluation of physicochemical/microbial properties and life cycle assessment (LCA) of PLA-based nanocomposite active packaging. *LWT—Food Science and Technology*, 75, 305–315.

Marambio-Jones, C., and Hoek, E. M. V. 2010. A review of the antibacterial effects of silver nanomaterials and potential implications for human health and the environment. *Journal of Nanoparticle Research*, 12, 1531–1551.

Mennini, T. 2013. Nanotech nanotechnology-enabled foods and food contact materials on the UK market. *Nutrafoods*, 12, 13749.

Morones, J. R., Elechiguerra, J. L., Camacho, A., Holt, K., Kouri, J. B., Ramírez, J. T., and Yacaman, M. J. 2005. The bactericidal effect of silver nanoparticles. *Nanotechnology*, 16, 2346–2353.

Mustafa, F., and Andreescu, S. 2020. Nanotechnology-based approaches for food sensing and packaging applications. *RSC Advances*, 10, 19309–19336.

NCES. 2020. *Plastic Packaging Waste Treatment in the European Union (EU-28) in 2006 and 2018 by Treatment Type (in Million Netric Tons)*. Statista—The Statistics Portal, United Kingdom, London.

Ndukwu, M. C., Ikechukwu-Edeh, C. E., Nwakuba, N., Okosa, I., Horsefall, I. T., and Orji, F. N. 2020. Nanomaterials application in greenhouse structures, crop processing machinery, packaging materials and agro-biomass conversion. *Materials Science for Energy Technologies*, 3, 690–699.

Nguyen, T. H. D., Lin, M., and Mustapha, A. 2015. Toxicity of graphene oxide on intestinal bacteria and Caco-2 cells. *Journal of Food Protection*, 78, 996–1002.

Oymaci, P., and Altinkaya, S. A. 2016. Improvement of barrier and mechanical properties of whey protein isolate based food packaging films by incorporation of zein nanoparticles as a novel bionanocomposite. *Food Hydrocolloids*, 54, 1–9.

Patel, M., Von Thienen, N., Jochem, E., and Worrell, E. 2000. Recycling of plastics in Germany. *Resources, Conservation and Recycling*, 29, 65–90.

Pauer, E., Wohner, B., Heinrich, V., and Tacker, M. 2019. Assessing the environmental sustainability of food packaging: An extended life cycle assessment including packaging-related food losses and waste and circularity assessment. *Sustainability (Switzerland)*, 11.

Persistence Market Research. 2014. *Global Market Study on Nano-enabled Packaging for Food and Beverages: Growth Estimated at a CAGR of 12.7% over 2014–2020*. Persistance Market Research, New York.

Perugini, F., Mastellone, M. L., and Arena, U. 2005. A life cycle assessment of mechanical and feedstock recycling options for management of plastic packaging wastes. *Environmental Progress*, 24, 137–154.

Rajendran, S., Scelsi, L., Hodzic, A., Soutis, C., and Al-Maadeed, M. A. 2012. Environmental impact assessment of composites containing recycled plastics. *Resources, Conservation and Recycling*, 60, 131–139.

Rawtani, D., Rao, P. K., and Hussain, C. M. 2020. Recent advances in analytical, bioanalytical and miscellaneous applications of green nanomaterial. *TrAC Trends in Analytical Chemistry*, 133, 116109.

Rey-Álvarez, B., Sánchez-Montañés, B., and García-Martínez, A. 2022. Building material toxicity and life cycle assessment: A systematic critical review. *Journal of Cleaner Production*, 341, 130838.

Rhim, J. W., Park, H. M., and Ha, C. S. 2013. Bio-nanocomposites for food packaging applications. *Progress in Polymer Science*, 38, 1629–1652.

Roberts, A. P., Mount, A. S., Seda, B., Souther, J., Qiao, R., Lin, S., Pu, C. K., Rao, A. M., and Klaine, S. J. 2007. In vivo biomodification of lipid-coated carbon nanotubes by Daphnia magna. *Environmental Science and Technology*, 41, 3028–3029.

Rossi, M., Passeri, D., Sinibaldi, A., Angjellari, M., Tamburri, E., Sorbo, A., Carata, E., and Dini, L. 2017. Nanotechnology for food packaging and food quality assessment. *Advances in Food and Nutrition Research*, 82, 149–204.

Rubilar, J. F., Cruz, R. M. S., Silva, H. D., Vicente, A. A., Khmelinskii, I., and Vieira, M. C. 2013. Physico-mechanical properties of chitosan films with carvacrol and grape seed extract. *Journal of Food Engineering*, 115, 466–474.

Salieri, B., Turner, D. A., Nowack, B., and Hischier, R. 2018. Life cycle assessment of manufactured nanomaterials: Where are we? *NanoImpact*, 10, 108–120.

Sarsar, V., Selwal, K., and Selwal, M. 2014. Nanosilver: Potent antimicrobial agent and its biosynthesis. *African Journal of Biotechnology*, 13, 546–554.

Schmidt, B., Petersen, J. H., Bender Koch, C., Plackett, D., Johansen, N. R., Katiyar, V., and Larsen, E. H. 2009. Combining asymmetrical flow field-flow fractionation with light-scattering and inductively coupled plasma mass spectrometric detection for characterization of nanoclay used in biopolymer nanocomposites. *Food Additives & Contaminants: Part A*, 26, 1619–1627.

Shen, L., Worrell, E., and Patel, M. K. 2010. Open-loop recycling: A LCA case study of PET bottle-to-fibre recycling. *Resources, Conservation and Recycling*, 55, 34–52.

Som, C., Berges, M., Chaudhry, Q., Dusinska, M., Fernandes, T. F., Olsen, S. I., and Nowack, B. 2010. The importance of life cycle concepts for the development of safe nanoproducts. *Toxicology*, 269, 160–169.

Song, H., Li, B., Lin, Q.-B., Wu, H.-J., and Chen, Y. 2011. Migration of silver from nanosilver—Polyethylene composite packaging into food simulants. *Food Additives & Contaminants: Part A*, 28, 1–5.

Souza, V. G. L., and Fernando, A. L. 2016. Nanoparticles in food packaging: Biodegradability and potential migration to food-A review. *Food Packaging and Shelf Life*, 8, 63–70.

Tong, T., Wilke, C. M., Wu, J., Binh, C. T. T., Kelly, J. J., Gaillard, J. F., and Gray, K. A. 2015. Combined toxicity of nano-ZnO and nano-TiO2: From single- to multinanomaterial systems. *Environmental Science and Technology*, 49, 8113–8123.

Tsagkaris, A. S., Tzegkas, S. G., and Danezis, G. P. 2018. Nanomaterials in food packaging: State of the art and analysis. *Journal of Food Science and Technology*, 55, 2862–2870.

Turkevich, L. A., Dastidar, A. G., Hachmeister, Z., and Lim, M. 2015. Potential explosion hazard of carbonaceous nanoparticles: Explosion parameters of selected materials. *Journal of Hazardous Materials*, 295, 97–103.

UN. 2015. *United Nations Sustainable Development Goals (SDGs)*. United Nations General Assembly (UNGA), New York.

Vigil, M., Pedrosa-Laza, M., Cabal, J. V. A., and Ortega-Fernández, F. 2020. Sustainability analysis of active packaging for the fresh cut vegetable industry by means of attributional & consequential life cycle assessment. *Sustainability*, 12, 1–18.

Vilarinho, F., Andrade, M., Buonocore, G. G., Stanzione, M., Vaz, M. F., and Sanches Silva, A. 2018. Monitoring lipid oxidation in a processed meat product packaged with nanocomposite poly(lactic acid) film. *European Polymer Journal*, 98, 362–367.

Wang, P., Lombi, E., Sun, S., Scheckel, K. G., Malysheva, A., McKenna, B. A., Menzies, N. W., Zhao, F. J., and Kopittke, P. M. 2017. Characterizing the uptake, accumulation and toxicity of silver sulfide nanoparticles in plants. *Environmental Science: Nano*, 4, 448–460.

Wang, Y., Wu, S., Zhao, X., Su, Z., Du, L., and Sui, A. 2014. In vitro toxicity evaluation of graphene oxide on human RPMI 8226 cells. *Bio-Medical Materials and Engineering*, 24, 2007–2013.

Westh, T. B., Hauschild, M. Z., Birkved, M., Jørgensen, M. S., Rosenbaum, R. K., and Fantke, P. 2015. The USEtox story: A survey of model developer visions and user requirements. *International Journal of Life Cycle Assessment*, 20, 299–310.

Zhang, H., and Selim, H. M. 2007. Colloid mobilization and arsenite transport in soil columns: Effect of ionic strength. *Journal of Environmental Quality*, 36, 1273–1280.

Chapter 18

Global Status of Nanotechnology Policies in the Packaging Sector

Christopher Igwe Idumah

List of Abbreviations

ACOP	Approved Code of Practice
Al	Aluminum
CAGR	Compound annual growth rate
CAS	Chinese Academy of Sciences
CEN	European Committee for Standardization
CO_2	Carbon dioxide
COSHH	Control of Substances Hazardous to Health Regulations
COVID-19	Severely Acute Respiratory Syndrome Coronavirus 2 (SARS-CoV-2)
DSEAR	Dangerous Substances and Explosive Atmosphere Regulations
EC	European Commission
EFSA	European Food Safety Authority
EHS	Environmental, Health, and Safety
EINECS	European Inventory of Existing Commercial Chemical Substances
EPA	U.S. Environmental Protection Agency
EU	European Union
FAO	Food and Agriculture Organization
FCS	Food contact substance
FDA	U.S. Food and Drug Administration (United States)
FIC	Food Information to Consumers
FSANZ	Australia/New Zealand
FSIS	U.S. Department of Agriculture's Food Safety and Inspection Service (United States)

DOI: 10.1201/9781003142287-18

HSE	Health and Safety Executive
IFCS	Intergovernmental Forum on Chemical Safety
ISO	International Organization for Standardization
KMFDS	Korea Ministry of Food and Drug Safety
METI	Ministry of Economic, Trade, and Industries
MEXT	Ministry of Education, Culture, Sports, Science, and Technology
MHLW	Ministry of Health, Labor, and Welfare
MOE	Ministry of the Environment
NANONET	Nanotechnology Researchers Network Center
NCNST	National Center for Nanoscience and Technology
NIFA	U.S. National Institute of Food and Agriculture
NIH/NIEHS	National Institutes of Health and National Institute of Environmental Health Sciences
NIMS	National Institute for Materials Science
NM	Nanomaterial
NNI	U.S. National Nanotechnology Initiative
NNSNIC	National Nanotechnology Strategy and Nanotechnology Innovation Centers
NONS	Notification of New Substances
NT	Nanotechnology
OECD	Organisation for Economic Co-operation and Development
PNC	Polymer nanocomposite
REACH	Registration, Evaluation, and Authorization and Regulation of Chemicals
RF	Russian Federation
RK	Republic of Korea
RS & RAE	Royal Society and the Royal Academy of Engineering
SA	South Africa
SAICM	Strategic Approach to International Chemicals Management
UNESCO	United Nations Educational, Scientific and Cultural Organization
UNEP	United Nations Environment Programme
UNITAR	United Nations Institute for Training and Research
UK	United Kingdom
US	United States
USDFDA	US Food and Drug Administration
USCPSC	Consumer Product Safety Commission
USEPA	United States Environmental Protection Agency
WHO	World Health Organization

18.1 Introduction

In previous decades, the advent of novel technologies founded on NMs has garnered great attention and high interest. NMs are elucidated as bio-persistent or insoluble and deliberately produced materials having a single or more outer sizes or inner architectures. Legislative parameters elucidating NMs essentially refer to a definition for deducing, identifying, and differentiating NMs from other materials. Hence, the EC released a publication, titled "Recommendation on the Definition of Nanomaterial," in a bid to facilitate continuity in interpreting the meaning of *NM* as pertaining

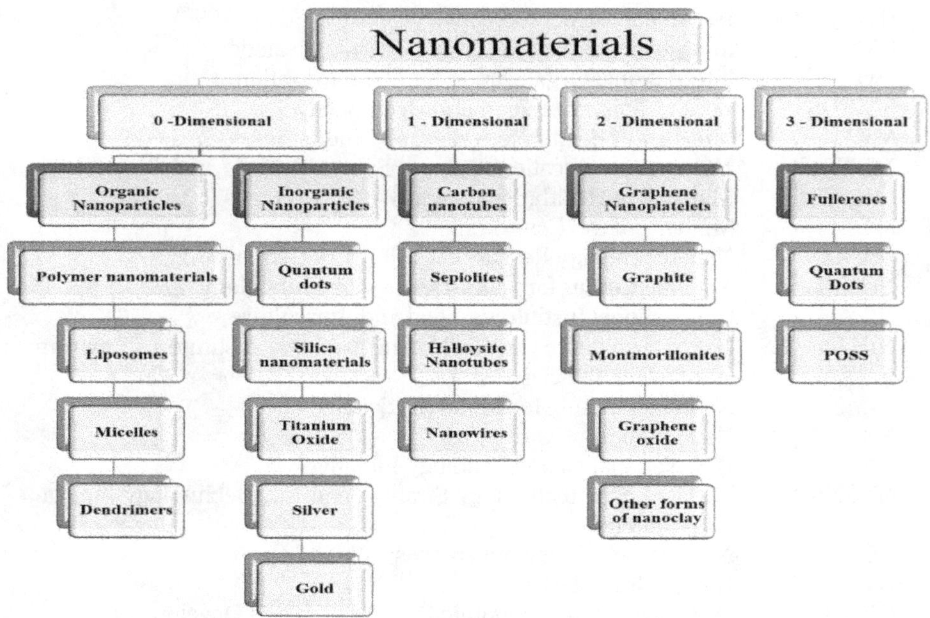

Figure 18.1 Classification of NMs.

to legislative, as well as policy, requirements concerning the EU (European Parliament and Council, 2011). However, this elucidation does not represent a legal instrument; it reference point that is vastly utilized across various regulatory segments and is adaptable to peculiar product legislating framework. The EC definition focuses on dimension, relative to size range between 1–100 nm, as the singularly defining attribute of the substrate (European Commission 2009a). Figure 18.1 elucidates the categories of NMs.

NMs are categorized as polymeric, metallic, ceramic, or composite materials (Idumah et al., 2020a). Novel attributes and performances of nanoscale particulates display new horizons for enhancement of products performances thereby considered as substitutes for newer materials cutting across a versatile range of fields, including pharmaceutically and biomedically inclined areas, construction, industrial, and information technology, as well as the electronics sector (Alghuthaymi et al., 2020; Idumah et al., 2020b), as elucidated in Figure 18.2. Hence, material attributes, such as shelf life, fire suppression, barrier behavior, flexibility, and recyclability, are enhanced (Alfei et al., 2020). Thus, due to their inherent multifunctionality, a broad range of NMs are utilized for various packaging products and numerous novel products that are continually being churned out in the market.

The nano-enhanced packaging market is expected to attain USD 105.06 billion by 2027, growing with a CAGR stable at 13.20% within the forecast duration of 2020 to 2027. Presently, the development of nanofacilitated packaging market is credited with the escalating global quest for packaged food products due to the ravaging COVID-19 pandemic. The rising quest from the food and beverage segment for efficient packaging material to reduce deterioration and extend product shelf life, along with food

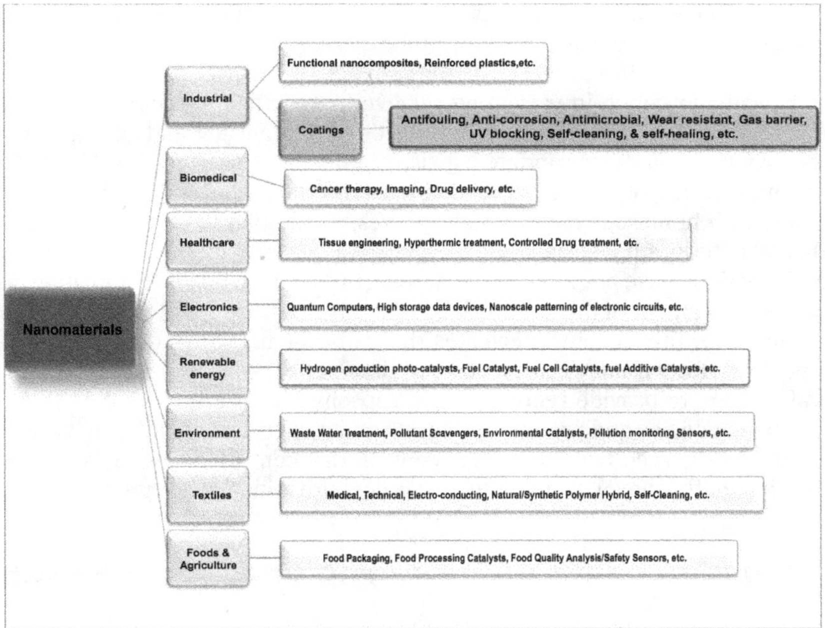

Figure 18.2 Application of NMs in various sectors.

records digitization, is anticipated to propel the nano-enhanced packaging market growth forecast between 2020 and 2027.

Therefore, steps are being taken to mitigate and ensure regulation in manufacturing and safe handling/usage of NMs, as well as NT, either through legislation or non-legal policies and guiding principles across the globe (Jildeh and Matouq, 2020). Globally, legislative activities are ongoing at ensuring NM regulation (Sharma et al., 2020). Presently, legislation in various countries is unique and appropriate enough to initiate the regulation of NMs and NT (European Parliament and Council, 2008). However, several proposals have been implemented by various concerned parties, like the European Parliament (European Parliament and Council, 2011). NMs are entailed by varying legislative activities. For instance, EU legislative activities on NMs include the Regulation of Active and Intelligent Materials and Articles (450/2009; European Commission, 2009b), the Regulation on Plastic Food Contact Materials and Articles (10/2011; European Commission, 2014), the Regulation on the Provision of FIC (1169/2011; Food Standards Agency, 2014), and the Regulation on Cosmetic Products (1223/2009).

Other legislative proposals are presently under review for improving oversight on NMs, for instance, annexes to the REACH, Novel Food Regulation (258/97), or the Regulation (1907/2006; European Commission, 2012). U.S. agencies on safety assessment, and other regulatory parastatals, have recommended different policies on the prospective human health challenges linked to the use of nanostructured materials in food.

Nanoscientists, using nanotoxicology, an aspect of toxicology and interdisciplinary field related to toxicity aspects of NMs, determine the potential risks of NMs in packaging (Indermun et al., 2020). Hence, there is a compelling quest to find a panacea to these challenges in order to expand the scope of knowledge on the use of these NMs relative to their toxicity, biocompatibility, and safety in packaging, especially in the food sector. From this perspective, the complexities, as well as limitations, in nanotechnological application in packaging relative to toxicity and accumulation is liable to be overcome through elucidation of the physiochemically and biologically affiliated behaviors through appropriate legislative guidelines and policies. NT, NMs, and nano-manufacturing have witnessed significant developments and applications across the globe, with numerous potential advantages, in addition to severe environmental, health, and deteriorating social ills. In response to this emerging lacuna caused by NT and NMs, nations have prepared national reports, based on policy formulation, for the effective management of these materials.

Therefore, this chapter elucidates the global status of nanotechnological policies and legislative framework on nano-packaging, especially in the food and agricultural sectors, to provide regulation and capacity building in the quest to develop national nano-policies and programs. Moreover, an overview of NMs, their uses, and the approaches geared toward societal control, risk control, and a social and ethical commitment of this novel, technologically successful initiative are presented.

18.2 Fundamentals of Packaging and Nano-Packaging

18.2.1 Fundamentals of Packaging

The manufacturing sector is under intense pressure to satisfy safety standards and regulations, as well as meet consumers' quest for safe and healthy products. Increasing awareness through ease of access to online media has induced consumers' desire for safe and adequately labeled products. In a bid to satisfy safety and authenticate products across the supply chain, manufacturers, traders, buyers, and regulatory agencies require a low-cost, novel, rapid, and permanent method for monitoring product packaging that will be highly efficient in comparison with the prevailing conventional barrier protection, since packaging is critical to transferring products to end users (Idumah et al., 2019a, 2020d).

Nanotechnological inculcation of packaging mainly alludes to three prospects: including it within products, embedding within the packaging material, and applying it during product processing. The successful implementation, commercialization, and responses to versatile nanotechnological applications are configured by the consumer's perception and assimilation of new technological concepts and applications (Mech et al., 2020a). Results emanating from versatile investigations demonstrate broad acceptance of nano-packaging by consumers (Chaudhary et al. 2020; Mech et al., 2020a). NT is anticipated to present a fulcrum for developing advanced nano-packaging materials to satisfy consumer's expectations.

Packaging materials facilitate minimizing critical interactions between packaging and products, thereby impacting consumers' health, as well as reducing waste materials, enhancing biodegradation, protecting against light and gases, and minimizing CO_2 release (Saqqa, 2020). Including NT in packaging fulfills essential packaging objectives, which entail preservation and protection against leakages and damages, as well as extended shelf life; containment during handling and transportation; convenience and consumer friendliness; and communication and marketing through facilitation of information about product quality (Idumah et al., 2019a, 2020c).

18.2.2 Functions of Packaging Materials

The main functions of packaging materials involve protecting products from interacting with external factors and other damaging factors, as well as containment especially in food packaging, in addition to providing consumers with supplements and nutritional information. Generally, some other functionalities include information on tampering and convenience, cost efficiency, customer and industrial satisfaction, and safety management, as well as eco-benign compliancy (Istiqola and Syafiuddin, 2020; Idumah et al., 2019a; Idumah et al., 2020c; Idumah and Odera, 2020).

18.2.2.1 Preservation and Protection

Product packaging especially pertaining to food products enhances resistance to food spoilage, exploration of the benefits of processing, elongation of product shelf life, as well as quality and safety enhancement. Therefore, packaging enables the protection of products from three major forms of external influences, namely, chemically, physically, and biologically affiliated parameters (Kumar et al., 2020). Pertaining to chemical protection, the minimization of compositional changes induced by environmental parameters, including gaseous exposure, especially to oxygen; loss or gain in moisture; or light, such as infrared, ultraviolet, or visible; are improved by chemically oriented protection (Fadiji et al. 2020). Varying materials utilized for packaging display prospects of improving chemical protection. Furthermore, the majority of metallic-oriented, as well as glass-based, packaging offer almost a total barricade toward both chemically and environmentally based parameters; however, a limited scope of packages are ultimately constituted of glass as well as metals (Das et al., 2020; Jafarzadeh and Jafari, 2020)

18.2.2.2 Minimization of Wastage and Containment

A very key function of materials packaging involves effective containment, as well as a reduction in waste, especially for food substrates prone to deterioration from varying factors. In numerous countries, food wastage has been ascribed to parameters such as poor storage systems and inefficient transportation systems. Generally, food packaging reduces food wastage by prolonging food's shelf life (Gholami et al., 2020).

18.2.2.3 Advertisement

Packaging is meant to attract the customer to a particular product. The extent of a product's packaging is attractive may determine how rapidly customers get attracted to a particular product and the level of sales garnered from the product. Hence, efficient packaging is likely to result in incremental sales in a competitive environment and may be positioned to improve the product image to the community while also identifying the product in a competitive situation.

18.2.2.4 Traceability

The objectives of traceability allude to enhancing the supply network, enabling food trace-back for improved health, and securing as well as enhancing quality, marketing, and differentiation with insignificant quality attributes. Here, manufacturers include special coding on package labels on their products, thereby enhancing the trackability of their products along the distribution channel. Various formats are utilized in presenting these codes. The Codex Alimentarius Commission opines that traceability is "the prospective of efficient supervising product through differing stages of processability, manufacturing, as well as dissemination" (Liu et al., 2020).

18.2.2.5 Convenience

Packaging is greatly influenced by convenience parameters such as ease of accessibility, disposal, handle, microwave ability, product appearance, as well as resealability. Hence, the essence of packaging is to minimize inconveniences attainable from product preservation (Zambrano-Zaragoza et al., 2020).

18.2.2.6 Tampering Exposure

Packaging also ensures the protected products are not criminally tampered with (Shafiq et al., 2020). In the food and pharmaceutical industries, specific packaging attributes are ensured to decrease susceptibility to criminal or malicious tampering, as well as deliberate adulteration of products. Tampering indicators include customized membranes, breakout closures, and customized bottle-lining printing, as well as graphical/textual composite canning. Other tamper-proof methods may include unique hologramic printing in the form of labels not susceptible to easy replication.

18.3 Materials Used in Food Packaging and Regulatory Policies

The design and constructional appearance play a key function in verifying product shelf life. The appropriate selection of packaging materials, as well as techniques, maintains product freshness as well as quality during distribution and storage (Camilleri, 2020). Conventionally used materials for packaging include glasses, metals such as tin-free steel, laminates, Al foils, and tin plates; paper and paperboards; plastics; and PNCs. Plastics have been utilized in both stiff and flexible forms. Nowadays, packages comingle various materials in order to exploit the combined material's aesthetic and functional attributes.

The USFSA undertakes the regulation of packaging materials under section 409 of the federal Food, Drug, and Cosmetic Act (USFDA, 2011). The key regulating technique is via product contact notification procedure which requires manufacturers to notify the agency before marketing FCS for new usage. FCS is defined as "any substrate that is intended to be used as a component material during manufacturing, packing, packaging, transportation or holding when usage is not meant to impact technically in such food" (USFDA, 2014a).

18.3.1 Glass

The use of glass packaging has been in existence for a very long while. Fabricating glassware entails heating comingled alumina (stabilizers), silica, sodium carbonate (the melt initiator), and limestone/calcium carbonate to elevated temperatures to the point where the materials undergo melting into a thickened liquid mass, which is then poured into molds. Recycled broken glass or cullet is also utilized to fabricate glass, accounting for about 60% of constituent raw materials. Glasswares utilized in food packaging undergo surface-coating to facilitate lubrication (Stefanini et al., 2020).

18.3.2 Metal

Metals are the most widely utilized form of packaging offering varying attributes including exceptional physical protection, barrier, formability, decorative ability, recyclability, as well as wide consumer appreciation. The main metals commonly utilized in packaging include Al and steel (Deshwal and Panjagari, 2020).

18.3.2.1 Al

Al is commonly utilized in fabricating cans, foil, as well as paper lamination or plastically affiliated packaging. Al is a low-weight, silvery-white metallic item garnered from bauxite ore and occurs in a mixture with oxygen (alumina). Manganese and magnesium are usually comingled with Al to enhance its strength. Unlike other metals, Al is highly resistant to corrosion, with its natural plating of Al oxide facilitating a highly efficient barricade to air, temperature, moisture, and chemical effects. Pristine Al is utilized in mild packaging in the form of cans for soft drinks and so on (Deshwal and Panjagari, 2020).

18.3.2.2 Laminates and Metalized Films

Lamination in terms of packaging entails foils of Al clinging to plastic film or paper in order to increase the protective behavior. This is facilitated by using thin gauges. Despite plastic lamination enhancement via heat sealability, the seal cannot entirely hinder moisture and air. Due to the high cost of laminated Al, it is usually utilized in packaging high-value products. Metalized film is a low-cost replacement for laminated-inclined packaging. Metalized sheets refer to plastics made up of a flat Al metallic layering (Ge et al., 2020).

18.3.2.3 Tinplate

Tinplate is fabricated using a lower carbon steel or blackish plate and is formed by coating both sides of the blackish plate using thin layers of tin. The coatability of the material is attained through immersing steel plates in molten tin (hot-immersed tin sheet) or through tin electro-layering on the steel sheet (Taroco et al., 2020).

18.3.2.4 Tin-Free Steel

Tin-free steel is additionally referred to as electrolytically inclined chromium or chrome oxide–coated steel and entails coating an organically inclined substrate to facilitate total corrosion repression. However, the chrome/chrome oxide posit tin-free steel is not good for welding. This attribute facilitates exceptional coating adhesion. Similar to a tin sheet, tin-devoid steel exhibit unique formability and strength, although marginally inexpensive compared to tin sheet. Tin-devoid steel is utilized in fabricating bottlecaps, food canning, canning ends, and trays, as well as back-away closures. More so, it is utilized in fabricating large containers like drums for the bulk storage of finished products (Taroco et al., 2020).

18.3.2.5 Plastic and NPC Packaging

Plastics are fabricated via condensation polymerization (polycondensation) or addition polymerization (polyaddition) of monomeric entities. Several benefits abound in plastics utilization for packaging in the form of architectures, sheets, and geometrical shapes, which facilitate design flexibility (Urban and Nakada, 2021).

PNCs are composed of a polymeric matrix in a continual or a discontinual phase (Idumah et al., 2015; Idumah and Hassan, 2015, 2016a, 2016b). PNC is a multiphase material resulting from the comingling of the matrix in the continuous phase and a nanoscale material as the discontinuous phase (Idumah, 2021a; Wang et al., 2020). Regarding the nanoscale substrate, the nanoscale phase is usually characterized into nanowhiskers or nanorods, nanospheres or nanoparticles, nanotubes and nanosheets, or nanoplatelets (Idumah et al., 2019b; Idumah, 2021b). In addition to enhancing mechanical and barrier behaviors, nanoparticles also inculcate active or smart behaviors into the packaging system. Polymeric materials utilized in fabricating PNC packaging materials include polypropylene, polyethylene, polyethylene terephthalate, polystyrene, and polyvinyl chloride (Abalansa et al., 2020; Schyns and Shaver, 2020).

PNC materials when utilized in packaging demonstrate durability while eliminating environmental waste challenges (Alghuthaymi et al., 2020). Hence, PNC packaging materials must offer inhibition to external factors such as gases, water vapor, solutes, and moisture, as well as facilitate conveying of certain active chemicals exhibiting specific attributes. For instance, in food packaging, packaging materials entail enhancing food quality and extending shelf life by resisting fungus and microbial development on the enclosed food substrate. The exhibition of attributes such as water vapor resistance, moisture, and mechanical behavior restrains product deterioration because of biological as well as physicochemical influences as well as quality maintenance during handling and storage (Alfei et al., 2020).

The basic functions of PNC packaging materials include prolonging the end life of the packaged products by restraining of unwanted variations caused by microbially affiliated activities, chemically induced spoilage, temperature variations, oxygen, wetness, light, external force, and stabilization of the standard and food items security from production commencement to consumption (Chaudhary et al., 2020). This is schematically elucidated in Figure 18.3.

Packaging achieves these functions by creating the appropriate physicochemical environment for products while restraining microorganisms, light, gases, and water vapor so as to sustain the standard and product safety while improving the end life of the packaged entities as depicted in Figure 18.4.

More so, in addition to inherent basic attributes such as mechanically, optically, and thermally affiliated attributes, materials appropriate for use in packaging must also hinder microbial development and contamination and suppress moisture gain or losses while positing as barrier against CO_2, water vapor permeability, oxygen, and other volatile entities such as flavors (Istiqola and Syafiuddin, 2020). PNCs attain effective packaging of products through various procedures, including intelligent/smart and active packaging as depicted in Figure 18.5. Figure 18.6 elucidates possible PNC packaging systems.

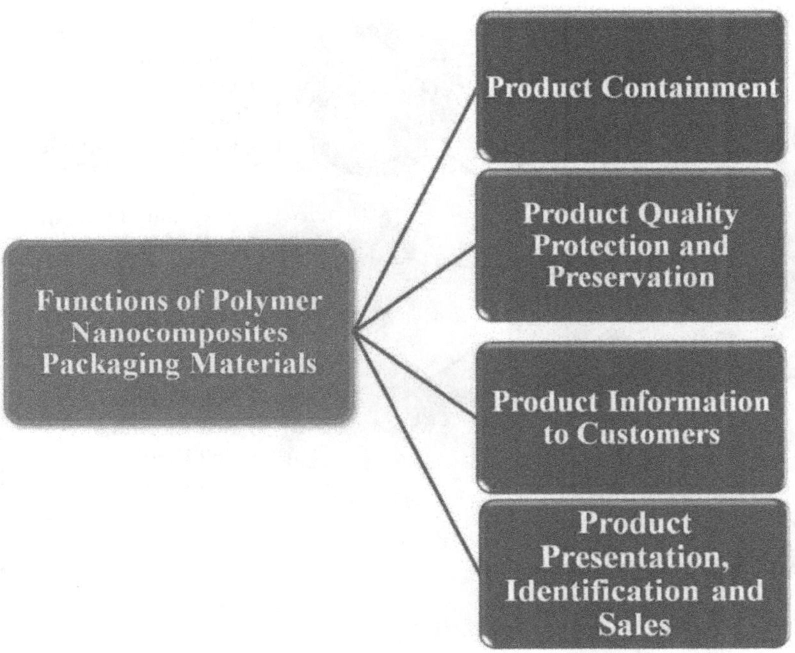

Figure 18.3 PNC packaging film functions.

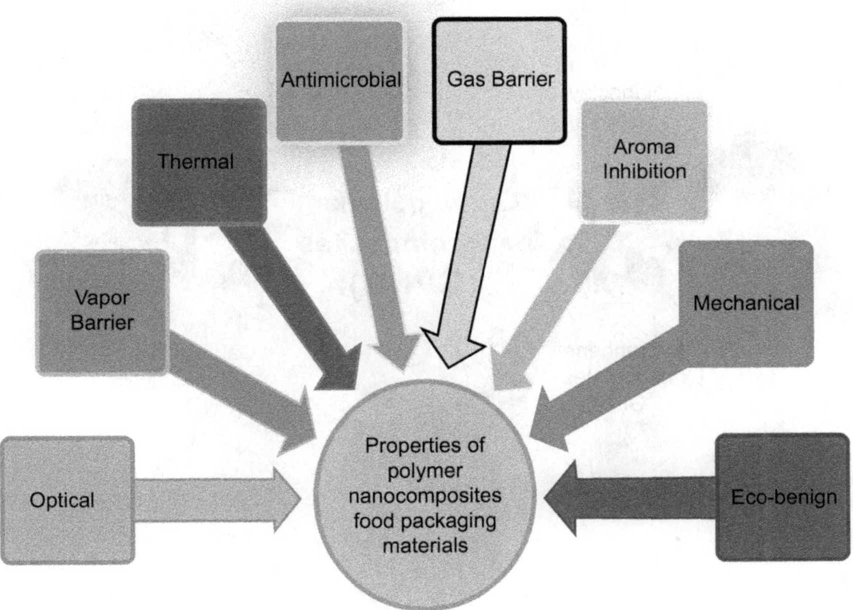

Figure 18.4 Attributes of PNC for product packaging.

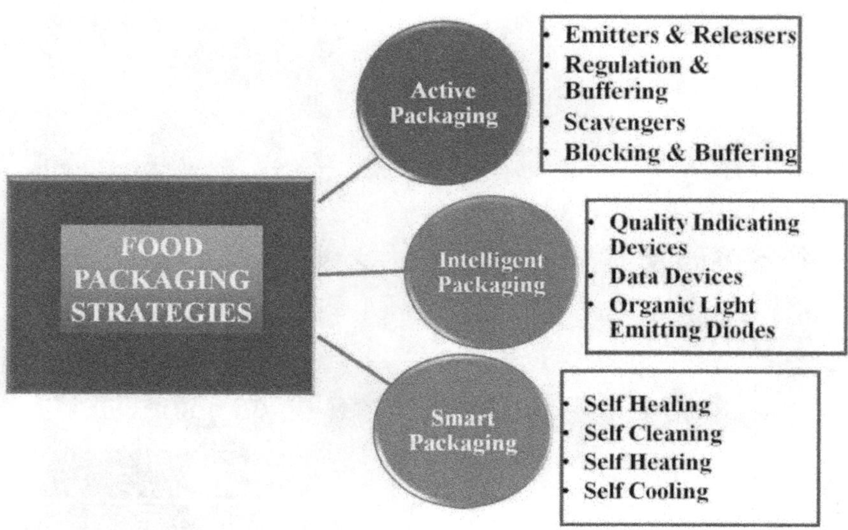

Figure 18.5 Packaging strategies.

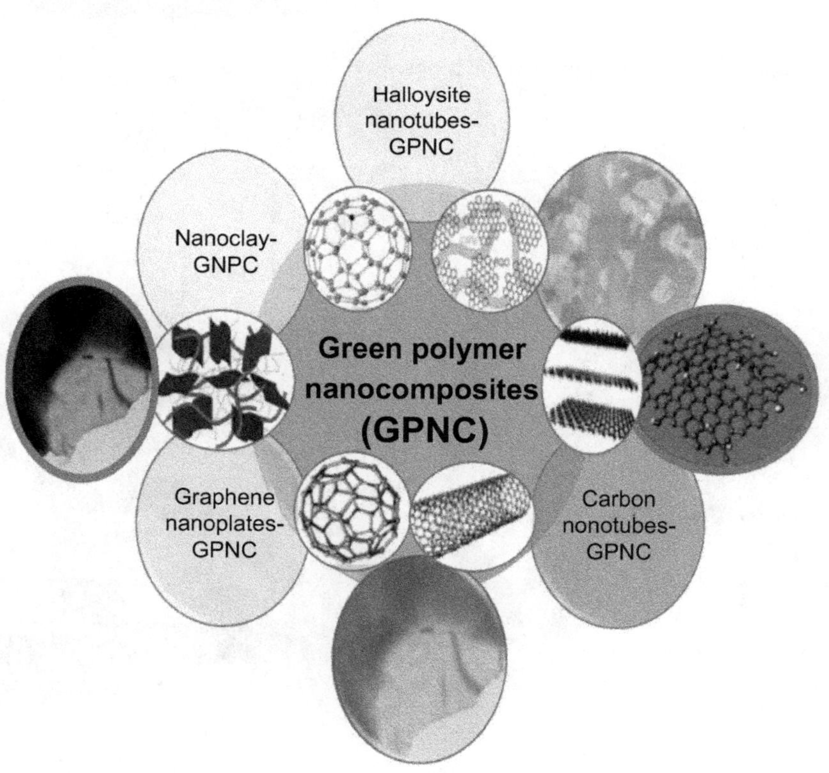

Figure 18.6 PNC packaging systems.

18.3.2.6 Nano-Packaging: Nanotechnologies and Electronics Packaging

Nowadays, NT is used in microelectronics packaging, involving nanoparticulate PNCs, exploiting superior mechanical, electrical, or thermal attributes of carbon derivatives such as graphene, carbon nanotubes, and so on, nanoclay derivatives like montmorillonites, halloysite nanotubes, and so on. PNC materials have been investigated for their high-*k* dielectrics, electrical conducting adhesives, conductive "inks," under-fill fillers, and solder enablement (Li and Qin, 2020; Mahajan, 2020).

18.3.2.7 Intelligent Nanosensor Packaging

NT has induced the evolution of nanosensors capable of being applied in the form of labeling or coatings to inculcate intelligent functioning to packaging of food products relative into maintaining packaging integrity through leaking detection, time—temperature variation indicating, and so on (Alizadeh et al., 2020; Faudzi et al., 2020). The utilization of NT-oriented printable inks has emerged (Oliveira et al., 2020). The availably conductive inks utilized for ink-jet printing relying on copper nanoparticles have also evolved (Li et al., 2020). Nano-barcodes using coatings/printing inks have demonstrated prospects for the authenticity traceability of packaged products (Ganzulenko and Petkova, 2020).

18.3.2.7.1 Nanocoatings as Intelligent Packaging for Surfaces In food packaging, nanoparticles, with biopolymeric matrices, susceptibly either improve the attributes of pristine polymers or enhance the functionalization of active as well as intelligent packaging (Esmaeili et al., 2020). Packaging can then be classified as "Improved", "Active," or "Intelligent" packaging, depending on the type of packaging material required for specific applications (Li et al., 2020). However, policies introduced by the EU regarding the use of "active" and "intelligent" packaging materials for foods, with the exception of titanium nitride in plastic bottles, have restrained the use of these terms (Mech et al., 2020b).

Nowadays, the packaging sector has focused on varying forms of nanoarchitectures and specific NMs, including nanoclay particulates as a result of their ease of accessibility, inexpensiveness, ease of processability, as well as high performance. Additionally, carbon derivatives entailing graphene nanosheets as well as carbon nanotubes have improved the development of packaging materials with superior attributes (Ramos et al., 2020; Yuan et al., 2020).

18.3.2.8 Paper-Based Packaging Materials

Paper packaging materials have demonstrated high prospects in the packaging sector. As a result of the inferior attributes of paperboard, the use of synthetic polymeric materials has developed and continues to develop in the packaging sector (Fengler and Gruber, 2020). Paper-packaging materials are eco-benign, which is attributable to their inherent recyclability and degradability. They are usually not very appropriate as packaging materials because of inherently inferior barrier attributes, especially against moisture (Nechita and Roman, 2020).

18.4 NT Policies Organizational Framework

Legislative parameters addressing NMs refer to the identification and distinguishing of NMs from other materials. A "Recommendation on Elucidation of NMs" to facilitate consistency in definition of "nanomaterial" for legislative and policy reasons in the EU has been published by the EC (European Parliament and Council, 2011). However, this elucidation may not be legally binding to the involved parties; however, it stands as a referring fulcrum globally applicable and adaptable to peculiar product legislation. Globally, the geometrical dimensional parameter is utilized as a reference for defining NMs, that is, dimensions within 1–100 nm. The dimension implies the external configuration of the constituting particulates of the material capable of being unbounded but liable to forming agglomerates and/or aggregates. This definition by EC is globally compliant to all NMs despite their source vis-à-vis manufactured, naturally occurring, or incidentally formed (European Parliament and Council, 2011).

Hence, an NM is a material having 50% or more of its constituting particulates, irrespective of being unbounded or partially agglomerated or aggregated, in numbering relying on particulate dimensional distribution expressing one or more external configurations within 1–100 nm. Before the release of the EC recommendation, NM definitions existed in certain specific regulations. These expositions are legally binding although different from the recommendation published by the EC in some areas. For instance, elucidation of "engineered nanomaterial" in FIC Regulatory Provision refers to a configuration consistent with the EC recommendation relative to external within proximity of 100 nm or less (European Parliament and Council, 2011).

NT facilitates novel lightweight materials exhibiting superior packaging barriers, protecting manufactured, agricultural, and food product quality during transportation (Ramos, et al., 2020). Nowadays, NMs have been utilized in various types of products packaging, as well as containers, as they demonstrate a novel alternative source of additives for enhancing polymer properties (barrier, mechanical, flame retardancy, heat resistance properties, and so on) for fabricating multifunctional packaging materials (Yuan et al., 2020).

18.4.1 NT Ethical Considerations

NT ethics have been elaborated from the national perspective of varying countries, and also within international agencies such as UNESCO, IFCS, and SAICM (Rahman, 2019). Most of the inputs discussed are focused on NT identification and analysis to enable the public, professional groups, and stakeholders to be conscious of the implicative aspects of this novel technology relative to potential benefits or deficits, along with novel linkages between technology, science, and society (OECD, 2013). The fundamental governing decision-making during NT development relative to public transparency and accountability is highly imperative in nano-policing and programming. The EU has also reported guidelines and principles involving a code of conduct in nano-policing. There exists an ongoing discussion regarding the "nano-divide", relative to whether nanotechnology will result in increasing or decreasing the margin between industrialized and developing countries. Inclusive factors considered in this debate relate to NM's developmental impact on the working classes in developing countries and influence the basic materials markets, as well as the economies of nations dependent on these NMs (European Parliament and Council, 2011).

18.4.2 IFCS

Globally, NT issues were initially discussed at the IFCS held in Dakar, Senegal 2008. Here, a pact of 21 recommendations, including reference to the importance of more research and risk evaluation, awareness creation, enhanced information networking, the development of a global code of conduct, and the utilization of precautionary ethics, was endorsed (SAICM. 2011).

18.4.3 ISO

ISO promulgated Technical Committee 229—Nanotechnologies. Presently, four working committees were established vis-à-vis: Measurement and characterization; Terminology and nomenclature; Material specifications; Health, safety, and nanotechnologies environmental aspects. TC229 had published 21 standards, while 23 standards have been developed since November 2011. Here, key documents addressed health as well as safety etiquette in occupational practices, particularly regarding terminologies, nanotechnologies, and definitions. ISO 21067 currently describes all packaging terms.

18.4.4 UNESCO

UNESCO has remained active in promoting ethical norms and principles relative to scientific and social transformation since 1970. Hence, UNESCO has actively participated with other international organizations in developing the ethical principles utilized in NT and manufactured NMs. A vital step in evolving a national nano-policy along with an implementation schedule involving the compilation of information that positions prepared NMs, also referred to as engineered NMs, into an appropriate chemicals controlling system facilitated by SAICM (SAICM, 2011). Similar to the National Chemicals Management Profile ("National Profile"), a "nanometric evaluation" entails the fundamental compilation of the present state of development, as well as NMs and nanotechnologies; an awareness of the national spectrum, and a compilation of critical issues impacting on a country's technical framework, institutional, legal, administrative, as well as chemical, management. These nano-evaluations are indicated as a vital baseline for national prioritizing and activities toward the provision of supporting capacity development as well as relevant technology-assisting programs. The drafting process of nano-evaluation depends on multiple stakeholders partnering with various governmental and nongovernmental organizations, research agencies, academia, communities, consumer organizations, and other societal interest groups.

18.5 NT and Nanoproduct Policies and Regulatory Framework

18.5.1 Insight into Global NT Policies for Nano-Packaging

Globally, varying NT policies, regulatory frameworks, commissions as well as initiatives have been pronounced by varying governments, private organizations, and the academic sector, in the EU, the US, and Asia aimed at ensuring fast evolvement and application of NT, economically inclined developmental promotion and sustenance, and maintenance of healthy global competitiveness, as well as improvement in innovative prowess (European Commission, 2004, 2007). Additionally, many of them have participated in putting forward regulatory frameworks for the enhanced protection of environmental parameters as well as human health. These research organizations, which are mainly financed by various governments across the globe, play key roles in conducting or affiliating with nanotechnological investigations, including fundamental studies on NT and applications, NM safety assessments, and the development of suitable regulatory frameworks.

More so, these organizations facilitate a long-term spectrum and forum for the interdisciplinary interaction of experts for promoting NT and information dissemination to the citizenry as well as elevating the NT industry. Pertaining to nanotechnological regulatory development, varying nanotechnologically oriented packaging companies are located in Asia (China), the US, Europe, and in other places across the globe. In the US, prevalent laws include the Occupational Safety, Toxic Substances Control Act, the Health Act, and the Food, Cosmetic Act, Drug Act, and notable environmental laws offer some level of legal acumen for NT regulation. An intergovernmental agency jointly promulgated by the FAO and the WHO referred as the Codex Alimentarius Commission, aimed at promoting and coordinating all nano-packaging related standards framework, undertaken by international governmental and nongovernmental organizations, is in place. Thus, in a bid to hinder regulatory lapses, the Codex Alimentarius Commission, involved in the development of global nano-packaging safe practice supervision, is in charge of nanoparticles and other nanoscale technology use in food and nano-packaging of agriculturally inclined products.

18.5.2 Global Efforts at Policing and Regulating NT for Nano-Packaging

Presently, efforts are continually being effected to establish an international regulation for NT, especially for nano-packaging. A significant number of government parastatals from varying governments across the globe have established standards as well as working regulatory frameworks aimed at defining and regulating NT usage. Hence, existing nanotechnological regulations in several nations including the US, the UK, Europe, and Asia are presented herein.

The EU (EC, EFSA Plastics Regulation (EC 10/2011) ensures the regulatory application of plastic substrates as well as items in close proximity to food (European Parliament and Council, 2011). Active and Intelligent Materials Regulation (EC 450/2009) formulate policies regulating the application of active, as well as intelligent, materials and items highly probable to be in close proximity to food (European Parliament and Council, 2011). In France, the Ministry of Agriculture/General Directorate on Food, EC, European Food Safety Authority, Grenelle II, Law Decree no 2012–232 Order of 6 August 2012 established a structure of compulsory declaration of all NMs. In Korea, the Korean Food and Drug Administration, Health and Functional Food Act gives authorization to the Ministry of Health and Welfare to regulate and license the manufacturing of health and functionalized food items.

18.5.3 NT Policing in the US, the UK, Germany, Japan, Netherlands, Norway, Poland, and Russia

In Germany, the Federal Ministry of Food, Agriculture and Consumer Protection, European Commission, European Food Safety Authority, Cellulose Film Directive (Directive 2007/42/EC ensure regulation of cellulose film materials and articles coming into contact with food. In Japan, the Ministry of Health, Labor, and Welfare, Food Sanitation Law (Law no. 233) authorizes food regulation by the Ministry of Health, Labor, and Welfare for reasons of health and safety. In Netherlands, the EC's and European Food Safety Authority's policies are in vogue. In Norway, the EC's, Norwegian Food Safety Authority's, and European Food Safety Authority's policies are in vogue. In Poland, the European Commission and the European Food Safety Authority ensure policy formulation, promulgation, and regulation. In the Russian Federation, the Federal Service for Surveillance on Consumer Rights Protection and Human Wellbeing and the Ministry of Health and Social Development of the Russian Federation ensure controlling of NMs is utilized for food products, the chemical industry, packaging materials, and agriculture. The USCPSC performs regulatory functions on consumer products, in addition to products having contact with food surfaces such as food containers, food cooking, the consumption, and the fabrication of articles. Moreover, the FDA ensures the regulation of products regarding substrate migration from substance released from the surface in contact with food.

The USFDA, a US government agency, is among the first parastatals across the globe to offer a definition for NT as well as nanometric products. Nevertheless, the FDA did not cut out its own exposition, although the organization took part in the development of the NNI definition of NT (USFDA, 2011). The NNI elucidates NT as the discernment and controlling of substrates at dimensions of about 1–100 nm, where specific insight improves new uses.

The description of NT by the NNI inculcates the (1) evolution of research and technology at the macromolecular, molecular, and atomic levels, within the length range of about 1–100 nm; (2) creation and usage of systems, architectures, and mechanisms exhibiting novel attributes and functions as a result of their minimal and/or intermediate dimension; and (3) capacity of controlling or manipulation on the atomic dimension (USFDA, 2011).

Regarding the dimensional limit (1–100 nm), there is a probability that novel therapeutic advantages are being garnered from products that are miniature compared to their traditional counterparts but occur beyond the 100-nm dimensional range limit of NT (USFDA, 2011). The FDA stipulates that it ensures the regulation of "products, and not technologies". The regulatory prospects of an application regarding nanotechnological products may unlikely take place until after the initial establishment of the NT as well as the development of its statutory categorization. The FDA mainly ensures the regulation of specific classes of products while anticipating that various nanotechnological entities may inculcate the regulatory boundaries among varying agencies within the FDA and will undergo regulation as "Combination Products" by the Combination Products office (USFDA, 2011).

FDA has conventionally ensured the supervision of varying products exhibiting particulate substrates in equal nanoparticulate dimension as the cells as well as molecular entities. Thus, FDA opines that a particular dimension is a null issue. On identification of emergent toxicological risks garnered from emerging products or manufacturing approaches, novel safety analysis will undergo necessitation (USFDA, 2011). Comingling of knowledge and policy between different governmental agencies is essential for developing a unitary regulation on NT usage. FDA has collaborated with the National Institutes of Health and NIH/NIEHS on nanotoxicity investigation (USEPA, 2013), and has directly contributed to toxicity of materials assessment. FDA created an interior FDA Nanotechnology Task Force for evaluating regulatory techniques encouraging the continual formation of innovative, safe, and effectual FDA-controlled products utilizing nanotechnological-oriented substrates. The Royal Society and the RS & RAE, which are commissioned by the UK government, have conducted independent assessments of nanotechnologies as well as evaluation of the benefits and challenges associated with these technological pathways (European Commission, 2004).

In addition, the HSE has additionally partaken in negotiating, agreeing, and enforcing regulations of produced nanoparticles (HSE, 2006). More so, the regulations garnered from HSE in dealing with toxic chemicals, flame hazards, and explosion tendencies of materials, such as the ACOP under the COSHH, guidance on COSHH risk evaluation, ACOP from the DSEAR, and the manipulation of precautions against the possibility of the explosion of combustible dusts, which is utilized in regulating NM application (HSE, 2004, 2006; IFST, 2006). Moreover, recommendations were put through such that chemicals such as nanoparticles or nanotubes could undergo treatment as new entities categorized under chemical safety regulations such as the NONS act as well as the REACH regulation from the European Commission (HSE, 2006).

In accordance with the European Inventory of Existing Commercial Chemical Substances (EINECS) under the NONS regulation, nanoscale substrates could be perceived as "existing" or "new" substrates relative to their production modes relative to top-down as well as bottom-up methods. Nevertheless, a "bottom-up" NM carbon fullerene, is known as an individual and lone carbon allotrope, which does not belong to the class of EINECS. Therefore, fullerenes as well as carbon nanotubes are classified as "novel" chemicals (HSE, 2006). In Japan, the evolving NT industry is seriously facilitated by Japanese government agencies including METI, the MEXT, the MHLW, and the MOE. The MEXT is in charge of research and development, forming academia, and building industry–government cooperative investigative spectrums in order to enhance promoting NT research and materials science.

METI evaluates testing techniques standardization for safety assessments of nanoparticles, while the MHLW ensures the development of the assessment techniques for the health impacts of NMs. Hence, the MEXT is in charge of conducting studies on fundamental sciences for public assimilation of NT, while METI, MHLW, and MOE work on the regulatory aspect. Moreover, NANONET, which was established by the NIMS, offers information facilitating the further evolution of NT.

In China, the NCNST was jointly established by the CAS and the Ministry of Education in March 2003. The NCNST is made up of various divisions, namely, (a) the nano-processability and nano-gadgets laboratory (lab), (b) the NM and nano-architectural lab, (c) the nanomedicine and nano-biotechnological lab, (d) the nano-structural properties evaluation and testing lab, and a (e) coordination lab (NCNST, 2005). The NCNST has created a public technological organization while also engaging in advanced studies in nanoscience. In 2005, a Commission on Nanotechnology Standardization in affiliation with the NCNST commenced with the responsibility of evolving national standards such as protocols, terminology, the safety essentials for particular dimensional measurements, nanomedicinal NMs and nano-processability, nano-gadgets, and nanobiotechnology. Relying on prevailing national standard of China, the commission participates in guidance as well as governing the evaluation and accreditation of nanoproducts, thereby facilitating the NT industrial sector at improving product quality, minimization of health risks, sustenance of safety, as well as fostering the development of new products (NCNST, 2005).

18.6 National Regulations and Legislative Policies for Nano-Packaging

18.6.1 International Agencies in Policing and Regulating Framework

The FAO of the United Nations as well as the WHO held a conference in 2009, focused on "Nanotechnologies in the Food and Agriculture Sectors: Potential Food Safety Implications" (FAO/WHO, 2010). In attendance were professionals from 13 countries across the globe, including the US, Canada, EU, FSANZ, China, Mexico, Switzerland, Brazil, Japan, Indonesia, the Republic of Korea, Malaysia, and South Africa (FAO/WHO, 2010). The meeting released insight into prevailing and emerging uses of nanotechnologies, including information on food safety implications, as well as potential risks and present the ability of mitigating these risks. Thus, the FAO and the WHO opined their intention at providing guidance as well as advice to global governments on critical challenges in food packaging and safety. Here, the participating nations agreed that NT offers significant advantages and privileges for developing novel products to enhance agriculture, water treatment, food processing, and preservation as well as nano-packaging. However, insight into food area was referred to the Codex Alimentarius Commission (FAO/WHO, 2013).

The FAO and the WHO generated a draft paper titled "State of the Art on the Initiatives and Activities Relevant to Risk Assessment and Risk Management of Nanotechnologies in the Food and Agriculture Sectors" for comments (FAO/WHO, 2012). The objective of this report was to initiate interaction with various countries on policies issued in the 2009 professionals meeting regarding nano-packaging and food safety regulation. This report presented an overview of the prevailing parameters in risk evaluation and NMs management in the agricultural and food sectors at both national and global levels. The prevailing status of NM uses in food packaging as a novel-technological panacea for minimizing food loss or enhancing traceability was expounded. Moreso, the emerging trends in nano-safety management prevailing in participating nations were elucidated.

The USEPA developed a N research policy in 2009, due to escalating use of NMs. The USEPA recognized NMs types commonly used in products, while six NMs were selected for testing. The Organisation for Economic Co-operation and Development (OECD) conducted the assessment and evaluated potential effects on the environment as well as human health. For food packaging applications, nano-silver and nanotitania were assessed as representative NMs (USEPA, 2013; FAO/WHO, 2013). Therefore, the USFDA brought out a draft policy for the public regarding FDA-regulated products composed of NMs or otherwise involved in NT applications (USFDA, 2011).

This policy was intended to assist manufacturers, as well as stakeholders, comprehend the potential effects, safety issues, and public health effects susceptible to emanate from NMs usage in FDA-regulated products. By comparison, the NNI issued a national policy intended for ensuring the development of NT and facilitating regulatory policymaking, thereby substituting the 2008 national policy (FAO/WHO, 2013). The policy focused on environmental, health, and safety challenges. Moreso, In June 2014, the USFDA issued "Guidance for Industry: Assessing the Effects of Significant Manufacturing Process Changes, Including Emerging Technologies, on the Safety and Regulatory Status of Food Ingredients and Food Contact Substances, Including Food Ingredients Constituting of Color Additives" (USFDA, 2014c). This report presents a pathway for food producers regarding ingredients, packaging substrates, and insights for end users.

Thus, the USFDA recommended that manufacturers should assess and fabricate an insightful toxicology evaluation. This report referred to the Code of Federal Regulations, Title 21 (21CFR 170.39). A substrate utilized in packaging materials for food, as well as processing equipment capable of migrating or susceptible to migrating into food will undergo exemption from regulation as a food inclusion as it constitutes a food inclusion at levels below the regulating threshold in case the substance meets specific criteria relating to (USFDA, 2014b). Hence, in March 2014, USEPA tried to prohibit the sale of plastic-oriented food containers composed of nano-silver manufactured by Pathway Investment Corp. as its products were not evaluated.

The EU presented "European Food Information to Consumers—Regulation (EU) No 1169/2011 on the offer of prepackaged food information to consumers on overall food labeling and nutritional labeling" in December 2014. However, this policy covered food ingredients only (FSA, 2014), combining two directives, 2000/13/EC and 90/496/EEC, into a single legislation (European Commission, 2013). Under this policy, all food additives containing engineered NMs must be revealed, outlining their ingredients with the names of such additives revealed by using "nano" in brackets (FAO/WHO, 2013). As this new policy does not include NMs used for packaging or containers, current legislations and regulations as per Commission Regulation (EU) No 202/2014 and 1282/2011, Commission Regulation (EC) No 975/2009 and 450/2009 are used only as guidelines and as a reference for producers and manufacturers.

In Africa, as per NMs, and NMs usage in food based applications especially food packaging, so far, there is no safety evaluation or particular regulation placed on the government website of any country in Africa (FAO/WHO, 2013). The safety of NMs was addressed in Oceania Food Standards. FSANZ in an article titled "Regulatory Approach to Nanoscale Materials", published in the *International Food Risk Analysis Journal* in 2011 (FAO/WHO, 2013). FSANZ identified and managed hazards propelled by chemically migrating from nano-food packaging, focused on the result emanating from industrial packaging evaluation, opining that present essentials required in packaging existing in the legal code are not enough (Tager, 2014; FSANZ, 2014).

The Japanese leadership in Asia created a nanotechnological research center in its Science and Technology Basic Plan for 2006–2010. Moreover, the Ministry of Health, Labor, and Welfare issued a 6-year plan (2009–2014) focused on the carcinogenicity of NMs, titled "Research Project on the Potential Hazards, and so on, of NMs" (FAO/WHO, 2013).

Relative to the RK, a policy was enacted on the safety evaluation of nano-oriented products in 2011. Recently, RK released a new research plan for NMs safety focused on "The First Master Plan on Management of NMs Safety" effectual between 2012 and 2016. The aim of this new study involves establishing a database for evaluation and analysis of NMs, in order to establish a system for assessing NM safety. The KMFDS is an agency involved in establishing data as well as a system for managing safety and

NM evaluation, especially in packaging. Presently, the KMFDS conducts a varieties of research regarding food nano-safety, especially in packaging, and has established new policies and safety guidelines pertaining to nano-packaging (Raj et al., 2020; Yu et al., 2020).

In Malaysia, the Nanotechnology Directorate issued strategic innovative and interdisciplinary research with about 20 projects focused on NT in the agriculture and food sectors emanated from this forum. The high focus of each project centered on health, safety, and environmental challenges. In compliance with future global safety standards relating to NT, Malaysia is encoding a distinct national NM pathway. Moreso, there are currently specific regulations guiding NT risk evaluation (Shekaftik et al., 2020; FAO/WHO, 2013).

In the RF, the Federal Service for Surveillance of Consumer Rights Protection and Human Well-Being has been in charge of NMs usage industrially, since 2007, resulting from guidelines stipulated by the Chief State Health Officer of the RF at July 23, 2007: N 54 “On the supervision of produce, received with use of nanotechnologies and containing nanomaterials.” About 50 standards and approaches for safety evaluation and risk assessment were issued and approved by the Chief State Health Officer of the RF. Approaches for risk assessment were developed by focusing on the recommendations and requirements stipulated by the OECD, the EFSA, and the FAO/WHO since 2012, especially for NMs used in fabricating packaging materials (FAO/WHO, 2013).

18.7 Packaging and Packaging Waste Directive 1994/62/EC

18.7.1 Packaging Directive 1994/62/EC

The Packaging Directive 1994/62/EC on packaging and packaging waste aims at inhibiting the release of packaging waste. Moreso, basic principles enhance packaging reuse, recycling, and other forms of waste package recovery and minimization of the disposal of these wastes (Fitch-Roy et al., 2020). The use of NMs in packaging has recently escalated. NT has demonstrated the capability of facilitating the minimization of packaging waste. This is expressed in the use of single-layer films instead of multilayer films to improve the material’s performance, implying that minimal packaging is essential in attaining the required performance. This is imperative because the directive states that the best approach for hindering the creation of packaging waste entails minimizing the overall packaging volume (Foschi and Bonoli, 2019).

Utilizing NMs in packaging posits potential health and environmental hazards to the human system. NMs in packaging materials for foods and drinks may unintentionally induce migration from the packaging material into foods/drinks, thereby increasing the prospects of NMs ingestion (Chaudhry et al., 2008). The European Food Safety Authority (EFSA, 2014) has studied prevailing developments in NT within its scope, hence published a policy on risk assessment involved in nanoscience and nanotechnologies applications within the food and feed chain in May 2011. This policy provided, *inter alia*, guidance on the physicochemical characterization requirements of engineered NMs utilized in food contact materials and testing methods for identifying and characterizing hazards emanating from the nano-attributes. This policy entails informing regulators about the types of NMs that may be potentially found in food packaging, including their chemical behaviors.

Very essential to NMs is the packaging legislation made on food contact materials under Regulation No 1935/2004 stating that food contact materials must maintain safety while lacking capability of transferring their constituents into food at levels likely to pose danger to human health, vary the food constitution in an unacceptable manner or degrade the taste as well as the odor of food products (European Commission, 2004). For instance, Regulation (EC) No 450/2009 involves additional parameters for active as well as intelligent packaging materials and products ensuring their safety in use, especially for nanoparticles (European Commission, 2004). This legislation entails a precautionary route in the presence of nanoparticles in packaging in separation from food via functional barrier not allowed, as a result of paucity of information as per nanoparticles potential in migrating through barriers.

The Packaging Directive covers all packaging materials positioned in the market in the EU as well as all packaging wastes, whether generated or released industrially, commercially, officially, in the shop, while servicing, within the household or at any other level, notwithstanding the material utilized. In Pursuance of Article 3 of the Packaging Directive, packaging implies all products fabricated from materials of any nature to be utilized for containment, handling, protection, delivery and goods outlay, from raw substrates to processed products, from the manufacturer to the end-user or the consumer. Herein, nonreturnable entities utilized for equal purposes will also be regarded as constituting packaging covering packaging materials composed of NMs. In pursuant to Article 5 for reuse, concerned nations may initiate packaging reuse systems. Furthermore, pursuant to Article 6 of the Packaging Directive focus is placed on recovery as well as recycling of packaging waste.

18.7.2 Packaging and Packaging Waste Directive

Herein, waste packaging materials composed of NMs generally or specific NMs are positioned for recovery via incineration, which requires some form of labeling mechanism to facilitate the identification of prevailing nano-wastes. On the other hand, incineration of nano-packaging materials may result in environmental hazards via flare gas emissions where NMs escape end-of-pipe controlling devices, or via elimination of filters where they are garnered.

18.7.3 Identification System

In a bid to garner the collection, reuse, and recovery, in addition to recycling, Article 8(2) entails that packaging must demonstrate the type of packaging material utilized based on Commission Decision 97/129/EC (1) (European Commission, 2014). This policy entails a voluntary number identification and abbreviation of plastics, for instance, polypropylene, polystyrene, fiberboard, and paper, including corrugated fiberboard, and zero-corrugated materials such as metals (steel and Al). Others include wood materials, textiles, glass, and composites. This system is meant for recyclers so as to enhance their identification of wastes at autonomously sorting facilities. This policy does not include NMs utilized in varying classes of packaging.

18.7.4 Reducing Adverse Effects of Packaging Wastes Disposal on the Environment

The Packaging Directive ensures that adverse environmental effects of harmful, and hazardous substrates on the environment during disposal of packaging wastes are reduced, including Articles 3, 9, 10, and 11. The influence of these articles on NMs is elucidated. Under Article 3(4), *[prevention]* 'is referred as — 'the reduction in quantity and harmfulness on the environment of materials and entities embedded in packaging and packaging wastes, as well as packaging and packaging waste at production levels including during marketing, distributing, using and eliminating stages'. Hence, this route to prevention invariably applies to NMs that are hazardous to the environment as absorbed in packaging as well as packaging waste.

The CEN standard on prevention (EN 13428:2004) is made up of two components (European Commission, 2004). The first part relates to hindrance via source minimization (weight reduction and/or volume of the packaging), while the second entails qualitative inhibition relative to decreasing the availability of harmful, as well as hazardous, substrates in packaging. The second component enables techniques for assessing and noting the availability of heavy metals and other harmful entities in packaging as well as their emission into the environment. Article 11 elucidates the maximum amounts of hexavalent chromium, lead, mercury, and cadmium allowable in packaging materials (European Commission, 2004).

18.7.5 NMs Included Under the Packaging and Packaging Waste Directive

In a bid to hinder the adverse effects of NMs and other substrates utilized in packaging, the inhibiting mechanism focuses on the prevalence of harm. Additionally, it is critically essential that parameters required for packaging abhor the presence of noxious as well as other hazardous entities in emissions of packaging waste through incineration or landfilling (NanoSafety Cluster (EU), 2020; Rowan and Laffey, 2021; Ibn-Mohammed et al., 2021).

18.8 Conclusion

Due to escalating market trend of packaging materials as a result of the advent of NT, there has been an escalating public outcry over the potential harm of NMs embedded in packaging materials to human health. Much research has revealed the possible emigrational affinity of NMs from the packaging material to products. It is imperative to emphasize the principles regulating policies and guidelines for NMs used in packaging materials. Global policies in packaging guide stakeholders and manufacturers regarding the potential harmful effects of NMs on packaging as they concern human lives and the environment. Various nations across the globe have actively participated in assessing the suitability of their policies and regulatory frameworks pertaining to nanotechnologies. This chapter has shown that, globally, many approaches have been applied in regulating nano-packaging and NT. Globally, nations have collaborated to ensure the adequate exchange of information aimed at guaranteeing strong and effective policies for protecting human lives as well as the environment, without hampering the development of critical products and their marketing strategies.

References

Abalansa S., El Mahrad B., Vondolia G., Icely J., Newton A. 2020. The marine plastic litter issue: A social-economic analysis. *Sustainability*, 12, 8677.

Alfei S., MarenGo B., Zuccari G. 2020. Nanotechnology application in food packaging: A plethora of opportunities versus pending risks assessment and public concerns. *Food Research International*, 137, 109664.

Alghuthaymi M., Abd-Elsalam K., Paraliker P., Rai M. 2020. Mono and hybrid nanomaterials: Novel strategies to manage postharvest diseases. pp. 287–317. *In*: Abd-Elsalam K. (ed.) *Micro and Nano Technologies, Multifunctional Hybrid Nanomaterials for Sustainable Agri-Food and Ecosystems*. Elsevier, Amsterdam.

Alizadeh A., Masoomian M., Shakooie M., Khajavi M., Farhoodi M. 2020. Trends and applications of intelligent packaging in dairy products: A review. *Critical Reviews in Food Science and Nutrition*. https://doi.org/10.1080/10408398.2020.1817847

Camilleri M.A. 2020. European environment policy for the circular economy: Implications for business and industry stakeholders. *Sustainable Development*, 1–9. https://doi.org/10.1002/sd.2113

Chaudhary P., Fatima F., Kumar A. 2020. Relevance of nanomaterials in food packaging and its advanced future prospects. *Journal of Inorganic and Organometallic Polymers and Materials*, 30, 5180–5192. https://doi.org/10.1007/s10904-020-01674-8

Das S., Chaudhari A.K., Dwivedy A.K., Upadhyay N., Singh V.K., Singh A., Dubey N.K. 2020. Nanoencapsulation technology: Boon to food packaging industries. *Nanomaterials and Environmental Biotechnology*, 17–40.

Deshwal G., Panjagari N. 2020. Review on metal packaging: Materials, forms, food applications, safety and recyclability. *Journal of Food Science and Technology*, 57, 2377–2392.

Esmaeili M., Ariaii P., Nasiraie L.R., *et al*. 2020. Comparison of coating and nano-coating of chitosan- *Lepidium sativum* seed gum composites on quality and shelf life of beef. *Food Measure*. https://doi.org/10.1007/s11694-020-00643-6

European Commission. 2004. Regulation (EC) No. 1935/2004 of the European Parliament and the Council of 27 October 2004 on materials and articles intended to come into contact with food and repealing directives 80/590/EEC and 89/109 EEC. *Official Journal of the European Union* 338, 4.

European Commission. 2007. Commission Directive 2007/19/EC of 2 April 2007 amending Directive 2002/72/EC relating to plastic materials and articles intended to come into contact with food and Council Directive 85/572/EEC laying down the list of simulants to be used for testing migration of constituents of plastic materials and articles intended to come into contact with foodstuffs. *Official Journal of the European Union* 97, 50.

European Commission. 2009a. Commission Regulation (EC) No 450/2009 of 29 May 2009 on active and intelligent materials and articles intended to come into contact with food. *Official Journal of the European Union* L135, 3–11.

European Commission. 2009b. Commission Regulation (EC) No 450/2009 of 29 May 2009 on active and intelligent materials and articles intended to come into contact with food. *Official Journal of the European Union* 135, 3.

European Commission. 2014. Commission Regulation (EU) No 202/2014 amending Regulation (EU) No 10/2011 on plastic materials and articles intended to come into contact with food. *Official Journal of the European Union* 62, 13.

European Parliament and Council. 2008. Regulation (EC) No 1272/2008 of the European Parliament and of the Council of 16 December 2008 on classification, labelling and packaging of substances and mixtures. *Official Journal of the European Union*, L353, 1–1355.

European Parliament and Council. 2011. Regulation (EU) No 1169/2011 of the European Parliament and of the Council of 25 October 2011 on the provision of food information to consumers. *Official Journal of the European Union*, L304, 18–63.

Fadiji T, Berry T, Coetzee C, Opara UL. 2020. Mechanical design and performance testing of corrugated paperboard packaging for the postharvest handling of horticultural produce. *Biosystems Engineering*, 171, 220–244.

Faudzi M., Sreekantan S., Mydin N., Romli A. 2020. Nanocomposite materials in food packaging: Opportunities, challenges and safety assessment. *In:* Siddiquee S., Gan Jet Hong M., Mizanur Rahman M. (eds.) *Composite Materials: Applications in Engineering, Biomedicine and Food Science*. Springer, Cham. https://doi.org/10.1007/978-3-030-45489-0_18

Fengler R., Gruber L. 2020. Mineral oil migration from paper-based packaging into food, investigated by means of food simulants and model substances. *Food Additives & Contaminants: Part A*, 37, 845–857.

Fitch-Roy O., Benson D., Monciardini D. 2020. Going around in circles? Conceptual recycling, patching and policy layering in the EU circular economy package. *Environmental Politics*, 29, 983–1003.

Food and Agriculture Organization of the United Nations/World Health Organization (FAO/WHO). 2010. *Expert Meeting on the Application of Nanotechnologies in the Food and Agriculture Sectors: Potential Food Safety Implications Meeting Report*. FAO/WHO, Rome, p. 107.

Food and Agriculture Organization of the United Nations/World Health Organization (FAO/WHO). 2012. *Joint FAO/WHO Seminar on Nanotechnologies in Food and Agriculture Meeting Report*. FAO/WHO, Rome, p. 12.

Food and Agriculture Organization of the United Nations/World Health Organization (FAO/WHO). 2013. *State of the Art on the Initiatives and Activities Relevant to Risk Assessment and Risk Management of Nanotechnologies in the Food and Agriculture Sectors: FAO/WHO Technical Paper Report*. FAO/WHO, Rome, p. 48.

Food Standards Australia New Zealand (FSANZ). 2014. *Summary of Responses to FSANZ's Industry Packaging Survey*. Parliament of Australia Website. Available from: www.aph.gov.au/~/media/Committees/clac_ctte/estimates/bud_1415/DoH/answers/SQ14-000586_Att2.pdf. Accessed 25 October 2020.

Foschi E., Bonoli A. 2019. The commitment of packaging industry in the framework of the European strategy for plastics in a circular economy. *Administrative Sciences*, 9, 18.

Ganzulenko Y., Petkova A. 2020. Testing a nano-barcodes marking technology for identification and protection of the mechanical products. *Journal of Physics: Conference Series*, 1582, 012032.

Ge C., Verma S.S., Burruto J., Ribalco N., Ong J., Sudhahar K. 2020. Effects of flexing, optical density, and lamination on barrier and mechanical properties of metallized films and aluminum foil centered laminates prepared with polyethylene terephthalate and linear low density polyethylene. *Journal of Plastic Film & Sheeting*. https://doi.org/10.1177/8756087920963532

Gholami R., Ahmadi E., Ahmadi S. 2020. Investigating the effect of chitosan, nanopackaging, and modified atmosphere packaging on physical, chemical, and mechanical properties of button mushroom during storage. *Food Science and Nutrition*, 8, 224–236.

Health and Safety Executive (HSE). 2004. *Nanotechnology Information Sheet HSIN1*. Available from: www.hse.gov.uk/pubns/hsin1.pdf

Health and Safety Executive (HSE). 2006. *Review of the Adequacy of Current Regulatory Regimes to Secure Effective Regulation of Nanoparticles Created by Nanotechnology*. Available from: www.hse.gov.uk/horizons/nanotech; www.nanomark.itri.org.tw/Eng2005; www.nanotechproject.org/consumerproducts2006

Ibn-Mohammed T., Mustapha K.B., Godsell J., Adamu Z., Babatunde A., Akintade D.A., Acquaye H., Fujii M.M., Ndiaye F.A., Koh Y. 2021. A critical analysis of the impacts of COVID-19 on the global economy and ecosystems and opportunities for circular economy strategies. *Resources, Conservation and Recycling*, 164, 105169.

Idumah C.I., Hassan A. 2015. Emerging trends in flame retardancy of biofibers, biopolymers, bio-composites, and bionanocomposites. *Reviews in Chemical Engineering*, 32, 115–148.

Idumah C.I., Hassan A. 2016a. Characterization and preparation of conductive exfoliated graphene nanoplatelets kenaf fibre hybrid polypropylene composites. *Synthetic Metals*, 212, 91–104.

Idumah C.I., Hassan A. 2016b. Recently emerging trends in thermal conductivity of polymer nano-composites. *Reviews in Chemical Engineering*, 32, 413–457.

Idumah C.I., Hassan A., Affam A. 2015. A review of recent developments in flammability of polymer nanocomposites. *Reviews in Chemical Engineering*, 31, 149–177.

Idumah C.I., Hassan A., Ihuoma D. 2019a. Recently emerging trends in polymer nanocomposites packaging materials. *Polymer-Plastics Technology and Engineering*, 58, 1054–1109.

Idumah C.I., Zurina M., Ogbu J., Ndem J., Igba C. 2020c. A review on innovations in polymeric nanocomposite packaging materials and electrical sensors for food and agriculture. *Composite Interfaces*, 27:1, 1–72.

Idumah C.I. 2021a. Recent advancements in self-healing polymers, polymer blends, and nanocomposites. *Polymers and Polymer Composites*, 29(4), 246–258.

Idumah C.I. 2021b. Novel trends in self-healable polymer nanocomposites. *Journal of Thermoplastic Composite Materials*, 34(6), 834–858.

Idumah C.I., Hassan A., Ogbu J., Ndem J.U., Nwuzor I.C. 2019b. Recently emerging advancements in halloysite nanotubes polymer nanocomposites. *Composite Interfaces*, 26, 751–824.

Idumah C.I., Obele C.M., Ezeani E.O., Hassan A. 2020a. Recently emerging nanotechnological advancements in polymer nanocomposite coatings for anti-corrosion, anti-fouling and self-healing. *Surfaces and Interfaces*, 21, 100734.

Idumah C.I., Obele M.C., Ezeani E.O. 2020b. Understanding interfacial dispersions in ecobenign polymer nano-biocomposites. *Polymer-Plastics Technology and Materials*. https://doi.org/10.1080/25740881.2020.1811312

Idumah C.I., Odera S.R. 2020. Recent advancement in self-healing graphene polymer nanocomposites, shape memory, and coating materials. *Polymer-Plastics Technology and Materials*, 59(11), 1167–1190.

Indermun S., Govender M., Kumar P., Choonara Y.E., Pillay V. 2020. Inorganic nanomaterials for enhanced therapeutic safety. *In:* Yata V., Ranjan S., Dasgupta N., Lichtfouse E. (eds.) *Nanopharmaceuticals: Principles and Applications Vol. 3. Environmental Chemistry for a Sustainable World*, vol. 48. Springer, Cham. https://doi.org/10.1007/978-3-030-47120-0_1.

Institute of Food Science and Technology (IFST). 2006. *Nanotechnology*. Available from: www.ifst.org/uploadedfiles/cms/store/ATTACHMENTS/Nanotechnology.pdf

Istiqola A., Syafiuddin A. 2020. A review of silver nanoparticles in food packaging technologies: Regulation, methods, properties, migration, and future challenges.

Journal of Chinese Chemical Society, 67, 1942–1956. https://doi.org/10.1002/jccs.202000179

Jafarzadeh S., Jafari S. 2020. Impact of metal nanoparticles on the mechanical, barrier, optical and thermal properties of biodegradable food packaging materials. *Critical Reviews in Food Science and Nutrition*. https://doi.org/10.1080/10408398.2020.1783200

Jildeh N.B., Matouq M. 2020. Nanotechnology in packing materials for food and drug stuff opportunities. *Journal of Environmental Chemical Engineering*, 8, 104338.

Kausar A. 2020. A review of high performance polymer nanocomposites for packaging applications in electronics and food industries. *Journal of Plastic Film & Sheeting*, 36, 94–112.

Kumar P., Mahajan P., Kaur R., Gautam S. 2020. Nanotechnology and its challenges in the food sector: A review. *Materials Today Chemistry*, 17, 100332.

Li K., Fina A., Marrè D., Carosio F., Monticellia O. 2020b. Graphite oxide nanocoatings as a sustainable route to extend the applicability of biopolymer-based film. *Applied Surface Science*, 522, 146471.

Li Y., Qin W. 2020. Foreword: Special section on "The Reliability of Advanced Microelectronic Packaging—Part I: Management of Thermal Effects". *IEEE Transactions on Components, Packaging and Manufacturing Technology*, 10, 1425–1426.

Li Z., Khuje S., Chivate A., Huang Y., Hu Y., An L., Shao Z., Wang J., Chang S., Ren S. 2020a. Printable copper sensor electronics for high temperature. *ACS Applied Electronic Materials*, 2, 1867–1873.

Liu W., Zhang M., Bhandari B. 2020. Nanotechnology—A shelf life extension strategy for fruits and vegetables. *Critical Reviews in Food Science and Nutrition*, 60, 1706–1721.

Mahajan R. 2020. Quiet revolutions: How advanced microelectronics packaging continues to drive heterogeneous integration. *In: 2020 19th IEEE Intersociety Conference on Thermal and Thermomechanical Phenomena in Electronic Systems (ITherm)*, Orlando, FL, USA, 1408–1412. https://doi.org/10.1109/ITherm45881.2020.9190247

Mech A., Rauscher H., Babick F., Hodoroaba V., Ghanem A., Wohlleben W., Marvin H., Weigel S., Brüngel R., Friedrich C.M., Rasmussen K., Loeschner K., Gilliland D. 2020b. *EUR 29876 EN, The NanoDefine Methods Manual*. European Commission, Luxembourg. Accessed 25 October 2020.

Mech A., Wohlleben W., Ghanem A., Hodoroaba V., Weigel S., Babick F., Brüngel R., Friedrich C., Rasmussen K., Rauscher H. 2020a. Nano or not nano? A structured approach for identifying nanomaterials according to the European Commission's definition. *Small*, 16, e2002228.

Nano-enabled Packaging Market-Global Industry Trends. n.d. Available from: www.databridgemarketresearch.com. Accessed 20 October 2020.

NanoSafety Cluster (EU). 2020. EU NanoSafety Cluster. Available from: www.nanosafetycluster.eu/. Accessed 25 October 2020.

National Center for NanoScience and Technology of China (NCNST). 2005. Available from: www.nanoctr.cn/e-view.jsp?tipid=1116405057331. Accessed 25 October 2020.

Nechita P., Roman M. 2020. Review on polysaccharides used in coatings for food packaging papers. *Coatings*, 10, 566.

OECD. 2013. *Regulatory Frameworks for Nanotechnology in Foods and Medical Products: Summary Results of a Survey Activity*, OECD Science, Technology and Industry Policy Papers, No. 4. OECD Publishing, Paris. http://dx.doi.org/10.1787/5k47w4vsb4s4-en

Oliveira S., Fasolin L., Vicente A., Fucinos P., Pastrana L. 2020. Printability, microstructure, and flow dynamics of phase-separated edible 3D inks. *Food Hydrocolloids*, 109, 106120.

Rahman N.A. 2019. Applications of polymeric nanoparticles in food sector. pp. 345–359. *In: Nanotechnology: Applications in Energy, Drug and Food*. Springer: Cham.

Raj A., Singh A., Shah P., Agrawal N. 2020. Safe dose of nanoparticles: A boon for consumer goods and biomedical application. *In:* Agrawal N., Shah P. (eds.) *Toxicology of Nanoparticles: Insights from Drosophila*. Springer, Singapore.

Ramos M., Fortunati E., Beltrán A., Peltzer M., Cristofaro F., Visai L., Valente A.J., Jiménez A., Kenny J.M., Garrigós M.C. 2020. Controlled release, disintegration, antioxidant, and antimicrobial properties of poly (lactic acid)/thymol/nanoclay composites. *Polymers*, 12, 1878.

Rowan N., Laffey J. 2021. Unlocking the surge in demand for personal and protective equipment (PPE) and improvised face coverings arising from coronavirus disease (COVID-19) pandemic—Implications for efficacy, re-use and sustainable waste management. *Science of the Total Environment*, 752, 142259. https://doi.org/10.1016/j.scitotenv.2020.142259

SAICM. 2011. *Nanomaterials: Applications, Implications and Safety Management in the SAICM Context*. Available from: www.saicm.org/documents/OEWG/Meeting%20documents/OEWG1%20INF8-Nano%20report.pdf. Accessed 25 October 2020.

Saqqa G.A. 2020. Nanotechnology in food packaging and food safety. *Journal of Advanced Research in Food Science and Nutrition*, 3, 24–33.

Schwirn K., Voelker D., Galert W., Quik J., Tietjen L. 2020. Environmental risk assessment of nanomaterials in the light of new obligations under the reach regulation: Which challenges remain and how to approach them? *Integrated Environmental Assessment and Management*, 16, 706–717.

Schyns Z., Shaver M. 2020. Mechanical recycling of packaging plastics: A review. *Macromolecular Rapid Communications*, 51, 2000415.

Shafiq M., Anjum S., Hano C., Anjum I., Abbasi B. 2020. An overview of the applications of nanomaterials and nanodevices in the food industry. *Foods*, 9, 148.

Sharma S., Rawat N., Kumar S., Mir Z., Gaikwad K. 2020. Nanotechnology for food: Regulatory issues and challenges. *In:* Sharma T.R., Deshmukh R., Sonah H. (eds.) *Advances in Agri-Food Biotechnology*. Springer, Singapore.

Shekaftik S., Ashtarinezhad A., Shirazi F., Hosseini A., Yarahmadi R. 2020. Assessing the risk of main activities of nanotechnology companies by the NanoTool method. *International Journal of Occupational Safety and Ergonomics*. https://doi.org/10.1080/10803548.2019.1693778

Silva A., Prata J., Walker T., Duarte A., Ouyang W., Barcelò D., Rocha-Santos T. 2021. Increased plastic pollution due to COVID-19 pandemic: Challenges and recommendations. *Chemical Engineering Journal*, 405, 126683.

Stefanini R., Borghesi G., Ronzano, A. *et al.* 2020. Plastic or glass: A new environmental assessment with a marine litter indicator for the comparison of pasteurized milk bottles. *International Journal of Life Cycle Assessment*. https://doi.org/10.1007/s11367-020-01804-x

Tager J. 2014. Nanomaterials in food packaging: FSANZ fails consumers again. *Chain Reaction*, 122, 16–17.

Taroco H.A., Garcia E.M., Duarte L.F., *et al.* 2020. Factorial design study of corrosion in Tin-Free Steel (TFS) tinplate of tomato paste. *Journal of Packaging Technology and Research*, 4, 135–144.

United States Environmental Protection Agency (USEPA). 2013. *Nanomaterials EPA Is Assessing*. The United States Environmental Protection Agency Website. Available from: www.epa.gov/nanoscience/quickfinder/nanomaterials.htm. Accessed 25 October 2020.

United States Food and Drug Administration (USFDA). 2011. *Draft Guidance for Industry: Dietary Supplements: New Dietary Ingredient Notifications and Related Issues*. 2011 July ed. Maryland. Available from: www.fda.gov/Food/Guidance-Regulation/GuidanceDocumentsRegulatoryInformation/DietarySupplements/ucm257563.htm. Accessed 25 October 2020.

United States Food and Drug Administration (USFDA). 2014a. *Guidance for Industry: Assessing the Effects of Significant Manufacturing Process Changes, Including Emerging Technologies, on the Safety and Regulatory Status of Food Ingredients and Food Contact Substances, Including Food Ingredients That Are Color Additives.* Available from: www.fda.gov/Food/GuidanceRegulation/GuidanceDocuments-RegulatoryInformation/IngredientsAdditivesGRASPackaging/ucm300661.htm. Accessed 24 October 2020.

United States Food and Drug Administration (USFDA). 2014b. *CFR-Code of Federal Regulations Title 21.* Maryland. Accessed 25 October 2020. Available from: www.fda.gov/MedicalDevices/DeviceRegulationandGuidance/Databases/ucm135680.htm.

United States Food and Drug Administration (USFDA). 2014c. *Guidance for Industry: Assessing the Effects of Significant Manufacturing Process Changes, Including Emerging Technologies, on the Safety and Regulatory Status of Food Ingredients and Food Contact Substances, Including Food Ingredients That Are Color Additives.* Available from: www.fda.gov/Food/GuidanceRegulation/GuidanceDocuments-RegulatoryInformation/IngredientsAdditivesGRASPackaging/ucm300661.htm. Accessed 25 October 2020.

Urban R.C., Nakada L.Y.K. 2021. COVID-19 pandemic: Solid waste and environmental impacts in Brazil. *Science of the Total Environment,* 755, 142471. https://doi.org/10.1016/j.scitotenv.2020.142471

Vanapalli K., Sharma H., Ranjan V., Samal B., Bhattacharya J., Dubey B., Goel S. 2021. Challenges and strategies for effective plastic waste management during and post COVID-19 pandemic. *Science of the Total Environment,* 750, 141514. https://doi.org/10.1016/j.scitotenv.2020.141514

Wang X., Tang Y., Zhu X., Zhou Y., Hong X. 2020. Preparation and characterization of polylactic acid/polyaniline/nanocrystalline cellulose nanocomposite films. *International Journal of Biological Macromolecules,* 146, 1069–1075.

Yoo J.E., Roev V., Bae J., Yoon D.-S., Kim S.-D., Lee E.-S. 2020. Multifunctional hybrid polymer nanocomposites for automotive-battery packaging. *Journal of Applied Polymer Science,* 137, 49059.

Yu Y., Bu F., Zhou H., Wang Y., Cui J., Wang X., Nie G., Xiao H. 2020. Biosafety materials: An emerging new research direction of materials science from the COVID-19 outbreak. *Materials Chemistry Frontiers,* 4, 1930–1953.

Yuan G.J., Xie J.F., Li H.H., Shan B., Zhang X.X., Liu J., Li L., Tian Z. 2020. Thermally reduced graphene oxide/carbon nanotube composite films for thermal packaging applications. *Materials,* 13, 317.

Zambrano-Zaragoza M.L., González-Reza R.M., Quintanar-Guerrero D., Mendoza-Muñoz N. 2020. Nano-films for food packaging. *In:* Hebbar U., Ranjan S., Dasgupta N., Kumar Mishra R. (eds.) *Nano-food Engineering. Food Engineering Series.* Springer, Cham. https://doi.org/10.1007/978-3-030-44552-2_10

Chapter 19

Future Scope of Nano-Based Methods for the Improvement of Postharvest Technologies and Increased Shelf Life of Minimally Processed Food

Cristian Josué Mendoza Meneses, Alma Karen Burgos Araiza, Betsie Martínez Cano, Ana Angélica Feregrino Pérez, and Manuel Toledano Ayala

List of Abbreviations

A. alternata	*Alternaria alternata*
Ag	Silver
A. niger	Aspergillus niger
B. cinerea	*Botrytis cinerea*
C. gloeosporioides	*Colletotrichum gloeosporioides*
CO_2	Carbon dioxide
Cu	Copper
DPPH	2,2-diphenyl-1-picrylhydrazyl
E. coli	*Escherichia coli*

 DOI: 10.1201/9781003142287-19

LDH	Layered double hydroxide
L. monocytogenes	*Listeria monocytogenes*
MPP	Minimally processed product
nm	Nanometers
PAL	Phenylalanine ammonia-lyase
PLA	Polylactic acid
PVA	Polyvinyl acid
S. aureus	*Staphylococcus aureus*
SiO_2	Silicon dioxide
TiO_2	Titanium dioxide
UV	Ultraviolet
ZnO	Zinc oxide

19.1 Introduction

Obtaining good quality horticultural products is of utmost importance due to its high demand in the population, which has increased in recent decades (Salama et al., 2021). However, postharvest losses of these products are a problem worldwide, and it is estimated that between 45% to 55% of world production of fruits and vegetables is lost or wasted along the value chain from production to consumption (Lufu et al., 2020).

The main cause of this is microbial damage, which, together with pests and other factors, causes 20% to 40% of world losses (González-Estrada et al., 2019; Ruffo-Roberto et al., 2019). A solution to these problems is nanotechnology, which arises with the purpose of maintaining quality integrity, improving sensory expression, extending the shelf life and preserving the product (Patel et al., 2018).

Nanotechnology and agroindustry are related to the inclusion of materials and compounds at nanometric scales in different sectors ranging from agriculture, livestock, aquaculture, and even related activities such as water filtration and food processing (He et al., 2019). This technology can be applied at any stage of agriculture from seed care to postharvest practices; therefore, nanotechnology can have the promise to manage current problems regarding the production, quality and processing of fresh horticultural products (Shukla et al., 2019).

Quality problems can be controlled with the application of nanotechnology-based control of growth, development and proliferation of phytopathogens. However, other causes of the loss of horticultural products are the humidity of the environment, degradation in processing and inherent physiological aspects of the product (Singh, 2020). The use of nanoscience in agroindustry is a widely used alternative since no harmful effects have been reported for its use in plants or in the consumption of products (Kadam and Shubhangi, 2020).

Nanotechnology is also used to increase the productivity of fruits and vegetables, in the same way, to minimize losses due to damage in processing (Salama et al., 2021). Even the MPPs are more vulnerable than products in their original form; some defects are due to poor handling of horticultural products, which modify the organoleptic characteristics (Prakash et al., 2018).

The advantages of nanotechnology applied in agroindustry are due to its different uses in nanopesticides (D. Kim et al., 2018), nanoemulsions (Pirozzi et al., 2020), nanoparticles (Naing and Kim, 2020), nanofilms (M. Zambrano-Zaragoza et al., 2020), nanocoatings (Park et al., 2017), nanocomposites (Anugrah et al., 2020) and nanofertilizers (Sharma et al., 2020).

19.2 Applications of Nanotechnology in Postharvest

Fruits and vegetables are of great importance for human nutrition, due to their high nutritional value and multiple health benefits (Yahia et al., 2017), However, these products are exposed to attack by microorganisms during their processing after harvest, which causes significant economic losses (Liu et al., 2018). Therefore, the preservation of fruits and vegetables through the implementation of postharvest treatments is crucial. For a long time, a great variety of traditional, natural and synthetic antimicrobial agents have been used to control the growth of microorganisms in the postharvest stage (González-Estrada et al., 2019).

Among the factors that promote losses during the harvest are injuries and damages, state of maturity, climate and harvesting method. Specifically, the factors that are involved in losses during postharvest are pretreatments (irradiation, heat treatment, chemical application, waxing, coatings, packaging) and storage conditions (range of airflow, temperature, humidity, gas composition, storage time; Lufu et al., 2020). Depending on agricultural practices, economic losses due to fungal attack in the postharvest chain range between 30% and 50%. Overall, 33% of the world's agricultural products is considered to be lost after being harvested (Ruffo-Roberto et al., 2019). In Figure 19.1, the physical and biological factors that influence the deterioration of fresh food are shown, which affect the sensory quality and the performance of horticultural products.

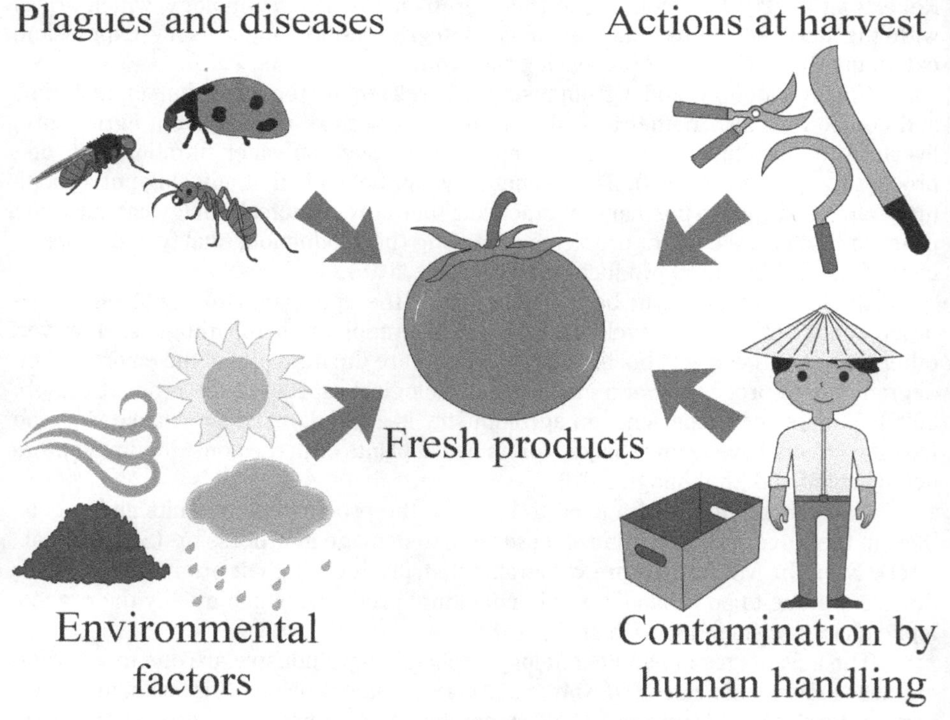

Figure 19.1 Physical and biological factors that affect the postharvest quality of fresh food.

Currently, environmental and health problems, together with the preference of consumers for chemical-free products, lead to the search for new, more efficient alternatives to control postharvest problems (Singh and Sharma, 2018). Nanotechnology is a versatile and smart tool that can sustainably achieve food security and increase agricultural productivity (Panpatte and Jhala, 2019), Furthermore, the use of nanomaterials is a promising alternative for the control of postharvest problems due to its bioavailability, protection that it exerts and better performance (Acevedo-Fani et al., 2017). A wide variety of nanoparticles of different natures have been studied that can easily penetrate the membrane of microorganisms and cause an inhibition of different physiological processes related to their growth and multiplication (Patra et al., 2013). In this sense, nanotechnology offers a novel alternative to develop different types of compound-bearing nanomaterials to control postharvest pathogens (González-Estrada et al., 2019).

In a study carried out by Mohammadi et al., (2015a) chitosan nanoparticles with the essential oil of *Cinnamomum zaylanicum* were used and their effect on *Phytophthora drechsleri* in cucumber was evaluated, and it was found that the severity and incidence of the disease were fully controlled, in addition to extending the useful life of the postharvest cucumber. In addition, the same type of nanoparticles added with pepper tree essential oil when used against *C. gloeosporioides* in avocado fruits was found to prevent and cure the infection (Chávez-Magdaleno et al., 2018). In another study, a coating made from nanochitosan added with Cu was reported to control decomposition and improve strawberry quality (Eshghi et al., 2014). The application of a mixture of chitosan and SiO_2 nanoparticles was also found to extend the shelf life of cherries (*Prunus pseudocerasus* L.) in the postharvest stage (Xin et al., 2017). However, there are different types of nanoparticles according to their composition and use in food; the most common are discussed next.

19.2.1 Nanochitosan

Chitosan is a safe, nontoxic polymer and can interact with polyanions to form complexes and gels. Chitosan nanoparticles measure between 50 and 100 nm and can be prepared by the interaction of oppositely charged macromolecules. Nanochitosan has the potential for the development of encapsulation or immobilization carriers, due to its favorable properties of biocompatibility, biodegradability, nontoxicity and antimicrobial potential (Zhao et al., 2011). The antimicrobial potential of chitosan nanoparticles is due to their unique physicochemical properties, high surface area and charge density that effectively interact with the surface of microorganisms (Pilon et al., 2015).

Studies have shown that the application of edible films and coatings with chitosan nanoparticles on fruits and vegetables significantly induced defense enzymes such as PAL, chitinase and glucanase, in addition to improving shelf life and preservation of bioactive components of strawberries (Eshghi et al., 2014). Chitosan nanoparticles can inhibit the growth of bacteria and fungi and hence are useful for managing microbial activity in apples (Pilon et al., 2015), bananas (Lustriane et al., 2018) and grapes (Castelo-Branco Melo et al., 2018).

Although there are some reports that show the benefits of applying nanochitosan to fruits and vegetables to minimize postharvest microbial contamination, more large-scale trials are still necessary to evaluate the form and frequency of application, since most of these tests have been developed in the laboratory.

19.2.2 Nanocomposites

Nanocomposites are a novel group of hybrid materials made up of natural polymers and inorganic solids (Pillai and Ray, 2012); incorporating nanofillers such as attapulgite, Ag, SiO_2 and TiO_2 into polymers makes it possible to improve their mechanical and barrier properties, which gives it many useful functional properties for different applications. Nanocomposites have functions such as CO_2 blocking, water vapor barrier and antimicrobial activity; in addition, it has been proven that they can remove ethylene.

The previously mentioned properties are suitable for the conservation of fresh agricultural products in the postharvest stage (Shi et al., 2018). Recently, the use of nanocomposites in postharvest storage has been studied with special interest. Different reports show that the use of nanocomposites improves the retention of nutrients and the protection of the taste of edible mushrooms (Donglu et al., 2016; Shi et al., 2018) Specifically, Pillai and Ray (2012) found that clay-loaded nanocomposite films improve the permeability and barrier properties and other useful qualities for food packaging and coating.

Ag nanocomposites, for their part, have powerful antifungal activity against *Rhizoctonia solani*, *A. flavus* and *A. alternata* (Saharan et al., 2013), while Cu and ZnO nanocomposites have a high antifungal effect against *A. alternata* and *B. cinerea*, in addition, when applied as a coating, they increase the shelf life and quality of strawberries (Al-Dhabaan et al., 2017) and tomatoes (Kaur et al., 2012). Youssef et al. (2019) evaluated the use of chitosan nanocomposites in Italia and Benitaka grapes and found it to use significantly reduces the development of gray mold and did not affect the quality of the grape.

Currently, the application of nanocomposites for active packaging in the food industry constitutes a promising alternative tool in technologies for the conservation of postharvest agricultural products.

19.2.3 Inorganic Nanoparticles

Inorganic nanoparticles are all those nanostructures that contain different types of metals and metal oxides, such as Ag, zinc, Cu, gold, ZnO, TiO_{-2} and magnesium oxide, among others. One of the functions of these materials is to make direct or slow-release contact to preserve food due to their antimicrobial potential against a large number of microorganisms (Alghuthaymi et al., 2020). The antimicrobial activity of inorganic nanoparticles can be attributed to their direct interaction with the surface of microbial cells, they interrupt transmembrane electron transfer and penetrate the cell, thus causing oxidation in cellular components and finally cell death (Dakal et al., 2016). Ag nanoparticles are the most studied due to their enormous antimicrobial potential; it is documented that they present this effect on *E. coli* and *S. aureus* (Sadeghnejad et al., 2014) and reduce mycelial growth of *A. niger* (Carrero-Sánchez et al., 2006). Therefore, Ag nanoparticle films were made to be used in food packaging as microbial agents (Azlin-Hasim et al., 2016).

A large number of inorganic nanoparticles have been studied and evaluated against different microorganisms; Cu nanoparticles inhibit the growth of *E. coli*, *S. aureus*, *Saccharomyces cerevisiae*, and *L. monocytogenes* (Cioffi et al., 2005), and its mechanism of action is attributed to multiple toxic effects, such as the generation of reactive oxygen species (Chatterjee et al., 2014). Zinc nanocrystals have been incorporated into a plastic matrix to inhibit the growth of different bacteria and fungi (Hong et al., 2016). In addition, TiO_2 nanoparticles have also been studied for their antimicrobial effect; however, the mechanism of action is not yet clear, but it has been attributed to an oxidative attack on the outer cell membrane of bacteria (Kubacka et al., 2014). ZnO_2 nanoparticles, when in direct contact with microorganisms, destroy microbial cell integrity (Sirelkhatim et al., 2015) and have been used for the postharvest storage of bananas, carrots and tomatos (Singh, 2020).

While Cu nanoparticles, for their part, when sprinkled on tomato fruits increase the content of antioxidant compounds and improve the quality of the fruit (López-Vargas et al., 2018). This has led to the development of packaging films for food enriched with inorganic nanoparticles that, by reducing the entry of oxygen and other gases, minimizing moisture loss and delaying the deterioration of food (Farhoodi, 2016).

19.2.4 Nanoemulsions

A nanoemulsion is a complex system made up of an oil phase, a surfactant, and water. It is a static colloidal solution with a droplet size of 20 to 200 nm and is capable of absorbing functional components within each droplet, which can reduce chemical degradation. In addition, they present a high kinetic and thermodynamic consistency against aggregation and coalescence and have a better solubilization of hydrophilic or hydrophobic components (Rashid et al., 2020).

Formulations from nanoemulsions have good storage stability in a wide temperature range and are relatively inexpensive and effective in agriculture and postharvest (Danish and Hussain, 2019). Nanoemulsions combine the antimicrobial activity that essential oils and nanoparticles by making a synergistic effect between them, the above has increased the interest of scientists for their use as a food preservative (Mohammadi et al., 2015b).

Limonene is a compound obtained from the essential oil of citrus fruits such as oranges, lemons, mandarins, limes and grapefruits. This compound has been shown to have antibacterial and antifungal properties against fungi that spoil food. This type of compound has the disadvantage of being hydrophobic, which makes its use in the development of edible coatings difficult; however, nanotechnology has provided an approach that involves the nanoencapsulation of phytochemical compounds in liposomes that allow their incorporation (Dhital et al., 2018).

There are several studies that involve nanoemulsions for the conservation of agricultural products in postharvest and show important results. *Zataria multiflora* essential oil with nanoparticles significantly inhibited the growth of the fungi *Alternaria solani* and *Rhizopus stolonifer* (Nasseri et al., 2016). The same type of essential oil with cadmium sulfide nanoparticles allows a controlled and sustained release for a month, and when applied to strawberries, the severity of the disease caused by *B. cinerea* was reduced (Mohammadi et al., 2015b).

Bernardos et al. (2015) found that an emulsion based on mesoporous silica nanoparticles with different essential oils can inhibit the growth of *A. niger* for 30 days, which suggests a better applicability of essential oils as food preservatives. Recently, the incorporation of nanoemulsions in edible coatings was studied in order to disperse, in low and effective doses, the active principles present in essential oils. Robledo et al. (2018) formulated a nanoemulsion of thymol and chitosan that they used as a coating on cherry tomatoes inoculated with *B. cinerea* and found that there is less growth of the fungus after 7 days in storage. A nanoemulsion made with pepper tree essential oil and bionanocomposites could improve avocado quality by inhibiting the growth of *C. gloeosporioides* (Chávez-Magdaleno et al., 2018). Therefore, it is considered that nanoemulsions can be an alternative fungicidal treatment to chemical fungicides to increase the postharvest life of fruits and vegetables.

19.2.5 Nanocoatings

To meet the consumer demand for fresh fruits and vegetables, scientists have sought alternatives to extend the food quality and shelf life by incorporating new bio-based packaging materials, such as biodegradable films and edible coatings. One of these alternatives here is the use of nano-edible coatings, which can increase the shelf life of food and control the flow of gases and moisture in food.

In particular, edible coatings or films consist of liquid solutions that are applied on the surface of the fruits, forming a semipermeable layer from biopolymers, with the aim of regulating humidity and gas exchange between the internal and external atmosphere (Ncama et al., 2018). These edible coatings must have certain characteristics: be semipermeable to water vapor, resistant to water and stable and resistant during the storage period, as well as low viscosity, transparency and be inexpensive (Saberi and Golding, 2018). In order to decide on the appropriate composition of a coating for a certain fruit, the essential factors that need to be taken into account are its cellular respiration process in the ripening stage and its production of ethylene (J. Zhang et al., 2017).

The use of edible coatings and films is an alternative for incorporating compounds that improve the organoleptic properties of fruits and vegetables while controlling microbial growth. In this context, nanomaterials can function as a vehicle to deliver additives and antimicrobial agents, resulting in better physical properties, greater thermal stability, resistance and tolerance to thermal stress (M. Zambrano-Zaragoza et al., 2018), which, may be important to minimize spoilage and maintain long-term storage of postharvest fruits and vegetables (González-Reza et al., 2018). In another study, the combination of nanoemulsions with certain nanostructures in an edible coating can be used to make a prolonged release of active ingredients (Salvia-Trujillo et al., 2015).

The incorporation of antimicrobial compounds in edible films and coatings is a widely used technique to extend the shelf life of food and maintain its quality (Dhital et al., 2018). This antimicrobial function is achieved by adding antimicrobial agents (such as nanoparticles) in the polymeric packaging system or by using nanoemulsions from antimicrobial compounds (I. Kim et al., 2020). Some essential oils that prevent microbial growth have been studied, such as oregano, thyme, cloves, cinnamon and lemongrass (Dhital et al., 2018). The different methods for preparing antimicrobial packaging are solution casting (for metallic particles), coating method with polymers and biopolymers, melt and injection extrusion (for plastics and metals in a fluid state; I. Kim et al., 2020).

The nanoparticles mostly used to modify the properties of edible coatings for fruits and vegetables in postharvest are those of titanium, zinc and Ag oxide (Al-Naamani et al., 2018; Koushesh-Saba and Amini, 2017; Shi et al., 2018). Nanocomposite films formed by nanoparticles of methylcellulose, pediocin and zinc oxide showed antimicrobial activity against *L. monocytogenes* and *S. aureus* (Shukla et al., 2019). A ZnO_2-chitosan nanocomposite was used as an antimicrobial polyethylene packaging film to protect okra (*Abelmoschus esculentus*), and it was found that, in addition to preserving its quality, it also inhibited antimicrobial growth (Al-Naamani et al., 2018).

Gorrasi and Bugatti, (2016) prepared edible active coatings based on pectin filled with LDH salicylate and managed to extend the shelf life of fresh apricots. H. Nguyen and Nguyen (2020) determined that a 0.2% nanochitosan coating preserves strawberry quality for up to 21 days in storage at 2°C. specifically, a coating made from a nanoemulsion of alginate and lemongrass oil on fresh-cut Fuji apples inactivated *E. coli* and improved the quality attributes of the product (Salvia-Trujillo et al., 2015). Electrospun zein nanofibers loaded with curcumin have a beneficial effect when used as a coating for apples infected with *Penicillium expansum* and *B. cinerea* by inhibiting the growth of fungi during storage (Yilmaz et al., 2016).

Recent advances in research on films and coatings have focused on developing nanocoatings, different extraction solvents (alcohol, acids, water), composites (mixture of hydrocolloids and lipids), multilayer coatings and edible coatings based on plant extracts; systems in which it has been sought to incorporate active agents such as antimicrobials, antioxidants, nutrients, emulsifiers, and flavorings (Ncama et al., 2018).

The rational use of nano-edible coatings made from nanocomposites and nanoemulsions can lead to reduced food spoilage, inhibit the growth of pathogenic microorganisms and significantly improve the shelf life of postharvest foods. Other less explored and investigational applications of nanomaterials in food packaging are the detection of pesticides, pathogens and toxins. Therefore, packaging made from nanomaterials is of great importance to control food quality (He et al., 2019). Figure 19.2 shows the benefits of the use of nanomaterials in harvested products in the prevention of physical and microbiological damage.

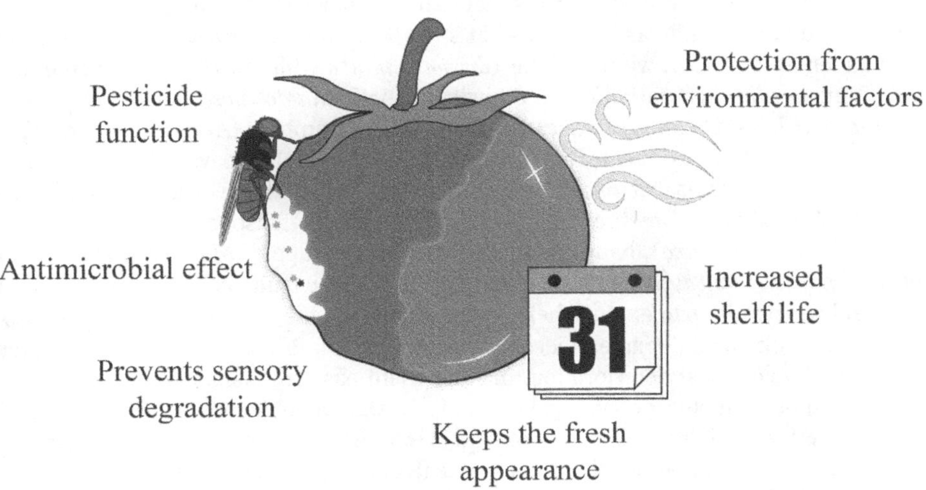

Figure 19.2 Benefits of the use of nanomaterials applied in fresh products in postharvest.

19.3 MPPs

A minimally processed product is defined as a fruit or vegetable that has been subjected to different physical processes without significantly altering its original properties (Shigematsu et al., 2018). These products are a sector of high current demand due to the preference of consumers for fresh foods with high nutritional content (Ali et al., 2018). Minimally processed foods emerged from new food trends: ready to use, ready to eat and ready to cook (Rai et al., 2019).

Horticultural products suffer damage by the action of microorganisms even in minimal processing, the main defects presented are darkening, changes in texture and loss of nutritional compounds and organoleptic characteristics (Putnik et al., 2017). Some of the actions during harvest that can produce contamination by microbial agents are handling and cutting activity, the cracks formed in the surface of the products produce lacerations that compromise the structural integrity of the fresh products (Corato, 2020).

The microorganisms most frequently found in minimally processed products are *L. monocytogenes* and *Salmonella spp*, although other bacteria such as *Cronobacter spp*, *Helicobacter pylori* and *S. aureus* can also occur; therefore, it is necessary to improve the washing of fresh products or use different methods to reduce the bacterial load, among these methods, the control of the refrigeration temperature stops the growth of microorganisms (Pina-Pérez et al., 2018; Sant'Anna et al., 2020).

Measures to minimize the problems that can occur in food processing start by evaluating the sensory characteristics of the product such as color, aroma, texture, smell and flavor; these organoleptic attributes can be evaluated by quantitative analyzes that indicate a parameter of the physiological state of the product (Possas et al., 2018). The arrival of this category of food has generated the development of new technologies for the packaging of these, such as flexible packaging, aseptic packaging, aluminum foil, metal cans and flexography (Rai et al., 2019).

There is a great variety of fruits that naturally have a very short shelf life or are considered very fragile, as is the case of strawberries (*Fragaria x ananassa* Duch.; V. Nguyen et al., 2020), winter jujube (*Zizyphus jujuba* Mill. cv. Dongzao; Kou et al., 2019), mango (*Mangifera indica*), red kiwifruit (*Actinidia deliciosa*; Xing et al., 2020), tangerine (*Citrus reticulata*), grapes (*Vitis vinífera*), carambolo (*Averrhoa carambola*), sweet cherry (*Prunus cerasus*; Emamifar and Bavaisi, 2020), papaya (*Carica papaya*), tomato (*Solanum lycopersicum*; H. Nguyen and Nguyen, 2020), avocado (*Persea americana*), plum (*Prunus domestica*), blueberry (*Cyanococcus*), raspberry (*Rubus idaeus*), guava (*Psidium guajava*), baby carrot (*Daucus carota*), apple (*Malus domestica*) and peach (*Prunus persica*; Saberi and Golding, 2018). These fruits have been involved in several investigations to extend their postharvest shelf life, due to their high susceptibility to mechanical damage, microbial decomposition, high respiration rate, water loss and physiological deterioration, making them easy to mishandle during collection, transport and storage (V. Nguyen et al., 2020). The objective of optimizing and prolonging the shelf life of this type of food is to reduce and control water loss, ripening, senescence and the development of rot (Saberi and Golding, 2018). Table 19.1, shows a list of nanomaterials used in postharvest to improve the quality and shelf life of fresh products.

Table 19.1 Nanomaterials Used in Postharvest Treatments to Improve the Quality and Shelf Life of MMPs

Method	*Functionality*	*Ingredient*	*Effect*	*Reference*
Nanoparticles	Antimicrobial, high solubility, yield and stability, antioxidant effect, prevents rust growth.	Nano-Ag	Increased the shelf life of pistachios, almonds, hazelnuts and walnuts.	H. Tavakoli et al., 2017
		Nano-Ag and clove oil	Maintains a better quality of lemon during storage.	S. Tavakoli et al., 2019
		Nano-Ag and PVA (film)	Extends the shelf life of grapes due to its breathability and antibacterial activity.	Deng et al., 2019
		Glyerosomes based on Nano-Ag polyvinylpyr-rolidone	Extends the shelf life of bell pepper for 12 days at 4°C.	Saravanakumar et al., 2020
		Nano-Ag impreg-nated packages	Cabbage and tomato were packed showing a bactericidal effect of nano-Ag on *Aeromonas hydrophila* and conserving moisture content, so they were kept fresh.	Singh and Sahareen, 2017
	Antimicrobial (against *Escherichia coli*, *Staphylococcus aureus*, *Listeria monocyto-genes*, *Bacillus cereus*, *Salmonella enteritidis*) UV protective, transparent, electrical conductivity.	Nano-ZnO, cinnamon oil in nitrogen-modified atmosphere	Inhibits microbial growth and preserved the sensory properties of canned pistachio for 12 weeks.	Kazemi et al., 2020
		Nano-ZnO and agar	Preserves the fresh appearance of green grapes for up to 21 days of storage at room temperature.	Kumar et al., 2019
		Nano-ZnO, chitosan and cellulose acetate phthalate (film)	Extends the shelf life of black grapes up to 9 days.	Indumathi et al., 2019

(*Continued*)

Table 19.1 (Continued)

Method	*Functionality*	*Ingredient*	*Effect*	*Reference*
		Nano-ZnO and pressurized argon (coating)	In fresh cut oranges, it effectively reduced the loss of mass and the increase of soluble solids, reduction of titratable acidity and ascorbic acid, inhibited the growth of aerobic bacteria and fungi.	Wu et al., 2021
		Nano-ZnO, polyurethane, chitosan and mahua oil	The shelf life of coated carrot pieces was extended up to 9 days.	Sarojini et al., 2019
	Antimicrobial.	Bag with Nano-Cu	Increases the shelf life of table grapes stored in bags with a nano-Cu-modified atmosphere.	Loyola and Arriola, 2017
	Antimicrobial, thermostability, inertia, nontoxic, photostable, cheap.	Nano-TiO_2, PLA, bergamot and nano-Ag (film)	Delays the loss of firmness, color, total acidity and vitamin C of packaged mangoes during storage for up to 15 days.	Chi et al., 2019
		Nano-TiO_2 and low-density polyethylene (packaging)	maintains overall quality of packaged strawberries, achieved lower decay rate and weight loss, delayed decline in firmness and titratable acid during storage.	Li et al., 2017b
	Stability, low toxicity, better insulating properties.	Nano-SiO_2 (packaging)	It significantly inhibits internal browning, delayed the decrease in total soluble solids, titratable acidity, ascorbic acid content and extractable juice in loquats.	Wang et al., 2020

Method	*Functionality*	*Ingredient*	*Effect*	*Reference*
		Nano-SiO_2 and potato starch (film)	In mushrooms (*Agaricus bisporus*) it reduced the browning index, delayed weight loss, prolonging its storage at 4°C.	Zhang and Wang, 2017
Bases for films with nanomaterials	Biodegradable, improves mechanical properties due to its hydrophilic properties and compatible structure, resistance to solvents.	PLA and nano-ZnO	The film showed greater permeability to water vapor and lower permeability to oxygen, better firmness, total phenolic content, color, sensory quality and inhibition of microorganisms, applied to freshly cut apple stored for 14 days at 4°C.	Li et al., 2017d
	Transparent, hygroscopic, biodegradable, reduces respiration rate.	Nanocellulose (coating)	The biodegradable film on guavas increased firmness, maintained the green color and reduced maturity for up to 13 days.	Francisco et al., 2020
	Barrier against water, high thermal stability, barrier against UV light, permeability to oxygen.	Sweet potato starch, montmorillonite Nanoclays and thyme essential oil	The film showed antimicrobial effect against *Escherichia coli* and *Salmonella typhimurium* on fresh spinach leaves during 5 days of storage.	Issa et al., 2017
	Antimicrobial effect in packaged fruits.	Polyvinyl alcohol/tea polyphenol composite	The films modified different aspects of the useful life of the fruits such as weight, firmness, acidity, soluble solids, which shows potential as a packaging material.	Lan et al., 2019

(Continued)

Table 19.1 (Continued)

Method	*Functionality*	*Ingredient*	*Effect*	*Reference*
Nanoemulsion	Antimicrobial against *Escherichia coli, Listeria monocytogenes, Salmonella typhimurium*, increases shelf life.	Oregano oil	Shows a decrease in the bacterial load in lettuce under storage at 4°C.	Bhargava et al., 2015
		Carvacrol and eugenol	Works as a control for *Salmonella enterica* inoculated in fresh spinach, showing a decrease in the total number of the pathogen.	Ruengvisesh et al., 2015
		Thymol	Treatment with thymol serves as an antagonist of foodborne bacteria, the effect is comparable with the action of chlorine for lettuce and blueberries inoculated with pathogens.	Li et al., 2017c
		Carvacrol	Decreases the number of pathogens after a 30-minute soak treatment in broccoli and mung bean seeds.	Landry et al., 2016
Nanocoating	Antimicrobial against *Escherichia coli, Listeria monocytogenes, Salmonella typhimurium*, rust and yeasts, increases shelf life, preservatives of sensory quality, retards deterioration.	Carvacrol and chitosan	It was determined that a coating and gamma irradiation reduces the number of pathogens in green beans, prolonging the shelf life up to 13 days at 4°C.	Severino et al., 2015

Method	Functionality	Ingredient	Effect	Reference
		Lemon oil and chitosan	The application of a coating with lemon oil reduces the microbial load of arugula and increases the shelf life from 3 to 7 days.	Sessa et al., 2015
		Carvacrol and chitosan	A pulsed light treatment applied in conjunction with a coating results in a surface decontaminant on cucumber slices, thus, it works as a substitute for chlorine.	Taştan et al., 2017
		Clove oil, cassava starch and glycerol	Treatment with different proportions of cassava starch and clove oil delays the deterioration of the quality of papaya stored for 15 days at 4°C.	Holsbach et al., 2019
		Cinnamon and pullulan oil	This treatment increased the shelf life of fresh strawberries, at the same time there was less loss of mass, firmness, soluble solids and acidity. There was also a lower bacterial load with respect to the control.	Chu et al., 2020

19.3.1 Postharvest Actions in the Handling of MPPs

Nanomaterials are defined as those with a length less than 100 nm (length, height or width; Ruffo-Roberto et al., 2019). These nanomaterials can be classified depending on their dimensions and their chemical nature. According to their dimensions, they are classified into nanoparticles, nanofibers, nanowires, nanofilms, thin films and nanocomposites. In relation to their chemical nature, they are classified into four categories: carbon, ceramic (metal oxides), metal and polymeric compounds (Paramo et al., 2020).

The main advantages of using nanomaterials are that they have been found to possess better antimicrobial characteristics and unusual properties, including high surface area, high reactivity, high strength, fine particle size and ductility. These properties give the containers mechanical resistance, thermal stability, elasticity and barrier performance (against gases such as oxygen, Co_2 and aromatic compounds; Huang et al., 2018). It has been found that the addition of nanoparticles to the materials used for the development of barriers, such as films and coatings, improves their properties such as rigidity and heat resistance (H. Nguyen and Nguyen, 2020).

19.3.1.1 Processing

The loss of moisture is one of the most recurrent problems during the processing and marketing of fresh fruits and vegetables, as it directly implies economic losses since this type of food is sold by weight. Some signs that show loss of moisture are wrinkling of the skin, diminished shine, deterioration of firmness and loss of fresh appearance (Saberi and Golding, 2018). V. Nguyen et al., (2020) found that dipping strawberries in 3% calcium chloride and then coating them with 0.2% nanochitosan caused a significant reduction in weight loss, preserved L-ascorbic acid, antioxidant capacity, total anthocyanin content and delayed production of malondialdehyde, during storage for 15 days at 4°C.

The addition of nano-TiO_2 and nano-SiO_2 to coatings made from chitosan has been reported to improve their tensile and water resistance. Tian et al. (2019) evaluated two coatings on *Ginkgo biloba* seeds: chitosan with nano-TiO_2 and chitosan with nano-SiO_2. In general, the addition of these nanomaterials improved the permeability of water vapor and gas, inhibited the appearance of mildew, maintained the firmness of the seeds and positively affected the antioxidant activity. For their part, R. Zhang et al. (2019) developed a coating from a mixture of 0.48% Konjac glucomannan, 0.6% carrageenan and 0.3% nano-SiO_2 for white fungi and found that this mixture showed the optimal values for transparency, oxygen transmission rate, transmission rate of CO_2 and water vapor transmission rate. The addition of nano-SiO_2 decreased gas permeability and delayed the effect of UV light.

19.3.1.2 Packing and Packaging

Packaging is the material that is in direct contact with food, which is why it is of vital importance in food processing. Nanomaterials designed for food packaging have many advantages over conventional materials, mainly in terms of their mechanical, thermal and barrier properties, as well as their low cost. Nanomaterials have demonstrated their potential for the preservation and storage of food (He et al., 2019), act as a gas and moisture barrier, improve sensory perceptions, prevent microbial spoilage and improve the shelf life of fresh produce (Flores-López et al., 2016).

The main reason for the development of packaging with innovative technologies is to avoid the development of microorganisms in the food contained (Rai et al., 2019). In the last decade, research focused on nanotechnology applied to containers and packaging has increased, either to improve them or to generate active, intelligent packaging made with bioactive materials (Ndukwu et al., 2020). Food packaging enables easy storage, handling, transport and protection against environmental contamination (Huang et al., 2018).

In a study, C. Zhang et al. (2018) found that packages made of PLA with nano-Ag particles improved the preservation of fresh strawberries, compared to pure PLA films, as it showed a lower loss of vitamin C, delaying the decrease in total phenols and DPPH in strawberries.

19.3.1.3 Storage

Some of the main methods used to extend the shelf life of postharvest fruits are refrigeration and conservation technology at low temperatures, short-wave UV radiation, chemical preservation, storage in modified atmospheres, heat treatments, the use of edible coatings, the application of biocontrol agents and wax coatings (V. Nguyen et al., 2020; Xing et al., 2020; C. Zhang et al., 2018). In general, these techniques seek to reduce respiration and perspiration rates, which causes a slower loss of moisture, preserving its optimum quality state for consumption for longer (Ncama et al., 2018).

Studies have shown that coatings made from sodium alginate can help extend the shelf life of some fruits such as melon, apple, papaya and strawberry (Kou et al., 2019). Emamifar and Bavaisi, (2020) evaluated the effect of a coating of 1.5% alginate and 1.25 g/L nano-ZnO in fresh strawberries, which showed the highest antioxidant and superoxide dismutase activity, as well as the lowest peroxidase activity, compared to other proportions studied. The incorporation of nano-ZnO in the coating improved its antimicrobial properties and extended the shelf life of this fruit by up to 20 days.

Similarly, the effect of a chitosan and nano-TiO_2 coating on mangos stored at 13°C was evaluated. The authors found that the decomposition index was 14.49% lower than the control without coating, the firmness of the fruit improved, in addition, the peroxidase and polyphenol oxidase activity, the total content of phenols and flavonoids were higher (Xing et al., 2020). Li et al. (2017a) evaluated the effect of a low-density polyethylene coating with nano-ZnO on cold tolerance and pectin metabolism of peaches stored at 2°C for 40 days. They found that the addition of nano-ZnO achieved the maintenance of the cell wall of the fruit since it inhibited pectin esterase and increased polygalacturonase and β-galactosidase, maintaining good quality during cold stress. In Figure 19.3, the postharvest control stages in the processing of minimally processed products are shown.

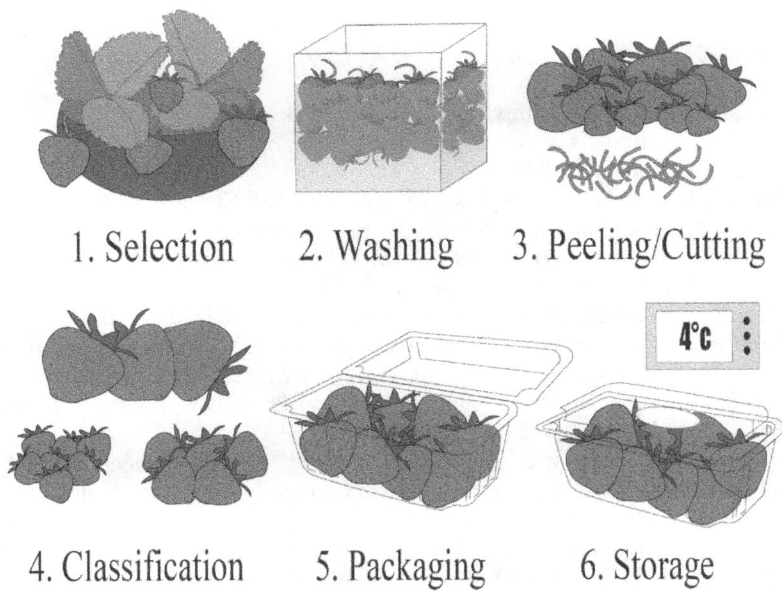

Figure 19.3 Control stages in the processing of minimally processed products.

19.4 Future Trends and Conclusion

One of the most important future perspectives for the application of nanotechnology in the food industry is the development of new packaging technologies; the main advances in this stage of the supply chain are intelligent or smart packaging, active packaging and sustainable or green packaging (Han et al., 2018). The technology is based on nanosensors for the detection of potential risks in food (Nile et al., 2020). The main alterations that are sought to prevent with this type of packaging are the change of temperature, presence of microorganisms and even freshness of the product for consumption; therefore, nanotechnology plays an important role for the optimization of the packaging.

Other aspects to be considered in future studies to increase the shelf life of fresh products are the cost and safety analyses of nanoparticles used in the processing of fresh food; recent studies have shown that nanoparticles can be used effectively on an individual basis; therefore, the combination of different types of nanoparticles is a possible topic of study to establish optimized models (Liu et al., 2020). Even the addition of nanoencapsulated with active agents to increase the shelf life of products is a field of opportunity for future research due to their biological implications (Hosseini and Jafari, 2020).

Information concentrated on nanotechnology applied in postharvest in fresh and MPPs indicates that it is possible to improve the quality and organoleptic characteristics of fruits and vegetables. Therefore, the field of agribusiness benefits from advances and combinations of methods to obtain greater results in terms of productivity and conservation of fresh horticultural products. This provides an overview of the use of nanotechnology throughout the value chain in obtaining products, from the care of the seeds to their commercialization.

References

Acevedo-Fani, A., Soliva-Fortuny, R., and Martín-Belloso, O. 2017. Nanoemulsions as edible coatings. *Current Opinion in Food Science, 15*, 43–49. https://doi.org/10.1016/j.cofs.2017.06.002

Al-Dhabaan, F. A., Shoala, T., Ali, A. A. M., Alaa, M., and Abd-Elsalam, K. 2017. Chemically-produced copper, zinc nanoparticles and chitosan—Bimetallic nanocomposites and their antifungal activity against three phytopathogenic fungi. *International Journal of Agricultural Technology, 13*(5), 753–769.

Alghuthaymi, M., Abd-Elsalam, K. A., Paraliker, P., and Rai, M. 2020. Chapter 13 — Mono and hybrid nanomaterials: Novel strategies to manage postharvest diseases. In K. A. Abd-Elsalam (Ed.), *Multifunctional Hybrid Nanomaterials for Sustainable Agri-Food and Ecosystems* (pp. 287–317). Elsevier. https://doi.org/10.1016/B978-0-12-821354-4.00013-3

Ali, A., Yeoh, W. K., Forney, C., and Siddiqui, M. W. 2018. Advances in postharvest technologies to extend the storage life of minimally processed fruits and vegetables. *Critical Reviews in Food Science and Nutrition, 58*(15), 2632–2649. https://doi.org/10.1080/10408398.2017.1339180

Al-Naamani, L., Dutta, J., and Dobretsov, S. 2018. Nanocomposite zinc oxide-chitosan coatings on polyethylene films for extending storage life of okra (Abelmoschus esculentus). *Nanomaterials, 8*(7), 479. https://doi.org/10.3390/nano8070479

Anugrah, D. S. B., Alexander, H., Pramitasari, R., Hudiyanti, D., and Sagita, C. P. 2020. A review of polysaccharide-zinc oxide nanocomposites as safe coating for fruits preservation. *Coatings, 10*(10), 988. https://doi.org/10.3390/coatings10100988

Azlin-Hasim, S., Cruz-Romero, M. C., Cummins, E., Kerry, J. P., and Morris, M. A. 2016. The potential use of a layer-by-layer strategy to develop LDPE antimicrobial films coated with silver nanoparticles for packaging applications. *Journal of Colloid and Interface Science, 461*, 239–248. https://doi.org/10.1016/j.jcis.2015.09.021

Bernardos, A., Marina, T., Žáček, P., Pérez-Esteve, É., Martínez-Mañez, R., Lhotka, M., Kouřimská, L., Pulkrábek, J., and Klouček, P. 2015. Antifungal effect of essential oil components against Aspergillus niger when loaded into silica mesoporous supports. *Journal of the Science of Food and Agriculture, 95*(14), 2824–2831. https://doi.org/10.1002/jsfa.7022

Bhargava, K., Conti, D. S., da Rocha, S. R. P., and Zhang, Y. 2015. Application of an oregano oil nanoemulsion to the control of foodborne bacteria on fresh lettuce. *Food Microbiology, 47*, 69–73. https://doi.org/10.1016/j.fm.2014.11.007

Carrero-Sánchez, J. C., Elías, A. L., Mancilla, R., Arrellín, G., Terrones, H., Laclette, J. P., and Terrones, M. 2006. Biocompatibility and toxicological studies of carbon nanotubes doped with nitrogen. *Nano Letters, 6*(8), 1609–1616. https://doi.org/10.1021/nl060548p

Castelo-Branco Melo, N. F., de MendonçaSoares, B. L., Marques Diniz, K., Ferreira Leal, C., Canto, D., Flores, M. A. P., Henrique da Costa Tavares-Filho, J., Galembeck, A., Montenegro Stamford, T. L., Montenegro Stamford-Arnaud, T., and Montenegro Stamford, T. C. 2018. Effects of fungal chitosan nanoparticles as eco-friendly edible coatings on the quality of postharvest table grapes. *Postharvest Biology and Technology, 139*, 56–66. https://doi.org/10.1016/j.postharvbio.2018.01.014

Chatterjee, A. K., Chakraborty, R., and Basu, T. 2014. Mechanism of antibacterial activity of copper nanoparticles. *Nanotechnology, 25*(13), 135101. https://doi.org/10.1088/0957-4484/25/13/135101

Chávez-Magdaleno, M. E., González-Estrada, R. R., Ramos-Guerrero, A., Plascencia-Jatomea, M., and Gutiérrez-Martínez, P. 2018. Effect of pepper tree (Schinus molle) essential oil-loaded chitosan bio-nanocomposites on postharvest control of colletotrichum gloeosporioides and quality evaluations in avocado (Persea americana) cv. Hass. *Food Science and Biotechnology, 27*(6), 1871–1875. https://doi.org/10.1007/s10068-018-0410-5

Chi, H., Song, S., Luo, M., Zhang, C., Li, W., Li, L., and Qin, Y. 2019. Effect of PLA nanocomposite films containing bergamot essential oil, TiO2 nanoparticles, and Ag nanoparticles on shelf life of mangoes. *Scientia Horticulturae, 249*, 192–198. https://doi.org/10.1016/j.scienta.2019.01.059

Chu, Y., Gao, C., Liu, X., Zhang, N., Xu, T., Feng, X., Yang, Y., Shen, X., and Tang, X. 2020. Improvement of storage quality of strawberries by pullulan coatings incorporated with cinnamon essential oil nanoemulsion. *LWT, 122*, 109054. https://doi.org/10.1016/j.lwt.2020.109054

Cioffi, N., Torsi, L., Ditaranto, N., Tantillo, G., Ghibelli, L., Sabbatini, L., Bleve-Zacheo, T., D'Alessio, M., Zambonin, P. G., and Traversa, E. 2005. Copper nanoparticle/polymer composites with antifungal and bacteriostatic properties. *Chemistry of Materials, 17*(21), 5255–5262. https://doi.org/10.1021/cm0505244

Corato, U. D. 2020. Improving the shelf-life and quality of fresh and minimally-processed fruits and vegetables for a modern food industry: A comprehensive critical review from the traditional technologies into the most promising advancements. *Critical Reviews in Food Science and Nutrition, 60*(6), 940–975. https://doi.org/10.1080/10408398.2018.1553025

Dakal, T. C., Kumar, A., Majumdar, R. S., and Yadav, V. 2016. Mechanistic basis of antimicrobial actions of silver nanoparticles. *Frontiers in Microbiology, 7*. https://doi.org/10.3389/fmicb.2016.01831

Danish, M., and Hussain, T. 2019. Nanobiofertilizers in crop production. In D. G. Panpatte & Y. K. Jhala (Eds.), *Nanotechnology for Agriculture: Crop Production & Protection* (pp. 107–118). Springer. https://doi.org/10.1007/978-981-32-9374-8_6

Deng, J., Chen, Q. J., Peng, Z. Y., Wang, J. H., Li, W., Ding, Q. M., Lin, Q. L., Liu, D. M., Wang, S. S., Shi, Y., Liu, Y., and Feng, S. S. 2019. Nano-silver-containing polyvinyl alcohol composite film for grape fresh-keeping. *Materials Express, 9*(9), 985–992. https://doi.org/10.1166/mex.2019.1592

Dhital, R., Mora, N. B., Watson, D. G., Kohli, P., and Choudhary, R. 2018. Efficacy of limonene nano coatings on post-harvest shelf life of strawberries. *LWT, 97*, 124–134. https://doi.org/10.1016/j.lwt.2018.06.038

Donglu, F., Wenjian, Y., Kimatu, B. M., Mariga, A. M., Liyan, Z., Xinxin, A., and Qiuhui, H. 2016. Effect of nanocomposite-based packaging on storage stability of mushrooms (Flammulina velutipes). *Innovative Food Science & Emerging Technologies, 33*, 489–497. https://doi.org/10.1016/j.ifset.2015.11.016

Emamifar, A., and Bavaisi, S. 2020. Nanocomposite coating based on sodium alginate and nano-ZnO for extending the storage life of fresh strawberries (Fragaria × ananassa Duch.). *Journal of Food Measurement and Characterization, 14*(2), 1012–1024. https://doi.org/10.1007/s11694-019-00350-x

Eshghi, S., Hashemi, M., Mohammadi, A., Badii, F., Mohammadhoseini, Z., and Ahmadi, K. 2014. Effect of nanochitosan-based coating with and without copper loaded on physicochemical and bioactive components of fresh strawberry fruit (Fragaria x ananassa Duchesne) during storage. *Food and Bioprocess Technology, 7*(8), 2397–2409. https://doi.org/10.1007/s11947-014-1281-2

Farhoodi, M. 2016. Nanocomposite materials for food packaging applications: Characterization and safety evaluation. *Food Engineering Review, 1*(8), 35–51. https://doi.org/10.1007/s12393-015-9114-2

Flores-López, M. L., Cerqueira, M. A., de Rodríguez, D. J., and Vicente, A. A. 2016. Perspectives on utilization of edible coatings and nano-laminate coatings for extension of postharvest storage of fruits and vegetables. *Food Engineering Reviews, 8*(3), 292–305. https://doi.org/10.1007/s12393-015-9135-x

Francisco, C. B., Pellá, M. G., Silva, O. A., Raimundo, K. F., Caetano, J., Linde, G. A., Colauto, N. B., and Dragunski, D. C. 2020. Shelf-life of guavas coated with biodegradable starch and cellulose-based films. *International Journal of Biological Macromolecules, 152*, 272–279. https://doi.org/10.1016/j.ijbiomac.2020.02.249

González-Estrada, R. R., Blancas-Benitez, F. J., Moreno-Hernández, C. L., Coronado-Partida, L., Ledezma-Delgadillo, A., and Gutiérrez-Martínez, P. 2019. Nanotechnology: A promising alternative for the control of postharvest pathogens in fruits. In D. G. Panpatte & Y. K. Jhala (Eds.), *Nanotechnology for Agriculture: Crop Production & Protection* (pp. 323–337). Springer. https://doi.org/10.1007/978-981-32-9374-8_15

González-Reza, R. M., García-Betanzos, C. I., Sánchez-Valdes, L. I., Quintanar-Guerrero, D., Cornejo-Villegas, M. A., and Zambrano-Zaragoza, M. L. 2018. The functionalization of nanostructures and their potential applications in edible coatings. *Coatings, 8*(5), 160. https://doi.org/10.3390/coatings8050160

Gorrasi, G., and Bugatti, V. 2016. Edible bio-nano-hybrid coatings for food protection based on pectins and LDH-salicylate: Preparation and analysis of physical properties. *LWT—Food Science and Technology, 69*, 139–145. https://doi.org/10.1016/j.lwt.2016.01.038

Han, J.-W., Ruiz-Garcia, L., Qian, J.-P., and Yang, X.-T. (2018). Food packaging: A comprehensive review and future trends. *Comprehensive Reviews in Food Science and Food Safety*, *17*(4), 860–877. https://doi.org/10.1111/1541-4337.12343

He, X., Deng, H., and Hwang, H. 2019. The current application of nanotechnology in food and agriculture. *Journal of Food and Drug Analysis*, *27*(1), 1–21. https://doi.org/10.1016/j.jfda.2018.12.002

Holsbach, F. M. S., Pizato, S., Fonteles, N. T., Souza, P. D. de, Pinedo, R. A., and Cortez-Vega, W. R. 2019. Evaluation of shelf life of Formosa papaya (Carica papaya L.) minimally processed using coating of cassava starch and essential clove oil. *Journal of Bioenergy and Food Science*, *6*(4), 78–96. https://doi.org/10.18067/jbfs.v6i4.269

Hong, J., Wang, L., Sun, Y., Zhao, L., Niu, G., Tan, W., Rico, C. M., Peralta-Videa, J. R., and Gardea-Torresdey, J. L. 2016. Foliar applied nanoscale and microscale CeO2 and CuO alter cucumber (Cucumis sativus) fruit quality. *Science of the Total Environment*, *563–564*, 904–911. https://doi.org/10.1016/j.scitotenv.2015.08.029

Hosseini, H., and Jafari, S. M. (2020). Introducing nano/microencapsulated bioactive ingredients for extending the shelf-life of food products. *Advances in Colloid and Interface Science*, *282*, 102210. https://doi.org/10.1016/j.cis.2020.102210

Huang, Y., Mei, L., Chen, X., and Wang, Q. 2018. Recent developments in food packaging based on nanomaterials. *Nanomaterials*, *8*(10), 830. https://doi.org/10.3390/nano8100830

Indumathi, M. P., Saral Sarojini, K., and Rajarajeswari, G. R. 2019. Antimicrobial and biodegradable chitosan/cellulose acetate phthalate/ZnO nano composite films with optimal oxygen permeability and hydrophobicity for extending the shelf life of black grape fruits. *International Journal of Biological Macromolecules*, *132*, 1112–1120. https://doi.org/10.1016/j.ijbiomac.2019.03.171

Issa, A., Ibrahim, S. A., and Tahergorabi, R. 2017. Impact of sweet potato starch-based nanocomposite films activated with thyme essential oil on the shelf-life of baby spinach leaves. *Foods*, *6*(6), 43. https://doi.org/10.3390/foods6060043

Kadam, S., and Shubhangi, B. 2020. Applications of nanotechnology in fruits and vegetables. *Food and Agriculture Spectrum Journal*, *1*(4), Article 4. https://doi.org/10.5281/zenodo.4152975

Kaur, P., Thakur, R., and Choudhary, A. 2012. An in vitro study of the antifungal activity of silver/chitosan nanoformulations against important seed borne pathogens. *International Journal of Scientific & Technology Research*, *1*(6), 83–86.

Kazemi, M. M., Hashemi-Moghaddam, H., Nafchi, A. M., and Ajodnifar, H. 2020. Application of modified packaging and nano ZnO for extending the shelf life of fresh pistachio. *Journal of Food Process Engineering*, *43*(12), e13548. https://doi.org/10.1111/jfpe.13548

Kim, D.-Y., Kadam, A., Shinde, S., Saratale, R. G., Patra, J., and Ghodake, G. 2018. Recent developments in nanotechnology transforming the agricultural sector: A transition replete with opportunities. *Journal of the Science of Food and Agriculture*, *98*(3), 849–864. https://doi.org/10.1002/jsfa.8749

Kim, I., Viswanathan, K., Kasi, G., Thanakkasaranee, S., Sadeghi, K., and Seo, J. 2020. ZnO nanostructures in active antibacterial food packaging: Preparation methods, antimicrobial mechanisms, safety issues, future prospects, and challenges. *Food Reviews International*. https://doi.org/10.1080/87559129.2020.1737709

Kou, X., He, Y., Li, Y., Chen, X., Feng, Y., and Xue, Z. 2019. Effect of abscisic acid (ABA) and chitosan/nano-silica/sodium alginate composite film on the color development and quality of postharvest Chinese winter jujube (Zizyphus jujuba Mill. Cv. Dongzao). *Food Chemistry*, *270*, 385–394. https://doi.org/10.1016/j.foodchem.2018.06.151

Koushesh-Saba, M., and Amini, R. 2017. Nano-ZnO/carboxymethyl cellulose-based active coating impact on ready-to-use pomegranate during cold storage. *Food Chemistry, 232*, 721–726. https://doi.org/10.1016/j.foodchem.2017.04.076

Kubacka, A., Diez, M. S., Rojo, D., Bargiela, R., Ciordia, S., Zapico, I., Albar, J. P., Barbas, C., Martins dos Santos, V. A. P., Fernández-García, M., and Ferrer, M. 2014. Understanding the antimicrobial mechanism of TiO 2 -based nanocomposite films in a pathogenic bacterium. *Scientific Reports, 4*(1), 4134. https://doi.org/10.1038/srep04134

Kumar, S., Boro, J. C., Ray, D., Mukherjee, A., and Dutta, J. 2019. Bionanocomposite films of agar incorporated with ZnO nanoparticles as an active packaging material for shelf life extension of green grape. *Heliyon, 5*(6), e01867. https://doi.org/10.1016/j.heliyon.2019.e01867

Lan, W., Zhang, R., Ahmed, S., Qin, W., and Liu, Y. 2019. Effects of various antimicrobial polyvinyl alcohol/tea polyphenol composite films on the shelf life of packaged strawberries. *LWT, 113*, 108297. https://doi.org/10.1016/j.lwt.2019.108297

Landry, K. S., Komaiko, J., Wong, D. E., Xu, T., McClements, D. J., and McLandsborough, L. 2016. Inactivation of salmonella on sprouting seeds using a spontaneous carvacrol nanoemulsion acidified with organic acids. *Journal of Food Protection, 79*(7), 1115–1126. https://doi.org/10.4315/0362-028X.JFP-15-397

Li, D., Li, L., Luo, Z., Lu, H., and Yue, Y. 2017a. Effect of nano-ZnO-packaging on chilling tolerance and pectin metabolism of peaches during cold storage. *Scientia Horticulturae, 225*, 128–133. https://doi.org/10.1016/j.scienta.2017.07.003

Li, D., Ye, Q., Jiang, L., and Luo, Z. 2017b. Effects of nano-TiO2-LDPE packaging on postharvest quality and antioxidant capacity of strawberry (Fragaria ananassa Duch.) stored at refrigeration temperature. *Journal of the Science of Food and Agriculture, 97*(4), 1116–1123. https://doi.org/10.1002/jsfa.7837

Li, J., Chang, J. W., Saenger, M., and Deering, A. 2017c. Thymol nanoemulsions formed via spontaneous emulsification: Physical and antimicrobial properties. *Food Chemistry, 232*, 191–197. https://doi.org/10.1016/j.foodchem.2017.03.147

Li, W., Li, L., Cao, Y., Lan, T., Chen, H., and Qin, Y. 2017d. Effects of PLA film incorporated with ZnO nanoparticle on the quality attributes of fresh-cut apple. *Nanomaterials, 7*(8), 207. https://doi.org/10.3390/nano7080207

Liu, J., Sui, Y., Wisniewski, M., Xie, Z., Liu, Y., You, Y., Zhang, X., Sun, Z., Li, W., Li, Y., and Wang, Q. 2018. The impact of the postharvest environment on the viability and virulence of decay fungi. *Critical Reviews in Food Science and Nutrition, 58*(10), 1681–1687. https://doi.org/10.1080/10408398.2017.1279122

Liu, W., Zhang, M., and Bhandari, B. 2020. Nanotechnology—A shelf life extension strategy for fruits and vegetables. *Critical Reviews in Food Science and Nutrition, 60*(10), 1706–1721. https://doi.org/10.1080/10408398.2019.1589415

López-Vargas, E. R., Ortega-Ortíz, H., Cadenas-Pliego, G., De Alba Romenus, K., Cabrera de la Fuente, M., Benavides-Mendoza, A., and Juárez-Maldonado, A. 2018. Foliar application of copper nanoparticles increases the fruit quality and the content of bioactive compounds in tomatoes. *Applied Sciences, 8*(7), 1020. https://doi.org/10.3390/app8071020

Loyola, N., and Arriola, M. 2017. Increasing the shelf life of post-harvest table grapes (Vitis vinífera cv. Thompson Seedless) using different packaging material with copper nanoparticles to change the atmosphere. *International Journal of Agriculture and Natural Resources, 44*(1), 54–63. https://doi.org/10.7764/rcia.v44i1.1640

Lufu, R., Ambaw, A., and Opara, U. L. 2020. Water loss of fresh fruit: Influencing preharvest, harvest and postharvest factors. *Scientia Horticulturae, 272*, 109519. https://doi.org/10.1016/j.scienta.2020.109519

Lustriane, C., Dwivany, F. M., Suendo, V., and Reza, M. 2018. Effect of chitosan and chitosan-nanoparticles on post harvest quality of banana fruits. *Journal of Plant Biotechnology*, *45*(1), 36–44. https://doi.org/10.5010/JPB.2018.45.1.036

Mohammadi, A., Hashemi, M., and Hosseini, S. M. 2015a. Nanoencapsulation of zataria multiflora essential oil preparation and characterization with enhanced antifungal activity for controlling Botrytis cinerea, the causal agent of gray mould disease. *Innovative Food Science & Emerging Technologies*, *28*, 73–80. https://doi.org/10.1016/j.ifset.2014.12.011

Mohammadi, A., Hashemi, M., and Hosseini, S. M. 2015b. Chitosan nanoparticles loaded with cinnamomum zeylanicum essential oil enhance the shelf life of cucumber during cold storage. *Postharvest Biology and Technology*, *110*, 203–213. https://doi.org/10.1016/j.postharvbio.2015.08.019

Naing, A. H., and Kim, C. K. 2020. Application of nano-silver particles to control the postharvest biology of cut flowers: A review. *Scientia Horticulturae*, *270*, 109463. https://doi.org/10.1016/j.scienta.2020.109463

Nasseri, M., Golmohammadzadeh, S., Arouiee, H., Jaafari, M. R., and Neamati, H. 2016. Antifungal activity of Zataria multiflora essential oil-loaded solid lipid nanoparticles in-vitro condition. *Iranian Journal of Basic Medical Sciences*, *19*(11), 1231–1237.

Ncama, K., Magwaza, L. S., Mditshwa, A., and Tesfay, S. Z. 2018. Plant-based edible coatings for managing postharvest quality of fresh horticultural produce: A review. *Food Packaging and Shelf Life*, *16*, 157–167. https://doi.org/10.1016/j.fpsl.2018.03.011

Ndukwu, M. C., Ikechukwu-Edeh, C. E., Nwakuba, N. R., Okosa, I., Horsefall, I. T., and Orji, F. N. 2020. Nanomaterials application in greenhouse structures, crop processing machinery, packaging materials and agro-biomass conversion. *Materials Science for Energy Technologies*, *3*, 690–699. https://doi.org/10.1016/j.mset.2020.07.006

Nguyen, H. V. H., and Nguyen, D. H. H. 2020. Effects of nano-chitosan and chitosan coating on the postharvest quality, polyphenol oxidase activity and malondialdehyde content of strawberry (Fragaria x ananassa Duch.). *Journal of Horticulture and Postharvest Research*, *3*(1), 11–24. https://doi.org/10.22077/jhpr.2019.2698.1082

Nguyen, V. T. B., Nguyen, D. H. H., and Nguyen, H. V. H. 2020. Combination effects of calcium chloride and nano-chitosan on the postharvest quality of strawberry (Fragaria x ananassa Duch.). *Postharvest Biology and Technology*, *162*, 111103. https://doi.org/10.1016/j.postharvbio.2019.111103

Nile, S. H., Baskar, V., Selvaraj, D., Nile, A., Xiao, J., and Kai, G. 2020. Nanotechnologies in food science: Applications, recent trends, and future perspectives. *Nano-Micro Letters*, *12*(1), 45. https://doi.org/10.1007/s40820-020-0383-9

Panpatte, D. G., and Jhala, Y. K. 2019. *Nanotechnology for Agriculture: Crop Production & Protection*. Springer. https://doi.org/10.1007/978-981-32-9374-8

Paramo, L. A., Feregrino-Pérez, A. A., Guevara, R., Mendoza, S., and Esquivel, K. 2020. Nanoparticles in agroindustry: Applications, toxicity, challenges, and trends. *Nanomaterials*, *10*(9), 1654. https://doi.org/10.3390/nano10091654

Park, J. H., Choi, S., Moon, H. C., Seo, H., Kim, J. Y., Hong, S.-P., Lee, B. S., Kang, E., Lee, J., Ryu, D. H., and Choi, I. S. 2017. Antimicrobial spray nanocoating of supramolecular Fe(III)-tannic acid metal-organic coordination complex: Applications to shoe insoles and fruits. *Scientific Reports*, *7*(1), 6980. https://doi.org/10.1038/s41598-017-07257-x

Patel, A., Patra, F., Shah, N., and Khedkar, C. 2018. Chapter 1 — Application of nanotechnology in the food industry: Present status and future prospects. In A. M. Grumezescu & A. M. Holban (Eds.), *Impact of Nanoscience in the Food Industry* (pp. 1–27). Academic Press. https://doi.org/10.1016/B978-0-12-811441-4.00001-7

Patra, P., Choudhury, S. R., Mandal, S., Basu, A., Goswami, A., Gogoi, R., Srivastava, C., Kumar, R., and Gopal, M. 2013. Effect sulfur and ZnO nanoparticles on stress physiology and plant (Vignaradiata) nutrition. In P. K. Giri, D. K. Goswami, & A. Perumal (Eds.), *Advanced Nanomaterials and Nanotechnology* (pp. 301–309). Springer. https://doi.org/10.1007/978-3-642-34216-5_31

Pillai, S., and Ray, S. S. 2012. Chitosan-based nanocomposites. *Natural Polymers, 2*(17), 33–68. https://researchspace.csir.co.za/dspace/handle/10204/7557

Pilon, L., Spricigo, P. C., Miranda, M., Moura, M. R. de, Assis, O. B. G., Mattoso, L. H. C., and Ferreira, M. D. 2015. Chitosan nanoparticle coatings reduce microbial growth on fresh-cut apples while not affecting quality attributes. *International Journal of Food Science & Technology, 50*(2), 440–448. https://doi.org/10.1111/ijfs.12616

Pina-Pérez, M. C., González, A., Moreno, Y., and Ferrús, M. A. 2018. Helicobacter pylori growth pattern in reference media and extracts from selected minimally processed vegetables. *Food Control, 86*, 389–396. https://doi.org/10.1016/j.foodcont.2017.11.044

Pirozzi, A., Del Grosso, V., Ferrari, G., and Donsì, F. 2020. Edible coatings containing oregano essential oil nanoemulsion for improving postharvest quality and shelf life of tomatoes. *Foods, 9*(11), 1605. https://doi.org/10.3390/foods9111605

Possas, A., Benítez, F. J., Savran, D., Brotóns, N. J., Rodríguez, P. J., and Posada-Izquierdo, G. D. 2018. Quantitative tools and procedures for shelf life determination in minimally processed fruits and vegetables. In F. Pérez-Rodríguez, P. Skandamis, & V. Valdramidis (Eds.), *Quantitative Methods for Food Safety and Quality in the Vegetable Industry* (pp. 223–254). Springer International Publishing. https://doi.org/10.1007/978-3-319-68177-1_11

Prakash, A., Baskaran, R., Paramasivam, N., and Vadivel, V. 2018. Essential oil based nanoemulsions to improve the microbial quality of minimally processed fruits and vegetables: A review. *Food Research International, 111*, 509–523. https://doi.org/10.1016/j.foodres.2018.05.066

Putnik, P., Roohinejad, S., Greiner, R., Granato, D., Bekhit, A. E.-D. A., and Bursać Kovačević, D. 2017. Prediction and modeling of microbial growth in minimally processed fresh-cut apples packaged in a modified atmosphere: A review. *Food Control, 80*, 411–419. https://doi.org/10.1016/j.foodcont.2017.05.018

Rai, M., Ingle, A. P., Gupta, I., Pandit, R., Paralikar, P., Gade, A., Chaud, M. V., and dos Santos, C. A. 2019. Smart nanopackaging for the enhancement of food shelf life. *Environmental Chemistry Letters, 17*(1), 277–290. https://doi.org/10.1007/s10311-018-0794-8

Rashid, F., Ahmed, Z., Ameer, K., Amir, R. M., and Khattak, M. 2020. Optimization of polysaccharides-based nanoemulsion using response surface methodology and application to improve postharvest storage of apple (Malus domestica). *Journal of Food Measurement and Characterization, 14*(5), 2676–2688. https://doi.org/10.1007/s11694-020-00514-0

Robledo, N., López, L., Bunger, A., Tapia, C., and Abugoch, L. 2018. Effects of antimicrobial edible coating of thymol nanoemulsion/quinoa protein/chitosan on the safety, sensorial properties, and quality of refrigerated strawberries (Fragaria × ananassa) under commercial storage environment. *Food and Bioprocess Technology, 11*(8), 1566–1574. https://doi.org/10.1007/s11947-018-2124-3

Ruengvisesh, S., Loquercio, A., Castell-Perez, E., and Taylor, T. M. 2015. Inhibition of bacterial pathogens in medium and on spinach leaf surfaces using plant-derived antimicrobials loaded in surfactant micelles. *Journal of Food Science, 80*(11), M2522–M2529. https://doi.org/10.1111/1750-3841.13085

Ruffo-Roberto, S., Youssef, K., Hashim, A. F., and Ippolito, A. 2019. Nanomaterials as alternative control means against postharvest diseases in fruit crops. *Nanomaterials*, *9*(12), 1752. https://doi.org/10.3390/nano9121752

Saberi, B., and Golding, J. B. 2018. Postharvest application of biopolymer-based edible coatings to improve the quality of fresh horticultural produce. In T. J. Gutiérrez (Ed.), *Polymers for Food Applications* (pp. 211–250). Springer International Publishing. https://doi.org/10.1007/978-3-319-94625-2_9

Sadeghnejad, A., Aroujalian, A., Raisi, A., and Fazel, S. 2014. Antibacterial nano silver coating on the surface of polyethylene films using corona discharge. *Surface and Coatings Technology*, *245*, 1–8. https://doi.org/10.1016/j.surfcoat.2014.02.023

Saharan, V., Mehrotra, A., Khatik, R., Rawal, P., Sharma, S. S., and Pal, A. 2013. Synthesis of chitosan based nanoparticles and their in vitro evaluation against phytopathogenic fungi. *International Journal of Biological Macromolecules*, *62*, 677–683. https://doi.org/10.1016/j.ijbiomac.2013.10.012

Salama, D. M., Abd El-Aziz, M. E., Rizk, F. A., and Abd Elwahed, M. S. A. 2021. Applications of nanotechnology on vegetable crops. *Chemosphere*, *266*, 129026. https://doi.org/10.1016/j.chemosphere.2020.129026

Salvia-Trujillo, L., Rojas-Graü, M. A., Soliva-Fortuny, R., and Martín-Belloso, O. 2015. Use of antimicrobial nanoemulsions as edible coatings: Impact on safety and quality attributes of fresh-cut Fuji apples. *Postharvest Biology and Technology*, *105*, 8–16. https://doi.org/10.1016/j.postharvbio.2015.03.009

Sant'Anna, P. B., Franco, B. D. de M., and Maffei, D. F. 2020. Microbiological safety of ready-to-eat minimally processed vegetables in Brazil: An overview. *Journal of the Science of Food and Agriculture*, *100*(13), 4664–4670. https://doi.org/10.1002/jsfa.10438

Saravanakumar, K., Hu, X., Chelliah, R., Oh, D.-H., Kathiresan, K., and Wang, M.-H. 2020. Biogenic silver nanoparticles-polyvinylpyrrolidone based glycerosomes coating to expand the shelf life of fresh-cut bell pepper (Capsicum annuum L. var. Grossum (L.) Sendt). *Postharvest Biology and Technology*, *160*, 111039. https://doi.org/10.1016/j.postharvbio.2019.111039

Sarojini, S., Indumathi, M. P., and Rajarajeswari, G. R. 2019. Mahua oil-based polyurethane/chitosan/nano ZnO composite films for biodegradable food packaging applications. *International Journal of Biological Macromolecules*, *124*, 163–174. https://doi.org/10.1016/j.ijbiomac.2018.11.195

Sessa, M., Ferrari, G., and Donsi, F. 2015. Novel edible coating containing essential oil nanoemulsions to prolong the shelf life of vegetable products. *Chemical Engineering Transactions*, *43*, 55–60. https://doi.org/10.3303/CET1543010

Severino, R., Ferrari, G., Vu, K. D., Donsì, F., Salmieri, S., and Lacroix, M. 2015. Antimicrobial effects of modified chitosan based coating containing nanoemulsion of essential oils, modified atmosphere packaging and gamma irradiation against Escherichia coli O157:H7 and Salmonella Typhimurium on green beans. *Food Control*, *50*, 215–222. https://doi.org/10.1016/j.foodcont.2014.08.029

Sharma, S., Rana, V. S., Pawar, R., Lakra, J., and Racchapannavar, V. 2020. Nanofertilizers for sustainable fruit production: A review. *Environmental Chemistry Letters*. https://doi.org/10.1007/s10311-020-01125-3

Shi, C., Wu, Y., Fang, D., Pei, F., Mariga, A. M., Yang, W., and Hu, Q. 2018. Effect of nanocomposite packaging on postharvest senescence of Flammulina velutipes. *Food Chemistry*, *246*, 414–421. https://doi.org/10.1016/j.foodchem.2017.10.103

Shigematsu, E., Dorta, C., Rodrigues, F. J., Cedran, M. F., Giannoni, J. A., Oshiiwa, M., and Mauro, M. A. 2018. Edible coating with probiotic as a quality factor for minimally processed carrots. *Journal of Food Science and Technology*, *55*(9), 3712–3720. https://doi.org/10.1007/s13197-018-3301-0

Shukla, P., Chaurasia, P., Younis, K., Qadri, O. S., Faridi, S. A., and Srivastava, G. 2019. Nanotechnology in sustainable agriculture: Studies from seed priming to postharvest management. *Nanotechnology for Environmental Engineering*, *4*(1), 11. https://doi.org/10.1007/s41204-019-0058-2

Singh, D., and Sharma, R. R. 2018. Chapter 1—Postharvest diseases of fruits and vegetables and their management. In M. W. Siddiqui (Ed.), *Postharvest Disinfection of Fruits and Vegetables* (pp. 1–52). Academic Press. https://doi.org/10.1016/B978-0-12-812698-1.00001-7

Singh, K. 2020. Nanotechnology: Advanced technique in postharvest management of horticultural crops. *Biotica Research Today*, *2*(6), 456–458.

Singh, M., and Sahareen, T. 2017. Investigation of cellulosic packets impregnated with silver nanoparticles for enhancing shelf-life of vegetables. *LWT*, *86*, 116–122. https://doi.org/10.1016/j.lwt.2017.07.056

Sirelkhatim, A., Mahmud, S., Seeni, A., Kaus, N. H. M., Ann, L. C., Bakhori, S. K. M., Hasan, H., and Mohamad, D. 2015. Review on zinc oxide nanoparticles: Antibacterial activity and toxicity mechanism. *Nano-Micro Letters*, *7*(3), 219–242. https://doi.org/10.1007/s40820-015-0040-x

Taştan, Ö., Pataro, G., Donsì, F., Ferrari, G., and Baysal, T. 2017. Decontamination of fresh-cut cucumber slices by a combination of a modified chitosan coating containing carvacrol nanoemulsions and pulsed light. *International Journal of Food Microbiology*, *260*, 75–80. https://doi.org/10.1016/j.ijfoodmicro.2017.08.011

Tavakoli, H., Rastegar, H., Taherian, M., Samadi, M., and Rostami, H. 2017. The effect of nano-silver packaging in increasing the shelf life of nuts: An in vitro model. *Italian Journal of Food Safety*, *6*(4). https://doi.org/10.4081/ijfs.2017.6874

Tavakoli, S. A., Mirzaei, S., Rahimi, M., and Tavassolian, I. 2019. Assessment of peppermint, clove, cumin essential oils and silver nano particles on biochemical and shelf life of citrus limon (L.). *Indian Journal of Biochemistry and Biophysics (IJBB)*, *56*(4), 269–275.

Tian, F., Chen, W., Wu, C., Kou, X., Fan, G., Li, T., and Wu, Z. 2019. Preservation of ginkgo biloba seeds by coating with chitosan/nano-TiO2 and chitosan/nano-SiO2 films. *International Journal of Biological Macromolecules*, *126*, 917–925. https://doi.org/10.1016/j.ijbiomac.2018.12.177

Wang, L., Shao, S., Madebo, M. P., Hou, Y., Zheng, Y., and Jin, P. 2020. Effect of nano-SiO2 packing on postharvest quality and antioxidant capacity of loquat fruit under ambient temperature storage. *Food Chemistry*, *315*, 126295. https://doi.org/10.1016/j.foodchem.2020.126295

Wu, D., Zhang, M., Xu, B., and Guo, Z. 2021. Fresh-cut orange preservation based on nano-zinc oxide combined with pressurized argon treatment. *LWT*, *135*, 110036. https://doi.org/10.1016/j.lwt.2020.110036

Xin, Y., Chen, F., Lai, S., and Yang, H. 2017. Influence of chitosan-based coatings on the physicochemical properties and pectin nanostructure of Chinese cherry. *Postharvest Biology and Technology*, *133*, 64–71. https://doi.org/10.1016/j.postharvbio.2017.06.010

Xing, Y., Yang, H., Guo, X., Bi, X., Liu, X., Xu, Q., Wang, Q., Li, W., Li, X., Shui, Y., Chen, C., and Zheng, Y. 2020. Effect of chitosan/nano-TiO2 composite coatings on the postharvest quality and physicochemical characteristics of mango fruits. *Scientia Horticulturae*, *263*, 109135. https://doi.org/10.1016/j.scienta.2019.109135

Yahia, E. M., Celis, M. E. M., and Svendsen, M. 2017. The contribution of fruit and vegetable consumption to human health. In *Fruit and Vegetable Phytochemicals* (pp. 1–52). John Wiley & Sons, Ltd. https://doi.org/10.1002/9781119158042.ch1

Yilmaz, A., Bozkurt, F., Cicek, P. K., Dertli, E., Durak, M. Z., and Yilmaz, M. T. 2016. A novel antifungal surface-coating application to limit postharvest decay on coated apples: Molecular, thermal and morphological properties of electrospun zein—Nanofiber mats loaded with curcumin. *Innovative Food Science & Emerging Technologies, 37*, 74–83. https://doi.org/10.1016/j.ifset.2016.08.008

Youssef, K., de Oliveira, A. G., Tischer, C. A., Hussain, I., and Roberto, S. R. 2019. Synergistic effect of a novel chitosan/silica nanocomposites-based formulation against gray mold of table grapes and its possible mode of action. *International Journal of Biological Macromolecules, 141*, 247–258. https://doi.org/10.1016/j.ijbiomac.2019.08.249

Zambrano-Zaragoza, M. L., González-Reza, R. M., Quintanar-Guerrero, D., and Mendoza-Muñoz, N. 2020. Nano-films for food packaging. In U. Hebbar, S. Ranjan, N. Dasgupta, & R. Kumar Mishra (Eds.), *Nano-food Engineering: Volume One* (pp. 287–307). Springer International Publishing. https://doi.org/10.1007/978-3-030-44552-2_10

Zambrano-Zaragoza, M. L., González-Reza, R., Mendoza-Muñoz, N., Miranda-Linares, V., Bernal-Couoh, T. F., Mendoza-Elvira, S., and Quintanar-Guerrero, D. 2018. Nanosystems in edible coatings: A novel strategy for food preservation. *International Journal of Molecular Sciences, 19*(3), 705. https://doi.org/10.3390/ijms19030705

Zhang, C., Li, W., Zhu, B., Chen, H., Chi, H., Li, L., Qin, Y., and Xue, J. 2018. The quality evaluation of postharvest strawberries stored in nano-Ag packages at refrigeration temperature. *Polymers, 10*(8), 894. https://doi.org/10.3390/polym10080894

Zhang, J., Cheng, D., Wang, B., Khan, I., and Ni, Y. 2017. Ethylene control technologies in extending postharvest shelf life of climacteric fruit. *Journal of Agricultural and Food Chemistry, 65*(34), 7308–7319. https://doi.org/10.1021/acs.jafc.7b02616

Zhang, R., and Wang, X. 2017. Process optimization of nano-SiO2/potato starch coatings and its improving effect for storage of agaricus bisporus by electrospraying. *Transactions of the Chinese Society of Agricultural Engineering, 33*(23), 291–299.

Zhang, R., Wang, X., Li, L., Cheng, M., and Zhang, L. 2019. Optimization of konjac glucomannan/carrageenan/nano-SiO2 coatings for extending the shelf-life of Agaricus bisporus. *International Journal of Biological Macromolecules, 122*, 857–865. https://doi.org/10.1016/j.ijbiomac.2018.10.165

Zhao, L.-M., Shi, L.-E., Zhang, Z.-L., Chen, J.-M., Shi, D.-D., Yang, J., and Tang, Z.-X. 2011. Preparation and application of chitosan nanoparticles and nanofibers. *Brazilian Journal of Chemical Engineering, 28*(3), 353–362. https://doi.org/10.1590/S0104-66322011000300001

Chapter 20

Nano-Based Methods for the Improvement of Postharvest Technologies and Increased Shelf Life of Minimally Processed Food

Mari Carmen López Pérez, Fabián Pérez Labrada, Yolanda González García, and Antonio Juárez Maldonado

List of Abbreviations

ADP	Adenosine diphosphate
APX	Ascorbate peroxidase
ATP	Adenosine triphosphate
CAT	Catalase
GPX	Glutathione peroxidase
GSNO	S-nitrosoglutathione
iATP	Intracellular adenosine triphosphate
LOX	Lipoxygenase
MAPK	Mitogen-activated protein kinase
MAPKK	Mitogen-activated protein kinase kinase
MAPKKK	Mitogen-activated protein kinase kinase kinase
MDA	Malondialdehyde

DOI: 10.1201/9781003142287-20

NM	Nanomaterial
NP	Nanoparticle
PLD	Phospholipase D
PPO	Polyphenoloxidase
PVA	Polyvinyl alcohol
RH	Relative humidity
ROS	Reactive oxygen species
SOD	Superoxide dismutase
TSS	Total soluble solid

20.1 Introduction

Agricultural production is a primary and fundamental activity for humanity and its development, since it is the main source of food for both humans and livestock. However, food obtained from agricultural activity is perishable due to its organic nature, and therefore, the extent of its use after harvest is limited. This requires the development of strategies and methods that allow increasing the use time, known as postharvest life, in such a way that the food remains in good quality for longer. Postharvest life considers physical characteristics such as firmness and color, and biochemical characteristics, such as aroma, sugar content, and organic acids, among others, that have the fruits or other organs of the plants that serve as food (Tang et al., 2020). These characteristics can be affected by different factors that influence the development of the crop such as the properties of the soil, the climate, some stress condition, and the culture management of the crop (Casajús et al., 2021), as well as by those factors that properly have to do with the process and manipulation from the harvest as the storage and packaging conditions (Maringgal et al., 2020) and by the physiological characteristics of each plant species (Chen et al., 2020).

There are different methods that have been developed to improve postharvest quality and increase shelf life, among which are the cultural practices carried out on crops before harvest (Ijaz et al., 2020), growth regulators, edible coatings, precooling methods, heat treatments, and modified atmosphere (Munhuweyi et al., 2020; Riva et al., 2020; Wang and Zhang, 2020). However, even with these tools available, there are still various problems in the postharvest. Nanotechnology, which takes advantage of the new characteristics presented by the so-called NMs (materials with at least one dimension between 1–100 nm), can be an alternative to improve the quality and postharvest life of agricultural products. These NMs, due to their physical and chemical characteristics (size, shape, porosity, characteristics of charges, free energy of their surface) function as biostimulants due to the series of biochemical, physiological, and genetic responses that they induce in plants (Juárez-Maldonado et al., 2019, 2021). It induces the production of secondary metabolites that have multiple functions in plants and that can also improve the quality of agricultural products and increase postharvest life. These NMs have been shown to have the ability to induce ROS production, as they can "stress" plants and thus activate their antioxidant defense system, resulting in an overproduction of enzymatic-type antioxidant defense compounds such as APX, GPX, CAT, SOD, and nonenzymatic component (ascorbic acid, glutathione, phenols, flavonoids). The accumulation of these antioxidant compounds in the different organs of a plant that serve as food will have a greater ability to maintain quality, since these compounds avoid the harmful effects of ROS on cell membranes that result in the loss of electrolytes and firmness and avoid oxidation of other important compounds (Li et al., 2017; Song et al., 2016; Taheri et al., 2020). In addition, they also prevent or reduce the activity of enzymes responsible for the loss of quality, such as PPO, which is responsible for "browning" (Karimirad et al., 2019; Lo'ay and Ameer, 2019).

These NMs have a diversity of positive functions in crops when they are applied during their development as they change the metabolism, cause an increase in biocompounds that improve the nutraceutical quality of food vegetables, as well as increase tolerance to stress (Juárez-Maldonado et al., 2019). Moreover, they have the great advantage that they can be applied directly to the fruits or other organs of the plant, after harvest, either directly or in conjunction with other compounds as in the case of Cu NP-chitosan (Meena et al., 2020), beeswax solid lipid NPs with xanthan gum and propylene glycol (Zambrano-Zaragoza et al., 2020), or even in the manufacture of containers, packages and/or bags, from NMs that can be organic (nanocellulose fibrils) or inorganic (montmorillonite, zinc oxide, silver; Basumatary et al., 2022). Finally, the effect obtained is an increase in the shelf life of foods, antimicrobial activity, and even an improvement in antioxidant activity.

This chapter presents a description of the problems in the postharvest handling of food obtained from crops, as well as the impact of different biotic or abiotic factors and the losses that are generated by this problem. Likewise, it describes how nanotechnology can influence food production, from the development of the crop and at harvest, and therefore be an alternative to increase the quality and postharvest life of food crops.

20.2 Importance of Postharvest in Vegetable Foods

Fruits and vegetables are an important source of nutrients such as carbohydrates, acids, minerals, polyphenols, vitamins, amino acids, aromatic compounds, carotenoids, fibers, and phytosterols among other bioactive compounds and, therefore, play a crucial role in the prevention of malnutrition and in reducing the risk of noncommunicable diseases (Oyedele et al., 2020; Ruiz-Rodríguez et al., 2020). The world production of fruits and vegetables has increased as a result of the growing demand of the population and the increase in the level of quality of life and awareness about the consumption of these (Maringgal et al., 2020). However, due to the perishable nature of these products, serious quality deterioration and losses will inevitably occur due to a variety of reasons during the pre-harvest and postharvest stages. So about a third of the fresh fruit produced in developing countries does not reach consumers in optimal conditions (Chen et al., 2020). Among the various factors that contribute to postharvest losses are poor handling and storage, improper packaging (Maringgal et al., 2020), and intrinsic physiological senescence (Chen et al., 2020). Therefore, maintaining the quality of food crops during storage has always been a challenge for scientists.

The ripening and senescence of the fruit are inevitable and irreversible physiological processes, during which the biochemical characteristics and the physiology of these organs are altered and influence various quality parameters of the fruit such as firmness, color, aroma, sugar content, and organic acids, among others (Tang et al., 2020). The ripening or senescence of the fruit is coordinated by a complex network of endogenous signals, such as those mediated by plant hormones, calcium-dependent signals, and transcription factors, where ethylene and abscisic acid are the main regulating hormones of this process in the climacteric fruits (Leng et al., 2014). While in non-climacteric fruits, they are generally not necessary for the coordination and completion of the ripening process (Li et al., 2016). During the ripening process, the fruits develop intrinsic and extrinsic changes, where the excessive softening derived from the loss of water results in an increase in susceptibility to pathogens, and leads to significant economic losses and postharvest deterioration (Posé et al., 2019). It has been reported that postharvest losses of crops can range from 10–40% due to mechanical, microbial, and physiological damage (Bose et al., 2021; Mossa et al., 2021).

The loss of water is the result of perspiration driven by the difference in the partial pressure of water vapor between the interior of the fruit and the external environment and is one of the main causes of postharvest deterioration (Jansasithorn et al., 2014). In general, most fruits are not marketed with a water loss of 5–10% of their initial weight, as this leads to a serious reduction in quality and wilting. Excessive water loss, in addition to causing marketable weight loss, promotes browning, loss of firmness, texture, flavor, accelerated senescence, susceptibility to cold damage, and membrane disintegration (Lufu et al., 2020). These modifications arise mainly from the degradation of pectin, the loss of sugars, the depolymerization of xyloglucan, a decrease in proteins, an increase in ROS and MDA production, and an increase in lignin, among others (Li et al., 2020; Xu et al., 2016).

The loss of water occurs through perspiration that strictly depends on the characteristics of the epidermis and environmental conditions. The epidermis of the fruit acts as a critical barrier to resist the movement of water both in liquid form and the vapors of the fruit tissue toward the outer surface and allows the fruit to maintain a high water content (Wei et al., 2019). Mechanical damage such as cuticular cracking, pitting, and bruising can lead to serious damage to the epidermis and excessive loss of water in the fruit (Jansasithorn et al., 2014). Furthermore, they can accelerate some physiological processes such as ethylene production and cause risks of microbial contamination, since spoilage pathogens can easily enter through dead or injured tissues and contaminate the rest of the fruit (Hussein et al., 2020). Hollow fruits such as sweet pepper (*Capsicum annuum* L.) are highly susceptible to postharvest water loss, which causes a deterioration in taste and texture (O'Donoghue et al., 2020).

Diseases are another important factor of cosmetic, organoleptic, and nutritional fruit damage, inducing a reduction in the perception of multifunctional quality. These are of biotic origin, such as fruit rot, generally associated with microbial pathogens and pests (Sharma et al., 2020). However, these diseases can be potentiated by other factors of abiotic origin, among these caused by physiological disorders such as stress due to high and low temperatures, strong winds, hail, and heavy rainfall (Burdon et al., 2013; Moscetti et al., 2016; Ramírez-Gil et al., 2021; Rojas et al., 2021). Synthetic fungicides and bactericides are used to control diseases, but they have adverse effects on the health of humans and ecosystems, in addition to favoring the development of resistance among microorganisms (Mossa et al., 2021). Postharvest diseases cause the loss of between 10–30% of total crop yield during handling, transport, storage, and marketing (Mossa et al., 2021). In addition to the decrease in quality and economic loss, infected fruits can increase the risk of serious diseases in human consumers since some agents can produce toxic secondary metabolites as in the case of *Penicillium expansum*, which produce patulin and citrinin in apple fruits (Yu et al., 2020).

The alteration of the metabolism of sugars is a common response against the attack of pathogens and the development of diseases in the fruit (Xu et al., 2021). In addition to this, ATP synthesis and energy status in plant cells are reduced, leading to the inability to meet energy needs for the maintenance of proper physiological metabolism, resulting in membrane degradation, cellular decompartmentalization, and electrolyte leakage (Chen et al., 2014). An imbalance of the ROS metabolism is also generated that causes irreversible damage to the organs and consequently the deterioration of the fruit (Tang et al., 2021). This process is associated with the enzymatic oxidation of phenolic compounds by the polyphenol oxidase enzyme that causes browning (Jiang et al., 2004). Among the main fungal pathogens that affect postharvest quality are *Botryosphaeria berengeriana*, *Penicillium italicum*, *Colletotrichum fructicola*, *Botrytis cinerea*, *Penicillum ulaiense*, and *Penicillum digitatum* (Tang et al., 2021). While among the bacterial ones *Pseudomonas spp*, *Listeria monocytogenes*, *Salmonella typhi*, *Staphylococcus aureus*, and *Escherichia coli* (Yi et al., 2020).

The postharvest life of horticultural products can also be affected by various pre-harvest factors, such as soil condition, climatic conditions, stress, and general crop management, as well as the time of day when the crops are harvested. Products could also influence postharvest life, since the diurnal cycle strongly influences various metabolic and physiological processes in plants (Casajús et al., 2021).

20.3 Postharvest Physiology

The vast majority of fruit and vegetable crops are prone to losing their nutraceutical (and economic) quality after harvest, to a greater or lesser extent. This is due to the fact that they have an active metabolism during harvest as well as a high metabolic activity (to a greater extent in climacteric fruits) once harvested (Gouda et al., 2020). In the case of maturation of climacteric fruits (e.g., *Solanum lycopersicum* L.), there is a high production of CO_2 and C_2H_4 (climacteric increase) while in non-climacteric fruits (e.g., *Fragaria x ananassa* Duch.) there are no significant changes in the production of these compounds (Gundewadi et al., 2018). The same authors point out that the respiration rate generates the decomposition of organic molecules causing the "softening" of the tissue, in climacteric fruits the ripening is controlled by C_2H_4 while in non-climacteric fruits it is generally hormones that are participants. Similarly, factors such as the solubilization of peptide substances, the transformation of starch to sugar, decrease in the content of organic acids, changes in the concentration of pigments, the production of volatile compounds, the content of phenolic compounds, alterations in the content of ROS, and modifications of the antioxidant system are related to the postharvest quality of horticultural crops (Gundewadi et al., 2018). It should be noted that the C_2H_4 produced during the growth of the plant stimulates the ripening of the fruit as well as the promotion of softening and the synthesis of pigments and loss of chlorophyll that occurs in greater proportion in climacteric fruits (Saltveit, 1999). Such conditions encourage producers to look for alternatives to minimize the reduction in the postharvest quality of their crops.

The fruits of the various fruit and vegetable crops tend to present an alteration in the coloration being a very important physiological disorder that causes economic losses. This phenomenon occurs in a greater proportion at the time of handling, harvesting, storing and/or processing fruits when the rupture of the tissue vacuoles releasing phenolic compounds into the cytoplasm occurs where the enzyme polyphenol oxidase oxidizes these compounds (a phenomenon known as "enzymatic browning"). This enzyme is characterized by containing copper which, under the presence of oxygen, hydroxylates monophenols to o-diphenols, and subsequently, these are oxidized to o-quinones (monophenolase and diphenolase activity, respectively), and finally a non-enzymatic polymerization occurs, causing brown, red and black pigments (Karimirad et al., 2018; Lo'ay and Ameer, 2019; Queiroz et al., 2008).

Furthermore, considering that the cellular respiration pathways (the Embden–Meyerhof–Parnas pathway, the tricarboxylic acid cycle, the mitochondrial electron transport, the cytochrome pathway, the pentose phosphate pathway, the glyoxylate pathway) have particular biological functions (oxidation of glucose to pyruvate, energy supply in fruits and vegetables postharvest during storage) in physiological alterations during the development of the plant fruit or at harvest, it can lead to damage in the shelf life (Aghdam et al., 2018; Lin et al., 2017; Mittler et al., 2011). A high respiration rate in fruit induces metabolic activity that accelerates the ripening, propitiating that the starch is degraded to glucose, therefore maintaining the content of total and reducing sugars could significantly control the sugars during storage (Meena et al., 2020). In this sense, the application of postharvest treatments would seek to maintain the quality of the crop; at a physiological level through the adequate supply of iATP, as well as a lower enzymatic activity PLD and LOX, maintaining the fluidity and integrity of the membrane (Aghdam et al., 2018). In accordance with Lin et al. (2017), the disruption of energy metabolism in fruit cell tissues (lower content of ATP and ADP), as well as an inhibition of H^+-ATPase, Ca^{2+}-ATPase and Mg^{2+}-ATPase activity during storage derived from the application of H_2O_2, could alter ionic homeostasis and integrity of the mitochondria by decreasing the compartmentalized energy, such events would cause the pericarp to browning. By contrast, harvest induces an analogous response to stress, causing morphological, physiological, biochemical, and molecular alterations. Here, protein kinases—MAPK, MAPKK, and MAPKKK—are of great importance in promoting responses to stress due to their vital participation in the sequential transduction of signals (channeling, integration, amplification); likewise, ROS play an important role as messengers and in the activation of MAPK (Gouda et al., 2020). Therefore, ROS production could stimulate the rapid development of the fruit depending on the stage it is in (Gouda et al., 2020). Gao et al. (2016) mentioned that the activation of the antioxidant system (enzymatic activity, SOD, CAT, peroxidase, APX) is essential to maintain the integrity of the membrane, acting as a mechanism to delay senescence and increase the shelf life of peaches (*Prunus persica* L.).

20.4 Nanotechnology in Agriculture Production

In the context of modern agriculture and its globalizing nature as well as the growing demand for food, nanotechnology has become a tool that may be viable to apply. Under this trend, the various aspects of nanotechnology (products with a nanometric scale between 1–100 nm) provide a new tool in crop production, within this are various technologies such as nanoformulations, nanosensors/nanobiosensors, nanodevices, NPs, nanofibers, nanoemulsions, and nanocapsules (Adisa et al., 2019; Bajpai et al., 2018; Mali et al., 2020; Sekhon, 2014; Wang et al., 2020a; Xing et al., 2020; Zulfiqar et al., 2019). About this, Abdel-Aziz et al. (2019) have reported the seed primer or the foliar application of nanofertilizers (nanochitosan or carbon nanotubes loaded with NPK) in French beans (*Phaseolus vulgaris* L.) to improve the not only the productivity of the crop but also the antioxidant system of the plant. Furthermore, Zhao et al. (2020b) have reported that the use of NMs (TiO_2, Ag, CuO, and graphene) generate a low environmental risk, which is why they are a good alternative for use in agriculture.

When talking about crop production and management, it is imperative to mention the losses that poor management can cause, especially fruit and vegetable crops (Mustafa et al., 2014), so it is necessary to establish new alternatives in postharvest quality such as the use of NMs (Zambrano-Zaragoza et al., 2013).

20.5 Postharvest Nanotechnology

Once harvested, the fruit presents a series of responses (high respiration rate, loss of water) that reduce its quality, so it is necessary to provide the best storage conditions. Various treatments (physical, chemical, gaseous) have been studied, and the use of edible coatings (biopolymers) have been tried. However, each of them carries a series of problems. Given this, nanotechnology is presented as a promising alternative in this area due to its particular characteristics, that is, shape, size, dimension, origin, and composition, that provide solubility and stability, derived from the stability in its zeta potential and its high density of surface charge that allow them to have nonspecific interactions with the surface charges of plant cell walls and membranes (Juárez-Maldonado et al., 2019; Manzoor et al., 2020; Wang et al., 2014). It has been documented that the responses observed in plants derive from the interaction between NMs with the different cell organelles, improving the physiological processes of plants, which allows the improvement postharvest life and maintain the nutritional quality of the crops (Mali et al., 2020). As an example, we can mention the use of NPs, whose photocatalytic and physicochemical activity stimulate the generation of ROS and activate the defense system of plants (Adisa et al., 2019; Yang et al., 2017). This could improve the postharvest life of the crops, although the behavior of the different crops varies depending on the type of NM and fruit (Ijaz et al., 2020).

The application of NMs (nanocoatings, NPs, nanonet, nanoencapsulates, nanopackings, etc.) can be carried out in pre-harvest or postharvest according to the characteristics of the crop or to the objectives that are sought (Adisa et al., 2019; Zulfiqar et al., 2019; Figure 20.1). In pre-harvest the applications can be on the seed, root, foliage, and/or fruits, with these applications the scientific evidence shows that there are a series of responses such as the increase in enzymatic and nonenzymatic antioxidant activity, increased ROS, and alterations at the morphological level, the preceding results in leaf tissue and/or fruit having a great capacity to preserve their useful life (Abdel-Aziz et al., 2019; Zulfiqar et al., 2019).

However, when the application of these nanotechnologies is carried out directly to the fruit, once harvested, the enzymes related to the degradation of the cell wall, a decrease in the browning of the fruit, as well as an increase in the content of ascorbic acid are observed (Karimirad et al., 2018). This enables the fruit to have a longer shelf life by reducing the production of ethylene (higher proportion of climacteric fruits; Nguyen et al., 2020; Sorrentino et al., 2007). The ROS load, antioxidants, and morphological modifications will be the reservoir to preserve the quality during the storage time of the fruits. The advantage will be in greater or lesser proportion depending on the NM used and the storage conditions (Liu et al., 2020).

20.5.1 Nanotechnology in Pre-Harvest to Improve Postharvest Quality

In accordance with Ijaz et al. (2020), the behavior of the crop during storage is closely related to factors that occurred during development in the field (handling, genetic material used, nutrition, temperature, RH, cultural practices, water potential, pests and diseases, stress, etc.). Faced with this situation, it has been sought to implement the application of nanotechnology during the growth and development of plants to provide physiological machinery that allows their postharvest quality to improve. In the case of some nanotechnological applications (Table 20.1), it is known that they can contribute to the improvement of the shelf life of crops, increasing the antioxidant system of plants and fruit (Achari and Kowshik, 2018).

Table 20.1 Preharvest Application of Nanotechnology in Crops

NM	*Crops*	*Storage conditions postharvest*	*Effect*	*Reference*
Cu NPs in Hydrogels of Chitosan-PVA	Jalapeño pepper (*Capsicum annuum* L. cv. Grande)	Room temperature (20 ± 1°C) and cold conditions (10 ± 1°C and 80% RH) for 30 days	Increase in antioxidant capacity, total phenols, and flavonoids.	Pinedo-Guerrero et al., 2017
Cu NPs	Tomato (*Solanum lycopersicum* L. cv. Huno F1)	Room temperature 24 ± 1°C. Refrigeration 4 ± 1°C and RH of 80%. 23 days.	Increase in the accumulation of bioactive compounds (total phenols, β-carotene, and ascorbic acid).	Hernández-Fuentes et al., 2017
Nano-calcium	Apple (*Malus domestica* L. cv. Red Delicious)	0°C, 90% RH for 4 months	Increased firmness, titratable acidity, total phenolic content, total antioxidant activity, and fiber content. Less reduction in weight loss, TSSs, and internal browning. Lower activity of polygalacturonase, pectin methyl esterase, and β-galactosidase enzymes.	Ranjbar et al., 2018
Nano-Chelate Zn	Apple (*Malus domestica* L. cv. Delbard Estival and cv. Kohanz)	4°C, 95% RH for 3 weeks	Decreased browning index and total color changes. Starch index delay. Greater firmness, higher content of ascorbic acid, total antioxidant activity, and SOD. Lower polyphenol oxidase activity.	Rasouli and Saba, 2018

(Continued)

Table 20.1 (Continued)

NM	*Crops*	*Storage conditions postharvest*	*Effect*	*Reference*
NPs composed of chitosan	Bell pepper (*Capsicum annuum* L. cv. California Wonder)	Stored in the dark at 11 ± 1°C and 70% RH for 21 days.	Only 30% reduced weight loss and 15% the change of color. Phenolic content, carotenoids, and capacity reduction increased. Microbiological activity was reduced 85%.	González-Saucedo et al., 2019
ZnO NPs photoactivated	Strawberries (*Fragaria x ananassa* Duch. cv. Darselect)	0 ± 5–7°C, 70–80 % RH for 8 days in a climate chamber.	Reduction of the incidence of *Botrytis cinérea*.	Luksiene et al., 2020

The application of NPs during the development of the crop or before the fruit harvest has been studied to know the impact that it could have on the postharvest life of the fruits. Hernández-Fuentes et al. (2017) reported alterations in the content of bioactive compounds in tomato fruit (*Solanum lycopersicum* L.) produced under stress conditions with NaCl and with applications of Cu NPs. The increase in these compounds could be derived from the exacerbation of ROS in the cellular tissues caused by NPs. González-Saucedo et al. (2019) reported an increase in the content of carotenoids in bell pepper (*Capsicum annuum* L.) treated with NPs composed of chitosan. Here, the induction of secondary metabolites might have occurred since these coatings can be a stress factor increasing, in turn, the antioxidant compounds and improving the storage quality of the fruit.

Ranjbar et al. (2018) mentioned that nano-calcium spraying on apple trees (*Malus domestica* L. cv. Red Delicious; 70 days after full flowering and up to a month before harvest) reduced the enzymatic activity of polygalacturonase, pectin methyl esterase, and β-galactosidase. These enzymes are related to the hydrolysis of the cell wall that causes softening in climacteric fruits, since the β-galactosidase enzyme consumes the galactose of the cell wall, by increasing its porosity (Mustafa et al., 2014; Ranjbar et al., 2018). The application of nano-calcium could stabilize the cell membrane by maintaining the structure, rigidity, and firmness of the cell wall of the fruit. This could be derived from a greater penetration and better absorption of NPs, and as a result, it could reduce respiration, metabolism, ethylene production, and hydrolysis of polysaccharides to monosaccharides. The spraying of Zn nano-chelate on apple trees (*Malus domestica* L. cv. Delbard Estival and cv. Kohanz) in 3–4 weeks prior to harvest induced an improvement in the postharvest quality of fruits by increasing the content of ascorbic acid and SOD as well as by reducing the polyphenol oxidase (Rasouli and Saba, 2018). In the case of spraying photoactivated ZnO NPs (7.5×10^{-3} M) on strawberries (*Fragaria × ananassa* Duch.) during the ripening stage of the fruit, they gave antifungal effect during the storage of this fruit derived from the production of ROS,

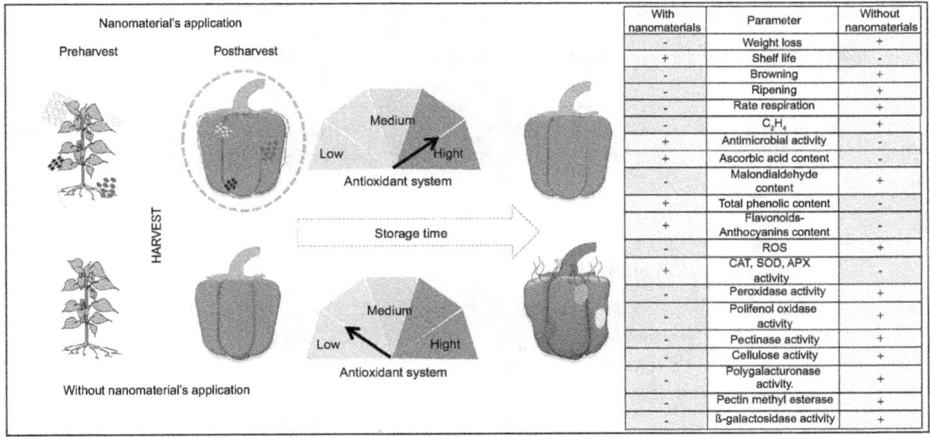

With nanomaterials	Parameter	Without nanomaterials
-	Weight loss	+
+	Shelf life	-
-	Browning	+
-	Ripening	+
-	Rate respiration	+
-	C_2H_4	+
+	Antimicrobial activity	-
+	Ascorbic acid content	-
-	Malondialdehyde content	+
+	Total phenolic content	-
+	Flavonoids-Anthocyanins content	-
-	ROS	+
+	CAT, SOD, APX activity	-
-	Peroxidase activity	+
-	Polifenol oxidase activity	+
-	Pectinase activity	+
-	Cellulose activity	+
-	Polygalacturonase activity.	+
-	Pectin methyl esterase	+
-	ß-galactosidase activity	+

Figure 20.1 Application of nanomaterials in vegetables to improve postharvest quality.

which can attack the cell wall and membrane of pathogens with a probable activation of enzymes of the fruit's defense system (Luksiene et al., 2020).

Neysanian et al. (2020) mentioned that when applying the Se NPs (3 and 10 mg L^{-1}) on tomato seedlings (*Solanum lycopersicum* L.), it generated oxidative stress, resulting in an increase in antioxidant capacity (catalase and peroxidase), as well as an increase in ascorbic acid, nonprotein thiols, soluble phenols, and higher enzymatic activity favoring postharvest longevity of the fruit. In the case of Cu NPs in chitosan-PVA hydrogels, when transplanted into jalapeño pepper plants (*Capsicum annuum* L.), an improvement in the antioxidant activity of the fruit could be observed as a response to the induced stress (Pinedo-Guerrero et al., 2017). Such methods can improve higher postharvest quality due to the higher antioxidant activity in the fruit (Figure 20.1).

20.5.2 Postharvest Application of Nanotechnology

Like the pre-harvest factors, there are factors that affect postharvest quality of crops such as pH,% sugar, weight loss, soluble solids content, and especially the storage conditions (Ijaz et al., 2020). Although the loss of weight and water content in crops can be considered as a process attributable to respiration and other metabolic processes after harvest, the changes that occur during postharvest ripening induce an increase in the softening of the fruit under higher storage temperatures (Malek et al., 2020; Melo et al., 2020). This softening is due to a high rate of degradation of the cell membrane by the activity of the enzymes phospholipase-D and pectin methyl esterase, polyphenol oxidase, and the production of MDA (Ashitha et al., 2020; Nguyen and Nguyen, 2020). In conjunction with this, a loss of fruit firmness occurs; the fruit may darken as a result of the oxidation of phenolic compounds to o-quinone and whose polymerization generates the formation of brown pigments (Ranjbar et al., 2018). Since nanotechnology can mitigate the loss of fruit quality by mitigating and/or minimizing the aforementioned processes, then it can be an effective tool in improving or increasing quality and postharvest life (Figure 20.1, Table 20.2).

Table 20.2 Postharvest Application of Nanotechnology in Crops

NM	*Crops—fruit*	*Storage conditions*	*Effect*	*Reference*
Sodium alginate film with Ag NPs	Carrots (*Daucus carota* L.) Pears (*Pyrus communis* L)	Room temperature 27° C for 10 days.	Weight loss reduction.	Fayaz et al., 2009
Nanocomposite packaging: polyethylene nano-Ag, nano-TiO_2 and montmorillonite.	Kiwifruit (*Actinidia deliciosa* L. cv. Qinmei)	Kiwis treated with 30 μL L^{-1} of ethylene for 24 h at 20°C in flasks. Packed, sealed and stored at 4 ± 1°C by 42 days.	Inhibition of weight loss, softening, content of TSSs. Lower concentration of malonaldehyde. Increase in the content of ascorbic acid and phenols.	Hu et al., 2011
Nanostructured bag of polyethylene /Ag_2O NPs	Apple (*Malus domestica* L. cv. Fuji)	5 and 15 ± 1°C dark room for 24 days.	Decreased weight loss. Microbial spoilage prevention.	Zhou et al., 2011
Agar-Ag NPs film	Lime (*Citrus aurantifolium* L.) Apple (*Pyrus malus* L.)	9 days.	Reduced weight loss. Increase in antimicrobial capacity.	Gudadhe et al., 2014
Chitosan/ nano-chitosan	Asiatic wild rice (*Zizania latifolia* (Griseb.) Turcz. ex Stapf.)	Stems was packaged in a low-density polypropylene plastic bag (400 × 280 × 0.02 mm) and sealed with a rubber band, stored at 1°C for 12 days.	Browning, lignification, and the increase in total color difference were retarded. Lower POD and MDA, inhibited the decrease in SOD, improved CAT and APX.	Luo et al., 2013
Dipping of solid lipid NPs + xanthan gum	Guava (*Psidium guajava* L.)	10°C and 85% RH for 30 days.	Lowest range of weight loss and preserved the best quality.	Zambrano-Zaragoza et al., 2013

NM	*Crops—fruit*	*Storage conditions*	*Effect*	*Reference*
Chitosan–surfactant nanostructure	Tomato (*Solanum lycopersicum* L.)	Stored at 15 ± 2°C and 70–80 % RH for 20 days.	The loss of firmness, decrease in titratable acidity, decrease in chlorophyll content, and increase in soluble solids content, were delayed, increase phenol content, lower level of respiration and weight loss.	Mustafa et al., 2014
Coating Carboxymethy-lcellulose-Ag NPs and guar gum-Ag NPs.	Kinnow (*Citrus reticulata* L. cv. Blanco)	4–10°C, 85–90% RH for 120 days.	Increased ascorbic acid, total phenols, and antioxidant activity. Lower incidence of fungal diseases.	Shah et al., 2015
Low density polyethylene film with Ag NPs	Carrot (*Daucus carota* L. cv. Brasília)	10 ± 2°C, 88% RH for 10 days.	Higher content of ascorbic acid, lower weight loss, and lower content of soluble solids.	Becaro et al., 2016
Nano-Zn	Strawberries (*Fragaria* x *anannasa* Duch. cv. Parous.)	Polystyrene boxes and stored at 1°C with 95% RH for 18 days.	Decrease in microbial load. Weight loss delay, firmness maintenance. Increase in anthocyanin, ascorbic acid, phenol, and antioxidant activity.	Sogvar et al., 2016
Chitosan/nano-silica	Loquat (*Eriobotrya japonica* Lindl. cv. Baiyu)	5°C for 40 days.	Delayed internal browning and weight loss, the decrease of TSSs, and titratable acidity was also inhibited, glucose and fructose, SOD, CAT, APX, PAL, PPO, LOX contents increased, decrease MDA and membrane permeability.	Song et al., 2016

(Continued)

Table 20.2 (Continued)

NM	*Crops—fruit*	*Storage conditions*	*Effect*	*Reference*
Ag NPs coating	Cherry tomatoes (*Solanum lycopersicum* L. cv. cerasiforme)	Room temperatura for 18 days.	Reduced weight loss, increased TSSs and ascorbic acid.	Gao et al., 2017
Solid lipid NPs from candeuba wax and xanthan gum	Guava (*Psidium guajava* L. cv. Media china)	10°C, ≈85% RH for 5 weeks.	Lower O_2 and CO_2 respiration rate. Less weight loss. Better retention of ascorbic acid and total phenols.	García-Betanzos et al., 2017
Polyethylene film with Ag NPs	Tomato (*Solanum lycopersicum* L. cv. Roma)	6°C for 21days in combination with 7% O_2 + 7% CO_2 + 86% N_2	Reduced weight loss. Reduced respiration rate.	Ghasemi-Varnamkhasti et al., 2016
Edible coatings (gelatin hydrogels added with curcumin-loaded zein NPs)	Grapes (*Vitis vinifera* L. cv. Benitaka)	25°C and 50% RH for 7 days.	There was no alteration in titratable acidity. The presence of the coating did not reduce the texture properties and in turn increased the intensity of the fruit's color.	Lemes et al., 2017
Nano-TiO_2—low-density polyethylene packaging	Strawberry (*Fragaria* x *ananassa* Duch. cv. Akihime)	4°C in the dark during 14 days.	Internal reduction of O_2 and increase of CO_2. Lower decomposition rate and weight loss, delay in the decrease of firmness and titratable acidity. ROS ($O_2^{\bullet -}$ and H_2O_2) decrease. Increased antioxidant activity. Retention in content of ascorbic acid and total phenols.	Li et al., 2017

NM	*Crops—fruit*	*Storage conditions*	*Effect*	*Reference*
Nano-ZnO + carboxymethyl cellulose	Pomegranate arils (*Punica granatum* L. cv. Malas)	4°C for 12 days.	Decreased total yeast + mold and total mesophilic bacteria, decreased weight loss, and greatest juice. Decrease soluble solids. Higher total anthocyanin, ascorbic acid, and antioxidant capacity.	Saba and Amini, 2017
Nano chitosan	Peach (*Prunus persica*. L. cv. Early Swelling)	4 ± 2°C for 7 weeks.	Weight loss reduction, improves fruit firmness.	Ali and Toliba, 2018
Edible NPs Coatings and Films: 1. Chitosan-Thyamol/Tripolyphosphate NPs, 2. Gelatin- Coconut fiber powder/titanium dioxide (TiO_2) NPs, 3. Chitosan-Methyl Cellulose/Silica (SiO_2) NPs, 4. Gelatin-Chitosan/(Ag/ZnO) NPs, Layer-by-layer, 5. Gelatin-Anthocyanin/kafirin NPs	Fruits: apples (*Malus Domestica* L. cv. Anna) and red grapes (*Vitis vinifera* L. cv. Flame Seedless) Vegetables: tomatoes (*Lycopersicon lycopersicum* L. cv. Suber Strain B') and weet green pepper (*Capsicum annuum* L. cv. California Wonde)	Fruit: 0–2°C, 90–95% RH (63 and 56 days Apple and grape) Vegetables: 8°C, 90–95% RH for 42 and 35 days for tomato and pepper.	Fruits and vegetables coated with NPs reduced weight loss, TSSs, firmness, total acidity, and ascorbic acid.	Bakhy et al., 2018
D- Limonene Liposomes by immersion for 10 min.	Strawberries (*Fragaria* x *ananassa* Duch. cv. Chandler)	Packed in sterile clamshell box stored at 4°C for 14 days.	Maintains the lower concentration of CO_2 and respiration, lower change in pH, and had higher anthocyanin content.	Dhital et al., 2018

(Continued)

Table 20.2 (Continued)

NM	*Crops—fruit*	*Storage conditions*	*Effect*	*Reference*
Chitosan-TiO_2 nanocomposite film	Cherry tomatoes (*Solanum lycopersicum* L.)	Packaged in a pouch low density polyethylene. Stored at 20°C and 85 % RH in a chamber.	Exhibited ethylene photodegradation activity and delayed the ripening process and changes in the quality of fruit.	Kaewklin et al., 2018
Chitosan NPs filled with chitosan oil of *C. aurantium*	Edible mushroom (*Agaricus bisporus*)	Packed in sealed polypropylene container 4°C for 20 days.	Color change rate slowdown, weight loss, and firmness. It promoted accumulation of phenolic compounds and ascorbic acid. Increased catalase and SOD and decreased polyphenol oxidase activity.	Karimirad et al., 2018
ZnO NPs coating	Fig (*Ficus carica* L.)	Room temperature for 6 days.	They delayed maturation, slowed weight loss, reduced color change, increased firmness, and inhibited the growth of microorganisms.	Lakshmi et al., 2018
Chitosan NPs	Cavendish banana (*Musa acuminata* AAA group)	25 ± 1°C for 15 days.	Shelf life was extended due to modifications in starch content, weight loss, pulp-peel ratio, and TSSs.	Lustriane et al., 2018
Carrageenan—ZnO NPs coating	Mango (*Mangifera indica* L. cv. Gedong Gincu)	20°C, 61% RH for 33 days.	Decreased transmission of water vapor, increased antimicrobial activity. He stood firm. Delayed discoloration.	Meindrawan et al., 2018

NM	Crops—fruit	Storage conditions	Effect	Reference
Chitosan NPs	Grapes (*Vitis labrusca* L.)	25°C for 12 days. 12°C for 24 days.	It delayed grape ripening, decreased weight loss, soluble solids, and reduced sugar content. Greater retention of moisture and preservation of valuable acidity values. Inhibition of microorganism growth.	Melo et al., 2018
Coating Thymol nanoemulsions incorporated in quinoa protein/ chitosan edible films	Cherry tomatoes (*Solanum lycopersicum* L.)	5°C for16 days.	Decreased growth of *Botrytis cinerea*	Robledo et al., 2018
Low density polyethylene film with TiO_2 NPs	Button mushrooms (*Agaricus* bisporus)	4°C for 15 days.	Reduction of weight loss, phenolic compounds, ascorbic acid. Lower browning index. Lower microbial count of fungi.	Almasi et al., 2019
Polylactic acid film + bergamot essential oil impregnated with nano-TiO_2 and nano-Ag.	Mango (*Magnifera indica* L.)	Room temperature for 15 days.	Less weight loss, delayed loss of firmness. Higher content of ascorbic acid.	Chi et al., 2019
Chitosan NPs	Cavendish banana (*Musa acuminata* AAA group)	22 ± 1°C	Slower skin discoloration.	Esyanti et al., 2019
Chitosan NPs filled with oil of *Cuminum cyminum*	Button mushroom (*Agaricus bisporus* cv. Sylvan A15)	Packed in sealed polypropylene container 4°C for 20 days.	Maintained color, firmness and acceptability. Inhibition of bacteria and fungi. Higher SOD and APX activity, antioxidant capacity and total phenolic content. PPO activity reduction.	Karimirad et al., 2019

(Continued)

Table 20.2 (Continued)

NM	*Crops—fruit*	*Storage conditions*	*Effect*	*Reference*
Nanocomposite coating (Soybean protein isolate + cinnamaldehyde + ZnO NPs)	Banana (Musaceae)	25°C, 40% RH for 7 days.	Less weight loss, more firmness. Reduction of the content of total soluble sugars and increase in titratable acidity.	Li et al., 2019
Ca NPs	Mango (*Mangifera indica* L. cv Hindi Be-Sennara)	6 ± 1°C, 95 ± 1% RH for 35 days.	Reduction of the internal browning index. Higher content of total phenolic compounds, flavonoids and lower activity of the PPO, PAL, LOX, and lower content of malonaldehyde. Reduction of pectinase and cellulase activity and less % leakage from the cell membrane. Higher antioxidant activity.	Lo'ay and Ameer, 2019
Multisystem coating based on polymeric nanocapsules containing *Thymus vulgaris* L. essential oil	Grapes (*Vitis vinifera* L)	25°C for 13 days.	Increased antioxidant capacity and maintained characteristics of color, firmness, titratable acidity, and TSS for longer time, acting as a barrier, which reduced the metabolism of fruits.	Pina-Barrera et al., 2019
Ag NP coating	Loquat (*Eriobotrya japonica* Lindl.)	Storage at 4 and 8°C. For 1 month.	Reduced loss of: weight, TSSs and sugars. Increased content of ascorbic acid, total phenols, antioxidants, and titratable acidity.	Ali et al., 2020
Chitosan NPs	Pomegranates arils (*Punica granatum* L. cv. Rabab)	Packed in polystyrene boxes, stored at 5°C for 18 days.	Reduction of microbial counts and higher content of total phenol, ascorbic acid in arils.	Amiri et al., 2021

NM	*Crops—fruit*	*Storage conditions*	*Effect*	*Reference*
Coating of chitosan + nano ZnO	Guava (*Psidium guajava* L.)	21 ± 1°C, 80 ± 2% RH for 15 days.	Inhibition of rot, delaying the ripening process, less weight loss.	Arroyo et al., 2020
Chitosan + nano SiO_2	*Blueberry (Vaccinium myrtillus* L.)	Room temperature for 8 days.	Inhibition of microbial populations. Increased anthocyanin content.	Eldib et al., 2020
Ag NPs	Orange (*Citrus* × *sinensis* L.)	Stored in bags at 25°C for15 days.	High efficacy against disease caused by *Penicillium digitatum*.	El khawaga, 2020
Carboxymethyl cellulose–based film containers enriched with ZnO NPs	Cherry tomatoes (*Solanum lycopersicum* L.)	Room temperature ~25°C, 45% RH for 10 days	Greater firmness of the fruit, reduced weight loss.	Guo et al., 2020
Nano-curcumin and Nano-rosemarinic acid	French basil (*Ocimum basilicum* L. cv. Grand Vert)	10°C for 3 weeks.	Less fresh weight loss (%), constant chlorophyll (during the first week), increase in the percentage of essential oil.	Hammam and Shoala, 2020
Chitosan NPs loaded with clove essential oil.	Pomegranate arils (*Punica granatum* L. cv. Malase Saveh)	5°C	It extended the shelf life of the aril for 54 days. Weight maintenance, TSSs, titratable acidity, pH, total phenol, and total anthocyanin content, as well as antioxidant activity and sensory quality.	Hasheminejad and Khodaiyan, 2020
Chitosan NPs with α-pineno	Bell peppers (*Capsicum annuum* L. cv. California)	12 ± 2°C for 21 days followed by 20 ± 2°C for 5 days.	Alterations in weight loss and preservation of physicochemical quality. Inhibition of *Alternaria alternata*.	Hernández-López et al., 2020

(Continued)

Table 20.2 (Continued)

NM	*Crops—fruit*	*Storage conditions*	*Effect*	*Reference*
Coating chitosan NPs	Bell pepper (*Capsicum annuum* L. cv. grossum)	5°C for12 days	Less weight loss (protection against water loss). Inhibition of pathogens.	Hu et al., 2020
Nanocomposite coating (chitosan combined with ZnO NPs)	Papaya (*Carica papaya.* L cv. California)	10°C for 12 days.	Suppression of microbial growth.	Lavinia et al., 2020
Natamycin-loaded zein NPs stabilized by carboxymethyl chitosan	Hongjia strawberries (*Fragaria* × *ananassa* Duch. cv. Benihoppe)	28°C 90% RH during 8 days.	Reduction of the occurrence of rot, mildew, and gray mold.	Lin et al., 2020
Edible coatings chitosan NPs.	Strawberries (*Fragaria* × *ananassa* Duch. cv. Camarosa)	Stored in PET containers under refrigeration at 4 ± 1°C for 8 days.	Less weight loss, more firmness. Maintenance of antioxidant compounds and antioxidant capacity.	Martínez-González et al., 2020
Edible coating of ZnO NPs	Mango (*Mangifera indica* L. cv. Golden Lily)	27 and 32°C for 7 days.	Maturation delay, weight loss slowdown, and increased firmness.	Malek et al., 2020
Edible coating of chitosan NPs	Strawberry (*Fragaria* × *ananassa* Duch.)	25 ± 2°C for 6 days 10 ± 2°C for12 days.	Antifungal action, reduction in weight loss, TSSs, maturity index, and moisture loss.	Melo et al., 2020
Nano-net of Cu-chitosan NPs	Tomato (*Solanum lycopersicum* L.)	27 ± 2°C, 55 ± 2% RH for 21 days.	Microbial decomposition reduction, physiological weight loss, respiration rate. Maintained fruit firmness. Delay of loss of titratable acidity, retention of TSSs, total and reducing sugars, lycopene, ascorbic acid. Inhibition of polyphenol oxidase.	Meena et al., 2020

NM	*Crops—fruit*	*Storage conditions*	*Effect*	*Reference*
Nano chitosan	Strawberry (*Fragaria* x *ananassa* Duch.)	2°C for 21 days.	Preservation of the quality index, reduction of weight loss, retention of firmness, valuable acidity, and ascorbic acid. Increase in total anthocyanin content. MDA production delay and inhibition of polyphenol oxidase activity.	Nguyen and Nguyen, 2020
$CaCl_2$+nano chitosan	Strawberry (*Fragaria* x *ananassa* Duch.)	4°C for 15 days.	Weight loss reduction, preservation of ascorbic acid, total anthocyanins. Malonaldehyde production delay.	Nguyen et al., 2020
Silver NPs polyvinylpyrrolidone based glycerosomes	Red and yellow bell pepper (*Capsicum annuum* L. var. grossum (L.) Sendt)	4°C and 15°C, 90–95 % RH for 15 days.	Extension of the shelf life of bell peppers (red and yellow).	Saravanakumar et al., 2020
Ag NPs	Apricot (*Prunus armeniaca* L. cv. Canino)	25 ± 3°C and 6 ± 2°C for 4 days.	Less weight loss, TSS, and MDA concentration. High content of total acidity and total carotenoids.	Shahat et al., 2020a
Chitosan NPs	Strawberry (*Fragaria* x *ananassa* Duch. cv. Camarosa)	25 ± 3°C for 6 days and 6 ± 2°C for 16 days.	Lower values for weight loss, decay, TSS, and MDA concentration; and higher levels of total acidity and anthocyanin and less of total aerobic, mold, and yeast	Shahat et al., 2020b
Chitosan-coated iron-oxide NPs	Peach (*Prunus persica* L.)	4°C for 25 days.	Reduction of weight loss and inhibition of microbial growth on the fruit surface.	Saqib et al., 2020

(Continued)

Table 20.2 (Continued)

NM	*Crops—fruit*	*Storage conditions*	*Effect*	*Reference*
Heracleum persicum essential oil-loaded chitosan NPs	Red sweet bell pepper (*Capsicum annuum* L.)	9 ± 1°C, 95% RH for 30 days.	Reduction in the change of color, firmness, and weight. Higher content of phenolic compounds, flavonoids, and ascorbic acid. Increased SOD, CAT, and reduced POD activity.	Taheri et al., 2020
Antimicrobial packaging films pullulan polymer and Solid lipid NPs containing 1% w/w cinnamaldehyde	Strawberrie (*Fragaria* x *ananassa* Duch.)	3°C for 10 days and 12°C for 6 days.	Reduction of yeast and mold population, reduction of metabolic rate, and maintenance of desirable color, delay in loss of firmness.	Trinetta et al., 2020
Coating of Ag NPs + hydroxypropyl-methylcellulose	Papaya (*Carica papaya* L. cv. Golden)	10°C for 7 days, 88.5% RH, followed by 20°C for 7 days, 88.5% RH	It reduced the incidence and severity of *Colletotrichum gloeosporioides* and weight loss.	Vieira et al., 2020
Nano SiO_2 packaging	Loquat (*Eriobotrya japonica* L. cv. Dawuxing and Qingzhong)	20 ± 1°C, 90% RH for 12 days.	Increased content of phenolic compounds, SOD, CAT, ascorbic acid. Inhibited internal browning, delayed decrease in TSSs.	Wang et al., 2020b
Chitosan/nano-TiO_2 Nanocomposite Coating	Mango (*Mangifera indica* L.)	13°C for 20 days.	Reduced decomposition index, respiration, TSSs and MDA, increased firmness, POD, PPO, phenol, and flavonoids.	Xing et al., 2020
Beeswax Solid Lipid NPs Nano-coatings + xanthan gum and propylene glycol	Strawberry (*Fragaria* x *ananassa* Duch. cv. camarosa)	4°C for 21 days.	Less weight loss (%), low decay rate and less loss of firmness.	Zambrano-Zaragoza et al., 2020

NM	Crops—fruit	Storage conditions	Effect	Reference
Treatment of S-nitrosoglutathione-chitosan NPs	Apple (*Malus domestica* L. cv. Fuji)	4°C for 4 days.	It preserved the quality of the freshly cut slices, higher polyphenol content, APX, POD, CAT, and SOD activity, maintained the surface color and weight, reduced the respiration rate and ROS.	Zhao et al., 2020a

The use of NMs on crops to increase their shelf quality has been widely documented (Table 20.2). The application of Si NPs can reduce lipid peroxidation by preserving the integrity of the membrane through the activation of the antioxidant system (El-Serafy, 2019), possibly due to electrostatic and hydrophobic interactions between NPs and fruit components (Malek et al., 2020; Melo et al., 2020). Bonilla-Bird et al. (2018) found in sweet potato (*Ipomoea batatas* L.) that the exposure of the roots to CuO NPs, once harvested, induced an increase in the Cu content in the peridermis and bark but not in the most internal tissues such as the pith; this suggests that this type of NPs can protect and increase the shelf life of sweet potato roots without compromising the consumer to excessive Cu consumption. By comparison, ZnO NPs based on polysaccharides (chitosan, alginate, carrageenan, cellulose, pectin) can reduce the ripening process in fruits due to the improved gas and water barrier properties (Anugrah et al., 2020).

According to Zhao et al. (2020a), the positive effect of GSNO–chitosan NPs in apples (*Malus domestica* L. cv. Fuji) could be due to its participation in the metabolism of GSNO and endogenous glutathione, as well as in the nitration process, and in the conversion of $O^{2\bullet-}$ in water by increasing the enzymatic activity of SOD. The antioxidant properties derived from the application of NPs could delay the browning of postharvest cultures due to the reduction in the oxidation process of the polyphenolic compounds that protect the cells from LOX (Karimirad et al., 2018).

In the same way, these NPs applied in postharvest on the fruits can prevent the activity of microorganisms since they have the ability to damage the cell membrane and even disintegrate the DNA of the microbes (Morones et al., 2005). This is probably due to the irruption in the transmembrane electron transfer, penetrating and oxidizing the envelope of the cellular components, as well as the production of secondary products as ROS that induce the formation of ions, efflux, and ultimately death of the microorganisms (Dakal et al., 2016). Sun et al. (2018) mentioned that the induction of ROS, the release of Zn^{2+} ions, and their eventual penetration into the cell wall, as well as their reaction with the content of the cytoplasm are the possible mechanisms of ZnO NP–based antimicrobial activity. Melo et al. (2018) reported that grapes (*Vitis labrusca* L.) immersed in chitosan NPs dispersed in a solution increased its antimicrobial activity. Eventually, the response of the plant depends on the type of culture, growth medium—storage and the NM used (surface coating, concentration, type, size, exposure time; Liu et al., 2020). But also, the particular characteristics of NMs (size, monodispersion, zeta potential, architecture) could facilitate the uniform and firm bond to the wax cuticular surfaces of fruits and vegetables (Meena et al., 2020).

In the case of nanocoatings (added with NPs), they form a semipermeable barrier on the surface of the fruit, inhibiting the decomposition symptoms, in addition to interfering with exposed macromolecules; reducing intracellular electrolyte leakage, weight loss, and water loss; and inducing lower respiration rate (Shah et al., 2015). This semipermeable film alters the gas exchange (water vapor, O_2, CO_2) between the fruit and the storage environment (Amiri et al., 2021; Melo et al., 2020). This microenvironment created between the coating and the surface of the fruit induces a lower rate of pigment oxidation and a longer postharvest life since, as mentioned, there is a reduction in O_2 accompanied by an increase in CO_2 (Melo et al., 2020). According to Sorrentino et al. (2007), the modification in the levels of O_2 and CO_2 could inhibit the degradation of chlorophyll on the surface of a fruit; in addition, a low O_2 and high CO_2 ratio can inhibit the production of ethylene and delay ripening as well as inhibit enzymes related to hydrolysis of starch.

Nanochitosan, having the ability to inhibit microbial mRNA, as well as protein synthesis, prolongs the shelf life (Ali and Toliba, 2018) and reduces the respiration and metabolism of the fruit (Chi et al., 2019). Amiri et al. (2021) reported that chitosan nanocoatings cause, in addition to less water loss, a reduction in the exposure to oxygen in the storage environment and eventually a reduction in the oxidative reactions of ascorbic acid in the fruit. Derived from the blockage of the pores of the fruit skin, the loss of moisture is less, and the respiratory rate is reduced (Amiri et al., 2021; Melo et al., 2018, 2020; Shah et al., 2015). The possible increase in the content of total phenols and anthocyanins derives from a decrease in their oxidation due to the chitosan-NP coating (Amiri et al., 2021). According to Melo et al. (2020), the increase in the antioxidant activity of chitosan NPs is given by the reaction between free radicals with the residual free amino groups (NH_2) that, in turn, can generate ammonium groups (NH^{+3}) by absorbing a hydrogen ion from the cell medium.

The application of Cu NP–chitosan forms an invisible and intangible nano-network on the surface of the fruit, being able to act as a potential barrier in all possible openings (stem scar, cuticle wax, lenticels, aquaporins) controlling microbial infection, loss of humidity, gas exchanges, and the frequency of the respiratory rate, in addition to considerably suppressing the metabolic activity that controls total and reducing sugars (Meena et al., 2020). The same authors point out that these NMs reduce lycopene synthesis during storage. This coating, according to electron microscopy studies, forms a smooth surface in contact with the fruit of strawberry (*Fragaria x ananassa* Duch.) and tomato (*Solanum lycopersicum* L.) that results in a reduction in respiration rate and weight loss (Meena et al., 2020; Nguyen et al., 2020). These types of coatings have, like the application of NPs, a high activity against the growth of microorganisms. Zambrano-Zaragoza et al. (2020) mentioned that beeswax solid lipid NPs with xanthan gum and propylene glycol when applied on strawberries (*Fragaria x ananassa* Duch. cv. camarosa) decreased fungal contamination. This was the result of the particle size and concentration of the polysaccharides, since its aggregation and formation significantly modified the atmosphere on the surface of the fruit.

One of the main mechanisms of these coatings with NPs is the inhibition of the respiratory rate and the reduction in the consumption of ascorbic acid in the fruit

(Chi et al., 2019), as well as reducing the degradation of phenolic compounds (Ali et al., 2020). This ascorbic acid, like its stereoisomer, isoascorbic acid, is used by fruits to minimize browning by reducing o-quinone to phenolic compounds and preventing polyphenol oxidase activity (Meena et al., 2020). In the case of containers, packaging, and/or bags, developed with organic (nanocellulose fibrils) or inorganic (montmorillonite, zinc oxide, silver) NMs based on biopolymers (derived from plants, animals, or microorganisms) and with the addition of NPs, it can have a functionality very similar to coatings, maximizing the useful life of the fruits, derived from the physicochemical and mechanical barrier and its antimicrobial and antioxidant activities (Basumatary et al., 2022).

20.6 Conclusion

The use of nanotechnology as a tool to improve postharvest quality and increased shelf life seems to be a good alternative, since it induces a series of positive responses that improve the characteristics of vegetable foods. A great advantage of NMs is that they can be applied in pre-harvest and thus induce beneficial responses observable during postharvest, improving the quality of vegetable foods at this stage. They can also be applied directly to the organs that serve as food, improving physical and even biochemical characteristics such as the content of bioactive compounds. In addition, these NMs can be used for the production of packages or bags in which vegetable foods are marketed and preserve their quality.

However, it is necessary to study at a multi-omic level the effects that NMs could have on the fruit–human consumption relationship for greater safety (Majumdar and Keller, 2021); for example, the nano-proteomic approach in the development of biomarkers to manage food quality (Agrawal et al., 2013) could focus on demonstrating the safety of fruits exposed to NMs.

References

Abdel-Aziz HMM, Hasaneen MNA, Omer AM. 2019. Impact of engineered nanomaterials either alone or loaded with NPK on growth and productivity of French bean plants: Seed priming vs foliar application. *South African Journal of Botany* 125: 102–108.

Achari GA, Kowshik M. 2018. Recent Developments on nanotechnology in agriculture: Plant mineral nutrition, health, and interactions with soil microflora. *Journal of Agricultural and Food Chemistry* 66: 8647–8661.

Adisa IO, Pullagurala VLR, Peralta-Videa JR, Dimkpa CO, Elmer WH, Gardea-Torresdey JL, White JC. 2019. Recent advances in nano-enabled fertilizers and pesticides: A critical review of mechanisms of action. *Environmental Science: Nano* 6: 2002–2030.

Aghdam MS, Jannatizadeh A, Luo Z, Paliyath G. 2018. Ensuring sufficient intracellular ATP supplying and friendly extracellular ATP signaling attenuates stresses, delays senescence and maintains quality in horticultural crops during postharvest life. *Trends in Food Science and Technology* 76: 67–81.

Agrawal GK, Timperio AM, Zolla L, Bansal V, Shukla R, Rakwal R. 2013. Biomarker discovery and applications for foods and beverages: Proteomics to nanoproteomics. *Journal of Proteomics* 93: 74–92.

Ali AA, Toliba AO. 2018. Effect of organic calcium spraying and Nano Chitosan fruits coating on yield, fruit quality and storability of peach cv 'early swelling.' *Current Science International* 7: 737–749.

Ali M, Ahmed A, Shah SWA, Mehmood T, Abbasi KS. 2020. Effect of silver nanoparticle coatings on physicochemical and nutraceutical properties of loquat during postharvest storage. *Journal of Food Processing and Preservation* 44: e14808.

Almasi H, Pourfathi B, Zonouzi RM. 2019. Studying of the effect of low density poly ethylene (LDPE) based antimicrobial nanocomposite packaging containing TiO_2 nanoparticles on the shelf-life extension of button mushroom (*Agaricus bisporus*) HADI. *Iranian Journal of Biosystems Engineering* 51: 195–209.

Amiri A, Ramezanian A, Mortazavi SMH, Hosseini SMH, Yahia E. 2021. Shelf-life extension of pomegranate arils using chitosan nanoparticles loaded with *Satureja hortensis* essential oil. *Journal of the Science of Food and Agriculture* 101(9): 3778–3786.

Anugrah DSB, Alexander H, Pramitasari R, Hudiyanti D, Sagita CP. 2020. A review of polysaccharide-zinc oxide nanocomposites as safe coating for fruits preservation. *Coatings* 10: 988.

Arroyo BJ, Bezerra AC, Oliveira LL, Arroyo SJ, Melo EA de, Santos AMP. 2020. Antimicrobial active edible coating of alginate and chitosan add ZnO nanoparticles applied in guavas (*Psidium guajava* L.). *Food Chemistry* 309: 125566.

Ashitha GN, Sunny AC, Nisha R. 2020. Effect of pre-harvest and postharvest hexanal treatments on fruits and vegetables: A review. *Agricultural Reviews* 41: 124–131.

Bajpai VK, Kamle M, Shukla S, Mahato DK, Chandra P, Hwang SK, Kumar P, Huh YS, Han YK. 2018. Prospects of using nanotechnology for food preservation, safety, and security. *Journal of Food and Drug Analysis* 26: 1201–1214.

Bakhy EA, Zidan NS, Aboul-Anean HED. 2018. The effect of nano materials on edible coating and films' improvement. *International Journal of Pharmaceutical Research and Allied Sciences* 7: 20–41.

Basumatary IB, Mukherjee A, Katiyar V, Kumar S. 2022. Biopolymer-based nanocomposite films and coatings: Recent advances in shelf-life improvement of fruits and vegetables. *Critical Reviews in Food Science and Nutrition* 62(7): 1912–1935.

Becaro AA, Puti FC, Panosso AR, Gern JC, Brandão HM, Correa DS, Ferreira MD. 2016. Postharvest quality of fresh-cut carrots packaged in plastic films containing silver nanoparticles. *Food and Bioprocess Technology* 9: 637–649.

Bonilla-Bird NJ, Paez A, Reyes A, Hernandez-Viezcas JA, Li C, Peralta-Videa JR, Gardea-Torresdey JL. 2018. Two-photon microscopy and spectroscopy studies to determine the mechanism of copper oxide nanoparticle uptake by sweetpotato roots during postharvest treatment. *Environmental Science and Technology* 52: 9954–9963.

Bose SK, Howlader P, Wang W, Yin H. 2021. Oligosaccharide is a promising natural preservative for improving postharvest preservation of fruit: A review. *Food Chemistry* 341: 128178.

Burdon J, Connolly P, de Silva N, Lallu N, Dixon J, Pak H. 2013. A meta-analysis using a logit non-linear mixed effects model for 'Hass' avocado postharvest performance data. *Postharvest Biology and Technology* 86: 134–140.

Casajús V, Perini M, Ramos R, Lourenco AB, Salinas C, Sánchez E, Fanello D, Civello P, Frezza D, Martínez G. 2021. Harvesting at the end of the day extends postharvest life of kale (*Brassica oleracea* var. sabellica). *Scientia Horticulturae* 276: 109757.

Chen T, Ji D, Zhang Z, Li B, Qin G, Tian S. 2020. Advances and strategies for controlling the quality and safety of postharvest fruit. *Engineering*. https://doi.org/10.1016/j.eng.2020.07.029

Chen Y, Lin H, Jiang Y, Zhang S, Lin Y, Wang Z. 2014. *Phomopsis longanae* Chi-induced pericarp browning and disease development of harvested longan fruit in association with energy status. *Postharvest Biology and Technology* 93: 24–28.

Chi H, Song S, Luo M, Zhang C, Li W, Li L, Qin Y. 2019. Effect of PLA nanocomposite films containing bergamot essential oil, TiO_2 nanoparticles, and Ag nanoparticles on shelf life of mangoes. *Scientia Horticulturae* 249: 192–198.

Dakal TC, Kumar A, Majumdar RS, Yadav V. 2016. Mechanistic basis of antimicrobial actions of silver nanoparticles. *Frontiers in Microbiology* 7: 1831.

Dhital R, Mora NB, Watson DG, Kohli P, Choudhary R. 2018. Efficacy of limonene nano coatings on postharvest shelf life of strawberries. *Lwt—Food Science and Technology* 97: 124–134.

El khawaga, MAY. 2020. The efficiency of biosynthesized silver nanoparticles by endophytic *Fusarium chlamydosporum* F25 against plants postharvest fungal pathogen. *Egyptian Academic Journal of Biological Sciences, G. Microbiology* 12: 29–43.

El-Serafy RS. 2019. Silica nanoparticles enhances physio-biochemical characters and postharvest quality of *Rosa hybrida* L. cut flowers. *Journal of Horticultural Research* 27: 47–54.

Eldib R, Khojah E, Elhakem A, Benajiba N, Helal M. 2020. Chitosan, nisin, silicon dioxide nanoparticles coating films effects on blueberry (*Vaccinium myrtillus*) quality. *Coatings* 10: 962.

Esyanti RR, Zaskia H, Amalia A, Nugrahapraja DH. 2019. Chitosan nanoparticle-based coating as postharvest technology in banana. *Journal of Physics: Conference Series* 1204: 012109.

Fayaz AM, Girilal M, Kalaichelvan PT, Venkatesan R. 2009. Mycobased synthesis of silver nanoparticles and their incorporation into sodium alginate films for vegetable and fruit preservation. *Journal of Agricultural and Food Chemistry* 57: 6246–6252.

Gao H, Zhang ZK, Chai HK, Cheng N, Yang Y, Wang DN, Yang T, Cao W. 2016. Melatonin treatment delays postharvest senescence and regulates reactive oxygen species metabolism in peach fruit. *Postharvest Biology and Technology* 118: 103–110.

Gao L, Li Q, Zhao Y, Wang H, Liu Y, Sun Y, Wang F, Jia W, Hou X. 2017. Silver nanoparticles biologically synthesised using tea leaf extracts and their use for extension of fruit shelf life. *IET Nanobiotechnology* 11: 637–643.

García-Betanzos CI, Hernández-Sánchez H, Bernal-Couoh TF, Quintanar-Guerrero D, Zambrano-Zaragoza ML. 2017. Physicochemical, total phenols and pectin methylesterase changes on quality maintenance on guava fruit (*Psidium guajava* L.) coated with candeuba wax solid lipid nanoparticles-xanthan gum. *Food Research International* 101: 218–227.

Ghasemi-Varnamkhasti M, Yoosefian SH, Mohammad-Razdari A. 2016. Investigation of active modified atmosphere and nanoparticle packaging on quality of tomatoes. *International Journal of Nutrition and Food Engineering* 10: 551–555.

González-Saucedo A, Barrera-Necha LL, Ventura-Aguilar RI, Correa-Pacheco ZN, Bautista-Baños S, Hernández-López M. 2019. Extension of the postharvest quality of bell pepper by applying nanostructured coatings of chitosan with *Byrsonima crassifolia* extract (L.) Kunth. *Postharvest Biology and Technology* 149: 74–82.

Gouda MHB, Zhang C, Wang J, Peng S, Chen Y, Luo H, Yu L. 2020. ROS and MAPK cascades in the postharvest senescence of horticultural products. *Journal of Proteomics & Bioinformatics* 13: 1–7.

Gudadhe JA, Yadav A, Gade A, Marcato PD, Durán N, Rai M. 2014. Preparation of an agar-silver nanoparticles (A-AgNp) film for increasing the shelf-life of fruits. *IET Nanobiotechnology* 8: 190–195.

Gundewadi G, Reddy VR, Bhimappa B. 2018. Physiological and biochemical basis of fruit development and ripening-a review. *Journal of Hill Agriculture* 9: 7–21.

Guo X, Chen B, Wu X, Li J, Sun Q. 2020. Utilization of cinnamaldehyde and zinc oxide nanoparticles in a carboxymethylcellulose-based composite coating to improve the postharvest quality of cherry tomatoes. *International Journal of Biological Macromolecules* 160: 175–182.

Hammam KA, Shoala T. 2020. Influence of spraying nano-curcumin and nano-rosemarinic acid on growth, fresh herb yield, chemicals composition and postharvest criteria of French basil (*Ocimum basilicum* L. var Grand Vert) plants. *Journal of Agricultural and Rural Research* 5: 1–22.

Hasheminejad N, Khodaiyan F. 2020. The effect of clove essential oil loaded chitosan nanoparticles on the shelf life and quality of pomegranate arils. *Food Chemistry* 309: 125520.

Hernández-Fuentes A, López-Vargas E, Pinedo-Espinoza J, Campos-Montiel R, Valdés-Reyna J, Juárez-Maldonado A. 2017. Postharvest behavior of bioactive compounds in tomato fruits treated with cu nanoparticles and NaCl stress. *Applied Sciences* 7: 1–14.

Hernández-López G, Ventura-Aguilar RI, Correa-Pacheco ZN, Bautista-Baños S, Barrera-Necha LL. 2020. Nanostructured chitosan edible coating loaded with α-pinene for the preservation of the postharvest quality of *Capsicum annuum* L. and *Alternaria alternata* control. *International Journal of Biological Macromolecules* 165: 1881–1888.

Hu Q, Fang Y, Yang Y, Ma N, Zhao L. 2011. Effect of nanocomposite-based packaging on postharvest quality of ethylene-treated kiwifruit (*Actinidia deliciosa*) during cold storage. *Food Research International* 44: 1589–1596.

Hu X, Saravanakumar K, Sathiyaseelan A, Wang MH. 2020. Chitosan nanoparticles as edible surface coating agent to preserve the fresh-cut bell pepper (*Capsicum annuum* L. var. grossum (L.) Sendt). *International Journal of Biological Macromolecules* 165: 948–957.

Hussein Z, Fawole OA, Opara UL. 2020. Harvest and postharvest factors affecting bruise damage of fresh fruits. *Horticultural Plant Journal* 6: 1–13.

Ijaz M, Zafar M, Afsheen S, Iqbal T. 2020. A review on Ag-nanostructures for enhancement in shelf time of fruits. *Journal of Inorganic and Organometallic Polymers and Materials* 30: 1475–1482.

Jansasithorn R, East AR, Hewett EW, Heyes JA. 2014. Skin cracking and postharvest water loss of jalapeño chilli. *Scientia Horticulturae* 175: 201–207.

Jiang Y, Duan X, Joyce D, Zhang Z, Li J. 2004. Advances in understanding of enzymatic browning in harvested litchi fruit. *Food Chemistry* 88: 443–446.

Juárez-Maldonado A, Ortega-Ortiz H, González-Morales S, Morelos-Moreno Á, Cabrera-de la Fuente M, Sandoval-Rangel A, Cadenas-Pliego G, Benavides-Mendoza A. 2019. Nanoparticles and nanomaterials as plant biostimulants. *International Journal of Molecular Sciences* 20: 1–19.

Juárez-Maldonado A, Tortella G, Rubilar O, Fincheira P, Benavides-Mendoza A. 2021. Biostimulation and toxicity: The magnitude of the impact of nanomaterials in microorganisms and plants. *Journal of Advanced Research*. https://doi.org/10.1016/j.jare.2020.12.011

Kaewklin P, Siripatrawan U, Suwanagul A, Lee YS. 2018. Active packaging from chitosan-titanium dioxide nanocomposite film for prolonging storage life of tomato fruit. *International Journal of Biological Macromolecules* 112: 523–529.

Karimirad R, Behnamian M, Dezhsetan S. 2019. Application of chitosan nanoparticles containing *Cuminum cyminum* oil as a delivery system for shelf life extension of *Agaricus bisporus*. *Lwt—Food Science and Technology* 106: 218–228.

Karimirad R, Behnamian M, Dezhsetan S, Sonnenberg A. 2018. Chitosan nanoparticles loaded *Citrus aurantium* essential oil: A novel delivery system for preserving the postharvest quality of *Agaricus bisporus*. *Journal of the Science of Food and Agriculture* 98: 5112–5119.

Lakshmi SJ, Bai RRS, Sharanagouda H, Ramachandra CT, Nadagouda S, Nidoni U. 2018. Effect of biosynthesized zinc oxide nanoparticles coating on quality parameters of fig (*Ficus carica* L.) fruit. *Journal of Pharmacognosy and Phytochemistry* 7: 10–14.

Lavinia M, Hibaturrahman SN, Harinata H, Wardana AA. 2020. Antimicrobial activity and application of nanocomposite coating from chitosan and ZnO nanoparticle to inhibit microbial growth on fresh-cut papaya. *Food Research* 4: 307–311.

Lemes GF, Marchiore NG, Moreira TFM, Da Silva TBV, Sayer C, Shirai MA, Gonçalves OH, Gozzo AM, Leimann FV. 2017. Enzymatically crosslinked gelatin coating added of bioactive nanoparticles and antifungal agent: Effect on the quality of Benitaka grapes. *Lwt—Food Science and Technology* 84: 175–182.

Leng P, Yuan B, Guo Y. 2014. The role of abscisic acid in fruit ripening and responses to abiotic stress. *Journal of Experimental Botany* 65: 4577–4588.

Li D, Ye Q, Jiang L, Luo Z. 2017. Effects of nano-TiO_2-LDPE packaging on postharvest quality and antioxidant capacity of strawberry (*Fragaria ananassa* Duch.) stored at refrigeration temperature. *Journal of the Science of Food and Agriculture* 97: 1116–1123.

Li H, Wu L, Tang N, Liu R, Jin Z, Liu Y, Li Z. 2020. Analysis of transcriptome and phytohormone profiles reveal novel insight into ginger (*Zingiber officinale* Rose) in response to postharvest dehydration stress. *Postharvest Biology and Technology* 161: 111087.

Li J, Sun Q, Sun Y, Chen B, Wu X, Le T. 2019. Improvement of banana postharvest quality using a novel soybean protein isolate/cinnamaldehyde/zinc oxide bionanocomposite coating strategy. *Scientia Horticulturae* 258: 108786.

Li L, Lichter A, Chalupowicz D, Gamrasni D, Goldberg T, Nerya O, Ben-Arie R, Porat R. 2016. Effects of the ethylene-action inhibitor 1-methylcyclopropene on postharvest quality of non-climacteric fruit crops. *Postharvest Biology and Technology* 111: 322–329.

Lin M, Fang S, Zhao X, Liang X, Wu D. 2020. Natamycin-loaded zein nanoparticles stabilized by carboxymethyl chitosan: Evaluation of colloidal/chemical performance and application in postharvest treatments. *Food Hydrocolloids* 106: 105871.

Lin Y, Lin Y, Lin H, Ritenour MA, Shi J, Zhang S, Chen Y, Wang H. 2017. Hydrogen peroxide-induced pericarp browning of harvested longan fruit in association with energy metabolism. *Food Chemistry* 225: 31–36.

Liu W, Zeb A, Lian J, Wu J, Xiong H, Tang J, Zheng S. 2020. Interactions of metal-based nanoparticles (Mbnps) and metal-oxide nanoparticles (monps) with crop plants: A critical review of research progress and prospects. *Environmental Reviews* 28: 294–310.

Lo'ay AA, Ameer NM. 2019. Performance of calcium nanoparticles blending with ascorbic acid and alleviation internal browning of 'Hindi Be-Sennara' mango fruit at a low temperature. *Scientia Horticulturae* 254: 199–207.

Lufu R, Ambaw A, Opara UL. 2020. Water loss of fresh fruit: Influencing pre-harvest, harvest and postharvest factors. *Scientia Horticulturae* 272: 109519.

Luksiene Z, Rasiukeviciute N, Zudyte B, Uselis N. 2020. Innovative approach to sunlight activated biofungicides for strawberry crop protection: ZnO nanoparticles. *Journal of Photochemistry and Photobiology B: Biology* 203: 111656.

Luo H, Jiang L, Bao Y, Wang L, Yu Z. 2013. Effect of chitosan/nano-chitosan composite coating on browning and lignification of fresh-cut *Zizania latifolia*. *Journal of Food Quality* 36: 426–431.

Lustriane C, Dwivany FM, Suendo V, Reza M. 2018. Effect of chitosan and chitosan-nanoparticles on post harvest quality of banana fruits. *Journal of Plant Biotechnology* 45: 36–44.

Majumdar S, Keller AA. 2021. Omics to address the opportunities and challenges of nanotechnology in agriculture. *Critical Reviews in Environmental Science and Technology* 51(22): 2595–2636.

Malek NSA, Rosman N, Mahmood MR, Khusaimi Z, Asli NA. 2020. Effects of storage temperature on shelf-life of mango coated with zinc oxide nanoparticles. *Science Letters* 14: 47–57.

Mali SC, Raj S, Trivedi R. 2020. Nanotechnology a novel approach to enhance crop productivity. *Biochemistry and Biophysics Reports* 24: 100821.

Manzoor A, Bashir MA, Hashmi MM. 2020. Nanoparticles as a preservative solution can enhance postharvest attributes of cut flowers. *Italus Hortus* 27: 1–14.

Maringgal B, Hashim N, Mohamed Amin Tawakkal IS, Muda Mohamed MT. 2020. Recent advance in edible coating and its effect on fresh/fresh-cut fruits quality. *Trends in Food Science & Technology* 96: 253–267.

Martínez-González MC, Bautista-Baños S, Correa-Pacheco ZN, Corona-Rangel ML, Ventura-Aguilar RI, Del Río-García JC, Ramos-García ML. 2020. Effect of nano-structured chitosan/propolis coatings on the quality and antioxidant capacity of strawberries during storage. *Coatings* 10: 90.

Meena M, Pilania S, Pal A, Mandhania S, Bhushan B, Kumar S, Gohari G, Saharan V. 2020. Cu-chitosan nano-net improves keeping quality of tomato by modulating physio-biochemical responses. *Scientific Reports* 10: 21914.

Meindrawan B, Suyatma NE, Wardana AA, Pamela VY. 2018. Nanocomposite coating based on carrageenan and ZnO nanoparticles to maintain the storage quality of mango. *Food Packaging and Shelf Life* 18: 140–146.

Melo CBNF, de MendonçaSoares BL, Diniz MK, Leal FC, Canto D, Flores MAP, Tavares-Filho H da CJ, Galembeck A, Stamford MTL, Stamford-Arnaud MT, Stamford TCM. 2018. Effects of fungal chitosan nanoparticles as eco-friendly edible coatings on the quality of postharvest table grapes. *Postharvest Biology and Technology* 139: 56–66.

Melo NFCB, Pintado MME, Medeiros JA da C, Galembeck A, Vasconcelos MA da S, Xavier VL, de Lima MAB, Stamford TLM, Stamford-Arnaud TM, Flores MAP, Stamford TCM. 2020. Quality of postharvest strawberries: Comparative effect of fungal chitosan gel, nanoparticles and gel enriched with edible nanoparticles coatings. *International Journal of Food Studies* 9: 373–393.

Mittler R, Vanderauwera S, Suzuki N, Miller G, Tognetti VB, Vandepoele K, Gollery M, Shulaev V, Van Breusegem F. 2011. ROS signaling: The new wave? *Trends in Plant Science* 16: 300–309.

Morones JR, Elechiguerra JL, Camacho A, Holt K, Kouri JB, Ramírez JT, Yacaman MJ. 2005. The bactericidal effect of silver nanoparticles. *Nanotechnology* 16: 2346–2353.

Moscetti R, Haff RP, Monarca D, Cecchini M, Massantini R. 2016. Near-infrared spectroscopy for detection of hailstorm damage on olive fruit. *Postharvest Biology and Technology* 120: 204–212.

Mossa A-TH, Mohafrash SMM, Ziedan E-SHE, Abdelsalam IS, Sahab AF. 2021. Development of eco-friendly nanoemulsions of some natural oils and evaluating of its efficiency against postharvest fruit rot fungi of cucumber. *Industrial Crops and Products* 159: 113049.

Munhuweyi K, Mpai S, Sivakumar D. 2020. Extension of avocado fruit postharvest quality using non-chemical treatments. *Agronomy* 10(2): 212. https://doi.org/10.3390/agronomy10020212

Mustafa MA, Ali A, Manickam S, Siddiqui Y. 2014. Ultrasound-assisted chitosan-surfactant nanostructure assemblies: Towards maintaining postharvest quality of tomatoes. *Food and Bioprocess Technology* 7: 2102–2111.

Neysanian M, Iranbakhsh A, Ahmadvand R, Ardebili ZO, Ebadi M. 2020. Comparative efficacy of selenate and selenium nanoparticles for improving growth, productivity, fruit quality, and postharvest longevity through modifying nutrition, metabolism, and gene expression in tomato; potential benefits and risk assessment. *PLoS ONE* 15: e0244207.

Nguyen HVH, Nguyen DHH. 2020. Effects of nano-chitosan and chitosan coating on the postharvest quality, polyphenol oxidase activity and malondialdehyde content of strawberry (*Fragaria x ananassa* Duch.). *Journal of Horticulture and Postharvest Research* 3: 11–24.

Nguyen VTB, Nguyen DHH, Nguyen HVH. 2020. Combination effects of calcium chloride and nano-chitosan on the postharvest quality of strawberry (*Fragaria x ananassa* Duch.). *Postharvest Biology and Technology* 162: 111103.

O'Donoghue EM, Somerfield SD, Chen RKY, Tiffin HR, Hunter DA, Brummell DA. 2020. Cell wall composition during expansion, ripening and postharvest water loss of red bell peppers (*Capsicum annuum* L.). *Postharvest Biology and Technology* 168: 111225.

Oyedele OA, Kuzamani KY, Adetunji MC, Osopale BA, Makinde OM, Onyebuenyi OE, Ogunmola OM, Mozea OC, Ayeni KI, Ezeokoli OT, Oyinloye AM, Ngoma L, Mwanza M, Ezekiel C. 2020. Bacteriological assessment of tropical retail fresh-cut, ready-to-eat fruits in south-western Nigeria. *Scientific African* 9: e00505.

Pina-Barrera AM, Alvarez-Roman R, Baez-Gonzalez JG, Amaya-Guerra CA, Rivas-Morales C, Gallardo-Rivera CT, Galindo-Rodriguez SA. 2019. Application of a multisystem coating based on polymeric nanocapsules containing essential oil of *Thymus vulgaris* L. to increase the shelf life of table grapes (*Vitis Vinifera* L.). *IEEE Transactions on Nanobioscience* 18: 549–557.

Pinedo-Guerrero ZH, Delia Hernández-Fuentes A, Ortega-Ortiz H, Benavides-Mendoza A, Cadenas-Pliego G, Juárez-Maldonado A. 2017. Cu nanoparticles in hydrogels of chitosan-PVA affects the characteristics of postharvest and bioactive compounds of jalapeño pepper. *Molecules* 22: 1–14.

Posé S, Paniagua C, Matas AJ, Gunning AP, Morris VJ, Quesada MA, Mercado JA. 2019. A nanostructural view of the cell wall disassembly process during fruit ripening and postharvest storage by atomic force microscopy. *Trends in Food Science & Technology* 87: 47–58.

Queiroz C, Lopes MLM, Fialho E, Valente-Mesquita VL. 2008. Polyphenol oxidase: Characteristics and mechanisms of browning control. *Food Reviews International* 24: 361–375.

Ramírez-Gil JG, Henao-Rojas JC, Morales-Osorio JG. 2021. Heliyon postharvest diseases and disorders in avocado cv. Hass and their relationship to pre-harvest management practices. *Heliyon* 7: e05905.

Ranjbar S, Rahemi M, Ramezanian A. 2018. Comparison of nano-calcium and calcium chloride spray on postharvest quality and cell wall enzymes activity in apple cv. red delicious. *Scientia Horticulturae* 240: 57–64.

Rasouli M, Saba MK. 2018. Pre-harvest zinc spray impact on enzymatic browning and fruit flesh color changes in two apple cultivars. *Scientia Horticulturae* 240: 318–325.

Riva SC, Opara UO, Fawole OA. 2020. Recent developments on postharvest application of edible coatings on stone fruit: A review. *Scientia Horticulturae* 262: 109074.

Robledo N, Vera P, López L, Yazdani-Pedram M, Tapia C, Abugoch L. 2018. Thymol nanoemulsions incorporated in quinoa protein/chitosan edible films; antifungal effect in cherry tomatoes. *Food Chemistry* 246: 211–219.

Rojas G, Fernandez E, Whitney C, Luedeling E, Cuneo IF. 2021. Adapting sweet cherry orchards to extreme weather events—Decision analysis in support of farmers' investments in Central Chile. *Agricultural Systems* 187: 103031.

Ruiz-Rodríguez LG, Zamora Gasga VM, Pescuma M, Van Nieuwenhove C, Mozzi F, Sánchez Burgos JA. 2020. Fruits and fruit by-products as sources of bioactive compounds. Benefits and trends of lactic acid fermentation in the development of novel fruit-based functional beverages. *Food Research International*: 109854.

Saba MK, Amini R. 2017. Nano-ZnO/carboxymethyl cellulose-based active coating impact on ready-to-use pomegranate during cold storage. *Food Chemistry* 232: 721–726.

Saltveit M. 1999. Effects of ethylene on the quality of fruits and vegetables. *Postharvest Biology and Technology* 15: 279–292.

Saqib S, Zaman W, Ayaz A, Habib S, Bahadur S, Hussain S, Muhammad S, Ullah F. 2020. Postharvest disease inhibition in fruit by synthesis and characterization of chitosan iron oxide nanoparticles. *Biocatalysis and Agricultural Biotechnology* 28: 101729.

Saravanakumar K, Hu X, Chelliah R, Oh DH, Kathiresan K, Wang MH. 2020. Biogenic silver nanoparticles-polyvinylpyrrolidone based glycerosomes coating to expand the shelf life of fresh-cut bell pepper (*Capsicum annuum* L. var. grossum (L.) Sendt). *Postharvest Biology and Technology* 160: 111039.

Sekhon SB. 2014. Nanotechnology in agri-food production: An overview. *Nanotechnology, Science and Applications* 7: 31–53.

Shah SWA, Jahangir M, Qaisar M, Khan SA, Mahmood T, Saeed M, Farid A, Liaquat M. 2015. Storage stability of kinnow fruit (*Citrus reticulata*) as affected by CMC and guar gum-based silver nanoparticle coatings. *Molecules* 20: 22645–22661.

Shahat M, Ibrahim M, Osheba A, Taha I. 2020a. Preparation and characterization of silver nanoparticles and their use for improving the quality of apricot fruits. *Al-Azhar Journal of Agricultural Research* 45: 38–55.

Shahat M, Ibrahim MI, Osheba AS, Taha IM. 2020b. Improving the quality and shelf-life of strawberries as coated with nano-edible films during storage. *Al-Azhar Journal of Agricultural Research* 45: 1–13.

Sharma RR, Nagaraja A, Goswami AK, Thakre M, Kumar R, Varghese E. 2020. Influence of on-the-tree fruit bagging on biotic stresses and postharvest quality of rainy-season crop of 'Allahabad Safeda' guava (*Psidium guajava* L.). *Crop Protection* 135: 105216.

Sogvar OB, Koushesh Saba M, Emamifar A, Hallaj R. 2016. Influence of nano-ZnO on microbial growth, bioactive content and postharvest quality of strawberries during storage. *Innovative Food Science and Emerging Technologies* 35: 168–176.

Song H, Yuan W, Jin P, Wang W, Wang X, Yang L, Zhang Y. 2016. Effects of chitosan/nano-silica on postharvest quality and antioxidant capacity of loquat fruit during cold storage. *Postharvest Biology and Technology* 119: 41–48.

Sorrentino A, Gorrasi G, Vittoria V. 2007. Potential perspectives of bio-nanocomposites for food packaging applications. *Trends in Food Science and Technology* 18: 84–95.

Sun Q, Li J, Le T. 2018. Zinc oxide nanoparticle as a novel class of antifungal agents: Current advances and future perspectives. *Journal of Agricultural and Food Chemistry* 66: 11209–11220.

Taheri A, Behnamian M, Dezhsetan S, Karimirad R. 2020. Shelf life extension of bell pepper by application of chitosan nanoparticles containing *Heracleum persicum* fruit essential oil. *Postharvest Biology and Technology* 170: 111313.

Tang J, Chen H, Lin H, Hung Y-C, Xie H, Chen Y. 2021. Acidic electrolyzed water treatment delayed fruit disease development of harvested longans through inducing the disease resistance and maintaining the ROS metabolism systems. *Postharvest Biology and Technology* 171: 111349.

Tang N, An J, Deng W, Gao Y, Chen Z, Li Z. 2020. Metabolic and transcriptional regulatory mechanism associated with postharvest fruit ripening and senescence in cherry tomatoes. *Postharvest Biology and Technology* 168: 111274.

Trinetta V, McDaniel A, Batziakas KG, Yucel U, Nwadike L, Pliakoni E. 2020. Antifungal packaging film to maintain quality and control postharvest diseases in strawberries. *Antibiotics* 9: 1–12.

Vieira ACF, de Matos Fonseca J, Menezes NMC, Monteiro AR, Valencia GA. 2020. Active coatings based on hydroxypropyl methylcellulose and silver nanoparticles to extend the papaya (*Carica papaya* L.) shelf life. *International Journal of Biological Macromolecules* 164: 489–498.

Wang A, Ng HP, Xu Y, Li Y, Zheng Y, Yu J, Han F, Peng F, Fu L. 2014. Gold nanoparticles: Synthesis, stability test, and application for the rice growth. *Journal of Nanomaterials* 2014.

Wang G, Zhang X. 2020. Evaluation and optimization of air-based precooling for higher postharvest quality: Literature review and interdisciplinary perspective. *Food Quality and Safety* 4: 59–68.

Wang L, Shao S, Madebo MP, Hou Y, Zheng Y, Jin P. 2020b. Effect of nano-SiO_2 packing on postharvest quality and antioxidant capacity of loquat fruit under ambient temperature storage. *Food Chemistry* 315: 126295.

Wang Y, Deng C, Cota-Ruiz K, Peralta-Videa JR, Sun Y, Rawat S, Tan W, Reyes A, Hernandez-Viezcas JA, Niu G, Li C, Gardea-Torresday JL. 2020a. Improvement of nutrient elements and allicin content in green onion (*Allium fistulosum*) plants exposed to CuO nanoparticles. *Science of the Total Environment* 725: 138387.

Wei X, Xie D, Mao L, Xu C, Luo Z, Xia M, Zhao X, Han X, Lu W. 2019. Excess water loss induced by simulated transport vibration in postharvest kiwifruit. *Scientia Horticulturae* 250: 113–120.

Xing Y, Yang H, Guo X, Bi X, Liu X, Xu Q, Wang Q, Li W, Li X, Shui Y, Chen C, Zheng Y. 2020. Effect of chitosan/Nano-TiO_2 composite coatings on the postharvest quality and physicochemical characteristics of mango fruits. *Scientia Horticulturae* 263: 109135.

Xu J, Zhao Y, Zhang X, Zhang L, Hou Y, Dong W. 2016. Transcriptome analysis and ultrastructure observation reveal that hawthorn fruit softening is due to cellulose/hemicellulose degradation. *Frontiers in Plant Science* 7: 1–14.

Xu W, Wei Y, Wang X, Han P, Chen Y, Xu F, Shao X. 2021. Molecular cloning and expression analysis of hexokinase genes in peach fruit under postharvest disease stress. *Postharvest Biology and Technology* 172: 111377.

Yang KY, Doxey S, McLean JE, Britt D, Watson A, Al Qassy D, Jacobson AR, Anderson A. 2017. Remodeling of root morphology by CuO and ZnO nanoparticles: Effects on drought tolerance for plants colonized by a beneficial pseudomonad. *Botany*: 1–12.

Yi L, Qi T, Ma J, Zeng K. 2020. Genome and metabolites analysis reveal insights into control of foodborne pathogens in fresh-cut fruits by *Lactobacillus pentosus* MS031 isolated from Chinese Sichuan Paocai. *Postharvest Biology and Technology* 164: 111150.

Yu L, Qiao N, Zhao J, Zhang H, Tian F, Zhai Q, Chen W. 2020. Postharvest control of *Penicillium expansum* in fruits: A review. *Food Bioscience* 36: 100633.

Zambrano-Zaragoza ML, Mercado-Silva E, Ramirez-Zamorano P, Cornejo-Villegas MA, Gutiérrez-Cortez E, Quintanar-Guerrero D. 2013. Use of solid lipid nanoparticles (SLNs) in edible coatings to increase guava (*Psidium guajava* L.) shelf-life. *Food Research International* 51: 946–953.

Zambrano-Zaragoza ML, Quintanar-Guerrero D, Del Real A, González-Reza RM, Cornejo-Villegas MA, Gutiérrez-Corte E. 2020. Effect of nano-edible coating based on beeswax solid lipid nanoparticles on strawberry's preservation. *Coatings* 10: 253.

Zhao H, Fan Z, Wu J, Zhu S. 2020a. Effects of pre-treatment with S-nitrosoglutathione-chitosan nanoparticles on quality and antioxidant systems of fresh-cut apple slices. *Lwt—Food Science and Technology*: 110565.

Zhao J, Lin M, Wang Z, Cao X, Xing B. 2020b. Engineered nanomaterials in the environment: Are they safe? *Critical Reviews in Environmental Science and Technology*: 1–36.

Zhou L, Lv S, He G, He Q, Shi B. 2011. Effect of PE/Ag_2O nano-packaging on the quality of apple slices. *Journal of Food Quality* 34: 171–176.

Zulfiqar F, Navarro M, Ashraf M, Akram NA, Munné-Bosch S. 2019. Nanofertilizer use for sustainable agriculture: Advantages and limitations. *Plant Science* 289: 110270.

Index

Note: Page numbers in *italic* indicate a figure and page numbers in **bold** indicate a table on the corresponding page.

O

P

For Product Safety Concerns and Information please contact our EU
representative GPSR@taylorandfrancis.com
Taylor & Francis Verlag GmbH, Kaufingerstraße 24, 80331 München, Germany

www.ingramcontent.com/pod-product-compliance
Lightning Source LLC
Chambersburg PA
CBHW082103220326
41598CB00066BA/4956

* 9 7 8 0 3 6 7 6 9 5 5 8 3 *